PEM FUEL CELL FAILURE MODE ANALYSIS

PEM FUEL CELL DURABILITY HANDBOOK

PEM Fuel Cell Failure Mode Analysis

PEM Fuel Cell Diagnostic Tools

PEM FUEL CELL FAILURE MODE ANALYSIS

EDITED BY
HAIJIANG WANG
HUI LI
XIAO-ZI YUAN

CRC Press
Taylor & Francis Group
Boca Raton London New York

CRC Press is an imprint of the
Taylor & Francis Group, an **informa** business

CRC Press
Taylor & Francis Group
6000 Broken Sound Parkway NW, Suite 300
Boca Raton, FL 33487-2742

First issued in paperback 2017

© 2012 by Taylor and Francis Group, LLC
CRC Press is an imprint of Taylor & Francis Group, an Informa business

No claim to original U.S. Government works

ISBN-13: 978-1-4398-3917-1 (hbk)
ISBN-13: 978-1-138-11819-5 (pbk)

Library of Congress Cataloging-in-Publication Data

PEM fuel cell failure mode analysis / editors, Haijiang Wang, Hui Li, and Xiao-Zi Yuan.
    p. cm.
  "A CRC title."
  Includes bibliographical references and index.
  ISBN 978-1-4398-3917-1 (alk. paper)
  1. Proton exchange membrane fuel cells--Reliability. 2. Proton exchange membrane fuel cells--Testing. 3. System failures (Engineering) I. Wang, Haijiang Henry. II. Li, Hui, 1964- III. Yuan, Xiao-Zi.

TK2933.P76P465 2011
621.31'2429--dc22
                                      2011007916

Visit the Taylor & Francis Web site at
http://www.taylorandfrancis.com

and the CRC Press Web site at
http://www.crcpress.com

Contents

Preface

State-of-the-art proton exchange membrane (PEM) fuel cell technology, which uses solid polymer membrane electrolyte and microporous gas diffusion electrodes to form a compact design that can generate much higher power density than traditional electrochemical power devices, has experienced three decades of technological development since the pioneering research work of Ballard Power Systems, a Canadian company located in Burnaby, British Columbia, in the 1980s. The first decade of this development was about concept demonstration. By using the Nafion® type of membrane as the electrolyte, gas diffusion electrodes made of a carbon fiber paper for the gas diffusion layer, and carbon-supported platinum nanoparticles for the catalyst, Ballard Power Systems successfully demonstrated a PEM fuel cell system with high-power density and high-energy density that had the potential to replace the internal combustion engine as the power-train for vehicles. To auto manufacturers, this was like sailors finally seeing the lighthouse after an endless journey in the dark, because at that time the auto giants had started to close their plants for battery-powered electric vehicles after about 30 years of technological development had resulted in failure due to the batteries' insufficient energy density. Huge subsequent investment by the auto companies led to the second stage of PEM fuel cell technology development: the technological leap. By the end of the 1990s, fuel cell-powered cars, buses, stationary generators, and so on had all shown superior performance. However, fuel cell developers had underestimated the challenges in the fuel cell commercialization process. The resulting "overstatement and under delivery" led to a cooldown in financial investment and public interest in the technology. The fuel cell companies had to change their strategies under these tight financial conditions. While continuing to overcome the technical challenges, they started to look at niche markets that might generate revenue in the short term, instead of focusing on fuel cell vehicle development, which is technically the most difficult technology and has additional challenges, such as creating a hydrogen infrastructure. Technological development in the last decade focused on addressing the technical challenges and improving fuel cell performance. This was the third stage: technological improvement. After 10 years of research and development, PEM fuel cell technology has improved substantially, even though the technical challenges have not yet been fully addressed. PEM fuel cell technology has quietly entered into the market-penetration, precommercial stage, especially in niche markets. It is predicted that fuel cell commercial products will be seen in about five years. The commercialization process follows the pattern of Concept Demonstration → Technology Leap → Technology Improvement → Market Penetration → Commercial Products.

The major technical challenges for PEM fuel cells in general are performance, reliability, durability, and cost. These challenges are interrelated. For instance, to reduce cost, cheap materials may be used but at the price of reliability and durability. To increase durability, thicker membranes may be used but performance will be lowered. Depending on the applications, the degree of these challenges is quite different. For example, for back-up power applications and applications for material handling, the cost of a state-of-the-art PEM fuel cell is quite competitive with conventional lead-acid battery technology; durability is also not a big problem. But for transportation applications, both cost and durability are major challenges.

To date, fuel cell researchers have invested a great deal of effort into PEM fuel cell durability research. Major failure modes and underlying failure mechanisms of components and fuel cell systems have been identified. The knowledge generated has been a great help in assisting fuel cell developers to increase durability from around 1000 h in the year 2000 to the current 3000 h or so for transportation applications, and it will continue to be essential for fuel cell researchers to address the remaining durability gap. The editors are extremely grateful to those experts in the field of PEM fuel cell durability who contributed to this handbook. We hope the book will help fuel cell researchers to overcome the technical challenges of PEM fuel cell technology and drive the technology into commercial products as soon as possible.

Editors

Haijiang Wang is a senior research officer, project manager of multiprojects, and the core competency leader of the Unit Fuel Cell Team in the National Research Council of Canada Institute for Fuel Cell Innovation (NRC-IFCI). He is currently leading a team of over 10 scientists to carry out research and development on novel fuel cell design and materials, as well as fuel cell diagnosis and durability. Dr. Wang received his PhD in electrochemistry from the University of Copenhagen, Denmark, in 1993. He then joined Dr. Vernon Parker's research group at Utah State University as a postdoctoral researcher to study electrochemically generated anion and cation radicals. In 1997, he began working with Natural Resources Canada as a research scientist to carry out research on fuel cell technology. In 1999, he joined Ballard Power Systems as a senior research scientist to continue his investigations. After spending five years with Ballard Power Systems, he joined NRC-IFCI in 2004. He is currently adjunct professor at five universities, including the University of British Columbia and the University of Waterloo. Dr. Wang has 25 years of professional research experience in electrochemistry and fuel cell technology. To date, he has published 115 journal papers, 3 books, 40 industrial reports, and 30 conference papers or presentations, and has been issued three patents.

Hui Li is a research council officer and project technical leader both for the PEM Fuel Cell Failure Mode Analysis project and PEM Fuel Cell Contamination Consortium at the National Research Council of Canada Institute for Fuel Cell Innovation (NRC-IFCI). Dr. Li received her BS and MSc in chemical engineering from Tsinghu University in 1987 and 1990, respectively. After completing her MSc, she joined Kunming Metallurgical Institute as a research engineer for four years and then took a position as an associate professor at Sunwen University (then a branch of Zhongshan University) for eight years. In 2002, she started her PhD program in electrochemical engineering at the University of British Columbia (UBC) under the supervision of Professor Colin Oloman. After obtaining her PhD in 2006, she carried out one term of postdoctoral research at the Clean Energy Research Centre (CERC) at UBC with Professor Colin Oloman and Professor David Wilkinson. In 2007, she joined the Low-temperature PEM Fuel Cell Group at NRC-IFCI, working on PEM fuel cell contamination and durability. Dr. Li has years of research and development experience in theoretical and applied electrochemistry and in electrochemical engineering. Her research is based on PEM fuel cell contamination and durability testing; preparation and development of electrochemical catalysts with long-term stability; catalyst layer/cathode structure; and catalyst layer characterization and electrochemical evaluation. Dr. Li has coauthored more than 30 research papers published in refereed journals and has one technology licensed to the Mantra Energy Group. She has also produced many industrial technical reports.

Xiao-Zi Yuan is a research officer and project leader of the Unit Cell Team at the Institute for Fuel Cell Innovation, National Research Council of Canada (NRC-IFCI). Dr. Yuan received her BS and MSc in electrochemical engineering from Nanjing University of Technology in 1991 and 1994, respectively, under the supervision of Professor Baoming Wei, and her PhD in material science from Shanghai

Jiaotong University in 2003, under the supervision of Professor Zifeng Ma. After graduating in MSc, she held a lecturer position at Nantong University for six years, and on completing her PhD she was an associate professor in the same university for one year. Beginning in 2005, she carried out a three-year postdoctoral research program at NRC-IFCI with Dr. Haijiang Wang. Dr. Yuan has over 16 years of R&D experience in applied electrochemistry, including over 10 years of fuel cell R&D (among these three years at Shanghai Jiaotong University, one year at Fachhochschule Mannheim, and six years, to date, at NRC-IFCI). Currently, her research focuses on PEM fuel cell design, testing, diagnosis, and durability. Dr. Yuan has published more than 50 research papers in refereed journals, produced two books and five book chapters, presented more than 30 conference papers or presentations, and holds five China patents.

Contributors

Laurent Antoni
Commissariat à l'Energie Atomique et aux
   Energies Alternatives
Grenoble, France

Olga A. Baturina
Naval Research Laboratory
Washington, DC

Peter Beckhaus
Zentrum für BrennstoffzellenTechnik
Center for Fuel Cell Technology
Duisburg, Germany

Zhongwei Chen
Department of Chemical Engineering
University of Waterloo
Waterloo, Ontario, Canada

Thorsten Derieth
Zentrum für BrennstoffzellenTechnik
Center for Fuel Cell Technology
Duisburg, Germany

Yannick Garsany
Naval Research Laboratory
Washington, DC

Benjamin D. Gould
Naval Research Laboratory
Washington, DC

Angelika Heinzel
Zentrum für BrennstoffzellenTechnik
Center for Fuel Cell Technology
Duisburg, Germany

Ryan Hsu
Department of Chemical Engineering
University of Waterloo
Waterloo, Ontario, Canada

Kui Jiao
Department of Mechanical and Mechatronics
   Engineering
University of Waterloo
Waterloo, Ontario, Canada

Lars Kühnemann
Zentrum für BrennstoffzellenTechnik
Center for Fuel Cell Technology
Duisburg, Germany

Bing Li
School of Automotive Studies
Tongji University
Shanghai, People's Republic of China

Xianguo Li
Department of Mechanical and Mechatronics
   Engineering
University of Waterloo
Waterloo, Ontario, Canada

Yinghao Luan
School of Chemical Engineering
East China University of Science and Technology
Shanghai, People's Republic of China

Jianxin Ma
School of Automotive Studies
Tongji University
Shanghai, People's Republic of China

Walter Mérida
Clean Energy Research Centre
University of British Columbia
Vancouver, British Columbia, Canada

Pucheng Pei
Department of Automotive Engineering
Tsinghua University
Beijing, China

Eric Pinton
Commissariat à l'Energie Atomique et aux
    Energies Alternatives
Grenoble, France

Sébastien Rosini
Commissariat à l'Energie Atomique et aux
    Energies Alternatives
Grenoble, France

Karen E. Swider-Lyons
Naval Research Laboratory
Washington, DC

Haijiang Wang
Institute for Fuel Cell Innovation
National Research Council of Canada
Vancouver, British Columbia, Canada

Daijun Yang
School of Automotive Studies
Tongji University
Shanghai, People's Republic of China

Xiao-Zi Yuan
Institute for Fuel Cell Innovation
National Research Council of Canada
Vancouver, British Columbia, Canada

Shengsheng Zhang
Institute for Fuel Cell Innovation
National Research Council of Canada
Vancouver, British Columbia, Canada

Yongming Zhang
School of Chemistry and Chemical
    Engineering
Shanghai Jiao Tong University
Shanghai, People's Republic of China

Junsheng Zheng
School of Automotive Studies
Tongji University
Shanghai, People's Republic of China

1

Haijiang Wang
*National Research Council
of Canada*

Hui Li
*National Research Council
of Canada*

Xiao-Zi Yuan
*National Research Council
of Canada*

Introduction

Fuel cells offer efficient, quiet, and virtually pollution-free energy conversion and power generation. Of the different types of fuel cells, proton exchange membrane (PEM) fuel cells are promising in being environmental friendly, and have several advantages over conventional energy-converting devices, including both high efficiency and power density, making them unique across a wide range of portable, stationary, and transportation power applications. However, several challenges still remain, including durability/reliability, cost, and performance, particularly for automotive and stationary applications. Durability has emerged as the top challenge. The 2015 US Department of Energy (DOE) lifetime requirements for transportation applications are 5000 h (cars) and 20,000 h (buses), and for on-site cogeneration systems 40,000 h. Currently, the lifetimes of fuel cell vehicles and stationary cogeneration systems are around 1700 h and 10,000 h, respectively (Payne, 2009). Clearly, intensive R&D is still needed to address the issues related to PEM fuel cell durability or degradation in order to achieve sustainable commercialization.

PEM fuel cells consist of many components, including catalysts, catalyst supports, membranes, gas diffusion layers (GDLs), bipolar plates, sealings, and gaskets. Each of these components can degrade or fail to function, thus causing the fuel cell system to degrade or fail. Component degradation includes, but is not limited to, catalyst particle ripening (particle coalescence), preferential alloy dissolution in the catalyst layer, carbon support oxidation (corrosion), catalyst poisoning, membrane dissolution, loss of sulfonic acid groups in the ionomer phase of the catalyst layer or in the membrane, bipolar plate surface film growth, hydrophilicity changes in the catalyst layer and/or GDL, and polytetrafluoroethylene (PTFE) decomposition in the catalyst layer and/or GDL. The durability of each component is affected by many internal and external factors, including the material properties, fuel cell operating conditions (such as humidification, temperature, cell voltage, etc.), impurities or contaminants in the feeds, environmental conditions (e.g., subfreezing or cold start), operation modes (such as start-up, shut-down, potential cycling, etc.), and the design of the components and the stack. In addition, the degradation processes of different components are often interrelated in a fuel cell system. It is therefore important to separate, analyze, and systematically understand the degradation phenomena of each component so that novel component materials can be developed and novel design for cells/stacks can be achieved to mitigate insufficient fuel cell durability. In this ongoing R&D, long-term durability tests are often required to evaluate the degradation mechanisms of various components and the corresponding fuel cell systems. However, it is generally impractical and costly to operate a fuel cell under its normal conditions for several thousand hours, and so accelerated test methods are required to facilitate rapid learning about key durability issues and thus enable the timely implementation of mitigation strategies.

This book provides a systematic review of PEM fuel cell durability and failure modes, progressing from component degradation to contamination-, environment-, operation-, and design-induced degradation, with each chapter covering one topic. Chapters 2 through 7 examine the degradation of various

1

fuel cell components, including degradation mechanisms, the effects of operating conditions (such as temperature, pressure, and humidification), mitigation strategies, and testing protocols. Chapter 8 discusses the effects of different contamination sources on the degradation of fuel cell components; Chapter 9 explains the relationship between external environment (mainly exposure to subfreezing conditions) and the degradation of fuel cell components and systems; Chapter 10 reviews the correlation between operational mode (such as start-up and shut-down) and the degradation of fuel cell components and systems; and finally, Chapter 11 presents how the design of fuel cell hardware (such as flow fields) relates to various failure modes.

Reference

Payne, T. 2009. DOE fuel cell R&D activities: Transportation, stationary, and portable power applications. In *Fuel Cells Durability & Performance*. US Brookline: The Knowledge Press Inc.

2

Catalyst Degradation

Shengsheng Zhang
National Research Council of Canada

Xiao-Zi Yuan
National Research Council of Canada

Haijiang Wang
National Research Council of Canada

2.1 Introduction

The catalyst is one of the most important components within a membrane electrode assembly (MEA) of a proton exchange membrane (PEM) fuel cell, as all the electrochemical reactions are realized with the aid of the catalyst and take place on its surface. As the heart of a PEM fuel cell, the MEA is generally composed of a PEM sandwiched between two electrodes with catalyst layers (CLs) on either side. Normally, a CL is spread on a gas diffusion layer (GDL). The catalyst-coated membrane (CCM) is a new form of MEA recently developed to improve the performance of PEM fuel cells, with CLs on both sides of the membrane. A carbon-supported Pt (or Pt-alloy) catalyst, together with a recast Nafion ionomer network, forms a complex composite CL with multiple interfaces, which is critical for the transport of reactant gas, protons, electrons, and water. Both the structure and the composition of the CL have been found to be important in improving cell performance. During the last few decades, catalyst durability has become a key issue, due to urgent demands for fuel cell applications. Research on this topic has shown severe catalyst degradation in both transportation and stationary applications, which hinders further development of this promising technology. Therefore, this chapter will mainly focus on catalyst degradation, including degradation of conventional carbon-supported Pt catalysts, Pt-based alloy catalysts, and non-Pt catalysts. After providing a brief introduction on the different failure modes of CLs in Section 2.2, the following sections will discuss catalyst degradation mechanisms and degradation testing procedures, and will describe degradation characterization and mitigation strategies.

2.2 CL Failure Modes and General Characterization Methods

2.2.1 The Catalyst and Its Degradation

For PEM fuel cell CLs, catalysts can be classified into three groups based on the active component: carbon-supported Pt; carbon-supported Pt-based catalysts in which Pt is alloyed with or modified by other metals such as Cr (Wells et al., 2008), Cu (Mani et al., 2008), Co (Qian et al., 2008), and Ru (Denis et al., 2008); and non-Pt-based catalysts such as nonnoble metal-based catalysts (Izhar et al., 2009) and organometallic complexes (Charreteur et al., 2008).

Although a variety of catalysts have been investigated so far, Pt-containing catalysts are still the most popular due to the low overpotential and high catalytic activity of Pt for the hydrogen oxidation reaction (HOR) and the oxygen reduction reaction (ORR). Pt has been the cathode electrocatalyst of choice since the first fuel cell experiments in 1839 by Grove (Steele et al., 2001). Generally, durability is one of the three essential requirements a successful PEM fuel cell catalyst should satisfy (the other two being performance and cost). Under normal conditions, the direct reason for Pt catalyst degradation is loss of active sites and surface area, due to either particle growth or mass loss during long-term operation. The degradation of adjacent materials such as the carbon support and ionomer may also have some impact on the Pt degradation process.

2.2.2 Carbon Support Degradation

Carbon black is the dominant material used in a PEM to support the catalyst (e.g., Pt), and thereby obtain a maximum catalyst utilization ratio and decrease the cost of fuel cells. Carbon degradation is prone to take place under prolonged operation of PEM fuel cells. The negative effect of carbon degradation on the catalyst is that it weakens the attachment of Pt particles to the carbon surface, and eventually leads to structural collapse, resulting in declines in catalyst active surface area and fuel cell performance. In a study by Sato et al. (2006), it was proven that no performance degradation occurred when a Pt-black catalyst was applied as the anode electrode catalyst during hydrogen starvation. Under the same conditions, however, a fuel cell with Pt/C catalyst experienced severe degradation. For more information on supports and their degradation, see Chapter 3.

2.2.3 Ionomer Degradation

Ionomer degradation/loss can be another critical factor leading to reduced CL performance after long-term operation. In a PEM fuel cell, hydrogen peroxide, OH radicals, or other contaminants produced from the fuel cell reactions can damage the functionality and integrity of the Nafion membrane by attacking its perfluorosulfonic acid (PFSA) structure. The same damage also occurs to the recast Nafion ionomer employed in the CL during the aging process (Zhang et al., 2009a; Young et al., 2010). However, our present grasp of ionomer degradation inside the CL is not thorough enough, given its high importance. The relationship between ionomer degradation and catalyst degradation is not distinctly understood and still needs more investigation.

2.2.4 CL Characterization Methods and Degradation Tests

With the development of fuel cell technology, investigative tools, including electrochemical and physical/chemical methods, have been employed to characterize catalyst performance, thereby greatly facilitating degradation research. The most frequently used tools can be classified into three categories: morphology observation, electrochemical diagnosis, and other analytical techniques. By comparing performance before and after operation, information about catalyst degradation can be obtained and analyzed accordingly. Table 2.1 summarizes the general investigative tools employed in the latest

TABLE 2.1 General Investigative Tools Employed in the Latest Catalyst Degradation Research

	Technologies	Characteristics	Information About the Catalyst Layer
Physical/Chemical methods	TEM (transmission electron microscope)	Morphology or Pt distribution analysis	Topography investigation and particle size distribution
	SEM or FEG-SEM or SEM-EDS (scanning electronic microscopy or field-emission gun with SEM or SEM with energy dispersive spectroscopy)		Topography investigation and element distribution analysis of cross-section of the MEA or CL
	AFM (atomic force microscopy)		Morphology of the CL surface
	Optical micrography		Dispersion of Pt/C catalyst on glassy carbon disk electrode
	EPMA (electron-probe microanalysis)		Characterize Pt content through cross-section of the MEA
	AAS (atomic adsorption spectroscopy)	Elemental content analysis	Investigate Pt content in Pt/C catalyst
	ICP or ICP-AES or ICP-MS (inductively coupled plasma combined with atomic emission spectrometry or mass spectrometry)		Investigate the amount of metal content
	UV (ultraviolet spectroscopy)		Detect the presence of Pt^{z+} ionic species ($z = 2,4$) in decantation solution of membrane
	XAS (X-ray absorption spectroscopy)	Atomic structure analysis	Give information for atomic structure of the catalyst, mainly of the surface atoms
	EXAFS (extended X-ray absorption fine structure) oscillations		Local structural parameters, coordination number, Pt-Pt bond distance and DW factor
	XPS (X-ray photoelectron spectroscopy)		Surface oxygen content or electronic structure change in other surface elements
	Laser Raman spectroscopy		Detect the degree of structural disorder of carbon
	XRD (X-ray diffraction)		Analyze Pt particle average size and crystallinity of alloy materials
Electrochemical methods	IV (polarization curve)	Potential or current scanning	Characterize cell performance by current density vs. potential under specialized conditions
	LSV (linear sweep voltammetry)		Obtain detailed information about the degradation mechanism of Pt by position shift of the potential peak
	CV (cyclic voltammetry)	CV based scanning	Determine the ECSA of Pt by hydrogen adsorption
	COSV (CO-stripping voltammetry)		Determine the ECSA of Pt by CO oxidation
	EIS (electrochemical impedance spectrometry)	Impedance	Characterize the polarization resistance (especially for ohmic resistance and charge transfer resistance)

Source: Reproduced from *J. Power Sources*, 194, Zhang, S. et al. A review of platinum-based catalyst layer degradation in PEM fuel cells, 588–600, Copyright (2009), with permission from Elsevier.

TABLE 2.2 General AST Conditions and Test Systems Feasible for Possible Pt Catalyst Degradation Investigation

Possible Causes of Pt Degradation	General AST Conditions	Available Test Systems
Pt particle growth	Potential cycling	Half-cell or single cell
	Potential holding	Half-cell or single cell
	Long-term operation under different conditions	Single cell or stack
Pt loss or migration	Potential cycling	Single cell or stack
	Potential holding	Single cell or stack
Contamination	Contamination from the air	Single cell or stack
	Contamination from the fuel	Single cell or stack
	Anion contamination	Half-cell or single cell or stack

catalyst degradation research. Among these technical methods, cyclic voltammetry (CV) is the most frequently employed tool to characterize electrochemical surface area (ECSA), which is a critical parameter to evaluate the amount of effective active sites for Pt in the electrode during degradation tests.

To investigate the durability of PEM fuel cell systems and their components, two basic degradation tests are commonly used: the steady/dynamic-state life test and the accelerated stress test (AST). Both tests are applicable to catalyst degradation investigations. Using the steady/dynamic-state life test, the catalyst failure mode in real steady-state or dynamic-state operating conditions can be identified via postanalysis. In an AST, the specified accelerated stressors can be conducted either *in situ* on a real fuel cell system or *ex situ* on an electrochemical half-cell system to facilitate identification of the durability issues. Compared with full cell or stack operation under steady/dynamic-state life testing, *ex situ* half-cell system testing in an aqueous acid solution (H_2SO_4 or $HClO_4$) has been applied widely, its advantages being simpler equipment and isolation of the catalyst from adjacent components; the latter avoids potentially confusing effects during degradation evaluation. The degradation rate and catalyst damage level under specific working conditions can be examined during and after ASTs, which is more helpful for explaining the probable failure mode. However, from a practical point of view, an actual fuel cell test is the ultimate evaluation method in catalyst innovation research. Table 2.2 presents the general AST conditions and test systems feasible for Pt catalyst degradation investigation.

2.3 Degradation of Platinum Catalyst

For PEM fuel cells to serve as a viable energy alternative, a major challenge is their relatively high cost. Part of this is due to the high prices of materials, including the catalyst, PEM, and bipolar plate. Effective use of these materials and the development of substitutes are the main methods for mitigating this critical disadvantage of PEM fuel cells in comparison with other current market technologies such as gasoline internal combustion engines. Effective use of Pt catalyst includes two basic strategies: lower the Pt loading and prolong the Pt use period. At present, Pt loading has been decreased to 0.2–0.4 mg cm^{-2}, and further lowering is anticipated (Gasteiger et al., 2005). However, extending the duration of Pt use still faces many problems, especially in long-term operating fuel cells.

2.3.1 Mechanisms of Pt Degradation

Under the harsh working conditions within fuel cell systems (such as strong acidic environments, oxidizing conditions, reactive intermediates, durative flow of liquids and gases, high electric currents, and large potential gradients), Pt catalyst tends to experience subtle changes and function losses during operation. These imperceptible changes accumulate and result in a gradual decline in power output during long-term operation of a PEM fuel cell. Generally, the direct reasons for Pt catalyst degradation include: (1) Pt particle agglomeration and particle growth, (2) Pt loss and redistribution, and (3) poisonous effects of contaminants. All of these factors will lead to a loss either of effective catalytic active sites or of electronic contact with conductors, resulting in apparent activity loss in the CL.

2.3.1.1 Pt Agglomeration and Particle Growth

As demonstrated by many researchers, agglomeration and particle growth of the Pt nanostructure is the dominant mechanism for catalyst degradation in PEM fuel cells. First of all, it is well known that nano-sized structural elements are able to show size-dependent properties different from the properties of bulk elements (Dingreville et al., 2005). Nanoparticles have the inherent tendency to agglomerate into bigger particles to reduce their high surface energy. As particles grow, their surface energy decreases and the growth process itself slows down. This phenomenon is occurring to any given random Pt catalyst system all the time.

However, during long-term real operating conditions or ASTs, the Pt particles in Pt/C catalysts can experience more severe agglomeration than normal. For example, Chung et al. (2009) described that in their study the average anode and cathode Pt particle size increased to 3.45 and 3.71 nm, respectively, from the original size of 1.35 nm after 1784 h of degradation testing at 80 mA cm^{-2}. The authors also detected that the growth in particle size occurred mostly during the initial stages of operation (40 h) and was possibly linked to the dissolution, migration, and redeposition of Pt. Examples of the particle size distribution of fresh and used Pt/C catalyst are compared in Figure 2.1. Pt particle agglomeration has generally been explained by "Ostwald ripening," in which small Pt particles dissolve in the ionomer phase and redeposit on larger particles that are a few nanometers apart (Ferreira et al., 2005). Virkar and Zhou (2007) proposed that Ostwald ripening also involves the transport of electrically charged species, whereby Pt is transported through the liquid and/or ionomer and the electrons through the carbon support. Figure 2.2 shows the process of Pt transport from small particles to a larger one (Virkar and Zhou, 2007). Furthermore, some other mechanisms have been proposed as responsible for Pt catalyst

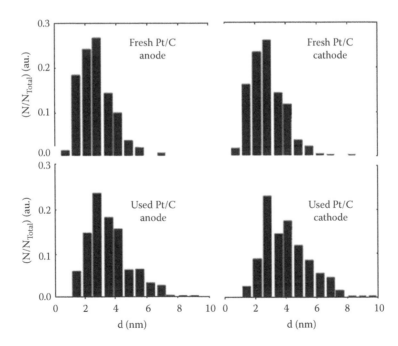

FIGURE 2.1 Particle size distribution of fresh and used Pt/C electrocatalysts, tested in constant-power mode, 0.12 W cm^{-2} at 333 K over 529 h. (Reproduced from Guilminot, E. et al. 2007b. *J. Electrochem. Soc.* 154: B96–B105. With permission from The Electro Chemical Society.)

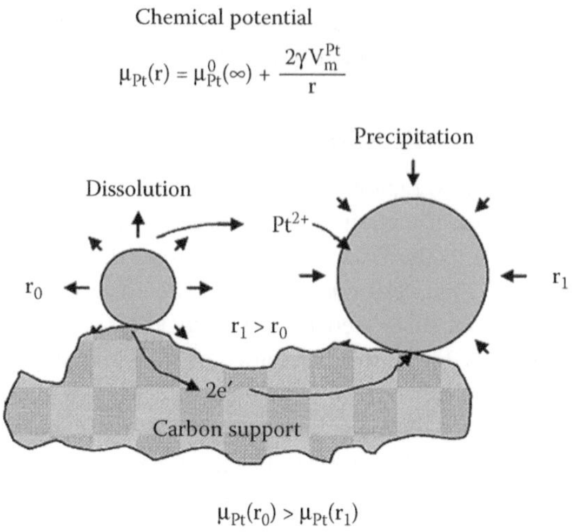

Chemical potential

$$\mu_{Pt}(r) = \mu_{Pt}^0(\infty) + \frac{2\gamma V_m^{Pt}}{r}$$

$$\mu_{Pt}(r_0) > \mu_{Pt}(r_1)$$

FIGURE 2.2 Transport of Pt from a smaller particle to a larger particle by coupled transport of Pt^{2+} through the liquid and/or ionomer and the transport of electrons through the carbon support. (Reproduced from Virkar, A. V. and Zhou, Y. 2007. *J. Electrochem. Soc.* 154: B540–B547. With permission from The Electro Chemical Society.)

agglomeration, including a combination of Pt particle coalescence and Pt solution/reprecipitation within the solid ionomer (More et al., 2006).

For all of these possible mechanisms, dissolution of Pt is an important step during the catalyst degradation process. According to Chung and coworkers' proposal (2009), at the anode, the following reactions take place:

$$Pt^0 \rightarrow Pt^{2+} + 2e^- \tag{2.1}$$

$$Pt^{2+} + H_2 \rightarrow Pt^0 + 2H^+ \tag{2.2}$$

At the cathode, the following reactions take place simultaneously:

$$Pt^0 + \frac{1}{2}O_2 \rightarrow PtO \tag{2.3}$$

$$PtO + 2H^+ \rightarrow Pt^{2+} + H_2O \tag{2.4}$$

$$Pt^{2+} + 2e^- \rightarrow Pt^0 \tag{2.5}$$

There is a balance between these reactions at the anode and cathode that is highly related to the Pt concentration. The lower the Pt ion concentration, the slower the degradation kinetics for the Pt/C catalyst. In addition, different electrode aging processes may reveal different dissolution reactions taking place at the anode and cathode. Using the rotating ring-disk electrode (RRDE) method and following different sweep protocols in a three-electrode system, Kawahara et al. (2006) proved that for a slow anodic triangular wave sweep, the Pt dissolution mechanism is expressed as

$$Pt \rightarrow Pt^{4+} + 4e^- \tag{2.6}$$

$$PtO_2 + 4H^+ \rightarrow Pt + 2H_2O \qquad (2.7)$$

while for a cathodic sweep, the Pt dissolution mechanism is expressed in Equation 2.8, with a charge transfer number of ~2.

$$PtO_2 + 4H^+ + 2e^- \rightarrow Pt^{2+} + 2H_2O \qquad (2.8)$$

Potential also plays an important role during the Pt degradation process. Higher potentials can accelerate Pt dissolvability. Wang et al. (2006a) suggested that the concentration of dissolved Pt increases monotonically as potential increases from 0.65 to 1.1 V, and then decreases at potentials higher than 1.1 V due to the formation of a protective oxide film; therefore, the potential limit for Pt catalyst degradation AST is generally lower than 1.0 V to avoid the possibility of oxide film formation and carbon support corrosion. Yoda et al. (2007) also proposed that Pt electrocatalysts are dissolved even under standard operating conditions. In addition, Pt agglomeration can be affected by many other operating conditions, such as temperature (Borup et al., 2006a; Cai et al., 2006) or relative humidity (RH) (Borup et al., 2006a,b; Xu et al., 2007). It has been proven that Pt catalyst degradation caused by the dissolution and redeposition process can be accelerated by increasing temperature and RH.

2.3.1.2 Pt Elemental Loss

Pt loss during operation is another major source of catalyst degradation, running parallel to Pt agglomeration and particle growth, and can be caused by many factors, such as Pt particles detaching from the support and dissolving into the electrolyte without redeposition (Mayrhofer et al., 2008), consequently being lost from the system. Several detection methods are available to evaluate Pt loss. Luo et al. (2006) conducted an experiment involving a 10-cell stack operated for 200 h at 60°C under ambient humidity and pressure. Pt content in the Pt/C CL was determined by atomic adsorption spectroscopy (AAS). The results showed Pt content of only 13.5% compared to the original value of 20%, which proved that serious Pt loss occurred during the aging process. By using inductively coupled plasma combined with mass spectrometry (ICP-MS), Ball et al. (2007a) also measured Pt loss from Pt/C catalyst in acidic electrolyte systems. After potential cycling, ppb levels of platinum were found in the electrolyte and a linear increase in dissolved Pt with cycle number was observed, which was in good agreement with Pt dissolution studies on Pt/C electrodes by Mitsushima et al. (2007) and Wang et al. (2006a).

2.3.1.3 Pt Migration

Another type of Pt loss arises when it migrates from its original position in the CL to elsewhere within the CL or to the PEM. Many research groups have reported the presence of Pt particles inside the PEM as well as Pt enrichment in the CL/PEM interface under different conditions (Ferreira et al., 2005; Xie et al., 2005; Guilminot et al., 2007a). Figure 2.3 shows Pt catalyst particles observed within the PEM near the cathode/PEM interface after degradation. According to Akita et al. (2006), these Pt particles originated from the dissolved Pt species, which diffused in the ionomer phase and subsequently precipitated in the ionomer phase of the CL or in the PEM. The Pt aggregates formed in the PEM have a well-ordered crystalline structure after the application of a potential AST, which indicates that Pt particles seem to form nuclei and grow atomically by the deposition of dissolved Pt ionic species.

A Pt band consisting of abundant Pt particles can be formed with the accumulation of Pt from migration. Precipitation occurs via the reduction of Pt ions by hydrogen that has crossed over from the anode, and thus it is called the "micrometer-scale diffusion process." According to a widely recognized hypothesis, the redistribution of Pt nanoparticles is a complex process involving (1) Pt dissolution, (2) formation of Pt^{z+} species, and (3) reduction to Pt particles by the crossover of H_2 from anode to cathode.

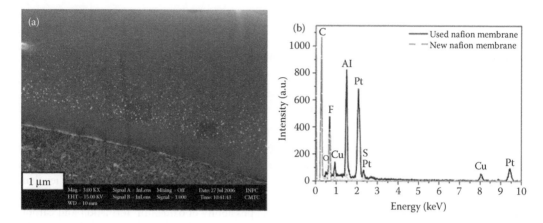

FIGURE 2.3   SEM image (a) and EDS spectra (b) of an aged PEM fuel cell MEA analysis location was ~6 μm from the cathode membrane interface in both cases. (Reproduced from Guilminot, E. et al. 2007a. *J. Electrochem. Soc.* 154: B1106–B1114. With permission from The Electro Chemical Society.)

For the first and second stages of Pt dissolution, Pt^{2+} is normally the main product following the reaction:

$$Pt_{(s)} \rightarrow Pt^{2+}_{(*)} + 2e^-$$ (2.9)

Except for Pt^{2+} formed under normal conditions, in the second stage faster electrooxidation of Pt into a higher valence state is also possible under higher potentials. Guilminot et al. (2007b) observed simultaneous Pt^{2+} and Pt^{4+} by ultraviolet (UV) detection, which proved the high mobility of Pt-containing species. The transport of Pt ions can also be facilitated by the presence of species containing counter ions, such as F^- or SO_x^- (Guilminot, 2007a).

In the third stage, the Pt^{z+} species are chemically and electrochemically reduced by H_2 that has crossed over the PEM and cathode to the Pt particles.

$$Pt^{z+}_{(*)} + \frac{z}{2}H_{2(g)} \rightarrow Pt_{(s)} + zH^+_{(*)}$$ (2.10)

The mechanism described above is a hypothesis based on the Pt dissolution and redeposition theory. Additional explanations have also been proposed based on further experimental results. For example, Chung et al. (2009) compared the degradation effects of OCV mode and constant-current mode at 80 mA cm^{-2}. The authors considered there to be a balance between chemical oxidation of Pt by oxygen and electrochemical reduction of PtO by electrons generated in constant-current mode. Due to this balance, the relative degradation of the catalyst and cell performance proceeded more slowly in the current generating mode. Conversely, in OCV mode, due to the higher availability of O_2, the rate of chemical oxidation was much faster than that of the electrochemical reduction of Pt^{2+}, causing rapid catalyst deactivation. This was also supported by the higher degradation rate observed when O_2 was fed into the cathode to replace the air feed under constant current conditions.

Concerning the direction and degree of Pt particle migration and redistribution, the results presented in the current literature are more complicated under different operating conditions. For example, when conducting a square-wave potential experiment between 0.87 and 1.2 V vs. RHE, Bi and Fuller (2008) found that Pt migrated into the PEM near the cathode. However, Ferreira et al. (2005) and More et al. (2006) observed Pt enrichment at the cathode/membrane interface after potential cycles between low potential (0.6 and 0.1 V) and 1.0 V, while Xie et al. (2005) and Guilminot et al. (2007b) observed Pt enrichment at the anode/membrane interface after long-term constant-current operation.

The main reason for the diversity of results is that Pt migration and redistribution is a complex process affected by many factors. Generally, the Pt migration profile across the CL/PEM interface and the Pt band position in the polymer membrane are thought to be dependent on potential control during cycling (Atrazhev et al., 2006; More et al., 2006), number of cycles accumulated (More et al., 2006), cell operating temperature (More et al., 2006; Bi et al., 2008), relative partial pressure of H_2 and O_2 (Li et al., 2006 ; Bi et al., 2007; Ohma et al., 2007), gas permeability of the membrane (Ohma et al., 2007), and RH (Xie et al., 2005; More et al., 2006). All of these factors will demonstrate varying degrees of influence on Pt migration behavior.

Regarding the ultimate effect of Pt migration on a PEM fuel cell, Bi et al. (2007) estimated that ~13% of the Pt initially in the cathode was transported into the membrane following 3000 potential cycles between 0.3 V and OCV. Undoubtedly, such migration would result in corresponding ECSA loss (Wang et al., 2006a). However, the exact contribution of Pt loss to total catalyst degradation is not yet known. More detailed explanations of Pt redistribution and its degrading effect on cell performance are therefore needed to further understand the specific characteristics of Pt catalyst behavior during operation.

2.3.1.4 Pt Catalyst Contamination

Another likely cause of severe degradation of the CL in PEM fuel cells is contamination. In general, contamination can be categorized into two groups based on the sources: the first includes gas contaminants from the fuel (such as CO, CH_4, H_2S, and NH_3) and the air (such as NO, NO_2, SO_2, SO_3, and some volatile organic compounds (VOCs)), and the second includes system-derived contaminants, such as trace amounts of metallic ions or silicon from system components (e.g., bipolar metal plates, membranes, and sealing gaskets). The above-mentioned contaminants may lead to degradation or failure of the operating fuel cell, mainly via any of the following three effects (Cheng et al., 2007): (1) deterioration of reaction kinetics, such as is caused by poisoning of reaction sites, (2) decrease in ionic conductivity of the membrane and/or ionomer, and (3) mass transport problems due to changes in the structure of the CL and GDL, or an inadequate hydrophilicity/hydrophobicity ratio. Poisonous effects vary based on the nature of the contaminant and on many other factors, such as the effective time, poison dosage, and so on. However, deterioration of the reaction kinetics, which is mainly caused by catalyst site poisoning, is the dominant degradation mechanism.

Contamination can have either reversible or irreversible effects on a PEM fuel cell, depending on the conditions. Even trace amounts of impurities such as sulfur containing species in the reactant gas are likely to reduce fuel cell performance due to kinetic losses, especially over long-term operation. Garzon et al. (2006) observed performance losses after just 4 h of exposure to 1 ppm of H_2S in hydrogen. When the test was continued with ultra-pure H_2 for many hours, no significant recovery occurred, indicating the irreversibility of the H_2S poisoning process. CV results demonstrated the presence of sulfur species chemisorbed onto the Pt surface. Even with levels of H_2S as low as 10 ppb, the authors found negative effects on PEM fuel cell performance. For the cathode side, some other sulfur-containing species such as SO_2 are also contaminants that can create irreversible effects in the MEA and have a strong negative impact on cell performance. Mohtadi et al. (2004) have reported that the rate of SO_2 poisoning at the cathode is strongly dependent on the bulk concentration of SO_2. Similarly, Wang et al. (2006b) demonstrated that the poisoning ability of SO_2 depended not only on the dosage, but also on its concentration in the air.

When ozone is in the cathode side, it has been suggested that the irreversible degradation effect arises from the accelerated dissolution of catalyst Pt particles and the corresponding degradation of the polymer (Franck-Lacaze et al., 2009).

However, for some contaminants in the feeding gas, such as CO in hydrogen and NO_2 in air, the negative influence on the fuel cell is reversible when the contaminating gas is switched to pure reactant gas after poisoning (Götz and Wendt, 1998; Mohtadi et al., 2004). The mechanism of recovery from the NO_2 effect is different from that from CO. CV spectra for clean and poisoned MEAs indicate that NO_2 does not act via catalyst surface poisoning. This could be because of an ionomer and/or catalyst–ionomer interface effect due to the formation of NH_4 from NO_2.

2.3.2 Effect of Operating Conditions on Pt Catalyst Degradation

With the knowledge of Pt catalyst degradation mechanisms introduced above, it is easy to understand that PEM fuel cell performance decline is a complicated event involving many parallel factors. The normal relevant operating conditions include temperature, RH, potential, and contaminant. The following subsections will provide more information on the investigation of current durability, as well as the specific research methods used to explore fuel cell degradation mechanisms.

2.3.2.1 Temperature

2.3.2.1.1 Low Temperature

The effect of temperature on catalyst durability was subjected to detailed experimental study by Borup et al. (2006), who concluded that the rate of Pt particle growth and performance degradation increased with temperature, as shown in Figure 2.4.

By employing advanced analytical techniques, more detailed quantitative information can be gained. For example, using a conventional electrochemical cell in combination with a quartz crystal microbalance, Dam and de Bruijn (2007) studied the influence of temperature and potentials on Pt thin-film dissolution in 1 M $HClO_4$ solution. They detected that the Pt catalyst dissolution rate was accelerated by increasing the temperature and the dissolution potential, which agrees with the Nernst equation. When temperature was increased from 60°C to 80°C, the Pt dissolution rate increased from 0.87 ng h^{-1} cm^{-2} to 1.58 µg h^{-1} cm^{-2} when exposed to a potential of 1.15 V. However, the amount of Pt dissolution was too small to measure at 40°C.

2.3.2.1.2 High Temperature

For high-temperature PEM fuel cells using commercially available H_3PO_4-doped poly-benzimidazole (PBI)-based MEAs (PEMEAS Fuel Cell Technologies), the degradation rate was also found to increase with increasing temperature (Sethuraman et al., 2008). When the operating temperature of the PEM fuel cell went from 240°C to 300°C, both the membrane and the kinetic charge-transfer resistances increased dramatically, mostly due to membrane and CL degradation. Similarly, Schmidt (2006)

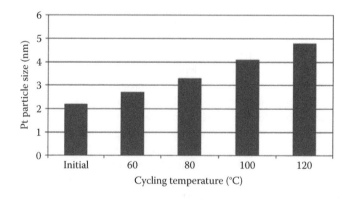

FIGURE 2.4 Platinum particle size after cycling from 0.1 to 0.96V as a function of operating cell temperature. (Reproduced from *J. Power Sources*, 163, Borup, R. L. et al. PEM fuel cell electrocatalyst durability measurements, 76–81, Copyright (2006), with permission from Elsevier.)

reported the durability modes of a Celtec®-P Series 1,000 MEA. Based on *in situ*, high-temperature operation experimental results and corresponding calculations, the author proposed that 55% of the cathode degradation resulted from increased mass transport overpotentials, 30% from reduced oxygen reduction kinetics, and 15% from increased ohmic cell resistance at the operating point of 0.2 A cm^{-2} at 160°C. This degradation trend is unhelpful for progress in the development of a high-temperature PEM fuel cell, which is an important direction for fuel cell technology due to its high-reaction kinetics and low water management requirements. To further facilitate the application of long-life fuel cells, more detailed evaluations of catalyst degradation under high temperatures, followed by the development of mitigation strategies, are urgently needed.

2.3.2.1.3 Subzero Conditions

By analyzing the performance and component properties of a PEM fuel cell stack under subzero conditions, Alink et al. (2008) concluded that although freeze/thaw cycling in a "wet" stack as well as cold start-ups can accelerate the reduction of surface areas in cross-sections, especially on the cathode side, neither the Pt concentration nor the absolute amount of Pt in the samples showed any correlation with subzero exposure. The electrode surface areas had already decreased during standard operation, and the performance degradation was mainly attributable to changes in wetting properties arising from distinct damage and increasing porosity in the electrode (Alink et al., 2008). Other investigations (Cho et al., 2003; Yan et al., 2006a) also proved that the possible delaminating of the CL from both the PEM and the GDL due to cycles of ice formation and melting may be the key issue leading to loss of thermal and electrical interfacial contact under subzero conditions.

2.3.2.2 Relative Humidity

With regard to the RH effect on catalyst degradation, Xie et al. (2005) used transmission electron microscopy (TEM) to show morphological changes and migrations of Pt in the anode and cathode CLs after high-humidity PEM fuel cell durability tests. During 1000-h testing, the authors found that cathode catalyst agglomeration occurred mainly in the first 500 h of operation. There were also significant morphological changes in the agglomeration, and migration of Pt particles occurred at the anode CL–Nafion membrane interface. Similar experiments also showed that when other operating conditions were kept constant during ASTs, the Pt degradation increased with the rising RHs of the feeding fuel and oxidant, as presented in Figure 2.5 (Borup et al., 2006). Bi et al. (2009) ascribed accelerated Pt degradation under high RH to increased Pt ion transport in the more abundant, larger water channel networks within the polymer electrolyte. According to their report, the observed loss rates for the cathode

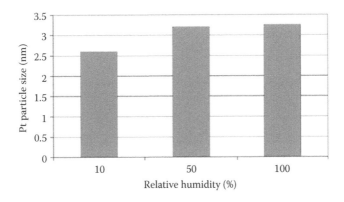

FIGURE 2.5 Platinum particle size after cycling from 0.1 to 0.96 V as a function of operating cell RH. (Reproduced from *J. Power Sources*, 163, Borup, R. L. et al. PEM fuel cell electrocatalyst durability measurements, 76–81, Copyright (2006), with permission from Elsevier.)

Pt mass and catalyst active surface area were reduced by about 3 and 2 times, respectively, when the RH was dropped from 100% to 50% at a cell temperature of 60°C. The degree of Pt oxidation was also found to increase significantly when RH went from 20% to 72%, but further increase was not apparent above 72% (Xu et al., 2007).

2.3.2.3 Potential Control/Load Cycling

PEM fuel cells are expected to be an excellent power source for automotive applications because they exhibit zero emissions, high-efficiency conversion, and high-power density. When used in an electric car, a fuel cell must have high durability under both frequent potential change and start/stop functions. This is the main reason why potential control and load cycling are the most frequently employed AST stressors for exploring catalyst durability in PEM fuel cells. Ordinary potential control protocols used to study catalyst degradation include: (1) potential cycling from low to high, (2) square- or triangular-wave potential control, and (3) steady-state potential control at specified voltage values.

As mentioned in Section 2.3.1.1, potential plays an important role during ASTs, since higher potentials can accelerate Pt dissolvability. Pt migration is the most likely negative effect of potential control/load cycling conditions. The most serious Pt migration is reported to occur at the cathode side of the MEA after potential cycling to OCV or higher voltages. For instance, according to Dam and Bruijn's (2007) *ex situ* experimental results, the Pt dissolution rate increased about 54-fold as the potential increased from 0.85 to 1.15 V at 80°C in 1 M $HClO_4$ solution; the dissolution rate became saturated at 80°C with potentials higher than 1.15 V, which they viewed as due to the formation of protected surface platinum oxide at high potentials.

Aside from potential value, load cycling profile is another important factor when investigating performance changes in PEM fuel cell catalyst degradation (Borup et al., 2006; Yan et al., 2006b). As an AST strategy, the load cycling profile of a fuel cell has been found to have a significant effect on its lifetime. This is especially notable in investigations simulating automotive drive cycles.

Under conditions of dramatic load changes, corrosion of the carbon support due to fuel starvation is also a reason for aggregation of Pt catalyst particles, ultimately resulting in reduced fuel cell durability. This aspect of degradation will be discussed in detail in Chapter 3.

2.3.2.4 Contamination

2.3.2.4.1 CO Contamination

In the fuel cell literature, the most extensively investigated catalyst degradation contaminant is CO. Basically, CO poisoning of Pt catalyst is understood to occur because CO competes with the adsorption of H_2 on Pt active sites at normal operating potentials; CO can adsorb preferentially and strongly on Pt catalytic active sites and thus block H_2. Even trace amounts of CO impurities from the reactant gas are likely to reduce fuel cell performance due to kinetic losses at the anode (Yamazaki et al., 2007). According to Ralph and Hogarth (2002), even 10 ppm CO penetrating the anode of a PEM fuel cell will seriously poison the Pt catalyst. In a quantitative investigation using fuel comprising H_2 mixed with 1% CO, the CO blocked 98% of the Pt active sites 25°C. The corresponding equations are

$$H_2 + 2Pt \rightarrow 2\left(\frac{H}{Pt}\right) \tag{2.11}$$

$$2\left(\frac{H}{Pt}\right) \rightarrow 2H^+ + 2e^- + 2Pt \tag{2.12}$$

$$CO + Pt \rightarrow \frac{CO}{Pt} \tag{2.13}$$

CO adsorbed on the platinum surface can be removed by its reaction with hydroxyl species that arise from the dissociation of water on the Pt surface at potentials above 700 mV. Therefore, the cell can recover from the poisoning effect of CO, under low current conditions (i.e., idling) when the CO supply is cut off. Figure 2.6 presents the performance revolution of the cell voltage when CO was introduced and then replaced in the anode gas supply.

$$H_2O + Pt \rightarrow \frac{OH}{Pt} + H^+ + e^- \tag{2.14}$$

$$\frac{CO}{Pt} + \frac{OH}{Pt} \rightarrow 2Pt + CO_2 + H^+ + e^- \tag{2.15}$$

2.3.2.4.2 Anion Contamination

Some anions, such as those containing Cl^-, are also potential contaminants, arising from catalyst preparation or from the fuel cell's feed stream. Schmidt et al. (2001b) found that the ORR activity of Pt catalysts with different anions decreased in the order of $ClO_4^- > HSO_4^- > Cl^-$, which is consistent with the increasing adsorption bond strength of the anions. By using TEM, CV, and electro probe microanalysis (EPMA), Matsuoka et al. (2008) also demonstrated the performance changes caused by Cl^-. The authors described the poison mechanism of Cl^- as two steps: (1) Cl^- promotes the dissolution of Pt and produces Pt ions ($(PtCl_4)^{2-}$ or $(PtCl_6)^{2-}$) in the inlet side of the cathode, and (2) crossover H_2 reduces the Pt-containing ions to metal Pt and deposits them in the PEM to form a Pt band.

2.3.2.4.3 Other Contaminants

Other contaminants, such as NH_3, SO_x, and NO_2 in the airstream, can also cause catalyst degradation. For example, NH_3 can damage cell performance by degrading the membrane and the ionomer in the CL, the extent of the damage depending on the level of impurity and duration of exposure (Uribe et al., 2002). As for SO_x contamination, it is generally believed that the S-containing species adsorbs on the

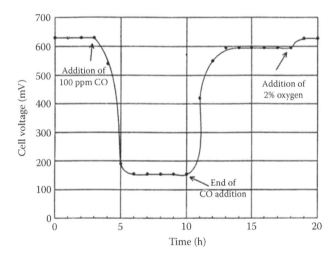

FIGURE 2.6 Effect of CO on a PEM fuel cell operated on H_2 at a current density of 100 mA cm^{-2}, $T = 70°C$, cathode and anode Pt loading 0.2 mg cm^{-2}. (Reproduced from *Electrochim. Acta*, 43, Götz, M. and Wendt, H. Binary and ternary anode catalyst formulations including the elements W, Sn, and Mo for PEMFCs operated on methanol or reformate gas, 3637–3644, Copyright (1998), with permission from Elsevier.)

active sites of a catalyst surface, occupying the polyatomic sites and thereby preventing the reactant oxygen from adsorbing on the catalyst surface. Contractor et al. (1979) and Loučka (1971) reported that SO_2 adsorbed on the Pt surface to produce linearly and bridged adsorbed S species. These two forms of chemisorbed S species on the Pt surface were responsible for catalyst poisoning in their studies.

2.3.2.4.4 Mixed Contaminants

Currently, combining a stressor with impurity mixtures has drawn considerable attention. Using a hydrogen mixture of H_2S (10 ppb), CO (0.1 ppm), CO_2 (5 ppm), and NH_3 (1 ppm), which simulated the preliminary maximum allowable fuel impurity specifications, Garzon et al. (2006) operated a 50 cm^2 fuel cell at 0.8 A cm^{-2} for 1000 h at 80°C. The authors attributed the total performance loss of 100 mV to NH_3 and H_2S impurities in the fuel supply, as apparently very little was attributable to CO and CO_2. The CV spectrum showed that about 40% of the Pt surface was covered with adsorbed H_2S during 1000 h of exposure; they explained the lack of CO adsorption from the multicomponent test as due to competitive adsorption effects.

Apparently, the effect of contaminants on catalysts is not yet thoroughly understood. To improve fuel cell lifetime and develop adequate mitigation strategies, it is important to gain a fundamental understanding of the main contamination mechanisms and their impact on fuel cell performance. More examples of contaminants' contributions to catalyst poisoning can be found in Chapter 8.

2.4 Degradation of Pt–M Catalysts

According to the understood decay mechanisms, a more affordable, durable catalyst that could also help lower the cost of PEM fuel cell production and increase cell durability requires the development of a catalyst with proper resilience to sintering and corrosion under universal working conditions. Currently, the majority of work in this area is focusing on developing new Pt–M electrocatalysts to stabilize Pt as well as lower the Pt loading in CLs.

2.4.1 Pt-Contained Catalysts for the HOR

As mentioned in Section 2.3.2.4, among the various durability problems, CO poisoning is one of the most likely environmental issues to deal with in stationary or cogeneration applications when reformed hydrogen is the fuel. CO is an inevitable impurity arising from the reforming process, and obviously can reduce the Pt catalyst's durability. Angelo et al. (2008) conducted a special experiment to characterize the effect of repeated cyclic exposure of the anode to a 2 ppm CO contaminant followed by removal of CO using air bleed, to simulate a repetitive adsorption process that may be experienced in a dynamically operated PEM fuel cell with low-level impurities in the anode feed stream. Result showed that the ECSA of the test cell anode—the representative parameter of the Pt catalyst—decreased significantly compared with the ECSA after the same cyclic operation without CO.

To improve the durability of PEM fuel cell catalysts exposed to CO poisoning, extensive research has been conducted over the past several decades on Pt catalysts alloyed with Ru or other metals (e.g., Sn, Re, Mo) to avoid serious activity losses. As an example, Götz and Wendt (1998) examined some Pt-alloy catalysts in a PEM fuel cell operated at 75°C. Figure 2.7 shows the performance when these catalysts were used at a poison concentration of 150 ppm CO in the anode gas feed. Among these alternative systems, Pt–Ru alloy catalysts have been the most widely investigated and a vast literature exists on the topic, including material fabrication techniques and tolerance mechanisms (Bacchuk and Li, 2001; Santo et al., 2006; Liao et al., 2008). The Pt/Ru ratio has been extensively investigated, with the optimum ratio suggested to be 50:50 for the highest CO oxidation activity (Gasteiger et al., 1993).

It has been demonstrated that improved durability under CO contamination by adding Ru to a Pt system results from either the ligand effect or the bifunctional mechanism. For the ligand effect, the

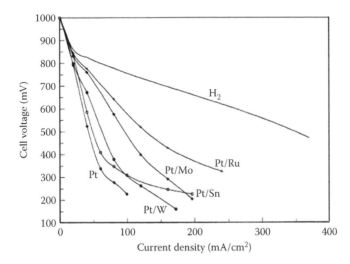

FIGURE 2.7 Polarization curves of cells operated on H_2/150 ppm CO with binary anode catalysts manufactured by the impregnation method. Noble metal loading at anode and cathode 0.4 mg cm^{-2}. ♦Pt, + Pt/Ru, ●Pt/W, *Pt/Mo, ○Pt/Sn. (Reproduced from *Electrochim. Acta*, 43, Götz, M. and Wendt, H. Binary and ternary anode catalyst formulations including the elements W, Sn and Mo for PEMFCs operated on methanol or reformate gas, 3637–3644, Copyright (1998), with permission from Elsevier.)

crucial issue is the modification of the Pt electronic properties by adding Ru, which is believed to be able to decrease the CO bond strength on the Pt surface. In the bifunctional mechanism, the major issue is posed by the adsorbates of hydrated oxides formed on Ru, which facilitate the oxidation of CO adsorbed on Pt sites at comparably negative potentials and leave free Pt sites for the anodic oxidation of H_2. However, although both of the two mechanisms have been proposed for many years, neither has been satisfactorily experimentally proven to explain the increased durability of Pt–Ru as compared to plain Pt catalyst.

Another aspect worth noting is that the CO-tolerance of a Pt–Ru/C alloy catalyst is not totally stable during long-term operation. It has been found that the performance of a PEM fuel cell using a Pt–Ru/C catalyst also gradually decreases, often with a serious voltage drop after an extended period. A significant loss in the CO-durability of Pt–Ru/C catalyst is frequently observed in ASTs with start-up and shutdown cycles and with potential cycling in different potential ranges (Inaba, 2009). This phenomenon is believed to result from the dissolution of Ru in the alloy catalyst. The same author also observed that loss of CO tolerance increased as the upper potential in potential cycling rose, especially when the upper potential was set at greater than 0.8 V. In another experiment, Yamada et al. (2007) detected that some of the commercially available Pt–Ru/C catalysts deteriorate even at potentials <0.4 V, at which potential the Ru should be thermodynamically stable. Some features of Pt–Ru catalyst, such as its crystallinity, the degree of alloying, and the presence of surface oxide, are also believed to play important roles in its stability.

In recent years, many other new Pt-based binary and ternary anode electrocatalysts, such as Pt–Sn/C, Pt–Mo/C, Pt–Ru–Mo/C, and Pt–Sn–SnO$_2$/C, have been presented for PEM fuel cells operated under reformate gas conditions. All of these alternatives are expected to further enhance anode durability under CO contamination conditions, based on different mechanisms. In the case of Sn, for example, the key effect in the anti-CO system is its ability to promote the formation of oxygen-containing species at low potentials and oxidize the poisoning CO on the Pt site to restore its activity. Sn can also weaken the Pt–CO bond to enhance diffusion and oxidation of the adsorbed CO. During this process, tin oxides might participate in catalysis as well (Jiang et al., 2005; Hung et al., 2010).

Novel CO-durable electrocatalysts based on organic metal complexes have also been explored. Okada et al. (2007) synthesized nitrogen and oxygen ligands containing transitional metal complexes and used the ligands as cocatalysts with Pt. Their results indicated that the composite catalysts, especially Pt–VO(salen)/C and Pt–Ni(mqph)/C provided high CO tolerance compared to Pt/C and Pt–Ru/C catalysts.

2.4.2 Pt-Containing Catalysts for the ORR

2.4.2.1 Pt–M Catalysts for the ORR

According to the model based on density functional theory (DFT), calculations proposed by Nørskov et al. (2004), a volcano-shaped relationship exists between the rate of the cathodic reaction and the oxygen adsorption energy during the ORR on a metal surface, as shown in Figure 2.8. Their model explains why Pt is the best elemental cathode material for the ORR. In addition, they predicted that metal alloys with a somewhat lower oxygen binding energy than Pt should have a higher catalytic activity toward oxygen reduction by making the oxygen on the Pt surface less stable. This is in good agreement with experimental results. During recent decades, numerous investigations have proven that Pt-M binary and ternary alloys show considerable enhancement in their catalytic activity for the ORR compared to Pt catalysts. The most attractive elements to alloy with Pt are selected from a variety of transition metals. For example, when Pt is alloyed with Cr, Mn, Co, and Ni, a three- to fivefold improvement in ORR kinetics can be achieved compared with a Pt catalyst (Mukerjee and Srinivasan, 1993). This enhancement of electrocatalysis has been ascribed to a combination of effects from different structural changes caused by alloying, such as modifications in the geometrical or electronic structure of the Pt metal and delayed onset of OH or O adsorption on Pt (Nørskov et al., 2008). Different base metals show different intrinsic activities when alloying with Pt. The particle size, shape, and composition as well as the method of preparing the alloy nanoparticles also play important roles in the activity of a Pt–M alloy catalyst.

2.4.2.1.1 Optimum Compositions of Pt–M Catalysts

A tremendous amount of investigation has been conducted on different Pt–M alloy systems to reveal the effects of Pt-alloy catalyst composition on fuel cell activity and stability. Numerous articles have been

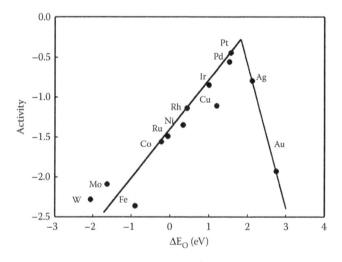

FIGURE 2.8   Trends in oxygen reduction activity plotted as a function of the oxygen binding energy. (Reproduced with permission from Nørskov, J. K. et al. Origin of the overpotential for oxygen reduction at a fuel-cell cathode. *J. Phys. Chem.* B 108: 17886–17892. Copyright (2004). The American Chemical Society.)

published discussing the optimal ratio of Pt to M; it is now generally accepted that Pt–M with an atomic ratio of 3:1 exhibits higher activity than other compositions. For example, Stamenkovic et al. (2006, 2007a) developed a series of well-characterized Pt_3M alloys (M = Ni, Co, Fe, Ti, V) that presented volcano-shaped values in electrocatalytic activity, with the maximum catalytic activity obtained for Pt_3Co. Figure 2.9 shows a high-resolution transmission electron micrograph for a Pt_3Co nanoparticle and the simulation results for a model Pt_3Co nanoparticle. Stamenkovic et al. also concluded that the catalyst's activity toward the ORR is determined by the strength of the oxygen–metal bond interaction, which in turn depends on the position of the metal d states relative to the Fermi level. In another paper (Stamenkovic et al., 2007b), the same group demonstrated that the Pt_3Ni (111) surface is 10 times more active for the ORR than the corresponding Pt (111) surface and 90 times more than current state-of-the-art Pt/C catalysts for PEM fuel cells, based on an unusual electronic structure (d-band center position) and arrangement of surface atoms in the near-surface region. Recently, considerable work has also been carried out on modification of Pt_3M catalysts with the aim of improving their catalytic activity and stability (Ma and Balbuena, 2009). Hopefully, novel catalysts with enhanced ORR activity and higher stability will be developed in the near future.

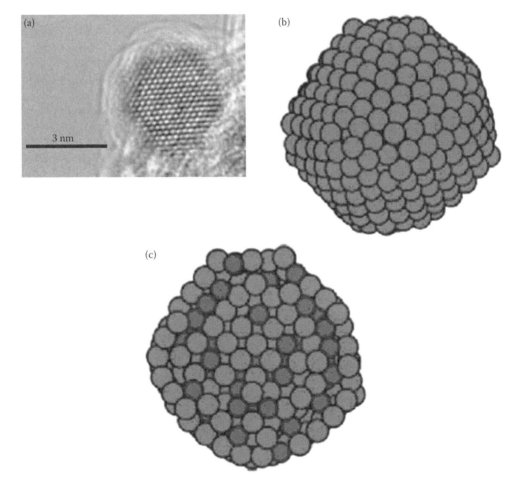

FIGURE 2.9 Nanoparticle catalysts: (a) high-resolution transmission electron micrograph of a Pt_3Co nanoparticle; external (b) and cross-section (c) snapshots of a model Pt_3Co nanoparticle of Monte Carlo (MC) simulations. (Reproduced by permission from Macmillan Publishers Ltd. *Nat. Mater.* Stamenkovic, V. R. et al. 2007b. Trends in electrocatalysis on extended and nanoscale Pt-bimetallic alloy surfaces. 6: 241–247. Copyright (2007).)

2.4.2.1.2 Durability of Pt–M Catalysts

Regarding stability, Greeley and Nørskov (2007) calculated that Pt atoms were more stable in a Pt = solute/Co = host system compared to pure Pt, by using DFT to estimate trends in the thermodynamics of binary surface alloy dissolution in acidic media. Franco and co-workers' (2009) simulation results also indicated that Pt_3Co was more stable than Pt, PtCo, or $PtCo_3$ for both low currents (0.1 A cm^{-2}) and intermediate currents (0.5 A cm^{-2}).

In practical operating systems, apart from factors related to Pt degradation in a Pt/C catalyst, an additional possible degradation source for Pt–M alloy catalysts is the presence of non-alloyed metal and its oxides on the catalyst surface, which could prevent molecular oxygen from reaching the Pt sites. Leaching and/or dissolution of the metal from the alloy surface in an acidic medium has been identified as one of the most important causes of electrocatalytic deactivation in PEM fuel cells. Kobayashi et al. (2009) proposed that the decrease in metallic Co atoms—Co oxidation into Co^{3+}—caused by Nafion and the electrochemical environment rather than oxidation by oxygen is probably a direct reason for cathode CL degradation in PEM fuel cells. The authors proposed that the possible reaction may involve the following processes: the dissolved Co is trapped by Nafion and oxidized to Co^{2+} by F, which has a higher electronegativity than oxygen; then the Co^{2+}-containing components will be oxidized to Co^{3+} by hydrogen peroxide, similar to the Fenton reaction in the case of Fe and Cu ions. These possible cationic impurities may contaminate the ionomer in the CL and the PEM at the same time (Kelly et al., 2005; Colón-Mercado and Popov, 2006). In a review paper, Gasteiger et al. (2005) concluded that, based on the readily occurring ion exchange with protonic sites in the ionomer of the CL and PEM, the possible contamination effects of the base-metal include: (1) higher membrane resistance, (2) higher ionomer resistance in the CL, (3) lower oxygen diffusion in the ionomer of the CL, and (4) membrane degradation.

Excessive base metal and incomplete alloying are undoubtedly the most likely sources of leaching because of intrinsic thermodynamic properties. During extended running, even well-alloyed metal may dissolve and leach from the surface under PEM fuel cell operating conditions. Antolini et al. (2006) proved that the stability of the second metal is linked to its content in the alloy. For alloys of $Pt_{1-x}M_x$ (M = Fe, Ni, and Co), when x was small, only the unalloyed base metal could be dissolved, but the alloyed second metal dissolved when x was increased to 0.6. A similar result was observed by Bonakdarpour et al. (2005) through electron microprobe and XPS measurement under simulated operating conditions. Figure 2.10 compares some of the nonnoble metal dissolution curves when alloyed with Pt at 0.8 V. Therefore, as proposed by Mukerjee et al. (2003), preleaching of the alloy to minimize the possible dissolution of metal from the system during operation is a necessary procedure in catalyst preparation. Stassi et al. (2009) also found that Pt–Co catalyst pre-leached with perchloric acid showed better stability than other catalysts in a Pt degradation test. Another method for stabilizing base metals on the surface layers of alloys is heat treatment, which reduces the dissolution of base metals by inducing a higher degree of alloying and crystalline structures. At the same time, heat treatment can also result in the formation of large catalyst nanoparticles. The increase in particle size relative to more conventional Pt/C catalysts is also expected to improve the electrocatalyst's stability. Bezerra et al. (2007) have reviewed the effects of heat treatment on the catalytic activity and stability of PEM fuel cells catalysts, including Pt–M/C. They presented the relationship between heat treatment and particle-size growth, alloying degree, and the catalyst surface morphology of different Pt–M/C catalysts, as well as the effects of these factors on the catalyst's properties.

2.4.2.1.3 Durability Test Results for Pt–Co/C Catalysts

Similar to Pt/C catalysts, the durability of Pt-alloy catalysts is usually characterized at two different levels: half-cell and full cell. Considerable work on this has been carried out during recent years. Both high and low stability have been reported for cathode alloy catalysts in PEM fuel cells using either half-cell or full cell systems. For instance, Tarasevich et al. (2007) compared the durability of Pt-Co/C and Pt/C catalysts using corrosion tests conducted in a 0.5 M H_2SO_4 solution in a half-cell. The temperature

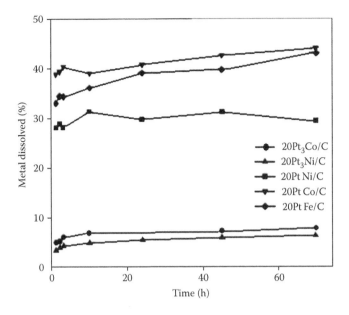

FIGURE 2.10 Non-noble metal dissolution data as a function of time for different Pt-alloy catalysts at a fixed potential of 0.8 V vs. NHE. (Reproduced from *J. Power Sources*, 155, Colón-Mercado, H. R. and Popov, B. N. Stability of platinum based alloy cathode catalysts in PEM fuel cells, 253–263, Copyright (2006), with permission from Elsevier.)

of the solution was periodically increased, up to 60°C, and the testing was performed in the presence of 10% H_2O_2 or air bubbling. The experimental results showed that the amount of dissolved Pt for Pt-Co/C (the alloy exists in two phases, Pt_3Co and PtCo) was lower than for Pt/C and did not rise with increased exposure to corrosion; however, this paper presented no results other than those from this purely chemical corrosion test. Many similar reports (e.g., Ye et al., 2006; Ball et al., 2007b) have also demonstrated that in a half-cell, Pt-alloy catalysts containing Co show significant improved durability in potential cycling compared to unalloyed Pt, an outcome ascribed to there being more active sites for fuel cell reactions due to special structural factors (e.g., smaller Pt–Pt bond distance) and surface electronic properties (e.g., inhibition of OH_{ads} formation).

Other than *ex situ* testing based on half-cell systems, *in situ* tests in full cells are conducted to further understand the durability issues for Pt-alloy based catalysts. Comparison of the results from different test systems can yield much useful information. Investigating the influence of carbon support corrosion, Pt dissolution, and sintering, Aricó et al. (2008) evaluated the stability of Pt/C and Pt-Co/C in both a gas-fed sulfuric acid electrolyte half-cell at 75°C and a PEM fuel cell at 130°C. Results in sulfuric acid revealed the Pt–Co alloy to be more stable than Pt after cycling and more sensitive to carbon support corrosion. In high-temperature PEM fuel cell testing, Pt–Co/C showed smaller sintering effects than Pt/C.

Yu et al. (2005) evaluated the durability of Pt-Co/C cathode catalyst in a dynamic fuel cell environment (potential cycling between 0.87 and 1.2 V) with continuous water fluxing on the cathode. The result indicated that Co dissolution neither detrimentally reduced cell voltage nor dramatically affected membrane conductance, even though the amount of Co dissolution was much larger than that of Pt. The performance loss of a Pt-Co/C MEA was as substantial as the Co dissolution in the first 400 cycles; however, the overall performance loss of the Pt-Co/C MEA was less than that of the Pt/C MEA. Zignani et al. (2008) reported that pure Pt/C showed higher electrochemical stability than the Pt–Co/C binary catalyst when they carried out durability tests of 30 h of constant potential operation at 0.8 V, repetitive potential cycling in the range 0.5–1.0 V, and thermal treatments. The authors ascribed the lower durability of Pt–Co/C during repetitive potential cycling to the dissolution/redeposition of Pt but not to Co loss, which meant that the ORR activities of both Pt and Pt–Co after repeated potential depended on particle size. The degree of alloying has also been regarded as a major factor influencing the stability of Pt-alloy catalysts in PEM fuel cells (Antolini et al., 2006).

On a large scale, an automotive application degradation test with a 160-cell stack was carried out by Haas and Davis (2009). They evaluated Pt–Co alloy catalyst durability with various designs (i.e., different carbon support and metal wt%) in a simulated drive-cycle test that included 640 air–air starts. Experimental results showed that in this voltage stress test, all Pt–Co alloy designs degraded quickly under simulated drive-cycle conditions relative to the Pt baseline. Pt–Co alloy stability was found only in a small voltage window between 0.6 and 0.9 V. The authors concluded that current alloys lack the expected stability for use in a realistic system approach, although they remain the most possible candidates for the future. Other than metal dissolution, changes in the structure and particle size of catalysts similar to Pt/C might be another degradation factor in Pt–M/C catalysts under the aggressive oxidative conditions of an operational fuel cell.

2.4.2.2 Novel-Structured Pt–M Catalysts

Other than Pt-based alloy catalysts, novel nanostructured Pt–M bimetal or tri-metal catalysts have also attracted research attention for their potential high activity and better durability in the ORR. Extensive R&D is being conducted to prepare such durable, high-activity catalysts.

Among diverse approaches, progress in creating a core–shell-structured catalyst is a promising breakthrough. For example, Koh and Strasser (2007) reported a "de-alloying" method achieved by applying a cyclic alternating current to carbon-supported Cu–Pt-alloy in the preparation of a novel catalyst with a core–shell nanoparticle structure. The catalyst exhibited MEA oxygen mass activities of 0.3–0.53 A mg^{-1} Pt at 0.9 V and 80°C. The authors demonstrated that the electrochemical dealloying of Cu from Pt–Cu bimetallics significantly altered the catalyst's surface activity and improved its durability under potential cycling conditions. Moreover, the larger size of the core–shell particles might be another factor making them intrinsically more stable than pure platinum. A group in Brookhaven National Laboratory (Brankovic et al., 2001; Zhang et al., 2007) has also developed different approaches to synthesize carbon-supported core–shell Pt–M nanoparticles. Their representative Pt–Cu elctrocatalysts were generated by the Pt galvanic displacement of Cu, preformed by underpotential deposition (UPD) (Zhang et al., 2007a). As shown in Figure 2.11, results of thin-film rotating disc electrode (RDE) measurements showed no ORR activity loss after 30,000 cycles of potential cycling between 0.6 and 1.1 V, compared with a more than 45% loss of ECSA on the Pt/C electrode. Reviews covering this aspect of the subject in detail can be found in references such as Bing et al. (2010).

2.4.2.3 Metal Oxide-Promoted Pt Catalysts

Recently, some metal oxides, including SnO_2, MnO_2, RuO_2, and Nb_2O_5 (Trogadas et al., 2007; Adzic et al., 2010; Parrondo et al., 2010), have also been applied to the carbon-supported Pt catalyst system to enhance its activity and durability. The mechanism of this performance enhancement varies according to the specific oxide.

For oxides such as TiO_2 or ZrO_2, the strong metal–support interaction and the anchor effect between the metal oxides and the adjacent Pt atoms are major reasons for the increased electrochemical stability (Liu et al., 2006, 2010; Bauer et al., 2010). In a report by Huang et al. (2010), after being subjected to continuous voltammetric cycles in the range of 0.6–1.4 V vs. RHE in an RRDE, the $Pt/Nb_xTi_{(1-x)}O_2$ catalyst showed nearly 10-fold higher ORR activity than the Pt/C catalyst. The main disadvantages of such catalysts are the relatively lower electronic conductivity and lower surface area. Under these conditions, the metal oxide is more like a support promoter than a catalyst promoter.

The addition of some metal oxides into the cathode Pt/C system has been studied to facilitate the supply of oxygen species and prevent Pt agglomeration, leading to enhanced fuel cell performance and durability. For example, CeO_2 was used as a promoter for Pt/C catalyst. Xu et al. (2003) reported that its

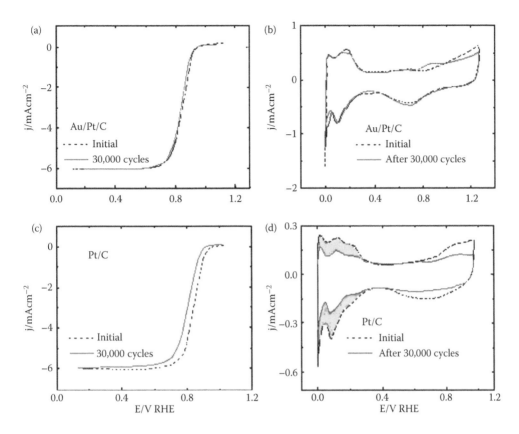

FIGURE 2.11 Polarization curves for the O_2 reduction reaction on Au/Pt/C (a) and Pt/C (c) catalysts on a rotating disk electrode, before and after 30,000 potential cycles. Voltammetry curves for Au/Pt/C (b) and Pt/C (d) catalysts before and after 30,000 cycles. The potential cycles were from 0.6 to 1.1 V in an O_2-saturated 0.1 M $HClO_4$ solution at room temperature. The shaded area in (d) indicates the lost Pt area. (From Zhang, J. et al. 2007a. Stabilization of platinum oxygen-reduction electrocatalysts using gold clusters. *Science* 315: 220–222. Reproduced with permission of The American Association for the Advancement of Science.)

oxygen storage ability (i.e., the ability to provide additional oxygen to the Pt active site) can improve cell performance at lower voltages. Furthermore, according to the potential cycling experiment conducted by Lim et al. (2010), the durability of Pt–CeO_2/C catalyst was much higher than commercial Pt/C in acidic solution. Compared with commercial Pt/C, the ECSA and particle size of the CeO_2 promoted catalyst showed no significant variation after the continuous application of potentials between 0 and 1.2 V in 0.5 mol L^{-1} H_2SO_4 solution. The authors ascribed this favorable performance to the CeO_2, which prevents the agglomeration and dissolution of Pt particles on the carbon support. Another advantage of oxide-promoted Pt catalysts is their lower price compared with traditional Pt/C. Therefore, although their catalytic efficiency is not as high as that of Pt/C, metal-oxide combined Pt/C catalysts are promising cathode material alternatives.

2.5 Degradation of Non-Pt Catalysts

In the quest to prepare efficient, durable, inexpensive catalysts as alternatives to Pt and Pt-based materials, non-Pt catalysts may be a feasible way to permanently resolve the cost issue in PEM fuel cell commercialization. Although there is still a long way to go in making them comparable with Pt-based catalysts, significant progress in non-Pt catalysts has been reported in recent years.

Two types of non-Pt catalysts are currently being explored: non-Pt metals and N-containing transition metal macrocycle catalysts.

2.5.1 Non-Pt Metal Catalysts

Non-Pt metal-based catalysts can be used as both anode and cathode components. The addition of tungsten carbide has been reported to improve the catalytic ability of some non-Pt catalysts. Izhar and coworkers (2009) investigated carbon-supported cobalt–tungsten and molybdenum–tungsten carbides and their activities as anode catalysts, using a single fuel cell and half-cell RDE. The maximum power densities of their 873 K-carburized CoWC/KB and MoWC/KB were 15.7 and 12.0 mW cm^{-2}, respectively, which were 14% and 11%, compared to a 20 wt% Pt/C catalyst.

In addition, non-Pt metal catalysts have also been investigated to improve the CO durability of the anode catalyst. For example, Schmidt et al. (2001a) prepared three types of Pd–Au/C catalyst with different compositions and proved that the surface of Pd–Au/C catalysts was less strongly poisoned by CO than that of Pt–Ru/C at a temperature of 60°C.

Compared with the anode, the substitution of the cathode Pt catalyst with a non-Pt material is likely to result in a greater reduction in Pt needed for PEM fuel cells, due to the greater Pt catalyst loading at the cathode. Such catalysts are typically based on noble metals other than Pt. For example, Fernández et al. (2005) prepared two novel catalysts, Pd–Co–Au and Pd–Ti, both of which showed considerable catalytic activity toward the ORR. Their study demonstrated the feasibility of developing non-Pt-alloy compositions that offer catalytic activity similar to that of Pt. Zhang et al. (2007b) also verified that heat treatment is necessary to enhance the catalytic activity toward the ORR of a carbon-supported Pd–Co alloy. However, these studies demonstrated few durability advantages in comparison with Pt-based catalysts.

2.5.2 N-Containing Transition Metal Macrocycle Catalysts

The possibility of lower cost means that the development of transition metal compound catalysts is more active than that of non-Pt metal catalysts. In the last few years, many researchers have attempted to improve the quality of non-Pt catalysts, with tremendous progress. For instance, Lefèvre et al. (2009a) synthesized an iron-based (Fe/N/C), Pt-free catalyst. During their experiments they achieved a great increase in site density when a mixture of carbon support, phenanthroline, and ferrous acetate was ball milled and then pyrolyzed twice, first in argon, then in ammonia. One special favorable step in the preparation was the introduction of a new material (pore filler) into the reaction system. The reported current density of the cathode made with this catalyst matched the performance of a Pt/C catalyst with a loading of 0.4 mg cm^{-2} in the kinetic region of a H_2/O_2 polarization curve (0.9 V iR-free), which is regarded as an important milestone for the development of N-containing transition metal macrocycles catalysts.

2.5.2.1 Catalytic Mechanisms for the ORR

Generally, the main feature of N-containing transition metal macrocyclic complexes involves structures such as phthalocyanines, porphyrines, and related derivatives (Lefèvre et al., 2000). Figure 2.12 presents an example of an Fe-based catalyst, phenanthroline structure suggested by time-of-flight secondary ion mass spectrometry (ToF SIMS). When such nitrogen atoms exist on the carbon support, adding Fe ions (even 50 ppm) induces catalytic activity. Increasing the Fe content increases the concentration of the catalytic sites until all nitrogen atoms of the phenanthroline type are coordinated with Fe.

Research into such catalysts started in the 1960s, when Jasinski (1964) discovered that cobalt phthalocyanine could catalyze the ORR in an alkaline medium. Fe and Co are the most commonly used metals for these catalysts. The origin of the electrocatalytic activity of N-containing non-Pt catalysts was generally recognized to be the N_4-chelates (or N_2-chelates) of transition metals, due to the simultaneous presence of metal precursors, active carbon, and a nitrogen source under pyrolysis conditions. Beck (1977) proposed that the mechanism of the ORR catalyzed by such catalysts was mainly involved by a modified

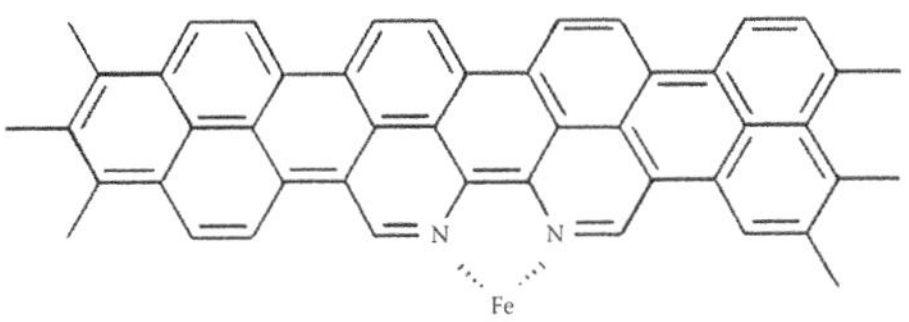

FIGURE 2.12 Proposed moiety of a high-temperature catalytic site. (Reproduced with permission from Lefèvre, M., Dodelet, J. P., and Bertrand, P. Molecular oxygen reduction in PEM fuel cells: Evidence for the simultaneous presence of two active sites in Fe-based catalysts. *J. Phys. Chem.* B 106: 8705–8713. Copyright (2002). The American Chemical Society.)

"redox catalysis": first, the oxygen was adsorbed on the catalyst metal center; after the reduction process, O_2 molecules and the reaction products (e.g., H_2O_2) remained strongly coordinated to the central metal ion of the chelates XMe^{II}; then, the potential-determining step, which regenerated the reduced chelates, was as follows:

$$(XMe^{III} \ldots O_2H)^+ + H^+ + 2e^- \rightarrow XMe^{II} + H_2O_2 \tag{2.16}$$

2.5.2.2 Synthesis

One possible approach for synthesizing such catalysts is to adsorb metal precursors and organic molecules onto carbon, following this with heat treatment in an inert atmosphere (Faubert et al., 1999). Another approach is based on the pyrolyzing of a metal precursor (a Fe or Co salt, a ferrocene derivative, etc.) adsorbed on carbon, and a nitrogen precursor that is either Co-adsorbed on carbon, or is an N-containing functionality at the carbon surface, or is an N-containing gas or vapor injected into the reactor (Lefèvre et al., 2000).

During the exploration of catalysts containing transition metal/nitrogen complex catalysts, many synthesis conditions have demonstrated considerable effects on catalysts' activity toward the ORR, including metal type and metal source (Lefèvre et al., 2005; Pylypenko et al., 2008; Wu et al., 2009), metal loading (Lalande et al., 1995; Zhang et al., 2009b), nitrogen resource (Wood et al., 2008; Wu et al., 2009), carbon support (Jaouen et al., 2003; Lefèvre et al., 2009b), and heat treatment (Biloul et al., 1996; Pylypenko et al., 2008). Different synthetic conditions may result in different activated sites for the ORR. Using ToF SIMS, Lefèvre et al. (2002) observed that both FeN_4/C and FeN_2/C catalytic sites were *ex situ* detected regardless of the precursors used, but the two sites were in different relative proportions, depending upon the precursor; meanwhile, for the Co-based catalyst, no similar dominant catalytic site such as $CoN_xC_y^+$ was found. The only catalytic site certainly present in Co-based catalysts is CoN_4/C (Lefèvre et al., 2005).

2.5.2.3 Durability

The stability of such macrocycle catalysts is totally different from that of Pt-based catalysts because of their different catalytic mechanisms. The former's durability is highly affected by factors such as synthesis conditions and catalyst types. For example, Biloul et al. (1996) tested the performance of Co porphyrin ($TpOCH_3PPCo$), its derivative (CF_3PPCo), and dihydrodibenzotetraazaannulene (CoTAA) as cathode catalysts under a constant current density of 100 mA cm^{-2} for fewer than 300 h. The durability order was observed to be $CF_3PPCo < TpOCH_3PPCo \ll CoTAA$. With respect to the temperature effect, heat treatment is widely accepted as a positive method for improving the stability of carbon-supported, N-containing transition metal macrocyclic complexes. Gouérec et al. (1997) studied the optimal temperature for preparing a stable CoTAA catalyst. According to the authors' observations, samples after heat treatment at 600°C under an N_2 or Ar atmosphere showed the best performance in terms of both

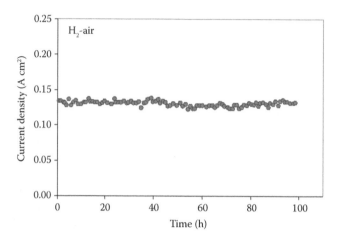

FIGURE 2.13   Long-term performance of an H$_2$-air fuel cell with Co-PPY-C composite cathode at 0.40 V. Co loading 6.0×10^{-2} mg cm^{-2}; cell temperature 80°C; flow rates of hydrogen and air at 5 mL s^{-1} and 9 mL s^{-1} (referred to as standard conditions), respectively. Anode and cathode gases humidified at 90°C and 80°C, respectively. Backpressure of gases 2.0 atm on both sides of the cell. (Reproduced by permission from Macmillan Publishers Ltd. *Nature*, Bashyam R. and Zelenay, P. 2006. A class of non-precious metal composite catalysts for fuel cells. 443: 63–66. Copyright (2006).)

activity and durability under 500 mV (vs. RHE) for 100 h. When the temperature was increased beyond this point, decomposition of the macrocycle was suggested by ToF-SIMS detection.

Recently, two durability tests of N-containing transition metal macrocycle catalyst were presented in *Nature* and *Science*. Both studies showed great improvement in catalyst activity and durability. According to representative work done by Bashyam and Zelenay (2006), shown in Figure 2.13, the durability of Co-polypyrrole-carbon (Co-PPY-C) catalysts synthesized via a nonpyrolytic method were comparable to that of an E-TEK 20 wt% Pt/C catalyst (Pt loading of 0.2 mg cm^{-2}) when it was used as the cathode catalyst in a PEM fuel cell during a 100-h life test. No obvious performance degradation was observed under fuel cell operating conditions. As mentioned previously, the ion-based catalyst prepared by Lefèvre et al. (2009a) showed a much higher initial activity. But the stability of this catalyst is not particularly high compared to that of the Co-based catalyst synthesized by Bashyam and Zelenay.

In summary, while none of the durability issues experienced with Pt catalysts were expected to arise with N-containing non-Pt catalysts during long-term operation, due to their unique structure and reaction mechanism these complexes do not so far show many fundamental advantages in chemical/electrochemical stability when used in PEM fuel cells. From a practical point of view, activity and stability are still significant challenges for state-of-the-art non-Pt catalysts in PEM fuel cell applications.

2.6 Summary

Catalyst degradation during long-term operation of PEM fuel cells under universal conditions is a complex process that includes many parallel mechanisms. Several factors can affect catalyst degradation, such as fuel cell operating conditions, preparation conditions, and catalyst type. Considerable effort has been made to understand the reasons behind electrocatalyst degradation and the possible solutions to this problem, based on half-cell, single cell, and stack tests. Currently, improvements to both traditional Pt-based electrocatalyst materials and non-Pt catalysts are important in moving fuel cell technology closer to mainstream applications. However, more work is needed to further understand failure modes and to develop corresponding mitigation methods to satisfy the demand for electrocatalysts with high durability in both transportation and stationary applications. From the viewpoint of the whole CL, more attention should also be paid to ionomer and interfacial degradation, to investigate the significant influences of fundamental mechanisms on catalyst degradation.

References

Adzic, R., Vukmirovic, M., and Sasaki, K. 2010. U.S. patent 7,704,918.

Akita, T., Taniguchi, A., Maekawa, J. et al. 2006. Analytical TEM study of Pt particle deposition in the proton-exchange membrane of a membrane-electrode-assembly. *J. Power Sources* 159: 461–467.

Alink, R., Gertersen, D., and Oszcipok, M. 2008. Degradation effects in polymer electrolyte membrane fuel cell stacks by sub-zero operation—An *in situ* and *ex situ* analysis. *J. Power Sources* 182: 175–187.

Angelo, M., Bender, G., Dorn, S. et al. 2008. The impacts of repetitive carbon monoxide poisoning on performance and durability of a proton exchange membrane fuel cell. *ECS Trans.* 16: 669–676.

Antolini, E., Salgado, J. R. C., and Gonzalez, E. R. 2006. The stability of Pt–M (M= first row transition metal) alloy catalysts and its effect on the activity in low temperature fuel cells, a literature review and tests on a Pt-Co catalyst. *J. Power Sources* 160: 957–968.

Aricó, A. S., Stassi, A., Modica, E. et al. 2008. Performance and degradation of high temperature polymer elecrtrolyte fuel cell catalysts. *J. Power Sources* 178: 525–536.

Atrazhev, V., Burlatsky, S. F., Cipollini, N. E., Condit, D. A., and Erikhman, N. 2006. Aspects of PEMFC degradation. *ECS Trans.* 1: 239–246.

Bacchuk, J. J. and Li, X. 2001. Carbon monoxide poisoning of proton exchange membrane fuel cells. *Int. J. Energy Res.* 25: 695–713.

Ball, S. C., Hudson, S. L., Leung, J. H., Russell, A. E., Thompsett, D., and Theobald, B. R. C. 2007a. Mechanisms of activity loss in PtCo alloy systems. *ECS Trans.* 11: 1247–1257.

Ball, S. C., Hudson, S. L., Thompsett, D., and Theobald, B. 2007b. An investigation into factors affecting the stability of carbons and carbon supported platinum and platinum/cobalt alloy catalysts during 1.2 V potentiostatic hold regimes at a range of temperatures. *J. Power Sources.* 171: 18–25.

Bashyam R. and Zelenay, P. 2006. A class of non-precious metal composite catalysts for fuel cells. *Nature* 443: 63–66.

Bauer, A., Lee, K., Song, C., Xie, Y., Zhang, J., and Hui, R. 2010. Pt nanoparticles deposited on TiO_2 based nanofibers: Electrochemical stability and oxygen reduction activity. *J. Power Sources* 195: 3105–3110.

Beck, F. 1977. The redox mechanism of the chelate-catalysed oxygen cathode. *J. Appl. Electrochem.* 7: 239–245.

Bezerra, C. W. B., Zhang, L., Liu, H. et al. 2007. A review of heat-treatment effects on activity and stability of PEM fuel cell catalysts for oxygen reduction reaction. *J. Power Sources* 173: 891–908.

Bi, W. and Fuller, T. F. 2008. PEM fuel cell Pt/C dissolution and deposition in Nafion electrolyte. *J. Electrochem. Soc.* 155: B215–B221.

Bi, W., Gray, G. E., and Fuller, T. F. 2007. PEM fuel cell Pt/C dissolution and deposition in Nafion electrolyte. *Electrochem. Solid-State Lett.* 10: B101–B104.

Bi, W., Sun, Q., Deng, Y., and Fuller, T. F. 2009. The effect of humidity and oxygen partial pressure on degradation of Pt/C catalyst in PEM fuel cell. *Electrochim. Acta* 54: 1826–1833.

Biloul, A., Gouèrec, P., Savy, M., Scarbeck, G., Besse, S., and Riga, J. 1996. Oxygen electrocatalysis under fuel cell conditions: Behaviour of cobalt porphyrins and tetraazaannulene analogues. *J. Appl. Electrochem.* 26: 1139–1146.

Bing, Y., Liu, H., Zhang, L., Ghosh, D., and Zhang, J. 2010. Nanostructured Pt-alloy electrocatalysts for PEM fuel cell oxygen reduction reaction. *Chem. Soc. Rev.* 39: 2184–2202.

Bonakdarpour, A., Wenzel, J., Stevens, D. A. et al. 2005. Studies of transition metal dissolution from combinatorially sputtered, nanostructured $Pt_{1-x}M_x$ (M = Fe, Ni; $0 < x < 1$) electrocatalysts for PEM fuel cells. *J. Electrochem. Soc.* 152: A61–A72.

Borup, R. L., Davey, J. R., Garzon, F. H., Wood, D. L., and Inbody, M. A. 2006a. PEM fuel cell electrocatalyst durability measurements. *J. Power Sources* 163: 76–81.

Borup, R. L., Davey, J. R., Garzon, F. H., Wood, D. L., Welch, P. M., and More, K. 2006b. PEM fuel cell durability with transportation transient operation. *ECS Trans.* 3: 879–886.

Brankovic, S. R., Wang, J. X., and Adzic, R. R. 2001. Metal monolayer deposition by replacement of metal adlayers on electrode surfaces. *Surf. Sci.* 474: L173–L179.

Cai, M., Ruthkosky, M. S., and Merzougui, B. 2006. Investigation of thermal and electrochemical degradation of fuel cell catalysts. *J. Power Sources* 160: 977–986.

Charreteur, F., Ruggeri, S., Jaouen, F., and Dodelet, J. P. 2008. Increasing the activity of Fe/N/C catalysts in PEM fuel cell cathodes using carbon blacks with a high-disordered carbon content. *Electrochim. Acta* 53: 6881–6889.

Cheng, X., Shi, Z., Glass, N. et al. 2007. A review of PEM hydrogen fuel cell contamination: Impacts, mechanisms, and mitigation. *J. Power Sources* 165: 739–756.

Cho, E., Ko, J., Ha, H. Y. et al. 2003. Characteristics of the PEMFC repetitively brought to temperatures below 0°C. *J. Electrochem. Soc.* 150: A1667–A1670.

Chung, C. G., Kim, L., Sung, Y. W., Lee, J., and Chung, J. S. 2009. Degradation mechanism of electrocatalyst during long-term operation of PEMFC. *Int. J. Hydrogen Energy* 34: 8974–8981.

Colón-Mercado, H. R. and Popov, B. N. 2006. Stability of platinum based alloy cathode catalysts in PEM fuel cells. *J. Power Sources* 155: 253–263.

Contractor, A. Q. and Lal, H. 1979. Two forms of chemisorbed sulfur on platinum and related studies. *J. Electroanal. Chem.* 96: 175–181.

Dam, V. A. T. and Bruijn, F. A. 2007. The stability of PEMFC electrodes—Platinum dissolution vs. potential and temperature investigated by quartz crystal microbalance. *J. Electrochem. Soc.* 154: B494–B499.

Denis, M. C., Lefèvre, M., Guay, D., and Dodelet, J. P. 2008. Pt–Ru catalysts prepared by high energy ball-milling for PEMFC and DMFC: Influence of the synthesis conditions. *Electrochim. Acta* 53: 5142–5154.

Dingreville, R., Qu, J., and Cherkaoui, M. 2005. Surface free energy and its effect on the elastic behavior of nano-sized particles, wires and films. *J. Mech. Phys. Solids* 53: 1827–1854.

Faubert, G., Côté, R., Dodelet, J.-P., Lefèvre, M., and Bertrand, P. 1999. Oxygen reduction catalysts for polymer electrolyte fuel cells from the pyrolysis of FeII acetate adsorbed on 3,4,9,10-perylenetetracarboxylic dianhydride. *Electrochim. Acta* 44: 2589–2603.

Franck-Lacaze, L., Bonnet, C., Besse, S., and Lapicque, F. 2009. Effects of ozone on the performance of a polymer electrolyte membrane fuel cell. *Fuel Cells* 9: 562–569.

Franco, A. A., Passot, S., Fugier, P. et al. 2009. Pt_xCo_y catalysts degradation in PEFC environments: Mechanistic insights I. Multiscale modeling. *J. Electrochem. Soc.* 156: B410–B424.

Fernández, J. L., Raghuveer, V., Manthiram, A., and Bard, A. J. 2005. Pd–Ti and Pd–Co–Au electrocatalysts as a replacement for platinum for oxygen reduction in proton exchange membrane fuel cells. *J. Am. Chem. Soc.* 127: 13100–13101.

Ferreira, P. J., la O', G. J., Shao-Horn, Y. et al. 2005. Instability of Pt/C electrocatalysts in proton exchange membrane fuel cells. *J. Electrochem. Soc.* 152: A2256–A2271.

Garzon, F. H., Rockward, T., Urdampilleta, I. G., Brosha, E. L., and Uribe, F. A. 2006. The impact of hydrogen fuel contaminates on long-term PEMFC performance. *ECS Trans.* 3: 695–703.

Gasteiger, H. A., Kocha, S. S., Sompalli, B., and Wagner, F. T. 2005. Activity benchmarks and requirements for Pt, Pt-alloy, and non-Pt oxygen reduction catalysts for PEMFCs. *Appl. Catal., B* 56: 9–35.

Gasteiger H. A., Marković, N., Ross, P. N., and Cairns, E. J. 1993. Methanol electrooxidation on well-characterized platinum–ruthenium bulk alloys. *J. Phys. Chem.* 97: 12020–12029.

Götz, M. and Wendt, H. 1998. Binary and ternary anode catalyst formulations including the elements W, Sn and Mo for PEMFCs operated on methanol or reformate gas. *Electrochim. Acta* 43: 3637–3644.

Gouérec, P., Biloul, A., Contamin, O. et al. 1997. Oxygen reduction in acid media catalysed by heat treated cobalt tetraazaannulene supported on an active charcoal: Correlations between the performances after longevity tests and the active site configuration as seen by XPS and ToF-SIMS. *J. Electroanal. Chem.* 422: 61–75.

Greeley, J. and Nørskov, J. K. 2007. Electrochemical dissolution of surface alloys in acids: Thermodynamic trends from first-principles calculations. *Electrochim. Acta* 52: 5829–5836.

Guilminot, E., Corcella, A., Chatenet, M. et al. 2007a. Membrane and active layer degradation upon PEMFC steady-state operation-I. Platinum dissolution and redistribution within the MEA. *J. Electrochem. Soc.* 154: B1106–B1114.

Guilminot, E., Corcella, A., Charlot, F., Maillard, F., and Chatenet, M. 2007b. Detection of Pt^{z+} ions and Pt nanoparticles inside the membrane of a used PEMFC. *J. Electrochem. Soc.* 154: B96–B105.

Haas, H. R. and Davis, M. T. 2009. Electrode and catalyst durability requirements in automotive PEM applications: Technology status of a recent MEA design and next generation challenges. *ECS Trans.* 25: 1623–1631.

Huang, S.-Y., Ganesan, P., and Popov, B. N. 2010. Electrocatalytic activity and stability of niobium-doped titanium oxide supported platinum catalyst for polymer electrolyte membrane fuel cells. *Appl. Catal., B* 96: 224–231.

Hung, W.-Z., Chung, W.-H., Tsai, D.-S., Wilkinson, D. P., and Huang, Y.-S. 2010. CO tolerance and catalytic activity of $Pt/Sn/SnO_2$ nanowires loaded on a carbon paper. *Electrochim. Acta* 55: 2116–2122.

Inaba, M. 2009. Durability of electrocatalysts in polymer electrolyte fuel cells. *ECS Trans.* 25: 573–581.

Izhar, S., Yoshida, M., and Nagai, M. 2009. Characterization and performances of cobalt–tungsten and molybdenum–tungsten carbides as anode catalyst for PEFC. *Electrochim. Acta* 54: 1255–1262.

Jaouen, F., Marcotte, S., Dodelet, J. P., and Lindbergh, G. 2003. Oxygen reduction catalysts for polymer electrolyte fuel cells from the pyrolisis of iron acetate adsorbed on various carbon supports. *J. Phys. Chem. B* 107: 1376–1386.

Jasinski, R. 1964. A new fuel cell cathode catalyst. *Nature* 201: 1212–1213.

Jiang, L., Sun, G., Zhou, Z. et al. 2005. Size-controllable synthesis of monodispersed sno2 nanoparticles and application in electrocatalysts. *J. Phys. Chem. B* 109: 8774–8778.

Kawahara, S., Mitsushima, S., Ota, K.-I., and Kamiya, N. 2006. Deterioration of Pt catalyst under potential cycling. *ECS Trans.* 3: 625–631.

Kelly, M. J., Fafilek, G., Besenhard, J. O., Kronberger, H., and Nauer, G. E. 2005. Contaminant absorption and conductivity in polymer electrolyte membranes. *J. Power Sources* 145: 249–252.

Kobayashi, M., Hidai, S., Niwa, H. et al. 2009. Co oxidation accompanied by degradation of Pt–Co alloy cathode catalysts in polymer electrolyte fuel cells. *Phys. Chem. Chem. Phys.* 11: 8226–8230.

Koh, S. and Strasser, P. 2007. Electrocatalysis on bimetallic surfaces: Modifying catalytic reactivity for oxygen reduction by voltammetric surface dealloying. *J. Am. Chem. Soc.* 129: 12624–12625.

Lalande, G., Tamizhmani, G., Côte, R. et al. 1995. Influence of loading on the activity and stability of heat-treated carbon-supported cobalt phthalocyanine electrocatalysts in solid polymer electrolyte fuel cells. *J. Electrochem. Soc.* 142: 1162–1168.

Lefèvre, M. and Dodelet, J.-P. 2000. O_2 reduction in PEM fuel cells: Activity and active site structural information for catalysts obtained by the pyrolysis at high temperature of Fe precursors. *J. Phys. Chem. B* 104: 11238–11247.

Lefèvre, M., Dodelet, J. P., and Bertrand, P. 2002. Molecular oxygen reduction in PEM fuel cells: Evidence for the simultaneous presence of two active sites in Fe-based catalysts. *J. Phys. Chem. B* 106: 8705–8713.

Lefèvre, M., Dodelet, J. P., and Bertrand, P. 2005. Molecular oxygen reduction in PEM fuel cell conditions: ToF-SIMS analysis of Co-based electrocatalysts. *J. Phy. Chem. B* 109: 16718–16724.

Lefèvre, M., Proietti, E., Jaouen, F., and Dodelet, J.-P. 2009a. Iron-based catalysts with improved oxygen reduction activity in polymer electrolyte fuel cells. *Science* 324: 71–74.

Lefèvre, M., Proietti, E., Jaouen, F., and Dodelet, J.-P. 2009b. Iron-based catalysts for oxygen reduction in PEM fuel cells: Expanded study using the pore-filling method. *ECS Trans.* 25: 105–115.

Li, J., He, P., Wang, K. P., Davis, M., and Ye, S. Y. 2006. Characterization of catalyst layer structural changes in PEMFC as a function of durability testing. *ECS Trans.* 3: 743–751.

Liao. M.-S., Cabrera, C. R., and Ishikawa, Y., A theoretical study of CO adsorption on Pt–Ru and Pt–M (M = Ru, Sn, Ge) clusters. *Surf. Sci.* 445: 267–282.

Lim, D.-H., Lee, W.-D., Choi, D.-H., and Lee, H.-I. 2010. Effect of ceria nanoparticles into the Pt/C catalyst as cathode material on the electrocatalytic activity and durability for low-temperature fuel cell. *Appl. Catal., B* 94: 85–96.

Liu, X., Chen, J., Liu, G., Zhang, L., Zhang, H., and Yi, B. 2010. Enhanced long-term durability of proton exchange membrane fuel cell cathode by employing $Pt/TiO_2/C$ catalysts. *J. Power Sources* 195: 4098–4103.

Liu, G., Zhang, H., Zhong, H., Hu, J., Xu, D., and Shao, Z. 2006. A novel sintering resistant and corrosion resistant Pt_4ZrO_2/C catalyst for high temperature PEMFCs. *Electrochim. Acta* 51: 5710–5714.

Loučka, T. 1971. Adsorption and oxidation of sulphur and of sulphur dioxide at the platinum electrode. *J. Electroanal. Chem.* 31: 319–332.

Luo, Z., Li, D., Tang, H., Pan, M., and Ruan, R. 2006. Degradation behavior of membrane-electrode-assembly materials in 10-cell PEMFC stack. *Int. J. Hydrogen Energy* 31: 1831–1837.

Ma, Y. and Balbuena, P. 2009. Surface adsorption and stabilization effect of iridium in Pt-based alloy catalysts for PEM fuel cell cathodes. *ECS Trans.* 25: 1037–1044.

Mani, P., Srivastava, R., and Strasser, P. 2008. Dealloyed Pt–Cu core–shell nanoparticle electrocatalysts for use in PEM fuel cell cathodes. *J. Phys. Chem. C* 112: 2770–2778.

Matsuoka, K., Sakamoto, S., Nakato, K., Hamada, A., and Itoh, Y. 2008. Degradation of polymer electrolyte fuel cells under the existence of anion species. *J. Power Sources* 179: 560–565.

Mayrhofer, K. J. J., Meier, J. C., Ashton, S. J. et al. 2008. Fuel cell catalyst degradation on the nanoscale. *Electrochem. Commun.* 10: 1144–1147.

Mitsushima, S., Kawahara, S., Ota, K., and Kamiya, N. 2007. Consumption rate of Pt under potential cycling. *J. Electrochem. Soc.* 154: B153–B158.

Mohtadi, R., Lee, W.-k., and Van Zee, J. W. 2004. Assessing durability of cathodes exposed to common air impurities. *J. Power Sources* 138: 216–225.

More, K. L., Borup, R., and Reeves, K. S. 2006. Identifying contributing degradation phenomena in PEM fuel cell membrane electrode assemblies via electron microscopy. *ECS Trans.* 3: 717–733.

Mukerjee, S. and Srinivasan, S. 1993. Enhanced electrocatalyssis of oxygen reduction reaction on platinum alloys in proton-exchange membrane fuel cells. *J. Electroanal. Chem.* 357: 201–224.

Mukerjee, S. and Srinivasan, S. 2003. O_2 reduction structure-related parameters for supported catalysts. In *Handbook of Fuel Cells—Fundamentals, Technology and Applications*, eds. W. Vielstich, A. Lamn, H. Gasteiger, pp. 502–503. Chichester: Wiley.

Nørskov, J. K., Bligaard, T., Hvolbæk, B., Abild-Pedersen, F., Chorkendorff, I., and Christensen, C. H. 2008. The nature of the active site in heterogeneous metal catalysis. *Chem. Soc. Rev.* 37: 2163–2171.

Nørskov, J. K., Rossmeisl, J., Logadottir, A., and Lindquist, L. 2004. Origin of the overpotential for oxygen reduction at a fuel-cell cathode. *J. Phys. Chem. B* 108: 17886–17892.

Ohma, A., Suga, S., Yamamoto, S., and Shinohara, K. 2007. Phenomenon analysis of PEFC for automotive use (1): Membrane degradation behavior during OCV hold test. *J. Electrochem. Soc.* 154: B757–B760.

Okada, T., Yano, H., and Ono, C. 2007. Novel CO tolerant anode catalysts for PEFC based on Pt and organic metal complexes. *J. New Mater. Electrochem. Syst.*, 10: 129–134.

Parrondo, J., Mijangos, F., and Rambabu, B. 2010. Platinum/tin oxide/carbon cathode catalyst for high temperature PEM fuel cell. *J. Power Sources* 195: 3977–3983.

Pylypenko, S., Mukherjee, S., Olson, T. S., and Atanassov, P. 2008. Non-platinum oxygen reduction electrocatalysts based on pyrolyzed transition metal macrocycles. *Electrochim. Acta* 53: 7875–7883.

Qian, Y., Wen, W., Adcock, P. A. et al. 2008. PtM/C catalyst prepared using reverse micelle method for oxygen reduction reaction in PEM fuel cells. *J. Phys. Chem. C* 112: 1146–1157.

Ralph, T. R. and Hogarth, M. P. 2002. Catalysis for low temperature fuel cells. *Platinum Met. Rev.*, 46: 117–135.

Santo, L., Colmati, F., and Gonzalez, E. R. 2006. Preparation and characterization of supported Pt-Ru catalyst with a high Ru content. *J. Power Sources* 159: 869–877.

Sato, Y., Wang, Z., and Takagi, Y. 2006. Effect of anode catalyst support on MEA degradation caused by hydrogen-starved operation of a PEFC. *ECS Trans.* 3: 827–833.

Steele, B. C. H. and Heinzel, A. 2001. Materials for Fuel-Cell Technologies. *Nature* 414: 345–352.

Schmidt, T. J. 2006. Durability and degradation in high-temperature polymer electrolyte fuel cells. *ECS Trans.* 1: 19–31.

Schmidt, T. J., Jusys, Z., Gasteiger, H. A., Behm, R. J., Endruschat, U., and Boennemann, H. 2001a. On the CO tolerance of novel colloidal PdAu/carbon electrocatalysts. *J. Electroanal. Chem.* 501: 132–140.

Schmidt, T. J., Paulus, U. A., Gasteiger, H. A., and Behm, R. J. 2001b. The oxygen reduction reaction on a Pt/carbon fuel cell catalyst in the presence of chloride anions. *J. Electroanal. Chem.* 508: 41–47.

Sethuraman, V. A., Weidner, J. W., Haug, A. T., and Protsailo, L. V. 2008. Durability of perfluorosulfonic acid and hydrocarbon membranes: effect of humidity and temperature. *J. Electrochem. Soc.* 155: B119–B124.

Stamenkovic, V. R., Fowler, B., Mun, B. S. et al. 2007a. Improved oxygen reduction activity on $Pt_3Ni(111)$ via increased surface site availability. *Science* 315: 493–496.

Stamenkovic, V. R., Mun B. S., Arenz, M. et al. 2007b. Trends in electrocatalysis on extended and nano-scale Pt-bimetallic alloy surfaces. *Nat. Mater.* 6: 241–247.

Stamenkovic, V., Mun, B. S., Mayrhofer, K. J. J. et al. 2006. Changing the activity of electrocatalysts for oxygen reduction by tuning the surface electronic structure. *Angew. Chem. Int. Ed.* 45: 2897–2901.

Stassi, A., Modica, E., Antonucci, V., and Aricò, A. S. 2009. A half cell study of performance and degradation of oxygen reduction catalysts for application in low temperature fuel cells. *Fuel Cells* 09: 201–208.

Tarasevich, M. R., Bogdanovskaya, V. A., Loubnin, E. N., and Reznikova, L. A. 2007. Comparative study of the corrosion behavior of platinum-based nanosized cathodic catalysts for fuel cells. *Prot. Met.* 43: 689–693.

Trogadas, P. and Ramani, V. 2007. $Pt/C/MnO_2$ hybrid electrocatalysts for degradation mitigation in polymer electrolyte fuel cells. *J. Power Sources* 174: 159–163.

Uribe, F. A., Gottesfeld, S., and Zawodzinski, J. T. A. 2002. Effect of ammonia as potential fuel impurity on proton exchange membrane fuel cell performance. *J. Electrochem. Soc.* 149: A293–A296.

Virkar, A. V. and Zhou, Y. 2007. Mechanism of catalyst degradation in proton exchange membrane fuel cells. *J. Electrochem. Soc.* 154: B540–B547.

Wang, X. P., Kumar, R., and Myers, D. J. 2006a. Effect of voltage on platinum dissolution. *Electrochem. Solid-State Lett.* 9: A225–A227.

Wang, W. T., Lee W.-k., and Van Zee, J. W. 2006b. A model for SO_2 impurity in air fed to a proton exchange membrane fuel cell. *ECS Trans.* 1: 131–137.

Wells, P. P., Qian, Y., King, C. R., Wiltshire, R. J. K., and Crabb, E. M. 2008. To alloy or not to alloy? Cr modified Pt/C cathode catalysts for PEM fuel cells. *Faraday Discuss.* 138: 273–285.

Wood, T. E., Tan, Z., Schmoeckel, A. K., O'Neill, D., and Atanasoski, R. 2008. Non-precious metal oxygen reduction catalyst for PEM fuel cells based on nitroaniline precursor *J. Power Sources* 178: 510–516.

Wu, G., Artyushkova, K., Ferrandon, M., Kropf, J., Myers, D., and Zelenay, P. 2009. Performance durability of polyaniline-derived non-precious cathode catalysts. *ECS Trans.* 25: 1299–1311.

Xie, J., Wood III, D. L., More, K. L., Atanassov, P., and Borup, R. L. 2005. Microstructural changes of membrane electrode assemblies during PEFC durability testing at high humidity conditions. *J. Electrochem. Soc.* 152: A1011–A1020.

Xu, H., Kunz, R., and Fenton, J. M. 2007. Investigation of platinum oxidation in PEM fuel cells at various relative humidities. *Electrochem. Solid-State Lett.* 10: B1–B5.

Xu, Z., Qi, Z., and Kaufman, A. 2003. Effect of oxygen storage materials on the performance of proton-exchange membrane fuel cells. *J. Power Sources* 115: 40–43.

Yamada, H., Shimoda, D., Matsuzawa, K., Tasaka, A., and Inaba, M. 2007. Stability of Pt–Ru/C catalysts: Effect of Ru content. *ECS Trans.* 11: 325–334

Yamazaki, O., Oomori, Y., Shintaku, H., and Tabata, T. 2007. Study on degradation of the Pt–Ru anode of PEFC for residential application (2) Analyses of the MEAs degraded in the anode performance. *ECS Trans.* 11: 287–295.

Yan, Q., Toghiani, H., Lee, Y.-W., Liang, K., and Causey, H. 2006a. Effect of sub-freezing temperatures on a PEM fuel cell performance, startup and fuel cell components. *J. Power Sources* 160: 1242–1250.

Yan, Q. G., Toghiani, H., and Causey, H. 2006b. Steady state and dynamic performance of proton exchange membrane fuel cells under various operating conditions and load changes. *J. Power Sources* 161: 492–502.

Ye, S., Hall, M., Cao, H., and He, P. 2006. Degradation resistant cathodes in polymer electrolyte membrane fuel cells. *ECS Trans.* 3:657–666.

Yoda, T., Uchida, H., and Watanabe, M. 2007. Effects of operating potential and temperature on degradation of eletctrocatalyst layer for PEFCs. *Electrochim. Acta* 52: 5997–6005.

Young, A. P., Stumper, J., Knights, S., and Gyenge, E. 2010. Ionomer degradation in polymer electrolyte membrane fuel cells. *J. Electrochem. Soc.* 157: B425–B436.

Yu, P., Pemberton, M., and Plasse, P. 2005. PtCo/C cathode catalyst for improved durability in PEMFCs. *J. Power Sources* 144: 11–20.

Zhang, S., Yuan, X.-Z., Hin, J. N. C., Wang, H., Friedrich, K. A., and Schultz, M. 2009a. A review of platinum-based catalyst layer degradation in PEM fuel cells. *J. Power Sources* 194: 588–600.

Zhang, L., Lee, K., C. Bezerra, W. B., Zhang, J., and Zhang, J. 2009b. Fe loading of a carbon-supported Fe–N electrocatalyst and its effect on the oxygen reduction reaction. *Electrochim. Acta* 54: 6631–6636.

Zhang, J., Sasaki, K., Sutter, E., and Adzic, R. R. 2007a. Stabilization of platinum oxygen-reduction electrocatalysts using gold clusters. *Science* 315: 220–222.

Zhang, L., Lee, K., and Zhang, J. 2007b. The effect of heat treatment on nanoparticle size and ORR activity for carbon-supported Pd–Co alloy electrocatalysts. *Electrochim. Acta* 52: 3088–3094.

Zigani, S. C., Antolini, E., and Gonzalez, E. R. 2008. Evaluation of the stability and durability of Pt and Pt-Co/C catalysts for polymer electrolyte membrane fuel cells. *J. Power Sources* 182: 83–90.

3

Catalyst Support Degradation

Zhongwei Chen
University of Waterloo

Ryan Hsu
University of Waterloo

3.1 Introduction

Cost and durability are major challenges for the commercialization of polymer electrolyte membrane fuel cells (PEMFCs) (Antolini et al., 2006; Borup et al., 2006; Lee et al., 2008). Great strides in PEMFC research have achieved long-term operation under near-ideal operating conditions, but the durability still needs to be improved under more realistic conditions for a wide range of operating conditions (Debe et al., 2006). For instance, the U.S. Department of Energy (DOE) has set a minimum lifetime target of 5000 h under external environmental conditions (–40°C to +40°C) for PEMFCs for automotive applications. This target is increased to 20,000 h for buses and 40,000 h for stationary applications. Therefore, detailed understanding of the failure modes and degradation mechanisms of fuel cell components and developing design improvements to mitigate or eliminate degradation are required to meet these stringent requirements.

The membrane electrode assembly (MEA) is the heart of a fuel cell stack and most likely to ultimately dictate stack life. Recent studies have shown that a considerable part of the cell performance loss is due to the degradation of the catalyst layer, in addition to membrane degradation. The catalyst layer in PEMFCs typically contains platinum/platinum alloy nanoparticles distributed on a catalyst support to enhance the reaction rate, to reach a maximum utilization ratio and to decrease the cost of fuel cells. The carbon-supported Pt nanoparticle (Pt/C) catalysts are the most popular for PEMFCs. Catalyst support corrosion and Pt dissolution/aggregation are considered as the major contributions to the degradation of the Pt/C catalysts (Bi et al., 2007; Asano et al., 2008). The effect of catalyst support and the interaction between catalyst support and Pt are investigated for the improvement of catalytic activity and stability of the electrocatalysts. In the typical Pt/C catalyst, the carbon support is prone to carbon oxidation that occurs during operation and can become detrimental to the structure of the catalyst over long periods of operation (Shao et al., 2007). Carbon corrosion has been known to weaken the interaction of Pt nanoparticles with carbon support, and eventually aggravate platinum agglomeration and the detachment of Pt nanoparticles from carbon support, resulting in decreases in the electrochemically active surface area (ECSA) of Pt thereby lowering the fuel cell performance (Schmittinger and Vahidi, 2008).

Recently, several mitigation strategies have been studied to enhance the lifetime and increase the stability of the catalyst support. These strategies may include alternative synthesis techniques, additional treatment steps, or new materials that are capable of withstanding the harsh conditions of an operating fuel cell for long durations. For example, the graphitization of carbon (in the form of graphene, carbon nanotubes (CNT), or carbon nanofibers (CNF) is a common treatment step that has successfully retarded the corrosion problems seen in carbon blacks (Shao et al., 2006). From a thermodynamic perspective, however, corrosion still occurs and catalysts made with graphitized carbon as the catalyst support still experience a reduction of ECSA after prolonged testing. Support materials other than carbon have also been tested as catalyst supports including indium tin oxides, tungsten oxides, tungsten carbides, doped titanium oxides, and zirconium oxides (Ioroi et al., 2005; Chhina et al., 2006, 2007a,b, 2008; Suzuki et al., 2007; Whitelocke et al., 2008; Antolini and Gonzalez, 2009). These mitigation strategies and novel materials are discussed in detail in this chapter to shed light on the direction of catalyst support research.

In this chapter, the following sections related to catalyst support degradation will be discussed in detail. Section 3.2 deals with the mechanisms, kinetics, and thermodynamics of support degradation; Section 3.3 discusses the effect of operating conditions on the support stability; Section 3.4 explains the degradation testing protocols, and Section 3.5 focuses on the mitigation strategies. The scope of this chapter will focus on traditional carbon black support degradation with further discussion on the reliability and stability of alternative catalyst supports. Failure modes dealing with the membrane, catalyst metals, or ancillary components of the fuel cell are beyond the scope of this chapter and thus will only be briefly mentioned. This chapter will be geared toward understanding the fundamental principles of catalyst support degradation and selectively reviewing peer-review journal articles which focus and discuss on the material stability and practical applications in PEMFCs.

3.2 Mechanisms, Thermodynamics, and Kinetics of Support Degradation

The durability has recently been recognized as one of the most important issues to be addressed before the large-scale commercialization of PEMFCs. Carbon-supported platinum (or platinum alloys) cathode catalysts degradation is considered as the major failure mode for fuel cell systems. Considerable effort has been made into detailing the mechanisms and thermodynamics of fuel cell degradation, including but not limited to, platinum contamination, platinum agglomeration/dissolution, and corrosion of the carbon support (Wu et al., 2008). The kinetics of the degradation over long-term operation has been modeled both theoretically and experimentally in numerous studies (Meng, 2008; Takeshita et al., 2008; Takeuchi and Fuller, 2008). Carbon support corrosion of cathode catalysts in PEMFCs is a major contributor to the overall catalyst degradation. The following sections provide background on the use of carbon blacks as a support for PEMFC catalysts and illustrate the commonly accepted mechanisms and kinetic models of the catalyst support degradation proposed by several research groups.

3.2.1 Carbon Catalyst Support

Currently, platinum is the only element having sufficient catalytic activity and stability for PEMFC applications. It is well known however, that platinum is scarce and expensive. To lower material costs for

PEMFCs, it is advantageous to disperse platinum as nanoparticles onto the supports to increase the active surface area (Antolini, 2009). In this manner, a good electronic network is formed providing one of the three critical pathways to the active metal sites. The second pathway, being the polymer electrolyte must be present at the active site to provide the pathway of protons. And finally, the third required pathway, the flow channel, is required for the mass transport of reactant gases (H_2/O_2). These three pathways are necessary for the formation of an active site and are termed as the triple phase boundary (TPB).

The support being the electronic pathway for the TPB must have good electronic conductivity to provide a pathway for electrons to reach the platinum nanoparticles. It is also beneficial for the support to be cheap, environmentally friendly, have high surface area, and be able to withstand the operating conditions of a fuel cell. The support surface properties are also of great importance when considering the interaction and dispersion between the support and the metal catalyst (Antolini, 2009).

Carbon black is favorable as a support material not only because of its high surface area and electronic properties, but it is also abundant, chemically inert, and environmental friendly (Bleda-Martinez et al., 2007). Carbon blacks are typically used as supports which are manufactured by the pyrolysis of hydrocarbons or oil fractions using oil furnaces or acetylene processes. Some of the most common carbon blacks used for platinum deposition in PEMFC catalysts are synthesized using the furnace method where the input materials are burned with limited air at about 1400°C (Dowlapalli et al., 2006). Vulcan XC-72 and Black Pearl 2000 represent these types of carbon blacks. These carbon blacks are easily made and abundant making them popular choices for carbon black supports for Pt/C catalysts (Cameron et al., 1990).

In the state-of-the-art PEMFCs, carbon black is normally used as a catalyst support material. Although carbon black has been widely utilized as a fuel cell catalyst support material, it is susceptible to corrosion in PEMFC operating conditions: high temperatures (50–90°C), low pH (<1), high operating voltages (0.6–1.2 V), and high water and oxygen concentrations. Especially during prolonged use, oxygen has the ability to react with the carbon support to produce CO or CO_2 which leaves the cell as a byproduct (Maillard et al., 2009). This oxidation of the carbon support reduces the electroconductive network of the electrocatalyst and may liberate platinum that has bonded onto the carbon support. This loss of ECSA is the reason why the carbon support corrosion of the cathode catalysts in PEMFCs is a major contributor to the catalyst degradation. Carbon support corrosion in the cathode of PEMFCs can occur because: (1) the cathode is held at relatively oxidative potentials and oxygen atoms are being generated by the catalyst particles; (2) furthermore, the cells are at elevated temperatures and carbon atoms are able to react with oxygen atoms and/or water to generate gaseous products such as CO and CO_2 that leave the cell (Guilminot et al., 2007). In the following section, the thermodynamics, mechanisms, and kinetics of carbon support corrosion will be discussed.

3.2.2 Mechanisms and Thermodynamics of Carbon Support Degradation

In a typical Pt/C catalyst, there are several modes of degradation that can occur through the oxidation of the carbon support. Figure 3.1 illustrates many of the mechanisms of performance loss in a PEMFC which are mainly due to the loss of TPB active sites. The first method (Figure 3.1a) in which TPB sites become inactive is from the loss of contact between the catalyst particles and the polymer electrolyte membrane. Detachment of the membrane, also known as membrane delamination, can be caused by the corrosion of the carbon support and can result in protons being unable to reach the platinum nanoparticles. This inevitably leads to lessening of the Pt utilization and catalytic activity of the MEA.

For a TPB site to exist, the catalytic metal must be bound onto the carbon support. The metal is usually deposited onto the support as dispersed platinum nanoparticles. To maximize the surface area and to increase the utilization of the active metal, these platinum particles usually are uniformly dispersed with particle diameters ranging from 3 to 8 nm for typical Pt/C catalysts. Poor adhesion to the carbon support or degradation of the support can lead to platinum particles being unanchored from the carbon. Unanchored platinum usually leaves the cell or migrates within the MEA by the assistance of water

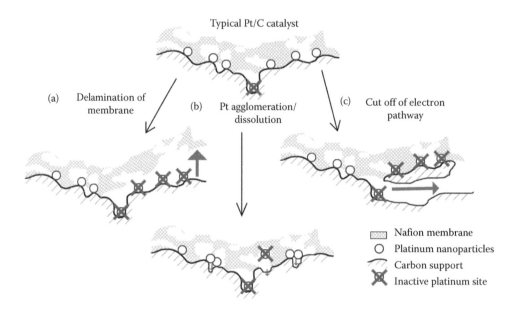

FIGURE 3.1   Degradation mechanisms for a PEMFC MEA by carbon corrosion for a typical Pt/C catalyst.

entering and leaving the cell (Figure 3.1b). Platinum agglomeration is a well-studied phenomenon that can also be aggravated by carbon corrosion. As carbon oxidation weakens the mechanical strength of the support between platinum nanoparticles, the platinum metals aggregate and become unevenly distributed on the support surface.

The carbon support not only anchors the polymer membrane and the platinum nanoparticles, but is the pathway for electrons to arrive at the active site. Destruction of active sites can occur by the corrosion of this pathway and separation of the carbon electrical network. This mechanism is illustrated in Figure 3.1c.

For other structures of carbon, for example, doped, graphitic, or nanostructured carbons, the degradation of the support can also lead to structural changes of the support that can lead to changes in electronic properties or other characteristics. For example, nitrogen-doped carbons have been known to have enhanced bond strength between adjacent carbons and platinum. Corrosion of the nitrogen-doped support can change the properties of these platinum–carbon interactions and lead to a more rapid decay of the active sites during operation.

These mechanisms of degradation all lead to performance degradation in the overall PEMFC. Ongoing degradation through carbon corrosion is a serious issue for the longevity of the fuel cell and in the severest cases cause catastrophic failure of the MEA.

Oxidation of carbon to CO_2 at standard conditions is thermodynamically possible at potentials greater than 0.207 V vs. the reversible hydrogen electrode (RHE) by Equation 3.1 (Tang et al., 2006).

$$C + 2H_2O \rightarrow CO_2 + 4H^+ + 4e^-, \ E^0 = 0.207 \, V \text{ vs. RHE} \tag{3.1}$$

$$C + H_2O \rightarrow CO + 2H^+ + 2e^-, \ E^0 = 0.518 \, V \text{ vs. RHE} \tag{3.2}$$

$$CO + H_2O \rightarrow CO_2 + 2H^+ + 2e^-, \ E^0 = -0.103 \, V \text{ vs. RHE} \tag{3.3}$$

Carbon can also oxidize to carbon monoxide through a two-electron pathway at 0.518 V at standard conditions and be further oxidized to carbon dioxide through the addition of water (shown in Equations 3.2 and 3.3, respectively).

As it now stands, there is no widely accepted mechanism which details the oxidation and CO_2 generation of surface carbon species. Stevens and Dahn (2005) proposed a four-step mechanism shown in Equation 3.4 that hypothesized this oxidation process and assumed water to be the source of oxygen (Stevens et al., 2005).

$$R\text{–}C_s\text{–}H \rightarrow R\text{–}C_s\text{–}OH \rightarrow R\text{–}C_s\text{=}O \rightarrow R\text{–}C_s\text{–}OOH \rightarrow R\text{–}H + CO_2 \tag{3.4}$$

The reduction of O_2 to H_2O (Equation 3.5) is the desired process in a H_2/O_2 PEMFC and produces twice as many electrons as the reduction to hydrogen peroxide (Equation 3.6). The H_2O_2 can be further reduced to H_2O depending on the catalyst type and its kinetics, as shown in Equation 3.7.

$$O_2 + 4H^+ + 4e^- \rightarrow 2H_2O_2, \ E_o = 1.229 \text{ V vs. RHE} \tag{3.5}$$

$$O_2 + 2H^+ + 2e^- \rightarrow H_2O_2, \ E_o = 0.67 \text{ V vs. RHE} \tag{3.6}$$

$$H_2O_2 + 2H^+ + 2e^- \rightarrow 2H_2O \tag{3.7}$$

The reduction to H_2O_2 is also disadvantageous to the MEA because peroxide radicals within the cell can attack and degrade the polymer electrolyte membrane and the carbon support. These attacks on the carbon support have been reported to proceed by the following reactions (Maass et al., 2008):

$$C + H_2O_2 \rightarrow C\text{–}O_{ad} + H_2O \tag{3.8}$$

$$C\text{–}O_{ad} + H_2O_2 \rightarrow CO_2 + H_2O \tag{3.9}$$

Equation 3.8 describes oxygen from the peroxide adsorbing onto the carbon surface to form a carbon–oxygen intermediate. This intermediate is reacted as in Equation 3.9 to release the carbon from the support and form CO_2 which is released from the cell as a gaseous effluent. Therefore, ensuring the catalyst is efficient in catalyzing the direct pathway of O_2 to H_2O is critical to the stability of the overall electrocatalyst.

3.2.3 Kinetics of Carbon Support Degradation

The rate at which the carbon support corrodes is heavily influenced by catalyst type and operating parameters of the cell. Several experiments suggested that the temperature and platinum loading can heavily influence carbon corrosion rates. Stevens and Dahn carried out a series of experiments to examine the effect of platinum loading and operating temperature on the thermal combustion of the carbon support (Stevens et al., 2005). Platinum was loaded onto Black Pearl 2000 carbon black with nominal loadings ranging from 5 to 80 wt% with platinum particle sizes ranging from 1.5 to 5.1 nm, respectively. The weights of the samples were recorded over time to determine the carbon loss due to corrosion of the carbon support. From their results they determined that the base carbon (carbon without any loaded platinum) displayed no weight loss over thousands of hours at a temperature of 195°C. By increasing the platinum loading, carbon corrosion rates were observed to increase. This was attributed to platinum particles catalyzing the carbon corrosion reaction, thus converting surface carbon to gaseous products (carbon monoxide (CO) and carbon dioxide (CO_2)). The temperature of the samples in the oven were also studied through the temperature range of 160–205°C and it was found that increasing temperature generally led to increased rates of carbon weight loss. Although the rates of degradation studied in this research do not accurately reflect the rates that occur in PEMFC operating conditions, it is proposed that a more thermally stable catalyst will also be a more electrochemically stable catalyst operating at PEMFC conditions.

Several kinetic models have been proposed to describe the rate of carbon corrosion in Pt/C electrocatalysts. Stevens et al. used a simple model to account for the fractional mass loss in carbon as a function of time, $\alpha(t)$ (Stevens et al., 2005):

$$\alpha(t) = \left[\frac{100}{100 - \%Pt}\right]\left[\frac{m_0 - m_t}{m_t}\right] \tag{3.10}$$

where %Pt is the mass percentage of platinum, m_0 is the initial dry mass of sample, and m_t is the mass of the sample at time t. For their modeling, they applied a first-order kinetic model to the degradation:

$$\frac{d\alpha}{dt} = k(1 - \alpha) \tag{3.11}$$

where k is the standard exponentially activated rate constant. In their reasoning, they describe that not all carbon within the sample may be adjacent to platinum particles to combust catalytically over an infinite period of time, and thus a slight modification to the equation yields:

$$\alpha = \alpha_{\max(1 - e^{-kt})} \tag{3.12}$$

where $\alpha_{\max}$ represents the maximum fraction of carbon that can combust catalytically over infinite time. By using the two parameters k and $\alpha_{\max}$, the degradation rates and reactivity of different carbon supports were compared.

Hu et al. also report the modeling of MEA and carbon corrosion based on four electrochemical and one chemical reactions taking place within the MEA (Hu et al., 2009). The two-dimensional model considers coupled transport of charged and noncharged species. The model was set up to solve for the local fuel starvation case and the start-up and shut-down case with the Butler–Volmer equation governing the kinetics of each half-cell reaction:

$$i_{RXN} = i_{0,RXN}\left[\prod_{Reactant\,k}\left(\frac{C_k}{C_{k,ref}}\right)^{\gamma k}\right]\left\{exp\left(\frac{\alpha_a F\eta}{RT}\right) - exp\left(\frac{\alpha_c F\eta}{RT}\right)\right\}\left\{\frac{S}{V}\right\}_{eff} \tag{3.13}$$

where C_k is the concentration for species (mol m^{-3}), F is Faraday's constant (96,487 C mol^{-1}), i is the volumetric current density (A m^{-3}), i_0 is the exchange current density (A m^{-2}), R is the universal gas constant (8.314 J mol^{-1} K^{-1}), S/V is the surface-to-volume ratio (m^2 m^{-3}), T is temperature (K), and the subscripts a, c, eff, and ref denote the anode, cathode, effective transport parameter, and reference state, respectively.

Owing to the numerous parameters and complexity of carbon support oxidation, many other papers (Meyers and Darling, 2006; Hu et al., 2007, 2008; Takeuchi and Fuller, 2007, 2008; Bi et al., 2008; Franco et al., 2008; Franco and Gerard, 2008; Gidwani et al., 2008) not reported here describe models such as the ones proposed to investigate the different parameters effecting carbon corrosion and fuel cell degradation.

3.3 Effect of Operating Conditions

In PEMFCs, understanding the relationships between the fuel cell's durability and performance with different operating conditions is an important task before the realization of their commercialization can occur. Much of the performance and durability testing with PEMFC benchmarking takes place under idealized conditions that will not accurately reflect those experienced in reality. Environmental real-world conditions may irreversibly aggravate problems in the fuel cell and exacerbate degradation mechanisms such as catalyst support corrosion. In the following section, the effects of several operating conditions to catalyst support degradation are reviewed. The discussion will focus on carbon black supports that have been most commonly studied in the literature; however, the concepts mentioned could be applicable to many noncarbon black supports as well.

3.3.1 Global Fuel Starvation

Fuel starvation is a general term that refers to the deprivation of fuel in parts of the electrode or the entire fuel cell. When insufficient fuel is being supplied to the cell for a certain operating voltage, gross fuel starvation occurs. However, even with sufficient fuel being supplied, localized fuel starvation can occur where insufficient fuel is provided to localized regions along the catalyst layer due to poor cell design, water blocking/flooding or poor operating conditions of the cell. The consequences of gross fuel starvation include carbon support corrosion. Several authors have experienced negative operating voltages as the anode potential increased to positive potentials. In these scenarios, the carbon support is being consumed in place of the fuel which leads to quick catalyst support degradation and decreased durability for the PEMFC (Meng, 2008; Takeshita et al., 2008; Takeuchi and Fuller, 2008).

Gross fuel starvation and its effects were confirmed by Ballard through experiments that replaced H_2 with N_2 at the anode of an operating fuel cell. The anode potential rose to >1.23 V vs. RHE the thereby allowing carbon corrosion to occur and the overall cell potential to achieve negative potentials. In experiments led by Taniguchi et al. cell potentials reached −2 V and the anode potential rose to 2.2 V as seen in Figure 3.2 (Taniguchi et al., 2004). Carbon corrosion was confirmed by the presence of CO_2 being detected in the effluent gas. Current densities at 0.7 V were recorded after 3 and 7 min of mimicked gross fuel starvation and resulted in 25 and 50% lower currents in Pt–Ru/C catalysts. Transmission electron microscopy (TEM) was used to observe changes in the average platinum particle sizes on the carbon support after simulating gross fuel starvation. The mean particle size increased from 2.6 to 5 nm at the anode and 2.8–4 nm at the cathode.

For fuel cell stacks consisting of multiple cells, gross fuel starvation can lead to several stacks operating with insufficient fuel, thus those respective stacks must degrade the carbon support to maintain the current being output by all the cells connected in series. To avoid this consequence, careful monitoring of the voltage and fuel distribution of each cell is necessary (Fowler et al., 2002). The implementation of such a control scheme certainly adds to the cost and complexity of the overall fuel cell stack.

It is important to understand the detrimental effects caused by gross fuel starvation as the anodic current that would have been supplied by the oxidation of hydrogen or the fuel is now being supplied by the oxidation of the carbon support to form carbonaceous species such as carbon monoxide and carbon dioxide. Gross fuel starvation results in irreversible damage to the carbon support and the loss of active site loss as the carbon corrodes.

3.3.2 Local Fuel Starvation

Carbon corrosion can also arise from a nonuniform distribution of fuel on the anode side (partial hydrogen coverage) and from crossover of reactant gas through the membrane. Local fuel starvations can cause this type of carbon corrosion. Because of its complexity and consequence to the durability of the fuel cell catalyst layer, local fuel starvation is both a widely studied and researched phenomena.

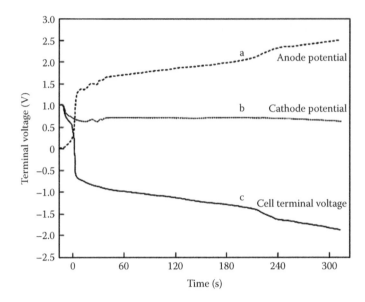

FIGURE 3.2 Time-dependent terminal voltages of the anode and cathode during a cell reversal experiment. (Reprinted from *Journal of Power Sources*, 130, Taniguchi, A. et al. Analysis of electrocatalyst degradation in PEMFC caused by cell reversal during fuel starvation, 42–49, Copyright (2004), with permission from Elsevier.)

Unlike gross fuel starvation happening when there is insufficient fuel being supplied to the electrode to generate the required current for the cell, local fuel starvation occurs when sufficient gas is supplied to the cell but the current distribution is not homogeneous and the fuel supplied is unevenly distributed across the membrane and electrode surfaces. This uneven distribution of gas in the cell with sufficient gas supply leads to uneven partial pressures and uneven current distribution. Postmortem thickness and morphology studies on a degraded MEA caused by carbon corrosion under a local H_2 starvation operation in a PEMFC were carried out using optical microscopy, SEM, and TEM (Natarajan and Van Nguyen, 2005a,b). Samples used for the postmortem studies were selected and indexed with the aid of a limiting current density distribution map that was premeasured from the degraded MEA using an electrochemical diagnostic technique. It is clear from the observations in this study that PEMFC operating conditions that cause local H_2 starvation will cause corrosion of the carbon support. This carbon corrosion resulted in a collapse of the electrode's porous structure, loss of mass transport in the electrode, and subsequent degradation of fuel cell performance. Further investigation is underway to fundamentally understand how carbon corrosion causes the collapse of the porous electrode structure and how it affects fuel cell durability (Natarajan and Van Nguyen, 2005a,b).

3.3.3 Load Cycling/Startup and Shutdown Cycling

Carbon corrosion and catalyst layer degradation may be aggravated by load cycling and startup conditions which may cause ambient air to replace hydrogen fuel in the anode (Roen et al., 2004; Paik et al., 2007). When the fuel cell is shut down or stopped, ambient air/oxygen may replace hydrogen in the gas channels through Brownian motion or via the cathode side by membrane crossover effects. When the fuel cell starts up, a transient fuel–air front is created in the anode layer that may lead to localized or complete fuel starvation (Ofstad et al., 2008). This resulted in dramatic carbon losses and performance decreases brought upon by decreased ECSA and structural changes due to carbon corrosion (see Figure 3.3).

To investigate further into these causes and effects, many reports have carried out detailed analysis to understand what reactions occur during start-up. Reiser et al. (2005) as well as Meyers and Darling

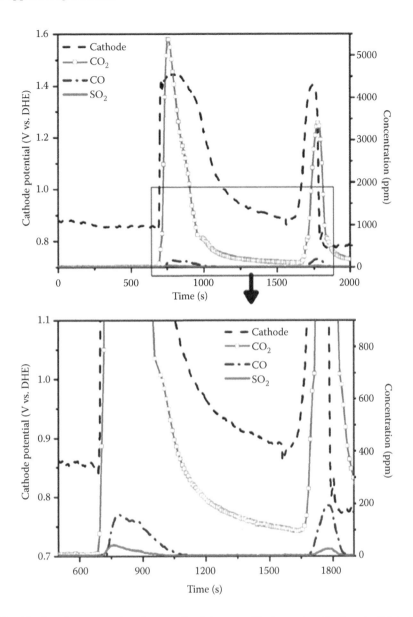

FIGURE 3.3 Monitored concentration changes of outlet gases with change in cathode potential during startup. (Reprinted from *Journal of Power Sources*, 192, Kim, J. et al. Relationship between carbon corrosion and positive electrode potential in a proton-exchange membrane fuel cell during start/stop operation, 674–678, Copyright (2009), with permission from Elsevier.)

(Meyers and Darling, 2006) modeled the behavior of the degradation assuming a maldistribution of hydrogen occurring at start-up. Their work has allowed them to determine the conditions that favored cathode carbon oxidation and the potential profile drops occurring between the electrode/electrolyte in the fuel–air anode. They concluded that this air–fuel boundary present in the anode might create locally favorable reverse-current conditions, which they called "the reverse-current decay mechanism" shown in Figure 3.4. This mechanism allowed for a reverse-current within the fuel cell and carbon cathode catalyst to degrade rapidly due to the cathode high interfacial potential that was calculated at 1.44 V vs. RHE.

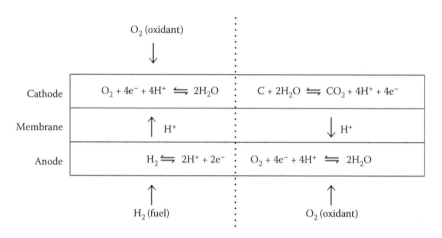

FIGURE 3.4 Typical half-reactions in H_2/O_2 PEM fuel cell (left) and "reverse-current" decay mechanism half-reaction with oxidant being introduced to the anode (right).

The effects of load cycling on the carbon support degradation must be carefully monitored during PEMFC operation. Mitigation techniques for sensing oxygen in the anode or a purging stage prior to operation may be implemented after long periods of downtime to avoid the problems associated with startup and increase the lifetime of these fuel cells.

3.3.4 Cold Start-Up

The U.S. DOE targets for 2010 for automotive applications intend to have PEMFCs capable of surviving temperatures of −40°C and starting operation from temperatures of −20°C. The main concern with cold temperature and cold temperature start-up is the formation of ice resulting in structural changes that might occur to the catalyst, MEA and/or gas diffusion layers (GDLs). Ice formation has also been said to impede oxygen transport to the catalyst sites and in severe cases, render entire cells inactive (Mao and Wang, 2007). A study by Cho et al. to examine the effect of cold temperatures was conducted using MEAs that were made to be cycled between 80°C and −10°C (Cho et al., 2004). In a previous study conducted by Cho et al. they hypothesized that the thermal cycling caused the pore sizes to increase in the catalytic layer that resulted in an increase in contact resistance due to structural changes within the cell catalyst layer (Cho et al., 2003).

Oszcipok et al. found similar results with Cho's in their investigation of starting up the MEA from −10°C (Oszcipok et al., 2005). They observed 5.4% loss in current at 450 mV per freeze–thaw cycle, and consecutive losses in ECSA after each start-up from −10°C. However, contradictory results were obtained by Knights et al. who experienced little performance loss for a fuel cell subjected to 55 freeze–thaw cycles (Knights et al., 2004).

Three primary mechanisms of MEA degradation occurring in cold start have been put forward by Wang et al.: (1) interfacial delamination between the membrane and cathode catalyst layer, (2) Pt particle growth and Pt dissolution in perfluorosulfonic acid (PFSA) ionomer, and (3) cathode catalyst layer structural damage and hence densification. The interfacial delamination and catalyst layer densification appear to be closely related to each other, and the key parameter to affect both is the ice volume fraction in the cathode catalyst layer after each cold-start step. Eliminating or minimizing these two degradation processes could improve the MEA cold-start durability greatly. However, there is not much research directly relating the effect of cold start-up with carbon support degradation. More work has to be done to understand the full effects of cold start-up on catalyst support degradation.

3.3.5 Relative Humidity

Stevens et al. conducted various studies to determine the effect of carbon type and humidification on the carbon degradation (Stevens et al., 2005). They conducted two sorts of studies in which (1) the catalyst durability was studied *ex situ* and degraded in an isothermal oven at 125°C and 150°C, and (2) the *in situ* MEA durability was studied by a 1.2 V potential hold for 50 h. For both carbons tested (Vulcan XC-72 and Black Pearl 2000) using both methods (*in situ* and *ex situ*), the carbon support was degraded faster with increased humidity introduced into the environment. This agreed with many reports in previous studies (Bi et al., 2009). Figures 3.5 and 3.6 show typical carbon corrosion rates (increase in the amount of CO_2 released from the outlet of the fuel cell) and cell performances at varying inlet humidity at different temperatures. Stevens et al. hypothesized that this increased degradation effect was due to an alternate pathway through a water–gas type of reaction.

Sun et al. (2007) and Hottinen et al. (2003) studied the humidity conditions and the effects of humidity on local fuel starvation and learned that when the fuel cell membrane was under-humidified the current densities increased monotonically across the flow channel due to the progressive hydration of the membrane and increased proton conductivity of the Nafion membrane. In a cell operating with an over-hydrated membrane, the current distribution decreased monotonically due to increased partial pressure and decreased as water content increased in the catalyst layer due to blockage of active sites or water flooding. Thus, water content within the membrane strongly affects the current distribution within the fuel cell and can aggravate localized fuel starvation.

Cai et al. also described studies to evaluate the effect of gas humidity on two Tanaka fuel cell catalysts (Pt/Vulcan and Pt/High surface area carbon (HSC)) at 250°C for up to 30 h (Cai et al., 2006). The Pt/Vulcan demonstrated 11% loss of its initial carbon mass when water was present but no observable mass changes when there was no water present. The Pt/HSC, however, demonstrated no changes in mass regardless of whether there was water or not. This effect was explained by both HSC having a

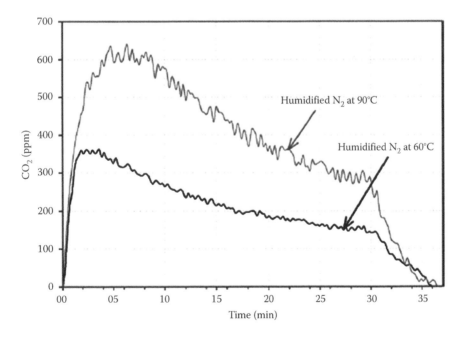

FIGURE 3.5 Monitored mass-spectrographic profiles of MEAs at different temperatures running corrosion tests at 1.4 V with humidified N_2. (Reprinted from *Journal of Power Sources*, 193, Lim, K. H. et al. Effect of operating conditions on carbon corrosion in polymer electrolyte membrane fuel cells, 575–579, Copyright (2009), with permission from Elsevier.)

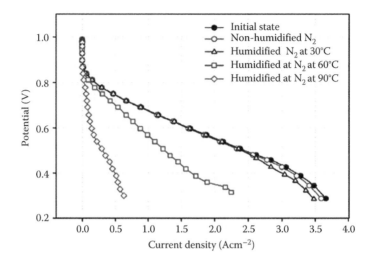

FIGURE 3.6 MEA polarization curves of before and after corrosion experiments using nonhumidified N$_2$, humidified N$_2$ at temperatures of 30°C, 60°C, and 90°C. (Reprinted from *Journal of Power Sources*, 193, Lim, K.H. et al. Effect of operating conditions on carbon corrosion in polymer electrolyte membrane fuel cells, 575–579, Copyright (2009), with permission from Elsevier.)

higher Brunauer–Emmett–Teller (BET) surface area than the Vulcan support and oxygen present in the system that adsorbed onto the carbon surface, thus causing mass increase that counteracted the loss of carbon.

The effects of relative humidity on the MEA performance are widely studied and the direct relationship between humidity increase and carbon corrosion are well noted in previous literature. The inevitability of humidity and water moisture in a fuel cell prompts for alternative operating methods or alternative catalyst supports which are more stable in a wide variety of PEMFC atmospheres.

3.3.6 Impurities

Feed stream impurities mainly in the hydrogen fuel fed to the anode electrode can cause both temporary and permanent performance degradation in PEMFCs. Steam reformation to produce hydrogen fuel inevitably leads to the presence of these impurities such as CO, H$_2$S, NH$_3$, organic sulfur–carbon, and carbon hydrogen compounds. The effects of these impurities in the fuel gas supply to the PEMFC have been well documented and studied in peer-review literature (Garzon et al., 2009).

The presence of CO in the hydrogen feed can poison the catalyst layer by binding onto platinum active sites rending them useless for fuel cell reactions. Many studies detail the effects and mechanism by which CO is able to degrade the PEMFC, as well as, prevention and recovery strategies (Angelo et al., 2008). In a process known as O$_2$/air bleeding, 0.5–1% of oxygen gas is fed into the fuel cell together with the anode feed stream in order to mitigate the poisoning effect of CO at the anode. The O$_2$ in the anode feed stream is able to oxidize adsorbed CO on the platinum particles freeing active sites for the hydrogen oxidation reaction (HOR) (Franco et al., 2009). This oxygen however, can result in further problems for the cathode carbon support if the cell undergoes cell reversal as described previously. In a report by Franco et al. they reported a 600-h test using CO as a competing degradation species method to O$_2$ in the anode caused by O$_2$ crossover from the cathode (Franco et al., 2009). Their simulated model and experimental results suggested that the effect of CO in the anode was strongly dependent on the current-cycle mode and could be utilized to reduce oxygen gas crossing over from the cathode and mitigate carbon support degradation caused by cell reversal. In order to capitalize on their findings, Franco et al. suggested the use of CO-tolerant catalysts at the anode.

The direct effect of contaminants entering the fuel cell gas feeds (both fuel and air streams) and affecting the corrosion of the support are not well reported although there is extensive study on the effect of performance overall by various impurities. Garzon et al. reported the effects of direct injection of various sulfur compounds, carbon monoxide, nitrogen oxides, ammonia, and salts into the fuel stream of the fuel cell (Garzon et al., 2009). In this work, they found that sulfur compounds such as H_2S and SO_2 could lead to severe performance losses in the fuel cell. Figure 3.7 demonstrates this loss in cell performance through MEA polarization curves after different periods of hydrogen sulfide exposure. Increased anode overpotentials were experienced leading to lower fuel cell efficiencies and higher cathode potentials. A consequence of high cathode potentials is the acceleration of platinum surface area loss and carbon support corrosion.

Other impurities have been experimented with in detail and have been reported to affect MEA and PEMFC performance; however, the exact interaction between the MEA and the impurity is not entirely known. The direct relationship with the carbon support may not be a factor when considering many of the impurities reported in literature which would degrade the MEA.

3.3.7 Oxidant Starvation

The effects of support degradation caused by air starvation are much less significant than that of fuel starvation. Similar to fuel starvation, oxidant (air) starvation can occur at the cathode whereby oxygen does not reach the catalyst layer to complete the electrochemical half-cell reaction of protons and oxygen to produce water. Common causes of oxidant starvation include a sudden change in the oxygen demand brought upon by start-up or a load change in the fuel cell. Water flooding at the cathode may also block the active sites at the cathode and therefore restrict gaseous oxygen from reacting. This water

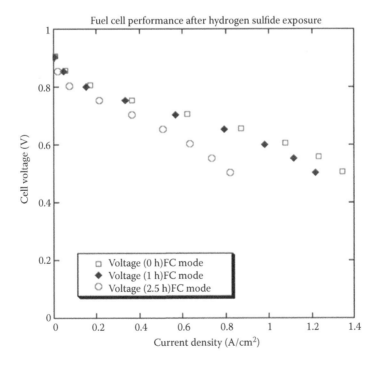

FIGURE 3.7 MEA polarization curves before and after 1 h and 2.5 h of hydrogen sulfide exposure. (Reprinted from Garzon, F. H. et al. 2009. *ECS Transactions* 25: 1575–1583. With permission from The Electrochemical Society.)

flooding typically comes about by three pathways: (1) water condensation from the humidified gas reactants, (2) water generation at the cathode active sites as a product from the electrochemical reaction, and (3) transport via electroosmotic drag associated with proton transport across the membrane. In experiments conducted by Taniguchi et al. air starvation led to a greater loss of ECSA, larger platinum particle sizes and decreased activity of MEAs compared to MEAs operating under normal conditions (Taniguchi et al., 2008). During oxidant starvation it was noticed that the cell terminal voltage rapidly changed to negative voltages, termed cell reversal. The electrode potential of the air-starved cathode electrode quickly dropped to less than 0 V while the anode potential decreased to about 0.1 V (Taniguchi et al., 2008).

Natarajan and Van Nguyen (2005a,b) also investigated cell performance over four hours using different oxidant flow rates at different temperatures. The reported results were comparable with Liu et al.'s results (2006a,b) indicating that during oxidant starvation uneven current distributions could be observed resulting in a drop in both the reaction rate and current density experienced downstream of the gas inlet channel.

Although the result of severe oxidant starvation can lead to damage of the carbon support, it is not often studied or focused on due to the relative by small effect it has when compared to fuel starvation. Oxidant starvation in PEMFCs is not considered a limiting factor in increasing the durability of PEMFCs.

3.4 Degradation Protocols

With the emphasis now on fuel cell durability as one of the main barriers holding back the commercialization of fuel cell technologies, it is important for researchers to know the tools and diagnostic methods available for accurate durability assessment. There is an arsenal of characterization techniques to evaluate catalyst degradation; however, there is no single approach that can be used to classify the severity of degradation nor is there one specific criteria used to assess the stability of the PEMFC catalyst. It is vital for the researcher to understand the wide variety of assessments available and to distinguish the capabilities and limitations of each testing method. The understanding of the individual degradation mechanisms of the catalyst support as well as the effect of the support degradation on the neighboring components of the cell is also important to synthesize new materials or develop new techniques to increase the longevity of the PEMFC. Fuel cell technology and energy organizations have established principle guidelines and set cost and durability targets based on the ongoing research that are meant to steer future endeavors toward high-stability and low-cost fuel cells. The following section outlines durability targets, diagnostic methods, and testing procedures that should be understood by those researchers in the field of PEMFC.

3.4.1 Fuel Cell Programs and Durability Targets

The U.S. DOE, New Energy and Industrial Technology Development Organization (NEDO), and the European Hydrogen and Fuel Cell Technology Platform (HFP) are among the main governing bodies recognizing the need to develop fuel cell technologies. Each of the organizations has identified PEMFC technology as an area for future development for vehicles and stationary power applications. Rollout plans and/or cost and durability targets have been set as guidelines for maturing technologies. The following sections detail more about the individual organizations and elaborate on the expectations set.

3.4.1.1 U.S. Department of Energy

The U.S. DOE Hydrogen, Fuel Cell & Infrastructure Technologies Program (HFCITP) was developed in recognition of America's growing dependence of foreign oil. The program teams up with industry, national laboratories, universities, government agencies, and other partners to overcome challenges to commercialization of hydrogen and fuel cell technologies. The DOE has set stringent targets for PEMFCs for stationary and automotive applications. They stress durability and cost as main barriers for commercialization. According to their targets set for 2010, the U.S. DOE aims at having fuel cell systems for automotive applications operate for at least 5000 h, which is equivalent to 150,000 driven miles, with less than 5% performance lost by the end of life. The projected cost per kilowatt for a fuel cell after mass manufacturing in 2005 is $75 kW^{-1}. The DOE expects that in order for fuel cells to be competitive with internal combustion engines the price per kW must be less than $50. DOE set targets for 2010 and 2015 are $25 kW^{-1} and $15 kW^{-1}, respectively (Borup et al., 2007).

For stationary fuel cell applications the durability requirements for fuel cells is more rigorous. By 2005, stationary PEMFCs have been able to undergo 20,000 h of operation however, market requirements will demand even greater lifetimes over a broad range of temperatures (−35–40°C). By 2011, U.S. DOE expects to see stationary fuel cells maintain lifetimes up to 40,000 h at a cost less than $750 kW^{-1} (Borup et al., 2007).

3.4.1.2 New Energy and Industrial Technology Development Organization

In Japan, NEDO is seeking development and advancement in PEMFCs for practical applications. The program collaborates with sectors of government, industry, and academia to promote hydrogen energy education and technology application in areas of automobiles, stationary power, and portable information devices. In response to growing environmental concerns and the promising advantages of PEMFCs, NEDO promotes the following four development areas under the Strategic Development of PEMFC Technologies for Practical Applications: (1) Development of technology on basic and common issues, (2) development of elemental technology, (3) development of basic production technology, and (4) development of technology for next-generation fuel cells. The NEDO program also recognizes the need for durable, highly efficient, and cost-effective fuel cell systems and by the full dissemination phase of PEMFCs (2020–2030) NEDO expects to see operating lifetimes for over 100,000 km (Borup et al., 2007).

Similar to the U.S. DOE, NEDO has set ambitious targets for PEMFCs for automotive and stationary power applications. By 2015, the targets set for automotive PEMFCs are as follows: Greater than 5000 h of operating life with at least 60% efficiency based on the lower heating value, an operation temperature of the cell in between 90°C and 100°C and a stack cost of approximately 10,000 Yen kW^{-1}. By 2020, operation cell temperatures are expected to increase to 100–120°C without a humidifier to avoid further complexity to the fuel cell system and must be able to operate with external temperatures of −40°C. Lifetime targets for degradation are expected to be less than 10% at end of life. For stationary power fuel cells, the set targets by NEDO are 40,000 h lifetime for 2010 and increased to 90,000 h by 2015.

3.4.1.3 European Hydrogen and Fuel Cell Technology Platform

The HFP focuses on the acceleration and development for cost reduction necessary to competitively market fuel cells for transportation, stationary, and portable power applications. Their Hyways Project Roadmap aims for a mass market rollout by 2020 for European class vehicles with efficiencies of at least 40% on the New European Drive Cycle (NEDC) at a cost of 100 Euros kW^{-1} and lifetimes of 5000 h for automobiles and 10,000 h for buses. The goals for stationary power systems mainly target residential power systems which the HFP hopes to see tangible market penetration by 2020. By 2009–2012, these systems are expected to maintain 34–40% electrical efficiency, a total fuel efficiency of 80%, >12,000 h of operation and cost less than 6000 Euro per system stack (Borup et al., 2007).

3.4.2 Diagnostic Methods

To test the degradation of fuel cell catalyst and assess the carbon support degradation effect on fuel cell performance, many diagnostic tools are available. These tools may test the morphology of the catalyst support directly or may evaluate the carbon corrosion indirectly through the fuel cell overall performance. Common parameters analyzed to evaluate the electrocatalyst degradation include measurement of the catalyst layer areas (cross-sectional and surface area), the ECSA, fuel cell current density, surface morphology, and elemental composition of material or effluent gas.

The morphology of the catalyst can be an important parameter when examining fuel cell catalyst degradation. Scanning electron microscopy (SEM) is often utilized to assess the thickness of catalyst layers and to determine surface morphology of the electrode. SEM (an example is shown in Figure 3.8) can also be useful in determining particle size distribution if the agglomerates or particles are relatively large (>20 nm) (Fujii et al., 2006). For smaller particle sizes or higher-resolution surface morphology analysis, TEM is often utilized. Platinum particle size distributions deposited on carbon supports can be seen through TEM as well as crystalline lattice parameters that are often conducted. SEM and TEM coupled with energy dispersive x-ray (EDX) analysis can become a powerful method to determine elemental composition of the surface of an electrode material.

Atomic force microscope (AFM) and optical micrography have also been used to characterize degradation of the catalyst layer. X-ray computer tomography (X-ray CT) was used by Garzon et al. (2007) and Lau et al. (2009) to conduct 3-D noninvasive scans of the catalyst layer before and after testing of the electrocatalyst. The morphology of the catalyst layer was observed through electron probe microanalyzers for surface images and cross-sectional elemental distribution of Pt. X-ray photoelectron spectroscopy (XPS) is a powerful tool in determining the structure of the support and changes in chemical bonding structure over corrosion periods. This characterization is utilized by Zhang et al.

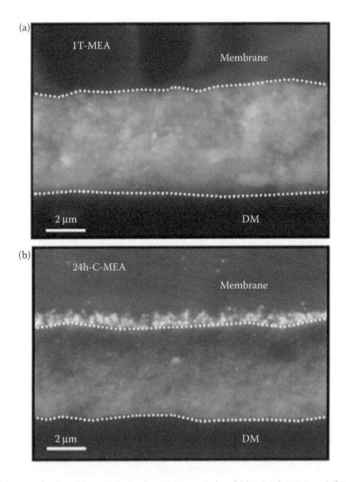

FIGURE 3.8 SEM cross-sectional images in backscattering mode of (a) initial MEA and (b) MEA after 24 h of electrochemical cycling. (Reprinted from Chen, S. et al. 2010. *Journal of the Electrochemical Society* 157: A82–A97. With permission from The Electrochemical Society.)

and illustrated in Figure 3.9 (Zhang et al., 2009). Such experimental techniques are useful for destructive and nondestructive methods for characterizing morphological changes to the catalyst layer through different stages of the fuel cell tests.

Electrochemical techniques are often used to evaluate the performance of the fuel cell catalysts and to observe changes of the catalyst over the operation life. Catalyst degradation can be inferred by decreases in performance and changes in the ECSA of the catalyst that can be measured by estimating the charge area under the hydrogen desorption peak in a typical cyclic voltammetry (CV) experiment in acidic conditions (Gallagher et al., 2007). Typically, to electrochemically test the degradation of the fuel cell electrocatalyst, two commonly used routes are taken: accelerated durability test (a.k.a. the accelerated stress tests (AST)) or the life test. The more reliable and accurate method to test the degradation of fuel cell electrocatalysts is to run the test according to real operating conditions of the fuel cell; however, these tests are time consuming, costly, and infeasible in some cases as real-world fuel cell lifetimes are expected to reach over 40,000 h of operation lifetimes. ASTs have become increasingly popular due to their simplicity and ability to assess catalyst degradation (Makharia et al., 2006). Furthermore, ASTs can be used to test full fuel cells (*in situ* testing) or electrochemical half-cells (*ex situ* testing). *In situ* testing is more useful for testing the system as a whole and assessing how different components interact with one another, whereas tests using an electrochemical half-cell are more useful for isolated electrochemical testing focused on the catalyst performance (Dross and Maynard, 2007). *Ex situ* testing will not be able to detect membrane degradation issues, stack problems, water flooding, fuel crossover, and temperature effects.

Therefore, knowing the variety of evaluation methods and their applicability to determining the stability of a catalyst support or MEA is extremely helpful. It should be noted that there are other protocols for assessing support degradation but these techniques may vary by circumstance and are not heavily reported upon. The following section will discuss protocols and testing procedures utilized by many research labs as standardized testing for catalyst support durability.

3.4.3 Degradation Testing Procedures

To test the durability of PEMFC catalyst under real-world conditions and time periods is infeasible if not impossible. Fuel cell lifetimes are expected to meet the DOE's target of 5000 h of operation for

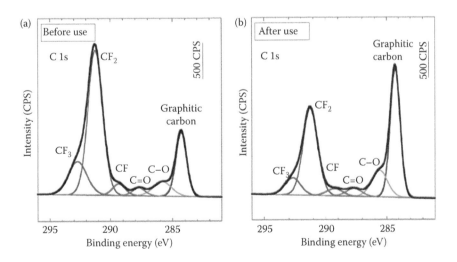

FIGURE 3.9 XPS of the C 1s spectra of MEAs constructed with Nafion membrane and Pt/C catalyst before and after 300 h of operation. (Reprinted from *Electrochimica Acta*, 54, Zhang, F. Y. et al. Quantitative characterization of catalyst layer degradation in PEM fuel cells by x-ray photoelectron spectroscopy, 4025–4030, Copyright (2009), with permission from Elsevier.)

automotive applications and 40,000 h for stationary application. Therefore, ASTs are developed to assess the degradation of these fuel cells under specified conditions. Several reports give detailed experimental conditions for assessing the durability of carbon-supported catalysts; however, the accuracy and effectiveness of these experiments to simulate real-world conditions is still in question. Table 3.1 gives commonly used examples of types of durability tests described by many peer-reviewed papers.

Lim et al. conducted durability experiments of Johnson Matthey Co. 40 wt% platinum catalysts with a focus on the effects of humidity, cell temperature, and gas-phase O_2 (Lim et al., 2009). Their degradation method was to hold the potential of their in-house-fabricated MEAs at 1.4 V for 30 min at various cell temperatures and varying humidity and measure the effluent CO_2 levels using online mass spectrometry.

Li et al. published a paper in which their degradation method of the catalyst support was to cycle the potential from 0.1 to 1.4 V at 10 mV s^{-1} scan rate (Liu et al., 2006a,b). They conducted a differential electrochemical mass spectroscopy (DEMS) study to analyze the CO_2 and platinum levels exiting the cathode during potential cycling. In their study, they claimed that the carbon support corrosion and the fuel cell degradation were further aggravated by multilayers of PtO and higher Pt oxides formed during cycling.

Zhang et al. conducted degradation tests by measuring the change in OCV with MEAs prepared from commercial Pt/C catalysts (Zhang et al., 2010). They reported ECSA losses of up to 60% after 254 h of operation. OCV degradation they observed was similar to OCV degradation trends reported in other literatures where the OCV greatly declined after a couple hours of operation and leveled off slowly with increased operation. The OCV can be recovered back to higher values through intermittent operation or after running a simple electrochemical recovery process. The thickness of the catalyst layer, as well as, the membrane was shown to decline with decreasing ECSA.

Due to the nature of PEMFC catalysts and their complexity usually involving precious metals or alloys, it is hard to evaluate the degradation method of the electrocatalyst and differentiate between carbon corrosion and other factors contributing to activity loss (e.g., platinum sintering or migration). It is therefore necessary in most cases to employ additional methodologies besides electrochemical measurements to determine the relative activity loss due to carbon corrosion. Such methods are mentioned in the previous sections; however, common techniques for characterizing carbon corrosion are SEM imaging of the MEA thickness and online CO_2 mass spectrometry.

In order to synchronize the efforts by various research groups and provide common metrics for evaluating the durability of fuel cell catalysts, the DOE has proposed specific conditions to evaluate and test the different components of the MEA. Tables 3.2 through 3.4 describe these metrics in full detail for different specific components, namely the electrocatalyst, the carbon support, and the MEA as a whole.

These degradation and testing procedures provide a standard form of measurement for accessing the stability of the MEA, the electrocatalyst, and the catalyst support. The metrics do not give an accurate measure of the lifetimes that the MEA would experience in reality, but rather give a comparative measure between different catalysts across different research laboratories.

TABLE 3.1 Commonly Used Carbon Corrosion Degradation Protocols

General Methods	Experiment Protocol	Reference Example
Potential holding	Hold cell at 1.4 V for 30 min using Johnson Matthey Co. 40 wt% platinum catalyst varying the gas inlet humidity and cell temperature	Ye et al. (2008); Lim et al. (2009); Ko et al. (2010)
Potential cycling	Cathode potentiodynamic cycling between 0.1 and 1.4 V vs. the anode electrode. 10 mV s^{-1} scan rate	Liu et al. (2006a,b)
Open circuit operation	Hold at open-circuit voltage (OCV) for 256 h with intermittent discontinues in operation 100 sccm min^{-1} and 50 sccm min^{-1} of air/H_2, respectively	Zhang et al. (2010)

TABLE 3.2 Electrocatalyst Potential Cycle and Metrics

Cycle	Step Change: 30 s at 0.7 V and 30 s at 0.9 V. Single Cell 25–50 cm^2	
Number	30,000 cycles	
Cycle time	60 s	
Temperature	80°C	
Relative humidity	Anode/cathode 100/100%	
Fuel/oxidant	Hydrogen/N$_2$	
Pressure	150 kPa absolute	
Metric	Frequency	Target
Catalytic activity[a]	Beginning and end of life	≤60% loss of initial catalytic activity
Polarization curve from 0 to ≥1.5 A/cm^{2b}	After 0, 1k, 5k, 10k, and 30k cycles	≤30 mV loss at 0.8 A cm^{-2}
ECSA/cycle voltammetry	After 1, 10, 30, 100, 300, 1000, 3000	≤40% loss of initial area

[a] Activity in A mg^{-1} at 150 kPa abs backpressure at 900 mV iR-corrected on H$_2$/O$_2$, 100% RH, 80°C.
[b] Polarization curve per the U.S. Fuel Cell Council (USFCC) "Single Cell Test Protocol" Section A6.

3.5 Mitigation Strategies

More of the current research regarding PEMFC catalyst supports focus on novel treatment processes, techniques, or different materials that allow for a more stable and active catalyst for PEMFC technologies. Based on the strict targets set upon by the DOE for fuel cell lifetimes, researchers have recognized the importance of stabilizing the support to impede corrosion mechanisms or to find novel materials that are able to avoid corrosion problems. This task of modifying the catalyst support has proven difficult for many laboratories who realized that an alteration of the catalyst stability will often affect other parameters to hinder its ability to succeed as a high-performing support. This section reviews possible strategies that have been used to either modify the carbon black as a more enduring support or replace it with another material able to prolong the lifetime of the overall catalyst.

TABLE 3.3 Catalyst Support Potential Cycle and Metrics

Cycle	Hold at 1.2 V for 24 h; Run Polarization Curve and ECSA; Repeat for Total 200 h Single Cell 25–50 cm^2	
Total time	Continuous operation for 200 h	
Diagnostic frequency	24 h	
Temperature	95°C	
Relative humidity	Anode/cathode 80/80%	
Fuel/oxidant	Hydrogen/nitrogen	
Pressure	150 kPa absolute	
Metric	Frequency	Target
CO$_2$ release	Online	<10% mass loss
Catalytic activity[a]	Every 24 h	≤60% loss of initial catalytic activity
Polarization curve from 0 to ≥1.5 A/cm^{2b}	Every 24 h	≤30 mV loss at 1.5 A cm^{-2} or rated power
ECSA/cycle voltammetry	Every 24 h	≤40% loss of initial area

[a] Activity in A mg^{-1} at 150 kPa abs backpressure at 900 mV iR-corrected on H$_2$/O$_2$, 100% RH, 80°C.
[b] Polarization curve per USFCC "Single Cell Test Protocol" Section A6.

TABLE 3.4 MEA Chemical Stability and Metrics

Test Condition	Steady State OCV Single Cell 25–50 cm²	
Total time	200 h	
Temperature	90°C	
Relative humidity	Anode/cathode 30/30%	
Fuel/oxidant	Hydrogen/air at stoics of 10/10 at 0.2 A cm⁻² equivalent flow	
Pressure, inlet kPa abs (bara)	Anode 250 (2.5), cathode 200 (2.0)	
Metric	Frequency	Target
F-Release or equivalent for nonfluorine membranes	At least every 24 h	No target—for monitoring
Hydrogen crossover (mA cm⁻²)[a]	Every 24 h	≤20 mA cm⁻²
OCV	Continuous	≤20% loss in OCV
High-frequency resistance	Every 24 h at 0.2 A cm⁻²	No target for monitoring

[a] Crossover current per USFCC "Single Cell Test Protocol" Section A3–2, electrochemical hydrogen crossover method.

3.5.1 Surface Modifications of Carbon Supports

Carbon supports typically undergo chemical or physical activation prior to platinum impregnation. The alteration of surface groups and functionalities on the carbon support can strongly influence the carbon–metal interaction that can directly affect the metal particle size, metal particle distribution, surface morphology of the carbon, and surface impurities that may be present. These parameters have been known to influence the catalytic metal stability and activity of the resulting catalyst. Common surface modifications strategies include chemical oxidation of the carbon or thermal activation to modify the surface structures.

3.5.1.1 Heat Treatments

One common approach to improve catalytic activity and durability of the catalyst is a heat-treatment process. Proper heat treatment of carbon support can increase the stability of Pt/C catalyst. Different types of heat treatments have been reported for fuel cell catalyst treatment, such as, oven/furnace heating, microwave heating, plasma thermal heating, and ultrasonic spray pyrolysis.

Continued efforts have been carried out to determine the effect of heat treatment on the individual components of the fuel cell catalyst. For a platinum-based catalyst, heat treatment has been known to have a significant impact on the metal particle size, particle size distribution, particle morphology, and dispersion on the support. A heat-treatment process could also be used to remove impurities present in the metallic catalyst (Bezerra et al., 2007). These effects help to increase the ECSA of the catalytic metal and generally improve the performance of these types of catalysts. Figure 3.10 illustrates the results obtained by Chen et al. who plotted the heat-treatment temperature of Vulcan XC-72R carbon black with specific corrosion rate seen in platinum alloy cathode catalysts at 1 V in H_3PO_4 at 180°C.

Separate studies carried out by Han et al. (2007) and Antolini et al. (2002) showed that typical Pt/C catalyst could be affected by heat treatments. In Han et al.'s study, they heat treated 20% Pt/C catalysts at several temperatures for varying durations and found that the average platinum particle size grew exponentially with increasing heating temperature and linearly with time. Antolini et al. showed that the heating rate was also an influencing factor in the heat-treatment process (Antolini et al., 2002). Platinum particles crystallinities, distribution and morphology were studied. Two different samples were treated at 15 and 45°C min⁻¹. The slow heating rate yielded more homogeneous particle sizes distribution and greater platinum crystallinity. The optimum pyrolyzation temperature was found in the range of 400–550°C.

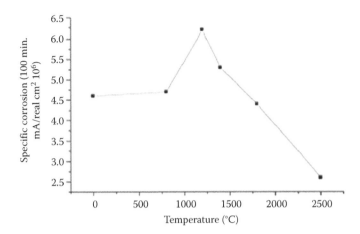

FIGURE 3.10 Effect of heat-treatment temperature versus specific corrosion rates of Vulcan XC-72R. Corrosion rates of the carbon black were tested at 1.0 V in H_3PO_4 at 180°C. (Reprinted from *Journal of Power Sources*, 173, Bezerra, C. W. B. et al. A review of heat-treatment effects on activity and stability of PEM fuel cell catalysts for oxygen reduction reaction, 891–908, Copyright (2007), with permission from Elsevier.)

Proper heat treatment of carbon support can increase the stability of Pt/C catalyst. For catalysts using carbon black as support for the catalytic metal, the heat-treatment process is understood to provide two main functions for the stability: (1) Removal of oxygenated functional groups and (2) graphitization of the carbon support surface. The carbon surface of most carbon blacks is functionalized with various oxygen-containing functionalities which indubitably affect the surface chemistry of the carbon support (Bleda-Martinez et al., 2006; Colmenares et al., 2009). Heat treatments can thermally decompose these surface oxides to H_2O, CO, CO_2 at temperatures from 100°C to 900°C. Studies show that the greater concentration of surface oxides on the carbon support leads to poorer platinum dispersion. By removing these acidic functionalities more π bonding formations (C=C) are present thus leading to better platinum dispersion, higher ORR activity and higher stability under fuel cell operating conditions.

Graphitization inevitably changes the electrochemical properties of the carbon and leads to a more corrosion-resistant support. This increased metal-support interaction is related to the increasing strength of the π sites on the support upon pregraphitization, which acts as anchoring centers for platinum (Stonehart, 1984; Prado-Burguete et al., 1991). Higher pyrolyzation temperatures are necessary for the graphitization using polymer and aromatic compounds (de Bruijn et al., 2008). At temperatures of 1630°C large aromatic molecules will condense into graphene, and at temperatures greater than 1730°C, aromatization occurs to form a graphitic structure. The temperature selected for the heat treatment plays an important role in the extent of graphitization and the number of defects formed within the surface of the carbon support. Graphitization yields a thermally, and electrochemically stable carbon support material which has been used extensively in carbon nanotubes and nanocrystalline graphene structures to improve the stability and durability of fuel cell carbon catalyst supports.

3.5.1.2 Surface Treatments and Coatings

Surface chemistry of the carbon support plays particular importance in the activity and stability of the resulting electrocatalyst in PEMFCs. Surface treatments typically include the use of oxidizing agents such as HNO_3, O_3, O_2, and H_2O_2 that can modify the functional sites on the carbon from basic to acidic. These groups usually exist in the form of oxides such as quinone, ether, anhydride, carbonyl, phenol, lactone, and carboxylic groups. Although the effects of these surface treatments may vary in the literature, it has been proposed by Torres et al. that HNO_3-treated carbons leave strong acid functionalities whereas carbons treated with O_3 and H_2O_2 result in weak acid functionalities (Torres et al., 1997). According to their temperature-programmed desorption results, carbons with more weak acid functionalities display a stronger interaction with the catalytic metal precursor (H_2PtCl_6) during impregnation than those with strong acid functionalities. This strong interaction would favor the dispersion of Pt on the carbon surface and hence yield higher catalytic activity. However, several other papers contest this by saying that basic sites are more likely to be the centers for adsorption of $PtCl_6^{2-}$ and that the oxidation treatment reduces the number of basic sites (Antonucci et al., 1994; Coloma et al., 1994; Roman-Martinez et al., 1995; Guerrero-Ruiz et al., 1998).

In a study made by Roman-Martinez et al. the relationship between support surface and the platinum precursor are studied (Roman-Martinez et al., 1995). It was found that there was a negative effect of the degree of surface oxidation on the final platinum dispersion seen after the resulting impregnation with two different platinum precursors H_2PtCl_6 and $[Pt(NH_3)_4]Cl_2$. This was a result of either the preferred anchorage of platinum precursor on the functional sites or a repelling electrostatic interaction by the oxidative groups. This discrepancy between reports suggests that different impregnation methods and surface functional groups on the carbon surface have not been fully explored with respect to the resulting platinum dispersion.

The treatment of carbon supports with HNO_3, however, has led some research groups to speculate that nitrogen functionalities are added during the activation process which benefit the anchorage of platinum metals to the carbon surface. Nitrogen functionalities have also been claimed to have inherent catalytic ability to the ORR and thus, HNO_3 is a common chemical pretreatment for the carbon support. Along these lines, nitrogen doping of the carbon support through the coating of carbon blacks has also been testified to and reported on in the next section—Novel Carbon Support Materials (Li et al., 2009).

Conductive polymers, such as polyacetylene, polythiophene, polypyrrole, polyisothianaphthene, polyethylene dioxythiophene, polyaniline, and so on, have interesting properties that make them suitable for use in PEMFCs (Heeger, 2001; Shirakawa, 2001). Their electroconductivity and noncarbon functionalities allow some of them to perform effectively as alternative carbon catalysts or with carbon supports to enhance their catalytic effects. Huang et al. utilized polypyrrole as a conductive polymer support for a platinum catalyst active for the ORR (Huang et al., 2009). Their results show significant resistance to carbon corrosion and improved conductivity over traditional Pt/C catalysts. They report that the platinum on polypyrrole catalyst (Pt/Ppy) has well-dispersed platinum particles of about 3.6 nm in diameter. CV scans up to 1.8 V revealed that there was little carbon support corrosion on the Pt/Ppy and a twofold increase in activity than Pt black at 0.9 V.

Various metals/metallic oxides have been coated onto carbons with the aim to improve the platinum tolerance to poisons, increase the platinum utilization and to avoid carbon corrosion of the catalyst support. Sn, Ru, Ti, and Co metals as well as their oxides have been reported on in fair amounts as being transition metals capable of increasing the platinum utilization by removing hydroxide species that would be present on the platinum such that the platinum can further catalyze the ORR. Although these materials have been coated onto the carbon as a support, there are alloying characteristics that occur with the impregnated platinum which result and thus will not be touched up in this section. However, many of these oxides and metals have been used as stand-alone catalyst supports without the use of a carbon substrate, and are discussed further in Section 3.5.3.

3.5.2 Novel Carbon Support Materials

With the introduction and expansion of the field of nanotechnology many novel materials and morphologies are being recognized as having similar but different properties than the bulk material. This observation has been extremely well noted for graphitized carbon structures which include CNTs, CNFs, graphene, and porous carbons. Most of these novel carbon structures have been noted to have higher chemical stability and mechanical strength, but also properties such as high aspect ratio, high electrical conductivity, and favorable surface properties. Because of this, these nanostructured carbons have been gaining tremendous popularity as alternatives for carbon blacks in a multitudinous variety of applications, especially for PEMFC catalyst supports. It is important to note, however, that although these catalyst supports differ in structure and morphology than typical carbon blacks, they are still considered catalyst supports that are based on carbon and due to this, are prone to degradation at normal fuel cell operating conditions (with the exception of conductive diamond). Still, the use of these nanostructured carbon materials has been ever increasing and likely to continue expanding because of their relative success. The following sections review multiple studies incorporating different carbon support materials that have been developed to improve the performance and durability over typical platinum on carbon black catalysts.

3.5.2.1 Graphitized Carbons—Carbon Nanotubes, Carbon Fibers, and Graphene

In light of the durability problems with conventional carbon black catalyst support materials, much research has been geared toward novel nanostructured carbon materials. CNTs (Baughman et al., 2002) and CNFs have been receiving a lot of attention due to their unique properties as PEMFC catalyst supports. Compared to carbon blacks, they offer higher electrical conductivity, form ordered catalyst layers, and are synthesized to be purer than many of the furnace burned carbon blacks that may, for example, contain organo-sulfur impurities that risk contaminating the platinum metal. Specific interactions between the platinum catalyst and CNT/CNF also favor the nanostructured support. The graphitized carbon layers of these supports offer greater mechanical stability and delocalized π bonds which significantly increase electroconductivity resulting in higher reactivity (Li et al., 2008).

Another advantage of the graphitized carbon support is that carbon corrosion rates have been shown to correlate with the degree of graphitization of the carbon support with increased electrochemical and thermal durability seen in CNFs, CNTs, graphene, and other graphitized carbon structures (Li et al., 2006). Although graphitization of carbon supports is successful in slowing down the kinetics of carbon corrosion, it does not change the fundamental thermodynamics of the degradation and thus are still prone to durability issues.

CNTs of various diameters and morphologies have been used extensively in research to improve the durability and performance of Pt/C catalysts. For CNTs, the number of cylindrical shells can have an outstanding impact on the catalytic activity experienced by Pt/CNT catalyst. Chen et al. conducted a comparative study of different carbon nanotube supports including single-walled carbon nanotubes (SWNT), double-walled carbon nanotubes (DWNT), multiwalled carbon nanotube (MWNT), and carbon black (Chen et al., 2007). In terms of performance, DWNTs exhibited the highest activity for the ORR followed by MWNTs, SWNTs, and then Pt/C. They found that the durability of the CNTs with greater graphitization was increased dramatically and the structure of the tubes led to increased performance and higher ECSA values. This result was attributed to better electrochemical properties of the CNTs such as higher electroconductivity and higher mass transport capabilities being provided by a network of CNT structures. Wu et al. (2005) and Tang et al. (2007) however, realized contradictory results when testing different morphologies of CNTs for the methanol oxidation reaction (MOR). They found that SWNT supported platinum catalysts outperformed DWNT and MWNT as catalyst supports and attributed this increased activity to the higher specific surface area and higher platinum dispersion on the tubes. These conflicting results from the authors could be due to different structures and morphologies of the nanotubes used. Some reports in the literature claim SWNTs to be semiconductors depending on the structure and chirality that would decrease electron transfer rates over MWNTs. If this is the case, it would promote a more thorough investigation to determine the relationship between various structures and chiralities of CNTs and their catalytic performances.

Besides the number of cylindrical shells contained in the CNTs, another important factor is the type or number of defects on the CNTs/CNFs. Bamboo-type MWNTs, as opposed to hollow-type CNTs, are a class of carbon nanotubes in which the graphite planes are at an angle to the principle axis of the nanotube. These types of tubes form cone-shaped compartments at varying lengths along the tube and as a result have inherently higher defected edges that are exposed down the surface of the tube. Bamboo-type MWNTs have been shown to have higher electron transfer in electrochemical processes compared to hollow CNTs, where the graphite planes are parallel to the principle nanotube axis. This improved electron transfer has been shown to lead to enhanced catalytic activity which is a direct results of the greater proportion of edge plane exposure and defects exhibited by the bamboo-MWNTs (Antolini, 2009).

The same parallels can be drawn when looking at different types of CNF structures. It was found that palladium supported on platelet type CNFs exhibited higher performances for the ORR than fish-bone-type CNFs which was a result of the greater edge to basal plane atoms of the platelet type CNFs. Tsuji et al. found similar results where they tested and compared the activity of platinum–ruthenium on different types of CNF structures: platelet, tubular, and herringbone, and found that catalysts prepared on platelet-type CNFs exemplified higher onset reduction potentials and current peak potentials for the ORR than the other structures (Tsuji et al., 2007).

Although it can be concluded from the above evidence that greater edge plane exposure and number of defected sites usually have a positive impact on catalytic activity of the carbon nanostructured supports, it is also known that the corrosion of carbon materials is initiated at these edge planes. This relationship between catalyst support enhancement and carbon corrosion needs to be kept in mind when developing novel nanostructured carbon catalysts. Unfortunately, the influence of defects in CNTs and CNFs on the durability of these structures is still not quite known and little research is published in this area.

Graphene-type carbon nanostructures, however, have not demonstrated similar promising results as carbon support structures compared with CNTs and CNFs (Antolini, 2009). Studies performed with platinum deposited onto graphene structures show low performance. Several explanations have been offered to explain why graphene sheets do not bode as well as their CNT/CNF counterparts. It has been argued that the delocalized electrons present in graphitized carbons are capable of three-dimensional mobility in CNT/CNF networks whereas on graphene sheets, electron mobility is hindered due to the planar, two-dimensional structure. This hindrance of mobility is exacerbated when the transport of gases and protons are considered to the catalyst active site to form a TPB. Platinum particles deposited on the graphene sheets become buried when layers of graphenes are stacked between one another—a problem not experienced by CNTs or CNFs that typically form a wire mesh network.

To further enhance the durability and performance of Pt/CNT structures, doping the CNTs has shown to have some interesting effects on the electrochemical properties of fuel cell catalysts (Acharya et al., 2009). A first principle study conducted by Yu-Hung Li et al. (2009) investigated the adsorption of platinum onto un-doped CNTs as well as CNTs doped with nitrogen or boron. The study confirmed the experimental results of many studies in that nitrogen-doped carbon nanotubes have increased reactivity for platinum adsorption on the surface of CNTs as well as higher binding energies when platinum adsorbs onto carbon atoms neighboring a nitrogen atom within the graphitic sheet. This study showed that the nitrogen incorporated into CNT structures simply mediated the enhancement of platinum adsorption onto the surface of CNTs by activating neighboring carbons and providing "donor-like" behavior due to an extra valence electron.

By the same study, boron was tested as a dopant for carbon nanotube structures and it was found that the binding energy of platinum bonded directly to boron within the graphitic sheet of the CNTs was higher than the binding energy of platinum adsorbed onto un-doped CNTs (Li et al., 2009). Unlike the case of nitrogen doping where the increased binding energy was due to the activation of the neighboring carbon atoms in the carbon graphite structure, the direct bond between platinum and boron is strengthened due to the hybridization of the platinum d-orbital with the boron p-orbital. In addition, it has been shown that the adsorption of platinum onto boron- and nitrogen-doped CNTs can be significantly increased with greater dopant content and different structures, for example, pyridinic nitrogen.

3.5.2.2 Porous Carbons

Recently, much study in the field of catalysis has sought interest in carbons with controllable pore diameters usually ranging from 2 to 50 nm. These mesoporous carbons can be classified into two categories, namely, ordered mesoporous carbons (OMCs) and disordered mesoporous carbons (DOMCs). OMCs can be synthesized by casting hard silica templates and impregnating a carbon source into the pores of the silica or by directly templating a triblock copolymer. OMCs are usually preferred over DOMCs because of their uniform pore diameters and interconnected pore chambers allowing for high specific surface area and better mass transport. DOMCs on the other hand are typically irregularly interconnected and exhibit wider pore distributions which results in lower conductivity.

As for optimal pore sizes, Yu et al. conducted studies on OMCs with pore sizes ranging from 10 to 1000 nm as a support for a PtRu catalyst used for the MOR (Yu et al., 2002; Chai et al., 2004) as seen in Figure 3.11. The optimal pore size observed for the catalyst was 25 nm leading to a 43% increase in activity compared to commercially available PtRu/C from E-TEK. They claimed that the pore size was optimal for high carbon support surface areas for catalyst dispersion and also large pore volumes for the formation of the triple phase boundary sites. The interconnected pore channels were also hypothesized to benefit the mass transport of gases to the reactant sites allowing for efficient flow of products and reactants.

Although the use of these OMCs have been utilized in recent years with great success as supports for PEMFC catalysts, not much work has been done on assessing the durability of these supports for long-term applications. And although these novel nanostructured carbons may display increased tolerance to the harsh operating conditions in fuel cells with respect to carbon blacks, carbon corrosion is still thermodynamically possible and not completely eliminated.

3.5.3 Noncarbon Supports

Several other materials are considered alternatives for carbon supports and are more durable and stable for a platinum support in fuel cell operating conditions. In light of the carbon corrosion problems faced by Pt/C catalysts today, many researchers have focused on novel materials which are able to operate efficiently in the harsh conditions of a PEMFC. An ideal catalyst support would be able to operate in high operating voltage, withstand low pH, be cost efficient, have high surface area, and be readily available. It would also be advantageous if the novel support material would be environmentally friendly and recyclable from an MEA. Much research has been documented reporting some of these materials such as tungsten oxide, indium tin oxide, tin oxide, tungsten carbide, W- and Nb-doped titania, and sulfated zirconia. It should be noted that although the supports presented in this section are reported as standalone noncarbon supports, it is common that some researchers will integrate the materials with a carbon support hoping to achieve added benefits of the noncarbon materials with the ease and conductivity of a traditional carbon support.

3.5.3.1 Tungsten Oxide

Tungsten oxide is an n-type semiconductor that has been used in electrochromic, photochromic, photocatalyst and gas-sensor applications. The inherent electrical conductivity of tungsten oxide stems

FIGURE 3.11 SEM images of the colloidal silica composite templates (insets) and the corresponding carbon replicas prepared by (a) volume templating and (b) surface templating, using 250 nm silica spheres. (Reprinted with permission from Yu, J. S. et al. 2002. Fabrication of ordered unifrom porous carbon networks and thir application to a catalyst supporter. *Journal of the American Chemical Society* 124: 9382–9383. Copyright (2002). The American Chemical Society.)

from its nonstoichiometric composition and oxygen vacancies in the crystal lattice. The good electrical conductivity and flexible material design allows tungsten oxide to be a good candidate as a PEMFC catalyst support alternative. However, what makes it attractive for PEMFC applications is the tolerance it exhibits to CO. Tungsten oxide has been used frequently as a CO-tolerant catalyst support especially for the MOR, ethanol oxidation reaction (EOR) and the ORR. CO tolerance is necessary when CO impurities are present in the inlet feed or during ethanol and methanol oxidation where CO may be an intermediate byproduct of the electrochemical reaction.

Chhina et al. conducted studies using tungsten oxide as platinum catalyst support materials (Chhina et al., 2007a,b). Thermogravimetric analysis and electrochemical testing showed that the tungsten oxide support material was much more resistant to oxidation than commonly used Vulcan XC-72R with and without deposited platinum. Tests showed that after repeated cycling tungsten oxide maintained its structural integrity much better than the widely used carbon support.

Maiyalagan et al. studied platinum supported on tungsten oxide nanorods that were synthesized using anodic alumnia membrane (Maiyalagan et al., 2008). Their catalyst showed good activity for the MOR, which was attributed to a synergistic effect between the platinum and WO_3 and the ability of their catalyst to avoid poisoning that could have otherwise occurred from oxidation of methanol intermediates.

The durability of WO_3 is a problem since it has been found that tungsten dissolution can be observed. This dissolution hinders WO_3 catalyst supports from its application in fuel cells. It has been found, however, that with the addition of Ti^{4+} into the WO_3 lattice framework, the stability improved and the catalyst experienced a decrease in ohmic resistance (Antolini and Gonzalez, 2009). Further research and understanding of these stability effects are necessary to adequately incorporate WO_3 as a catalyst support for PEMFC applications.

3.5.3.2 Indium Tin Oxide

Since the first report of transparent conducting CdO films in 1907, vast research and development has been carried out on transparent conductive oxides to tune their properties so that they can be effectively utilized in numerous applications (Hosono, 2007). Conductive oxide films have been found to be well suited for applications such as liquid crystal displays, transparent electrodes of solar cells, photodetectors, and smart windows (Wan et al., 2004).

Indium tin oxide (ITO) is a Sn(IV)-doped In_2O_3 n-type semiconductor on which a great deal of research has been conducted. The creation of electron degeneracy is fulfilled by introducing a tin dopant that allows for the simultaneous occurrence of high optical transparency (~90%) and high electron conductivity (Tahar et al., 1998a,b). In comparison, the electrical conductivity of ITO is much greater (1000 S cm^{-1}) whereas graphitized carbon exhibits electron conductivity of 727 S cm^{-1}. To meet economical and technological demands, most research reporting the synthesis of indium tin oxide describe the use of dc and rf sputtering, spray pyrolysis, chemical vapor deposition, and vacuum evaporation, although the use of a sol–gel technique has been recently reported (Mattox, 1991; Tahar et al., 1998a,b).

Chhina et al. conducted thermal and electrochemical stability studies on an ITO oxidation-resistant support as a replacement for carbon supports for fuel cell applications (Chhina et al., 2006). Chinna et al. compared their Pt/ITO catalyst with HiSpec4000, a commercially available Johnson Matthey 40 wt% platinum catalyst deposited onto Vulcan XC-72R carbon black. As a further means of comparison, their group synthesized platinum supported onto Vulcan XC-72R catalyst in-house using the same procedure as their Pt/ITO to impregnate the platinum onto the support. This allowed for a more accurate measurement between the two supports since the same platinum deposition method was used. Results of the ITO catalyst showed improved stability both electrochemically and thermally especially at elevated voltages and temperatures. Thermogravimetric analysis revealed that at 1000°C in an air-flowing atmosphere, both carbon-supported catalysts lost their support mass significantly (~55 wt% loss in mass), whereas the Pt/ITO catalyst maintained its mass with only 0.7 wt% loss. To determine the catalysts' stability, 100 cycles of electrochemical potential step cycling were conducted with electrode coated with the three individual supported catalysts starting with a potential hold at 0.6 V for 60 s and then 1.8 V for 20 s. These electrochemical tests were conducted in 0.5 M H_2SO_4 at 30°C in a deoxygenated environment with the electrode rotating at 2000 rpm. Each catalyst was tested thrice for repeatability and the limits of error and normalized activity for the first 30 cycles for the three catalysts are plotted in Figure 3.12. It should be noted that cyclic voltammograms were obtained during the testing, before and after every 10 potential step cycles and hence, there is a slight increase in activity for cycles 11 and 21 where the activity would increase due to the CV. Based on the results, Chhina et al. concluded that the electrochemical stability of the ITO support was far greater than that of Vulcan XC-72R and led to a more durable catalyst at high-potential sweeps (Chhina et al., 2006). However, even though the stability was far greater for the Pt/ITO catalyst, the activity and ECSA were poor. XRD of the Pt/ITO showed that platinum particles deposited on the ITO support was around 13 nm, where the support particles themselves averaged 38 nm in diameter. Typically, carbon black particles range from 50 to 100 nm and platinum crystalline sizes range between 3 and 8 nm. Chhina et al. claimed that the ITO particle size was too small to optimally disperse platinum nanoparticles on the surface. Decreased platinum particle sizes tend to show greater catalytic activity in PEMFC catalysts as a result of increased ECSA. Chhina et al. proposed that with synthesis methods able to control the particle sizes of ITO, possibly a sol–gel method, it would help limit the agglomeration seen in the platinum deposits in the resulting Pt/ITO catalyst and help increase the activity to acceptable levels (Chhina et al., 2006).

Owing to the high electrical conductivity and the known stability of ITO, this material is particularly interesting for fuel cell catalyst supports. It has been proposed as an alternative material for the cathode support in PEMFCs that could avoid the corrosion problems experienced by conventional carbon supports. However, the application for PEMFC catalyst supports is novel and there are few reported works geared specifically toward fuel cell applications.

3.5.3.3 Titanium Oxide

Titanium dioxide or titania (TiO_2) has been considered a promising alternative catalyst support in low-temperature fuel cells because of its excellent corrosion resistance, good stability in acid, and tolerance to high potentials (Diebold, 2003; George et al., 2008). As an electrocatalyst support, TiO_2 deposited

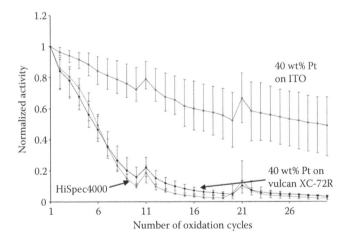

FIGURE 3.12 Normalized activity of three separate 40 wt% platinum catalysts. Average of three separate tests and limits of error are plotted. (Reprinted from *Journal of Power Sources*, 161, Chhina, H. et al. An oxidation-resistant indium tin oxide catalyst support for proton exchange membrane fuel cells, 893–900, Copyright (2006), with permission from Elsevier.)

with catalytic metals have shown good ORR performance for PEMFCs mainly attributed to good dispersion of catalytic metals on the surface of the support and to the metal–support interaction (von Kraemer et al., 2008). TiO_2 also exhibits proton conduction across the surface of the material that is greatly beneficial for a PEMFC catalyst support. This allows for greater utilization of the catalytic metal in the catalyst as more TPBs are formed between the catalyst and the support (García et al., 2007; Shao et al., 2009). Furthermore, TiO_2 is cheap, readily available, and nontoxic, making it widely accessible for use in PEMFCs.

TiO_2 is known to exist in mainly three crystalline forms, each with distinctive properties: anatase, rutile, and brookite. Anatase, the metastable state, and rutile, the stable form, are the only forms of interest for TiO_2 applications (Diebold, 2003; Antolini and Gonzalez, 2009). Pure stoichiometric TiO_2 exists in a distorted octahedral configuration and has a low electrical conductivity of only 10^{-13} S cm^{-1} at 298 K; however, substochiometric compositions or added n-type dopants such as vanadium, niobium, and tantalum can create oxygen deficiencies within the lattice structure thereby drastically increasing the conductivity of these materials.

The high electroconductivity (e.g., Ti_4O_7 having an electrical conductivity of 1000 S cm^{-1}), wide diversity of structures, and previously mentioned properties allow them to be used as PEMFC catalyst supports (Chhina et al., 2009a,b,c). Although Ti_4O_7 and Ebonex® are stable in acid, Chen et al. reported that they could still be oxidized under extensive polarization at high potentials (Chen et al., 2002). For this reason, they claimed that doping titania was the preferred method of enhancing the electrical conductivity and stability of these materials.

Chhina et al. examined the stability of such a catalyst support by comparing Pt/C and Nb-doped titania after the addition of 10 wt% Pt (Chhina et al., 2009a,b,c). They found that their Pt/Nb-doped titania was slightly outperformed by traditional Pt/C during MEA test studies; however, after a potential hold at 1.4 V for 60 h the performance of the Pt/Nb-doped titania remained nearly unaffected compared to Pt/C which showed a 50% drop in catalyst layer thickness from SEM images. Figure 3.13 shows the MEA polarization curves of commercial HiSpec4000 before and after 20 h of 1.4 V potential hold as well as the Pt/Nb–TiO_2 catalyst before and after the same durability test.

Nanostructured TiO_2 has also been utilized as PEMFC catalyst supports with dimensions well below 100 nm. Like other TiO_2 structures, these materials also inherit high stability, high surface areas, and moderate electrical conductivity. Titanium oxide nanotubes are of particular interest for fuel cell catalyst supports due to their abilities to form porous networks, disperse catalytic metals upon the surface

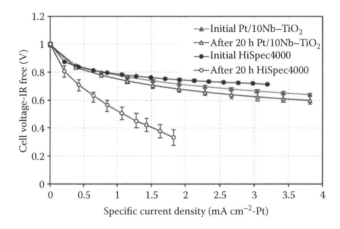

FIGURE 3.13 MEA polarization curves depicting Pt/Nb-doped titania and commercial HiSpec4000 catalyst before and after a 1.4 V potential hold for 20 h. (Reprinted from Chhina, H. et al. 2009b. *Journal of the Electrochemical Society* 156: B1232–B1237. With permission from The Electrochemical Society.)

of the material, as well as, allow for better mass transport, all of which increase the catalytic ability of the catalyst. Kang et al. evaluated these titanium oxide nanotubes as supports for catalytic metals for ORR (Kang et al., 2008a,b). They synthesized arrays of nanotube titanium oxide structures of approximately 1 μm long and 120 nm in diameter grown on titanium substrates. Deposition of bimetallic catalysts was carried out using a dual gun sputtering technique which deposited either Pt–Ni (Kang et al., 2008a) or Pt–Co (Kang et al., 2008b) onto the surface of the nanotubes. HRTEM determined that the Pt–Ni particles ranged in diameter between 5 and 10 nm (shown in Figure 3.14), while the Pt–Co particles were between 3 and 4 nm. HRTEM also confirmed the tube structure of the titanium oxide nanotubes as well as the crystalline phases of the Co and Pt. It was remarked that the TiO_2 catalyst support was of polycrystalline anatase phase. The initial ORR performance with $PtNi/TiO_2NT$ was poor; however, an annealing procedure at 400°C in hydrogen reducing environment allowed for the ORR activity to be significantly improved (Kang et al., 2008a). No comparisons to similar carbon structures were made. $Pt_{70}Co_{30}/TiO_2NT$ was also assessed for its ability to catalyze the ORR (see Figure 3.15). When compared with PtCo deposited on a TiO_2 film, the ORR activity of the $Pt_{70}Co_{30}/TiO_2NT$ displayed a significant activity increase (Kang et al., 2008b). The nanotube structure of the titanium oxide was thought to provide better mass transport of oxygen to the catalysts, as well as, a higher surface area for better dispersion of the catalytic metal on the support.

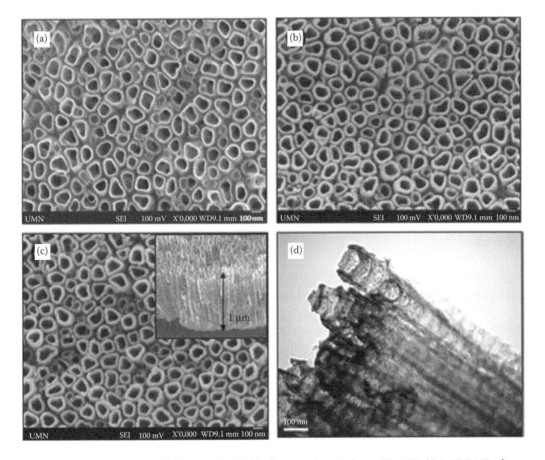

FIGURE 3.14 SEM images of (a) bare TiO_2NT, (b) platinum deposited onto TiO_2NT, (c) Pt/TiO_2NT after an annealing treatment at high temperature and, (d) a HRTEM of annealed $PtNi/TiO_2NT$. (Reprinted from Kang, S. et al. 2008a. *Journal of the Electrochemical Society* 155: B1058–B1065. With permission from The Electrochemical Society.)

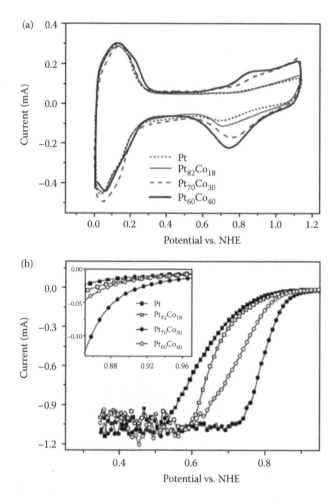

FIGURE 3.15 (a) Cyclic voltammograms and (b) ORR performance of Pt/TiO$_2$NTs, and Pt–Co/TiO$_2$NT catalysts. (Reprinted from Kang, S. et al. 2008b. *Journal of the Electrochemical Society* 155: B1128–B1135. With permission from The Electrochemical Society.)

Many structures of TiO$_2$ and substoichiometric TiO$_2$ exist which seem suitable for PEMFC catalyst supports. The added benefit of proton conductivity and improved durability in the catalyst support is advantageous over traditional carbon black supports; however, the performance of Pt/TiO$_2$ catalysts are lacking according to published research. More directed research with these types of catalyst supports might prove effective for high durability, low Nafion-loaded MEA applications.

3.5.3.4 Sulfur ZrO$_2$

Many metals such as titanium, zirconium, tantalum, and niobium have been known to provide high corrosion resistance in acid media. This is mainly due to their ability to passivate on the surface of the bulk material by forming metallic oxides (Antolini and Gonzalez, 2009). Thus, these oxide materials have potential as corrosion-resistant materials in PEMFCs. Zirconium oxide (ZrO$_2$) can be easily modified with sulfonic acid groups (SO$_x$) to increase the acidity and proton conductivity (5×10^{-2} S cm^{-1} at 60–150°C) of the material. In fact, sulfated-ZrO$_2$ (S-ZrO$_2$) is the strongest solid acid among well-known super acids ($H_o < -16$). For these reasons, much research has been conducted with S-ZrO$_2$ in fuel cells as a catalyst for conversion of organic compounds (Hino et al., 1979; Song and Sayari, 1996; Yadav and Nair, 1999), additive for Nafion composite membranes and liquid electrolytes (Zhai et al., 2006; Navarra et al., 2008), and catalyst supports for electrocatalysts (Zhang et al., 2007).

There are several advantages for the use of S-ZrO_2 as a catalyst support in PEMFC applications. Because of its hydrophilicity, it has been suggested that this type of fuel cell catalyst would be well suited for low-relative humidity conditions and possibly simplify fuel cell components to operate without the use of a humidifier. Due to the proton conductivity across the surface of the material, less Nafion ionomer needs to be cast to form the TPBs. Platinum utilization increases as the S-ZrO_2 support acts as both the platinum and proton conductor and better gas diffusion to the catalyst site results from the decreased blockage of Nafion ionomer (Liu et al., 2006a,b). It is believed that within porous carbon catalyst supports, platinum deposited within the pores may not have proton conductivity due to the perfluorosulfonated ionomer unable to penetrate into the pores. Thus, a TPB which is necessary for a catalyst active site will not be formed. Therefore, the S-ZrO_2 support has an additional benefit over porous carbon material supports in that by using the S-ZrO_2 as a support for platinum catalysts, the surface of the support can act as a proton conductor and platinum deposited anywhere on the surface of the support will provide immediate access to the electron and proton pathways thereby requiring less Nafion. Thus the use of S-ZrO_2 in fuel cell MEA components may potentially lower the cost of materials substantially, as the catalytic metals and membrane materials are among the most costly in a PEMFC. However, like most metallic oxides, the downside of their use stems from their relatively low electron conductivity and low surface areas that results in poor platinum dispersion.

To overcome the poor dispersion of platinum on S-ZrO_2, Suzuki et al. proposed an ultrasonic spray pyrolysis method to disperse platinum onto commercial S-ZrO_2 (Suzuki et al., 2007). The platinum precursor ($H_2PtCl_6 \cdot 6H_2O$) and S-ZrO_2 supports were suspended in solution and mists of the solution were generated using an ultrasonic atomizing unit and allowed to pyrolyze at 650°C with air flow. With this synthesis technique, platinum particle sizes of ~8 nm were deposited. Single cell electrochemical tests showed that, compared to Pt/C, the Pt/S-ZrO_2 catalyst exhibited a slightly poorer ORR performance when Nafion was incorporated as a binding agent into the catalyst layer for both catalysts. However, when Nafion was not incorporated into the catalyst layer, but only in the electrolyte membrane, the cell performance of Pt/S-ZrO_2 was much greater. This substantiated the claim by Suzuki et al. that Pt/S-ZrO_2 could be used as a catalyst to reduce Nafion ionomer in the MEA (Suzuki et al., 2007).

The improved corrosion resistance demonstrated by S-ZrO_2 over carbon supports shows much promise for it to be a replacement for carbon as a catalyst support, and utilizing the ultrasonic spray technique, platinum particle sizes were adequate for ORR catalysis. However, like many of the other novel noncarbon supports, much more study is required to optimize this support and to determine the real-world performance of the catalyst on these types of supports.

3.5.3.5 Carbides

Although carbide materials contain carbon, their structure and properties differ drastically from traditional carbon blacks, and thus for the purpose of this text, they will be considered here as noncarbon materials. Boron carbide has been used considerably since the 1960s for catalyst supports in phosphoric acid and alkaline fuel cells (Shao et al., 2009). The platinum deposited onto these supports was shown to be more resistant to sintering than platinum on graphite or platinum black. There was also an enhanced catalytic activity effect by the platinum that was deposited onto the carbide support. By the 1980s, silicon carbide and titanium carbide were also tested as phosphoric acid supports in fuel cell catalysts. Silicon carbide showed poor electroconductivity and titanium carbide had low surface area (<1 m^2 g^{-1}) making both of them disadvantageous for electrocatalyst supports (Antolini and Gonzalez, 2009). Of all the transition metal carbides, tungsten carbides shows the biggest promise in PEMFC applications as it has an electronic density near the Fermi level most resembling platinum. This has attracted the use of tungsten carbides for either a catalyst or catalyst support for many researchers.

In addition to having inherent catalytic properties, tungsten carbides have shown to be resistant in both acid and alkaline media at anodic potentials. The electrical conductivity of WC is about 10^5 S cm^{-1} at 20°C, and the active sites are resistant to poisoning species such as carbon monoxide, numerous hydrocarbons, and hydrogen sulfide (McIntyre et al., 2002). Different phases of tungsten carbide allow for the material to be highly flexible with facile methods to change the surface properties and chemical composition. Tungsten carbide has little catalytic activity for the MOR on its own; however, in combination with a platinum catalyst, the support–catalyst system has a synergistic effect for the MOR and improves performance greatly (Zellner and Chen, 2005). Thus there are several reports that describe the use of Pt/WC to enhance MOR activity (Hwu et al., 2001; Ganesan et al., 2005, 2007; Zellner and Chen, 2005).

As for the stability of these types of support materials, according to Zellner and Chen W_2C and WC were the most important forms of tungsten carbides. They tested the stability of both materials in 0.5 M H_2SO_4 (Zellner and Chen, 2005) and confirmed that W_2C was thermodynamically unstable at low temperatures and that WC was stable at potentials lower than 0.6 V. This was in agreement with studies conducted by Chhina et al., who performed thermal and electrochemical stability of Pt/WC catalysts (Chhina et al., 2007a,b). Their group found that WC was a more stable catalyst support than Vulcan XC-72R after potential cycling between 0.6 and 1.8 V. The activity of the platinum deposited onto WC support remained after 100 cycles while the Pt/C catalyst was completely degraded after 20 cycles. Still the initial activity of the Pt/C catalyst was higher than that of the Pt/WC catalyst. This was attributed to the extremely low surface area of the WC support and poor particle size distribution of platinum. XRD analysis showed that the particle sizes of WC were ~36 nm, while the platinum deposited onto its surface was ~30 nm. Therefore, they concluded that better methods to synthesize WC with higher surface areas could drastically improve the catalytic activity. The same group conducted a follow-up study utilizing three different methods for WC synthesis to improve the platinum dispersion. Commercial WC was synthesized through a direct carburization of the tungsten metal with carbon or graphite at high temperatures (1400–2000°C) in hydrogen or a vacuum (Antolini and Gonzalez, 2009). With this method, however, additional phases of amorphous tungsten carbides may be generated. Chhina et al. presented three alternative synthesis routes of a WC catalyst support to increase the support surface area by aqueous tungstate dispersion on carbon, an incipient wetness technique to disperse tungstate on carbon and a DC magnetron sputtering of tungsten on carbon (Chhina et al., 2008). Of the three methods, DC magnetron sputtering showed to be the optimal method as the resulting WC support contained an outer shell of WC and a porous inner support of carbon. This outer layer was uniformly distributed with tungsten and had a relatively high surface area. Oxidation cycles of 0.6 and 1.8 V showed that the durability of the Pt/WC catalyst remained stable after 100 cycles, whereas commercial HiSpec4000 catalyst lost its activity in several cycles. The plot of the normalized activity of the Pt/WC over 30 cycles is shown in Figure 3.16. It is also important to note that tungsten carbide can be oxidized to form substoichiometric tungsten carbide which remains slightly conductive (Chhina et al., 2007a,b). Thus, these tungsten carbide supports have attracted much attention for PEMFC applications; however, better methods of platinum dispersion are required to improve the electrochemically active surface area and performance.

3.5.3.6 Conductive Diamond

Pure diamonds are nonconductive and generally have low specific surface area. However, by doping diamonds with a dopant such as nitrogen or boron, doped diamonds exhibit the electrical conductivity and chemical stability necessary for PEMFC applications. Several experiments have shown that these conductive diamonds exhibit a phenomenal stability being able to endure an extremely wide potential window. For instance, platinum and platinum alloy metals have been experimentally deposited onto doped diamond supports and were shown to have excellent activity for the ORR, MOR, EOR, and oxidation of ethylene glycol (Sp taru et al., 2008). The doped diamonds were found to act similarly to the commonly used sp^2 carbon, which illustrates their potential as fuel cell catalyst supports (Hupert et al., 2003). Moreover, no microstructural changes were found in nanocrystalline boron-doped diamonds when the platinum dispersed catalysts were stressed to 1.6 V.

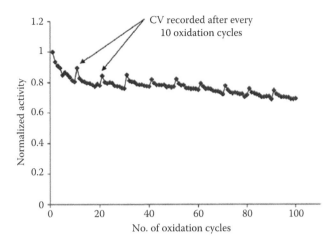

FIGURE 3.16 Normalized catalytic activity of 40 wt% platinum deposited onto WC support at 80°C in N_2. (Reprinted from *Journal of the Electrochemical Society*, 179, Chhina, H. et al. High surface area synthesis, electrochemical activity, and stability of tungsten carbide supported Pt during oxygen reduction in proton exchange membrane fuel cells. 50–59. Copyright (2008), with permission from Elsevier.)

The use of doped conductive diamonds as PEMFC catalyst supports is still unperfected as there are still several disadvantages when compared with typical sp^2 carbons. Diamonds typically exhibit low conductivity and low surface areas which is unfavorable for dispersing the catalytic metal uniformly onto the support. In several experiences the catalytic metal nanoparticles deposited onto the diamond surface were too large for PEMFC applications. Some researchers reported that using techniques such as magnetron sputtering and electrochemical deposition, platinum particle sizes between 10 and 300 nm were obtained (Montilla et al., 2002; Salazar-Banda et al., 2006). Finally, doping nanocrystalline diamonds with boron is not easily controllable to realize a homogenous dispersion.

To increase the surface area of conductive diamond supports, a technique called vacuum annealing is utilized in place of doping that anneals un-doped nanocrystalline diamonds to make a conductive diamond. These diamonds, also termed nanodiamonds, are advantageous as catalyst supports because they have high surface areas created by the crevices and surface boundaries between the nanocrystallites. These surface defects acting in favor of platinum deposition however cripple the stability of the material compared to pure diamond.

Much of the current research with conductive diamonds in PEMFC research is to develop platinum deposition techniques that can result in more uniform and smaller particle sizes on the conductive surfaces of diamond. However, conductive diamonds for catalyst support applications have often been used to examine the intrinsic properties of catalytic metals because of their inertness to electrochemical processes, lack of surface corrosion, and oxide formation. They remain strong candidates for fuel cell applications where catalyst integrity and durability are high priorities, and where typically carbon supports may fail due to the harsh operating conditions and high operating voltages.

3.6 Summary

Catalyst support degradation has been identified as a crucial PEMFC degradation mechanism, especially in automotive applications. Catalyst support degradation during long-term fuel cell operation under universal conditions is a complex process that occurs through many parallel mechanisms. Thermodynamically, carbon corrosion can occur at potentials greater than 0.207 V. However, studies have shown that increased voltages, humidity, load cycling, and impurities can cause structural and chemical changes to the carbon support or cause operational changes such as local/gross fuel starvation which accelerates carbon corrosion even further.

Extensive effort has been put into the study on methods to assess the structural damage brought upon by support degradation. The most popular methods include electrochemical tests such as CV and potential holding. Physical characterization can include TEM, and online mass spectrometry; however, SEM, AFM, XPS, EDX, and X-ray CT have been used as well to assess catalyst layer damage and structural changes. Typically, electrocatalysts will be operated in real time or operated through some AST to simulate long-time operation and a final or online assessment can be carried out. Several goals as well as protocols have been outlined by U.S. DOE, HFP, and NEDO to standardize support degradation testing and prioritize the needs of the fuel cell research industry.

The preferred strategy to reduce the negative influence of carbon corrosion on fuel cell performance is to use alternative, more stable materials as catalyst supports. Several strategies are reported to reduce the degradation that has been experienced in the previous literature. Further treatments steps to the carbon black such as heat treatments or oxidation treatments are commonly used to increase the graphitization or add additional functionalities to the surface of the support. This has resulted in slower kinetics of the corrosion, better dispersion of the catalyst metal and stronger Pt-support bonds to increase the performance and/or stability of catalysts in PEMFC environments. It has been proposed that new types of carbon with higher graphite components, such as CNTs, CNFs, and graphene, show greater stability than normal carbon black as support materials for PEMFCs; however, these are still prone to degradation. Other species of materials, such as tungsten carbide, tin-oxide, zirconia oxide and tungsten oxide, and integrating these materials with a carbon support to achieve added benefits of the noncarbon materials and maintains the ease and conductivity of a traditional carbon support have also demonstrated potential as novel catalyst supports in PEMFCs. However, little work has been published on the long-term stability of carbon substitutes used in PEMFCs. In order to meet the targets set out by DOE and increase the operating lifetimes for automotive and stationary PEMFCs, more research must be conducted to understand and mitigate fuel cell degradation.

References

Acharya, C. K. et al. 2009. Effect of boron doping in the carbon support on platinum nanoparticles and carbon corrosion. *Journal of Power Sources* 192: 324–329.

Angelo, M. et al. 2008. The impacts of repetitive carbon monoxide poisoning on performance and durability of a proton exchange membrane fuel cell. *Proton Exchange Membrane Fuel Cells 8, Pts 1 and 2* 16: 669–676.

Antolini, E. 2009. Carbon supports for low-temperature fuel cell catalysts. *Applied Catalysis B: Environmental* 88: 1–24.

Antolini, E. et al. 2002. Study on the formation of Pt/C catalysts by non-oxidized active carbon support and a sulfur-based reducing agent. *Journal of Materials Science* 37: 133–139.

Antolini, E. and Gonzalez, E. 2009. Ceramic materials as supports for low-temperature fuel cell catalysts. *Solid State Ionics* 180: 746–763.

Antolini, E. et al. 2006. The stability of Pt-M (M = first row transition metal) alloy catalysts and its effect on the activity in low temperature fuel cells. *Journal of Power Sources* 160: 957–968.

Antonucci, P. et al. 1994. On the role of surface functional groups in Pt carbon interaction. *Journal of Applied Electrochemistry* 24: 58–65.

Asano, K. et al. 2008. Degradation of Pt catalyst layer in PEFCs during load cycling under pressurized conditions. *Proton Exchange Membrane Fuel Cells 8, Pts 1 and 2* 16: 779–785.

Baughman, R. et al. 2002. Carbon nanotubes—The route toward applications. *Science* 297: 787.

Bezerra, C. W. B. et al. 2007. A review of heat-treatment effects on activity and stability of PEM fuel cell catalysts for oxygen reduction reaction. *Journal of Power Sources* 173: 891–908.

Bi, W. and Fuller, T. F. 2008. Modeling of PEM fuel cell Pt/C catalyst degradation. *Journal of Power Sources* 178: 188–196.

Bi, W. et al. 2007. PEM Fuel Cell Pt/C Dissolution and deposition in Nafion electrolyte. *Electrochemical and Solid-State Letters* 10: B101.

Bi, W. et al. 2009. The effect of humidity and oxygen partial pressure on degradation of Pt/C catalyst in PEM fuel cell. *Electrochimica Acta* 54: 1826–1833.

Bleda-Martinez, M. et al. 2006. Chemical and electrochemical characterization of porous carbon materials. *Carbon* 44: 2642–2651.

Bleda-Martinez, M. et al. 2007. Polyaniline/porous carbon electrodes by chemical polymerisation: Effect of carbon surface chemistry. *Electrochimica Acta* 52: 4962–4968.

Borup, R. et al. 2006. PEM fuel cell electrocatalyst durability measurements. *Journal of Power Sources* 163: 76–81.

Borup, R. et al. 2007. Scientific aspects of polymer electrolyte fuel cell durability and degradation. *Chemical Reviews* 107: 3904–3951.

Cai, M. et al. 2006. Investigation of thermal and electrochemical degradation of fuel cell catalysts. *Journal of Power Sources* 160: 977–986.

Cameron, D. et al. 1990. Carbons as supports for precious metal catalysts. *Catalysis today* 7: 113–137.

Chai, G. et al. 2004. Ordered porous carbons with tunable pore sizes as catalyst supports in direct methanol fuel cell. *Journal of Physical Chemistry B* 108: 7074–7079.

Chen, G. et al. 2002. Development of supported bifunctional electrocatalysts for unitized regenerative fuel cells. *Journal of the Electrochemical Society* 149: A1092.

Chen, Z. et al. 2007. Durability and activity study of single-walled, double-walled and multi-walled carbon nanotubes supported Pt Catalyst for PEMFCs. *ECS Transactions* 11: 1289–1299.

Chen, S. et al. 2010. Platinum-alloy cathode catalyst degradation in proton exchange membrane fuel cells: Nanometer-scale compositional and morphological changes. *Journal of the Electrochemical Society* 157: A82–A97.

Chhina, H. et al. 2006. An oxidation-resistant indium tin oxide catalyst support for proton exchange membrane fuel cells. *Journal of Power Sources* 161: 893–900.

Chhina, H. et al. 2007a. *Ex situ* evaluation of tungsten oxide as a catalyst support for PEMFCs. *Journal of the Electrochemical Society* 154: B533–B539.

Chhina, H. et al. 2007b. Thermal and electrochemical stability of tungsten carbide catalyst supports. *Journal of Power Sources* 164: 431–440.

Chhina, H. et al. 2008. High surface area synthesis, electrochemical activity, and stability of tungsten carbide supported Pt during oxygen reduction in proton exchange membrane fuel cells. *Journal of Power Sources* 179: 50–59.

Chhina, H. et al. 2009a. Characterization of Nb and W doped titania as catalyst supports for proton exchange membrane fuel cells. *Journal of New Materials for Electrochemical Systems* 12: 177–185.

Chhina, H. et al. 2009b. *Ex situ* and *in situ* stability of platinum supported on niobium-doped titania for PEMFCs. *Journal of the Electrochemical Society* 156: B1232–B1237.

Chhina, H. et al. 2009c. Transmission electron microscope observation of Pt deposited on Nb-doped titania. *Electrochemical and Solid-State Letters* 12: B97.

Cho, E. et al. 2003. Characteristics of the PEMFC repetitively brought to temperatures below 0°C. *Journal of the Electrochemical Society* 150: A1667.

Cho, E. et al. 2004. Effects of water removal on the performance degradation of PEMFCs repetitively brought to <0°C. *Journal of the Electrochemical Society* 151: A661.

Colmenares, L. C. et al. 2009. Model study on the stability of carbon support materials under polymer electrolyte fuel cell cathode operation conditions. *Journal of Power Sources* 190: 14–24.

Coloma, F. et al. 1994. Preparation of platinum supported on pregraphitized carbon blacks. *Langmuir* 10: 750–755.

de Bruijn, F. A. et al. 2008. Review: Durability and degradation issues of PEM fuel cell components. *Fuel Cells (Weinheim, Germany)* 8: 3–22.

Debe, M. et al. 2006. High voltage stability of nanostructured thin film catalysts for PEM fuel cells. *Journal of Power Sources* 161: 1002–1011.

Diebold, U. 2003. The surface science of titanium dioxide. *Surface Science Reports* 48: 53–229.

DOE cell component accelerated stress test protocols for PEM fuel cells. 2010.

Dowlapalli, M. et al. 2006. Electrochemical oxidation resistance of carbonaceous materials. *ECS Transactions* 1: 41–50.

Dross, R. and Maynard, B. 2007. *In-situ* reference electrode testing for cathode carbon corrosion. *ECS Transactions* 11: 1059–1068.

Fowler, M. et al. 2002. Incorporation of voltage degradation into a generalised steady state electrochemical model for a PEM fuel cell. *Journal of Power Sources* 106: 274–283.

Franco, A. A. and Gerard, M. 2008. Multiscale model of carbon corrosion in a PEFC: Coupling with electrocatalysis and impact on performance degradation. *Journal of the Electrochemical Society* 155: B367–B384.

Franco, A. A. et al. 2008. Carbon catalyst-support corrosion in polymer electrolyte fuel cells: mechanistic insights. *ECS Transactions* 13: 35–55.

Franco, A. A. et al. 2009. Impact of carbon monoxide on PEFC catalyst carbon support degradation under current-cycled operating conditions. *Electrochimica Acta* 54: 5267–5279.

Fujii, Y. et al. 2006. Degradation investigation of PEMFC by scanning electron microscopy and direct gas mass spectroscopy. *ECS Transactions* 3: 735–741.

Gallagher, K. G. et al. 2007. The effect of transient potential exposure on the electrochemical oxidation of carbon black in low-temperature fuel cells. *ECS Transactions* 11: 993–1002.

Ganesan, R. et al. 2007. Platinized mesoporous tungsten carbide for electrochemical methanol oxidation. *Electrochemistry Communications* 9: 2576–2579.

Ganesan, R. and Lee, J. 2005. Tungsten carbide microspheres as a noble-metal-economic electrocatalyst for methanol oxidation. *Angewandte Chemie* 117: 6715–6718.

García, B. et al. 2007. Low-temperature synthesis of a PtRu/Nb~ 0~.~ 1Ti~ 0~.~ 9O~ 2 electrocatalyst for methanol oxidation. *Electrochemical and Solid State Letters* 10: 108.

Garzon F. H. et al. 2007. Micro and nano x-ray tomography of PEM fuel cell membranes after transient operation, *ECS Transactions* 11: 1139–1149.

Garzon, F. H. et al. 2009. The impact of impurities on long term PEMFC performance. *ECS Transactions* 25: 1575–1583.

George, P. P. et al. 2008. Selective coating of anatase and rutile TiO_2 on carbon via ultrasound irradiation: Mitigating fuel cell catalyst degradation. *Journal of Fuel Cell Science and Technology* 5: 041012(1–9).

Gidwani, A. et al. 2008. CFD study of carbon corrosion in PEM fuel cells. *ECS Transactions* 16: 1323–1333.

Guerrero-Ruiz, A. et al. 1998. Study of some factors affecting the Ru and Pt dispersions over high surface area graphite-supported catalysts. *Applied Catalysis A: General* 173: 313–321.

Guilminot, E. et al. 2007. Membrane and active layer degradation upon PEMFC steady-state operation. *Journal of the Electrochemical Society* 154: B1106.

Han, K. et al. 2007. Heat treatment and potential cycling effects on surface morphology, particle size, and catalytic activity of Pt/C catalysts studied by 13C NMR, TEM, XRD and CV. *Electrochemistry Communications* 9: 317–324.

Heeger, A. 2001. Semiconducting and Metallic Polymers: The Fourth Generation of Polymeric Materials (Nobel Lecture) Copyright (c) The Nobel Foundation 2001. We thank the Nobel Foundation, Stockholm, for permission to print this lecture. *Angewandte Chemie (International ed. in English)* 40: 2591.

Hino, M. et al. 1979. Solid catalyst treated with anion. 2. Reactions of butane and isobutane catalyzed by zirconium oxide treated with sulfate ion. Solid superacid catalyst. *Journal of the American Chemical Society* 101: 6439–6441.

Hosono, H. 2007. Recent progress in transparent oxide semiconductors: Materials and device application. *Thin Solid Films* 515: 6000–6014.

Hottinen, T. et al. 2003. Effect of ambient conditions on performance and current distribution of a polymer electrolyte membrane fuel cell. *Journal of Applied Electrochemistry* 33: 265–271.

Hu, J. et al. 2007. Modelling of carbon corrosion in a PEMFC caused by local fuel starvation. *ECS Transactions* 11: 1031–1039.

Hu, J. et al. 2008. Modeling and simulations on mitigation techniques for carbon oxidation reaction caused by local fuel starvation in a PEMFC. *ECS Transactions* 16: 1313–1322.

Hu, J. et al. 2009. Modelling and simulations of carbon corrosion during operation of a Polymer Electrolyte Membrane fuel cell. *Electrochimica Acta* 54: 5583–5592.

Huang, S. et al. 2009. Development of conducting polypyrrole as corrosion-resistant catalyst support for polymer electrolyte membrane fuel cell (PEMFC) application. *Applied Catalysis B: Environmental* 93: 75–81.

Hupert, M. et al. 2003. Conductive diamond thin-films in electrochemistry. *Diamond and Related Materials* 12: 1940–1949.

Hwu, H. et al. 2001. Potential application of tungsten carbides as electrocatalysts. 1. Decomposition of methanol over carbide-modified W (111). *Journal of Physical Chemistry B* 105: 10037–10044.

Ioroi, T. et al. 2005. Sub-stoichiometric titanium oxide-supported platinum electrocatalyst for polymer electrolyte fuel cells. *Electrochemistry Communications* 7: 183–188.

Kang, S. et al. 2008a. Effect of annealing PtNi nanophases on extended TiO nanotubes for the electrochemical oxygen reduction reaction. *Journal of the Electrochemical Society* 155: B1058–B1065.

Kang, S. et al. 2008b. The effectiveness of sputtered PtCO catalysts on TiO nanotube arrays for the oxygen reduction reaction. *Journal of the Electrochemical Society* 155: B1128–B1135.

Kim, J. et al. 2009. Relationship between carbon corrosion and positive electrode potential in a proton-exchange membrane fuel cell during start/stop operation. *Journal of Power Sources* 192: 674–678.

Knights, S. et al. 2004. Aging mechanisms and lifetime of PEFC and DMFC. *Journal of Power Sources* 127: 127–134.

Ko, Y.-J. et al. 2010. Effect of heat-treatment temperature on carbon corrosion in polymer electrolyte membrane fuel cells. *Journal of Power Sources* 195: 2623–2627.

Lau, S. et al. 2009. Noninvasive, multiscale 3D x-ray characterization of porous functional composites and membranes, with resolution from MM to sub 50 NM. *Journal of Physics: Conference Series* 152: 012059 (1–9).

Lee, H. et al. 2008. Improvements of electrical properties containing carbon nanotube in epoxy/graphite bipolar plate for polymer electrolyte membrane fuel cells. *Journal of Nanoscience and Nanotechnology* 8: 5464–5466.

Li, Y. et al. 2009. A first-principles study of nitrogen-and boron-assisted platinum adsorption on carbon nanotubes. *Carbon* 47: 850–855.

Li, L. and Xing, Y. C. 2006. Electrochemical durability of carbon nanotubes in noncatalyzed and catalyzed oxidations. *Journal of the Electrochemical Society* 153: A1823–A1828.

Li, L. and Xing, Y. C. 2008. Electrochemical durability of carbon nanotubes at 80°C. *Journal of Power Sources* 178: 75–79.

Lim, K. H. et al. 2009. Effect of operating conditions on carbon corrosion in polymer electrolyte membrane fuel cells. *Journal of Power Sources* 193: 575–579.

Liu, Z. et al. 2006a. Behavior of PEMFC in starvation. *Journal of Power Sources* 157: 166–176.

Liu, G. et al. 2006b. A novel sintering resistant and corrosion resistant Pt_4ZrO_2/C catalyst for high temperature PEMFCs. *Electrochimica Acta* 51: 5710–5714.

Maass, S. et al. 2008. Carbon support oxidation in PEM fuel cell cathodes. *Journal of Power Sources* 176: 444–451.

Maillard, F. et al. 2009. Carbon materials as supports for fuel cell electrocatalysts. *Carbon Materials for Catalysis*: 429–480.

Maiyalagan, T. and Viswanathan, B. 2008. Catalytic activity of platinum/tungsten oxide nanorod electrodes towards electro-oxidation of methanol. *Journal of Power Sources* 175: 789–793.

Makharia, R. et al. 2006. Durable PEM fuel cell electrode materials: Requirements and benchmarking methodologies. *ECS Transactions* 1: 3–18.

Mao, L. and Wang, C. 2007. Analysis of cold start in polymer electrolyte fuel cells. *Journal of the Electrochemical Society* 154: B139.

Mattox, D. 1991. Sol-gel derived, air-baked indium and tin oxide films. *Thin Solid Films* 204: 25–32.

McIntyre, D. et al. 2002. Effect of carbon monoxide on the electrooxidation of hydrogen by tungsten carbide. *Journal of Power Sources* 107: 67–73.

Meng, H. 2008. Numerical analyses of non-isothermal self-start behaviors of PEM fuel cells from subfreezing startup temperatures. *International Journal of Hydrogen Energy* 33: 5738–5747.

Meyers, J. P. and Darling, R. M. 2006. Model of Carbon corrosion in PEM fuel cells. *Journal of the Electrochemical Society* 153: A1432–A1442.

Montilla, F. et al. 2002. Electrochemical oxidation of benzoic acid at boron-doped diamond electrodes. *Electrochimica Acta* 47: 3509–3513.

Natarajan, D. and Van Nguyen, T. 2005a. Current distribution in PEM fuel cells. Part 1: Oxygen and fuel flow rate effects. *AIChE Journal* 51: 2587–2598.

Natarajan, D. and Van Nguyen, T. 2005b. Current distribution in PEM fuel cells. Part 2: Air operation and temperature effect. *AIChE Journal* 51: 2599–2608.

Navarra, M. et al. 2008. Properties and fuel cell performance of a Nafion-based, sulfated zirconia-added, composite membrane. *Journal of Power Sources* 183: 109–113.

Ofstad, A. B. et al. 2008. Carbon corrosion of a PEMFC during shut-down/start-up when using an air purge procedure. *ECS Transactions* 16: 1301–1311.

Oszcipok, M. et al. 2005. Statistic analysis of operational influences on the cold start behaviour of PEM fuel cells. *Journal of Power Sources* 145: 407–415.

Paik, C. et al. 2007. Influence of cyclic operation on PEM fuel cell catalyst stability. *Electrochemical and Solid-State Letters* 10: B39.

Prado-Burguete, C. et al. 1991. Effect of carbon support and mean Pt particle size on hydrogen chemisorption by carbon-supported Pt ctalysts. *Journal of Catalysis* 128: 397–404.

Reiser, C. et al. 2005. A reverse-current decay mechanism for fuel cells. *Electrochemical and Solid-State Letters* 8: A273.

Roen, L. M. et al. 2004. Electrocatalytic corrosion of carbon support in PEMFC cathodes. *Electrochemical and Solid-State Letters* 7: A19–A22.

Roman-Martinez, M. et al. 1995. Metal-support interaction in Pt/C catalysts. Influence of the support surface chemistry and the metal precursor. *Carbon* 33: 3–13.

Salazar-Banda, G. et al. 2006. On the changing electrochemical behaviour of boron-doped diamond surfaces with time after cathodic pre-treatments. *Electrochimica Acta* 51: 4612–4619.

Schmittinger, W. and Vahidi, A. 2008. A review of the main parameters influencing long-term performance and durability of PEM fuel cells. *Journal of Power Sources* 180: 1–14.

Shao, Y. et al. 2009. Novel catalyst support materials for PEM fuel cells: Current status and future prospects. *Journal of Materials Chemistry* 19: 46–59.

Shao, Y. et al. 2006. Durability study of Pt/ C and Pt/ CNTs catalysts under simulated PEM fuel cell conditions. *Journal of the Electrochemical Society* 153: A1093.

Shao, Y. et al. 2007. Understanding and approaches for the durability issues of Pt-based catalysts for PEM fuel cell. *Journal of Power Sources* 171: 558–566.

Shirakawa, H. 2001. The discovery of polyacetylene film: The dawning of an era of conducting olymers (Nobel Lecture) Copyright (c) The Nobel Foundation 2001. We thank the Nobel Foundation, Stockholm, for permission to print this lecture. *Angewandte Chemie (International ed. in English)* 40: 2574.

Song, X. and Sayari, A. 1996. Sulfated zirconia-based strong solid-acid catalysts: Recent progress. *Catalysis Reviews* 38: 329–412.

Sp taru, N. et al. 2008. Platinum electrodeposition on conductive diamond powder and its application to methanol oxidation in acidic media. *Journal of the Electrochemical Society* 155: B264.

Stevens, D. and Dahn, J. 2005. Thermal degradation of the support in carbon-supported platinum electro-catalysts for PEM fuel cells. *Carbon* 43: 179–188.

Stevens, D. et al. 2005. *Ex situ* and *in situ* stability studies of PEMFC catalysts. *Journal of the Electrochemical Society* 152: A2309.

Stonehart, P. 1984. Carbon substrates for phosphoric acid fuel cell cathodes. *Carbon* 22: 423–431.

Sun, H. et al. 2007. Effects of humidification temperatures on local current characteristics in a PEM fuel cell. *Journal of Power Sources* 168: 400–407.

Suzuki, Y. et al. 2007. Sulfated-zirconia as a support of Pt catalyst for polymer electrolyte fuel cells. *Electrochemical and Solid-State Letters* 10: B105.

Tahar, R. et al. 1998a. Electronic transport in tin-doped indium oxide thin films prepared by sol–gel technique. *Journal of Applied Physics* 83: 2139.

Tahar, R. et al. 1998b. Tin doped indium oxide thin films: Electrical properties. *Journal of Applied Physics* 83: 2631.

Takeshita, T. et al. 2008. Analysis of Pt catalyst degradation of a PEFC cathode by TEM observation and macro model simulation. *Proton Exchange Membrane Fuel Cells 8, Pts 1 and 2* 16: 367–373.

Takeuchi, N. and Fuller, T. 2007. Modeling of transient state carbon corrosion for PEMFC electrode. *ECS Transactions* 11: 1021–1029.

Takeuchi, N. and Fuller, T. 2008. Modeling and investigation of design factors and their impact on carbon corrosion of PEMFC electrodes. *Journal of the Electrochemical Society* 155: B770–B775.

Tang, J. et al. 2007. High performance hydrogen fuel cells with ultralow Pt loading carbon nanotube thin film catalyst. *Journal of Physical Chemistry C* 111: 17901–17904.

Tang, H. et al. 2006. PEM fuel cell cathode carbon corrosion due to the formation of air/fuel boundary at the anode. *Journal of Power Sources* 158: 1306–1312.

Taniguchi, A. et al. 2004. Analysis of electrocatalyst degradation in PEMFC caused by cell reversal during fuel starvation. *Journal of Power Sources* 130: 42–49.

Taniguchi, A. et al. 2008. Analysis of degradation in PEMFC caused by cell reversal during air starvation. *International Journal of Hydrogen Energy* 33: 2323–2329.

Torres, G. et al. 1997. Effect of the carbon pre-treatment on the properties and performance for nitrobenzene hydrogenation of Pt/C catalysts. *Applied Catalysis A: General* 161: 213–226.

Tsuji, M. et al. 2007. Fast preparation of PtRu catalysts supported on carbon nanofibers by the microwave-polyol method and their application to fuel cells. *Langmuir* 23: 387–390.

von Kraemer, S. et al. 2008. Evaluation of TiO_2 as catalyst support in Pt–TiO_2/C composite cathodes for the proton exchange membrane fuel cell. *Journal of Power Sources* 180: 185–190.

Wan, Q. et al. 2004. Single-crystalline tin-doped indium oxide whiskers: Synthesis and characterization. *Applied Physics Letters* 85: 4759.

Wang C. Y. ,Yang X. G. ,Tabuchi Y. , Kagami F. 2010. Cold-start durability of membrane-electrode assemblies, *Handbook of Fuel Cells: Fundamentals Technology and Applications*. Wiley Online Library. DOI: 10.1002/9780470974001.f500057a

Whitelocke, S. A. and Kalu, E. E. 2008. Catalytic activity and stability of tungsten oxide electrocatalyst for fuel cell applications. *AIChE Annual Meeting, Conference Proceedings*, Philadelphia, PA, Nov. 16–21, 2008: 117/1-/8.

Wu, G. et al. 2005. Remarkable support effect of SWNTs in Pt catalyst for methanol electrooxidation. *Electrochemistry Communications* 7: 1237–1243.

Wu, J. et al. 2008. A review of PEM fuel cell durability: Degradation mechanisms and mitigation strategies. *Journal of Power Sources* 184: 104–119.

Yadav, G. and Nair, J. 1999. Sulfated zirconia and its modified versions as promising catalysts for industrial processes. *Microporous and Mesoporous Materials* 33: 1–48.

Ye, S., Hall, M., He, P. 2008. PEM fuel cell catalysts: The importance of catalyst support. *ECS Trans.* 16 (2), 2101–2113.

Yu, J. et al. 2002. Fabrication of ordered uniform porous carbon networks and their application to a catalyst supporter. *Journal of the American Chemical Society* 124: 9382–9383.

Zellner, M. and Chen, J. 2005. Potential application of tungsten carbides as electrocatalysts: Synergistic effect by supporting Pt on C/W (110) for the reactions of methanol, water, and CO. *Journal of the Electrochemical Society* 152: A1483.

Zhai, Y. et al. 2006. Preparation and characterization of sulfated zirconia (SO42-/ZrO2)/Nafion composite membranes for PEMFC operation at high temperature/low humidity. *Journal of Membrane Science* 280: 148–155.

Zhang, F. Y. et al. 2009. Quantitative characterization of catalyst layer degradation in PEM fuel cells by X-ray photoelectron spectroscopy. *Electrochimica Acta* 54: 4025–4030.

Zhang, S. S. et al. 2010. Effects of open-circuit operation on membrane and catalyst layer degradation in proton exchange membrane fuel cells. *Journal of Power Sources* 195: 1142–1148.

Zhang, Y. et al. 2007. Investigation of self-humidifying membranes based on sulfonated poly (ether ether ketone) hybrid with sulfated zirconia supported Pt catalyst for fuel cell applications. *Journal of Power Sources* 168: 323–329.

4

Membrane Degradation

Yinghao Luan
*East China University of
Science and Technology;
National Engineering
Research Center of Ultra
Fine Powder*

Yongming Zhang
*Shanghai Jiao Tong
University*

4.1 Introduction

In a polymer electrolyte membrane fuel cell (PEMFC), the most essential component is the membrane electrode assembly (MEA), which consists of gas diffusion layers (GDLs), catalyst layers (CLs), and a polymer electrolyte membrane. The polymer electrolyte membrane, or ionomer membrane, transports protons from anode to cathode and also serves as a separator to prevent reactant gases (oxygen in the cathode side and hydrogen in the anode side) from contacting each other. In the development of PEMFCs, especially for automotive applications, insufficient durability remains one of the major factors impeding the commercialization of fuel cell technology. The performance of PEMFCs gradually degrades owing to the deterioration of cell components, particularly the MEA, and membrane deterioration is one of the most serious contributing factors to MEA degradation. Membrane deterioration includes chemical, mechanical, and thermal degradation that can lead to membrane thinning and pinhole formation. The presence of pinholes will then result in direct contact between the oxygen and hydrogen, leading to exothermic combustion, cell short circuit, and ultimately catastrophic degradation of fuel cell performance. For automobile applications, fuel cells are often required to function under extreme conditions, such as cold start-up, shutdown, and unplanned disruptions. Therefore, the ionomer membranes must be sufficiently robust to survive all possible conditions of service to which the fuel cell may be exposed. The required properties of the polymer electrolyte membrane include: (1) sufficient ionic conductivity, (2) ability to maintain flexibility in a dried state, (3) homogeneity, (4) impermeability toward gases, (5) chemical stability and resistance to redox reactions, (6) thermal and hydrolytic stability, (7) ability to form thin membranes, and (8) good mechanical stability and strength (Lee et al., 2006).

The goal of this chapter is to provide a thorough review of membrane degradation, including a fundamental understanding of the various mechanisms of membrane degradation, and the development of mitigation strategies. In addition, the most commonly used membrane degradation testing procedures (protocols) are also discussed.

4.2 Types of Membrane

4.2.1 Perfluorosulfonic Membranes

At present, perfluorosulfonic ionomer (PFSI) is the most widely used membrane material in well-developed PEMFCs because of its excellent chemical and thermal stability, high ionic conductivity, and good mechanical strength.

The first PFSI was discovered in the early 1960s by DuPont, and later became known as the now famous Nafion®. The use of Nafion as a separator membrane in a chlor-alkali cell was demonstrated by Grot in 1964 and achieved great success in the 1970s. In 1966, Grot and Selman approached General Electric (GE) regarding the use of this polymer in fuel cells (Grot, 2007).

4.2.1.1 Manufacturing of PFSIs

PFSIs are chemically converted from the copolymerization of tetrafluoroethylene (TFE) and perfluorinated vinyl ether monomers containing a sulfonyl fluoride functional group that can be converted into a sulfonated group in hot aqueous alkali.

Steps in the manufacturing of Nafion PFSI are shown in Figure 4.1 (Connolly and Gresham, 1966; Grot, 2007).

The typical acid capacity of Nafion membranes used in present-day fuel cells is 0.9–1.01 meq g^{-1}, which signifies 0.9–1.01 millimol sulfonic acid groups per gram of dry ionomer polymer, and so the

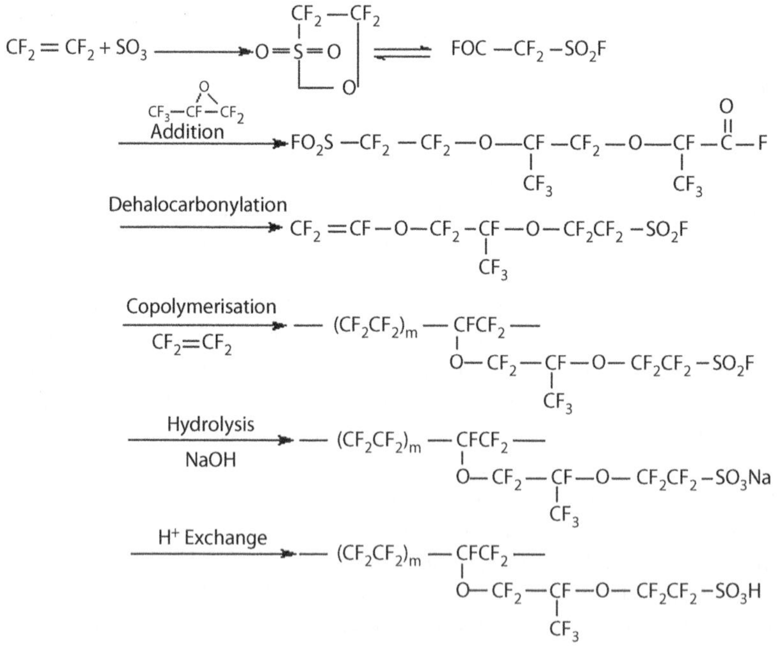

FIGURE 4.1 Steps in the manufacturing of Nafion PFSI.

average number of m (the average number of continuous repeating $-CF_2CF_2-$ units) in Figure 4.1 is 4–6. Because perfluorosulfonic polymers are phase separated at the molecular level in polar solvents at normal temperatures, resulting in colloidal dispersions or sols rather than solutions, with a fluorocarbon core and ionic groups at the polymer–liquid interface (Szajdzinska-Pietek et al., 1994; Gebel and Loppinet, 1996; Cirkel et al., 1999; Cirkel and Okada, 2000; Jiang et al., 2001; Lee et al., 2004), it is very hard to obtain true solutions to measure the actual molecular weight. Lousenberg (2005) developed a size exclusion chromatography method incorporating static light scattering detection to measure Nafion PFSI molecular weight in aqueous dispersions. The initial apparent mass distributions of all dispersions were broad and bimodal, with a high-molar-mass shoulder consistent with a postulated aggregate structure. The apparent weight-average molar masses ranged from 1.3×10^6 to 3.9×10^6 g mol^{-1}. When the dispersions were heated at or above 230°C, the aggregate structure was broken down, resulting in monomodal mass distributions. Simultaneously, the weight-average molar masses were reduced to approximately 2.5×10^5 g mol^{-1}, and the polydispersities were approximately 1.7–1.8. Assuming that the Nafion ionomer molecular weight is 2.5×10^5 g mol^{-1} with a value of $m = 4$–6 in Figure 4.1, approximately 1000–1200 $-CF_2CF_2-$ units and 210–250 sulfonic groups containing perfluorinated vinyl ether units can be estimated per average polymer chain. Such calculations are important in developing a full understanding of the chemical degradation pathways of PFSI molecules.

Nafion ionomer is not the only perfluorosulfonic membrane material to be considered for PEMFCs. There are many other commercial perfluorinated ionomers, such as Asahi Glass (Flemion®), Asahi Kasei (Aciplex®), 3M (3M polymer), and Solvay Solexis (Hyflon®), all of which share some structural similarities with the polytetrafluoroethylene-based Nafion but use different perfluorinated vinyl ethers, as shown in Table 4.1.

4.2.1.2 Fabrication of PFSI Membranes

The biggest application for fluorinated ionomers currently is as unreinforced membranes for fuel cells, or polytetrafluoroethylene (PTFE) fiber-reinforced composite membranes for electrolytic baths. The membranes can be fabricated by the extrusion method or the solution-cast method.

Extrusion of the ionomer material in the precursor form is the earliest and most commonly used method for membrane fabrication. The synthesized polymer, occurring in a fluffy form, is first extruded and chopped into pellets, and then the pellets are fed into an extruder to produce the precursor polymer membrane. Due to the high perfluorinated vinyl ether content of the precursor polymer, its melting point is fairly low, with a correspondingly low extrusion temperature, typically 280°C. After the precursor polymer has been fabricated for the membrane, it is hydrolyzed in a hot aqueous solution of sodium hydroxide or potassium hydroxide, optionally also containing some dimethyl sulfoxide (DMSO) to increase the rate of hydrolysis. Acid-type membranes are exchanged for sodium-type or potassium-type products using nitric or hydrochloric acid. After the acid exchange, the membrane is rinsed with deionized water, followed by drying of the product using warm air (Grot, 2007).

The extrusion method is usually used to produce thick membranes. Since membrane resistance is normally proportional to membrane thickness, thinner membranes are desirable for fuel cell applications. To fabricate membranes thinner than about 50 μm, solution-cast perfluorosulfonate ionomer offers some advantages compared to the extrusion method. DuPont produces some Nafion membranes, such as NRE211 and NRE212 (25 and 50 μm thickness, respectively) on a commercial scale. The process is described by Curtin et al. (2004), as shown in Figure 4.2.

In Figure 4.2, first a base backing film (1) is unwound and measured for thickness (2). Then polymer dispersion is applied (3) to the base backing film, and both materials enter a drying section (4) to form the composite membrane. The composite membrane/backing film is then measured for total thickness (5), with the membrane thickness being the difference from the initial backing film measurement. The membrane is inspected for defects (6), protected with a coversheet (7), and wound on a master roll (8). Membrane production takes place in a clean room environment (9). Master rolls are slit into product rolls, which are individually sealed and packaged for shipment. This process has several key

TABLE 4.1 Molecular Structures of Commercial PFSIs and Corresponding Perfluorinated Vinyl Ethers

Polymer Name	Molecular Structure of Ionomer	Perfluorinated Vinyl Ether	Company
Nafion Flemion Aciplex	$-(\text{CF}_2\text{CF}_2)_m-\text{CFCF}_2-$ $\text{O}-\text{CF}_2-\text{CF}-\text{O}-\text{CF}_2\text{CF}_2-\text{SO}_3\text{H}$ CF_3	$\text{CF}_2=\text{CF}-\text{O}-\text{CF}_2-\text{CF}-\text{O}-\text{CF}_2\text{CF}_2-\text{SO}_2\text{F}$ CF_3	DuPont Asahi Glass Asahi Kasei
3M polymer	$-(\text{CF}_2\text{CF}_2)_m-\text{CFCF}_2-$ $\text{O}-\text{CF}_2\text{CF}_2\text{CF}_2\text{CF}_2-\text{SO}_3\text{H}$	$\text{CF}_2=\text{CF}-\text{O}-\text{CF}_2-\text{CF}_2-\text{CF}_2-\text{CF}_2-\text{SO}_2\text{F}$	3M
Hyflon Dow polymer	$-(\text{CF}_2\text{CF}_2)_m-\text{CFCF}_2-$ $\text{O}-\text{CF}_2\text{CF}_2-\text{SO}_3\text{H}$	$\text{CF}_2=\text{CF}-\text{O}-\text{CF}_2-\text{CF}_2-\text{SO}_2\text{F}$	Solvay Solexis Dow

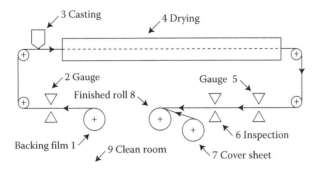

FIGURE 4.2 Solution-casting process for Nafion membranes. (Reprinted from *Journal of Power Sources*, 131, Curtin, D. E. et al. Advanced materials for improved PEMFC performance and life, 41–48, Copyright (2004), with permission from Elsevier.)

advantages: (a) pre-qualification of large dispersion batches for quality (e.g., free of contamination) and expected performance (e.g., acid capacity); (b) increased overall production rates for the H^+ membrane from solution casting as compared to from polymer extrusion followed by chemical treatment; and (c) improved thickness control and uniformity, including capability to produce very thin membranes (Curtin et al., 2004).

4.2.2 Non-PFSI Membranes

PFSI membranes have long been used in the most state-of-the-art PEMFCs owing to their excellent thermal, chemical and mechanical stability, and their high ionic conductivity. However, such membranes are too expensive to be used commercially, and they also have the disadvantages of high methanol permeability when used in direct methanol fuel cells (DMFCs) and lower conductivity at high temperatures because of reduced water content. Many researchers have tried to develop alternative ionomer membranes, such as partially fluorinated membranes and hydrocarbon-based membranes that are generally made of polyaromatic or polyheterocyclic repeat units, including polysulfones (PSU), poly(ether sulfones) (PES), poly(ether ketones) (PEK), poly(phenyl quinoxaline) (PPQ), polybenzimidazole (PBI), and so on. Such materials and membranes are summarized very well in other articles (Rozière and Jones, 2003; Collier et al., 2006), and therefore are not introduced in this chapter.

4.2.3 High-Temperature Ionomer Membranes

State-of-the-art of PEMFC technology is mainly based on PFSI membranes operating at a typical temperature of 80°C. But recent research into H_2/air PEMFCs has focused on the need to develop fuel cells that can operate at temperatures higher than 100°C. There are several compelling reasons for operating PEMFCs at higher temperatures: (1) the electrochemical kinetics for both cathode and anode reactions are enhanced; (2) water management can be simplified because only a single water phase needs to be considered; (3) the cooling system is simplified due to the increased temperature gradient between the fuel cell stack and the coolant; (4) waste heat can be recovered as a practical energy source; (5) CO tolerance is dramatically increased, thereby allowing fuel cells to use lower-quality reformed hydrogen (Li et al., 2003; Zhang et al., 2006).

At operating temperatures above 100°C, the major drawback of PFSI membranes is that the proton conductivity of the membranes drops because of water evaporation and alterations in the mechanical properties. As a result, several proton-conducting membranes have been developed for high-temperature operation. Li et al. (2003) classified the existing high-temperature polymer membranes into three groups: modified PFSI membranes, alternative sulfonated polymers and their inorganic composite membranes, and acid-based complex membranes. DMFCs and H_2/O_2(air) cells based on modified PFSI membranes have been successfully operated at temperatures up to 120°C under ambient pressure, and up to 150°C under pressures of 3–5 atm. The sulfonated hydrocarbons and their inorganic composites exhibit promising conductivity and thermal stability at temperatures above 100°C. Acid-based polymer membranes, especially H_3PO_4-doped PBI, have been demonstrated for DMFCs and H_2/O_2(air) PEMFCs at temperatures up to 200°C under ambient pressure. Such membranes and corresponding technologies are well reviewed by Li et al. (2003) and Zhang et al. (2006).

4.3 Mechanisms of Membrane Degradation

Membrane degradation can result in loss of the electrolyte, loss of separator functionality, and severe fuel cell failure. In following sections, three membrane degradation modes—chemical, mechanical, and thermal—are introduced.

4.3.1 Chemical Degradation

4.3.1.1 Chemical Degradation Mechanisms of PFSI Membranes

The bond strength for C–F is about 485 kJ mol^{-1}, higher than for C–H bonds (typically 350–435 kJ mol^{-1}) and C–C bonds (typically 350–410 kJ mol^{-1}) (Li et al., 2003). Even with its highly stable C–F bonds, the PFSI membrane still exhibits appreciable chemical degradation. HF, CO_2, SO, SO_2, SO_4^{2-}, trifluoroacetic acid, H_2SO_2, H_2SO_3, and other fluoro-containing sulfonic acids and carboxylic acids (such as HOOC–CF(CF$_3$)–O–CF$_2$CF$_2$–SO$_3$H and CF$_3$CF$_3$–O–CF$_2$CF$_2$–SO$_3$H), which are products involved in perfluorosulfonic membrane degradation in PEMFCs, have been found in the venting water of an operating fuel cell (Healy et al., 2005; Teranishi et al., 2006; Chen and Fuller, 2009a). It is widely accepted that chemical degradation of the PFSI membrane proceeds via the reaction of radical species.

Radicals (such as •OH or •OOH) generated in the MEA are generally considered to be the direct cause of membrane chemical degradation. The presence of radicals in perfluorosulfonic membranes has been detected in the degradative process by electron spin resonance (ESR) spectroscopy (Almeida and Kawano, 1998; Bosnjakovic and Schlick, 2004; Endoh et al., 2004; Kadirov et al., 2005; Endoh, 2006; Danilczuk et al., 2007; Lund et al., 2007; Danilczuk et al., 2009). Many research groups have found the formation of hydrogen peroxide (H_2O_2) in the venting water of fuel cells with or without the cells being on load (Inaba et al., 2004; Inaba et al. 2004, 2006; Liu and Zuckerbrod, 2005; Mittal et al., 2006b; Teranishi et al., 2006; Chlistunoff and Pivovar, 2007; Sethuraman et al., 2008). H_2O_2 can easily generate radical species, especially in the presence of metal ions, such as ferrous ions, so H_2O_2 is strongly believed to be generated in an operating fuel cell and has been claimed to play a role in membrane degradation processes.

Oxygen can be electrochemically reduced either directly to water without the intermediate formation of H_2O_2 (direct four-electron reduction), or to H_2O_2 (series two-electron reduction). Using the rotating ring disk electrode (RRDE) technique, numerous groups have demonstrated that H_2O_2 can be electrochemically formed through a two-electron reduction of O_2 on catalytic sites such as platinum particles, as shown in following equations, and is affected by many factors such as temperature, gas concentration, potential, catalyst properties, and other fuel cell operating parameters (Kinoshita, 1990; Guo et al., 1999; Antoine and Durand, 2000, 2001; Antoine et al., 2001; Paulus et al., 2002; Inaba, 2004; Liu and Zuckerbrod, 2005; Ramaswamy et al., 2008; Sethuraman et al., 2008).

$$O_2 + 2e + 2H^+ \rightleftharpoons H_2O_2 (E_0 = 0.695 \text{ V}) \tag{4.1}$$

This mechanism can produce H_2O_2 at the cathode side under normal fuel cell operating conditions where current is generated with an external load. Since membrane chemical degradation is more severe under open-circuit voltage (OCV) conditions than loading conditions, and the cathode potential (approximately 1 V) under OCV conditions is more positive than the potential required for the electro-chemical formation of H_2O_2, some researchers propose that the two-electron oxygen reduction mechanism can be considered an insignificant pathway in membrane degradation (Liu et al., 2009).

Laconti et al. (2003) proposed that H_2O_2 can be formed at the anode side via the reaction of crossover O_2 with H_2 on the anode platinum catalyst, through the following three steps:

$$\text{Step 1}: H_2 \rightarrow 2H \left(\text{on platinum catalyst}\right) \tag{4.2}$$

$$\text{Step 2}: H + O_2 \rightarrow HOO \tag{4.3}$$

$$\text{Step 3}: HOO + H \rightarrow H_2O_2 \tag{4.4}$$

According to their experimental results, Endoh (2006) concluded that the degradation mechanism of the MEA under OCV tests was via H_2O_2, chemically formed at the cathode and chemically or electro-chemically generated at the anode, then decomposing to produce hydroxyl radicals.

Inaba et al. (2006) proposed that gas crossover and catalytic combustion at the electrodes are key factors in membrane degradation; H_2O_2 is most probably formed by gas crossover and the resulting catalytic combustion.

Some researchers have proposed that H_2O_2 alone would account at best for a small fraction of PFSI membrane degradation in a fuel cell (Miyake et al., 2006; Mittal et al., 2006b, 2007; Vogel et al., 2008). Mittal et al. (2006b, 2007) found that the fluoride release rate (FRR) in their experiments decreased proportionally with decreasing hydrogen crossover rate, and that the FRR was much lower at OCV when N_2 and H_2O_2 were used as reactants compared to when H_2 and O_2 were used. They believed in the possibility of an alternative membrane degradation mechanism whereby harmful species are formed by a mechanism other than H_2O_2 decomposition in the presence of impurities or the catalyst, and proposed that molecular H_2 and O_2 react on the surface of the platinum catalyst to form the membrane-degrading species. This proposal is in agreement with work by Vogel et al. (2008), who found that no degradation occurred via the attack of •OH radicals on Nafion ionomer in the absence of ferric/ferrous ions in UV-illuminated H_2O_2 aqueous solution, or in *ex situ* exposure of membranes to aqueous H_2O_2 or its vapor at elevated temperatures.

But interestingly, using a gas-phase H_2O_2 exposure method, Hommura et al. (2008) found that PFSI membranes could be decomposed by H_2O_2 alone, and that the number of carboxyl groups resulting from ionomer main-chain degradation increased in proportion to the time exposed to gaseous H_2O_2; ferrous ions could work as a catalyst for the degradation reaction but they were not indispensable species for degradation. Delaney and Liu (2007) tested PFSI stability in vapor-phase H_2O_2 and aqueous Fenton's reagent. They found that ionomer degradation followed different mechanisms under each condition, and they suggested that chain scission dominated in vapor-phase H_2O_2 tests, while unzipping of end groups dominated under aqueous Fenton's reagent conditions. They also found iron contamination to be critical: even a small amount of iron was more effective in the vapor-phase tests than in the solution tests. Because H_2O_2 is found in the gas phase in operating PEMFCs, ionomer degradation in fuel cells might result from a combination of vapor- and liquid-phase degradation, especially under OCV or very-low-power conditions.

According to a detailed thermochemical and kinetic analysis, Coms (2008) from General Motors (GM) proposed that the hydroxyl radical is the only oxidant capable of abstracting a hydrogen atom from a carboxylic acid intermediate and thereby propagating the PFSI main-chain unzipping process in the highly oxidizing environment of an operating PEMFC, while both H_2O_2 and •OOH radicals most often function as reducing agents. It is a good reason to explain the experimental results from Hommura and Delaney's work mentioned above that PFSI degraded more severely in H_2O_2 vapor than in liquid phase, because fewer reactive hydroxyl radicals are scavenged in the gas phase due to lower H_2O_2 concentration.

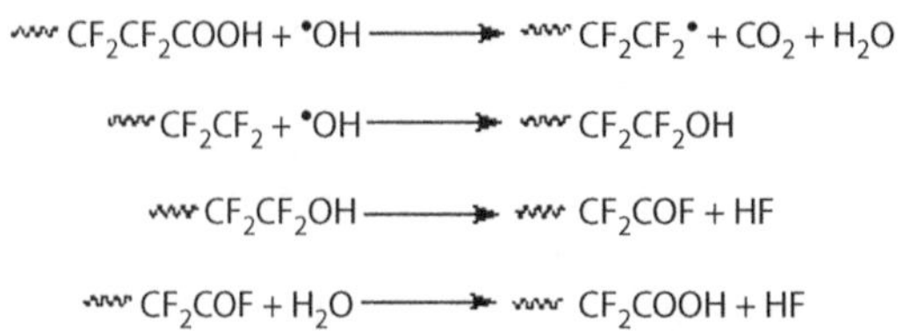

$$\text{⌁ CF}_2\text{CF}_2\text{COOH} + {}^{\bullet}\text{OH} \longrightarrow \text{⌁ CF}_2\text{CF}_2{}^{\bullet} + CO_2 + H_2O$$

$$\text{⌁ CF}_2\text{CF}_2 + {}^{\bullet}\text{OH} \longrightarrow \text{⌁ CF}_2\text{CF}_2\text{OH}$$

$$\text{⌁ CF}_2\text{CF}_2\text{OH} \longrightarrow \text{⌁ CF}_2\text{COF} + HF$$

$$\text{⌁ CF}_2\text{COF} + H_2O \longrightarrow \text{⌁ CF}_2\text{COOH} + HF$$

FIGURE 4.3 End-group unzipping degradation mechanism of main chain of PFSI, proposed by Curtin et al. (Redrawn from Curtin, D. E. et al. 2004. *Journal of Power Sources* 131: 41–48.)

From the above review it can be concluded that whether or not H_2O_2 is involved in the PFSI membrane degradation mechanism in PEMFCs, its radical species are involved in the PFSI degradation process. For PFSI membranes in fuel cells, three main possible degradation pathways exist: end-group unzipping, chain scission, and side-group attack.

4.3.1.1.1 End-Group Unzipping

Although PFSI membranes have been commercialized for more than 40 years, because of the sophisticated molecular structure and complicated electrochemical environment in fuel cells, the degradation mechanisms of these membranes are still hotly contested. Among those reported in the literature, "end-group unzipping," originated by reactive radical attack, seems the one generally recognized by researchers as most relevant. The PFSI end-group unzipping mechanism comes from the work of Curtin et al. (2004) at DuPont. Radicals play an important part in this degradation mechanism, as shown in Figure 4.3: hydrogen is abstracted by a hydroxyl radical from a carboxylic acid end group to give a fluorocarbon radical, carbon dioxide, and water. The fluorocarbon radical can react with the hydroxyl radical again to form an intermediate that rearranges to an acid fluoride and one equivalent of hydrogen fluoride (HF). Hydrolysis of the acid fluoride generates a second equivalent of HF and another carboxylic end group subject to further degradation (Curtin et al., 2004; Hommura et al., 2008). FTIR measurement of PFSI membranes proved that the number of carboxyl groups in the ionomers increased in proportion to the time exposed to gaseous H_2O_2 (Hommura et al., 2008).

Based on Curtin et al.'s unzipping mechanism along with well-known thermochemical and kinetic parameters, Coms (2008) proposed a modified unzipping mechanism, shown in Figure 4.4. After a

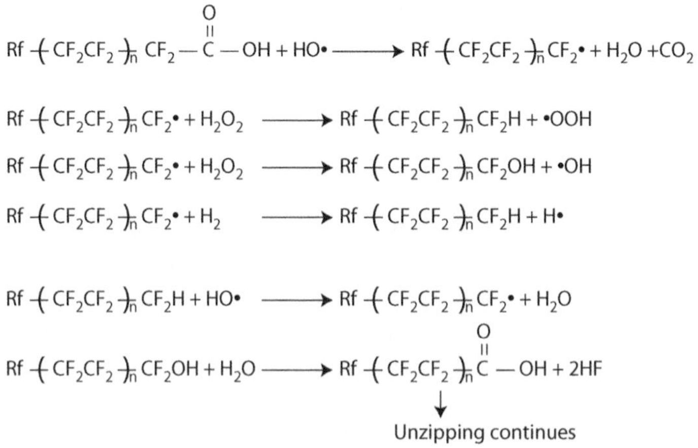

$$Rf \text{—}(CF_2CF_2)_n\, CF_2 \text{—} \overset{O}{\overset{\|}{C}} \text{—} OH + HO^{\bullet} \longrightarrow Rf \text{—}(CF_2CF_2)_n CF_2{}^{\bullet} + H_2O + CO_2$$

$$Rf \text{—}(CF_2CF_2)_n CF_2{}^{\bullet} + H_2O_2 \longrightarrow Rf \text{—}(CF_2CF_2)_n CF_2H + {}^{\bullet}OOH$$

$$Rf \text{—}(CF_2CF_2)_n CF_2{}^{\bullet} + H_2O_2 \longrightarrow Rf \text{—}(CF_2CF_2)_n CF_2OH + {}^{\bullet}OH$$

$$Rf \text{—}(CF_2CF_2)_n CF_2{}^{\bullet} + H_2 \longrightarrow Rf \text{—}(CF_2CF_2)_n CF_2H + H^{\bullet}$$

$$Rf \text{—}(CF_2CF_2)_n CF_2H + HO^{\bullet} \longrightarrow Rf \text{—}(CF_2CF_2)_n CF_2{}^{\bullet} + H_2O$$

$$Rf \text{—}(CF_2CF_2)_n CF_2OH + H_2O \longrightarrow Rf \text{—}(CF_2CF_2)_n \overset{O}{\overset{\|}{C}} \text{—} OH + 2HF$$

↓
Unzipping continues

FIGURE 4.4 Modified end group unzipping degradation mechanism of main chain of PFSI, proposed by Coms. (Redrawn from Coms, F. D. 2008. *ECS Transactions* 16: 235–255.)

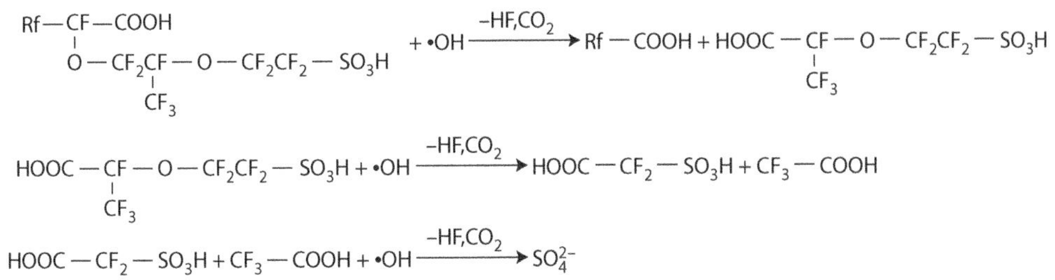

$$\underset{\overset{\displaystyle |}{\underset{\overset{\displaystyle |}{CF_3}}{O-CF_2CF-O-CF_2CF_2-SO_3H}}}{Rf-CF-COOH} + \cdot OH \xrightarrow{-HF,CO_2} Rf-COOH + \underset{\overset{\displaystyle |}{CF_3}}{HOOC-CF-O-CF_2CF_2-SO_3H}$$

$$\underset{\overset{\displaystyle |}{CF_3}}{HOOC-CF-O-CF_2CF_2-SO_3H} + \cdot OH \xrightarrow{-HF,CO_2} HOOC-CF_2-SO_3H + CF_3-COOH$$

$$HOOC-CF_2-SO_3H + CF_3-COOH + \cdot OH \xrightarrow{-HF,CO_2} SO_4^{2-}$$

FIGURE 4.5 Secondary unzipping degradation propagation reaction. (Redrawn from Xie, T. and Hayden, C. A. 2007. *Polymer* 48: 5497–5506.)

comprehensive thermochemical analysis of PFSIs and the reactive oxygen species formed in an operating fuel cell, Coms believed that the hydroxyl radical is the only oxidant capable of abstracting a hydrogen atom from a carboxylic acid intermediate and thereby propagating the PFSI main-chain unzipping process. The first step forms H_2O, CO_2, and perfluororadicals. In Curtin's mechanism, after perfluororadicals are formed they react with hydroxyl radicals to form perfluoroalcohol. While Coms believed that the concentrations of the two kinds of radicals are very low, the most likely reaction for the perfluororadicals and the perfluoroalcohol is with H_2O_2 to form hydrofluorocarbon or perfluoroalcohol. The perfluoroalcohol then hydrolyzes to form perfluorocarboxylic acid, and hydrofluorocarbon is converted back to perfluororadical via reaction with hydroxyl radical. In this mechanism, the perfluororadical can also react with hydrogen gas to form the potentially damaging hydrogen radical as a by-product.

Xie et al. mechanistically proposed a secondary unzipping degradation propagation reaction of PFSIs, as shown in Figure 4.5. During the chain end-group unzipping reaction, the side chain can be cleaved from the polymer to form HOOC–CF(CF_3)–O–CF_2CF_2–SO_3H. The newly generated carboxylic acid group on the main chain may continue the main-chain unzipping process, while the cleaved side chain may react with hydroxyl radical to form CO_2, HF, and sulfate ions (Xie and Hayden, 2007).

PFSIs are sometime produced with hydrogen-containing initiators, and the end groups may hydrolyze during polymerization or chemical conversion of sulfonyl groups to sulfonic groups. For example, when a PFSI precursor is synthesized in water with a persulfate initiator, hydrogen-containing groups may be generated, as shown in Figure 4.6 (Bro and Sperati, 1959; Pianca et al., 1999). It has been observed that during industrial extrusions of perfluoropolymers manufactured by aqueous emulsion polymerization with $K_2S_2O_8$ as initiator, the absorptions of the carboxylic groups decrease with formation of a band at 1884 cm^{-1}, attributable to the $-CF_2-COF$ end groups (Pianca et al., 1999). For PFSIs, the $-COF$ end groups, if they exist, can be chemically converted to $-COOH$ groups when the precursor is hydrolyzed in aqueous alkali, followed by conversion to H-type ionomer in acid. Researchers at DuPont also found that unstable groups such as $-CF=CF_2$, $-CF_2H$, $-CONH_2$, $-CF_2CH_2OH$, $-COF$ and $-CF=CF(CF_2)CF_3$ may be generated under polymerization conditions after thermo-extrusion of tetrafluoroethylene/perfluorinated vinyl ethers (Morgan and Sloan, 1986; Imbalzano and Kerbow, 1988). PFSI precursor

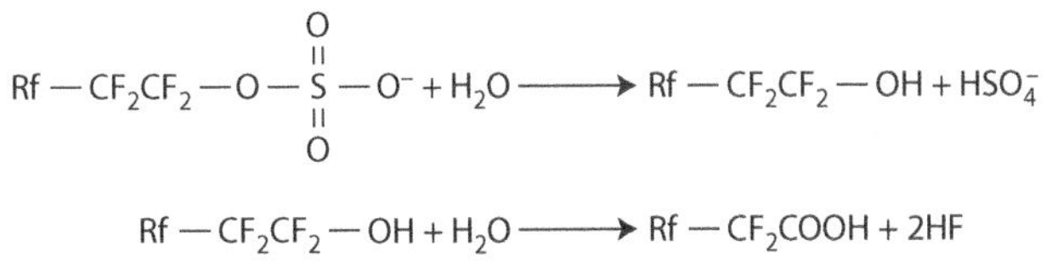

$$Rf-CF_2CF_2-O-\overset{\overset{\displaystyle O}{\|}}{\underset{\underset{\displaystyle O}{\|}}{S}}-O^- + H_2O \longrightarrow Rf-CF_2CF_2-OH + HSO_4^-$$

$$Rf-CF_2CF_2-OH + H_2O \longrightarrow Rf-CF_2COOH + 2HF$$

FIGURE 4.6 Generation of initial carboxylic groups in PFSIs. (Redrawn from Pianca, M. et al. 1999. *Journal of Fluorine Chemistry* 95: 71–84.)

polymer is one such kind of copolymer, and so the above-mentioned unstable groups might be formed after thermo-extrusion for pellets and membranes, and thereby exist in the product membranes.

So, although PFSIs contain no hydrogen in the main chain, the hydrogen-containing or other unstable end groups are susceptible to reactive radical attack, which is generally believed to be the principal degradation mechanism of perfluorosulfonic membranes in fuel cells (Curtin et al., 2004; Wang et al., 2008).

4.3.1.1.2 Chain Scission

There is considerable experimental evidence indicating that chain scission reactions occur in the degradation process of PFSI membranes. Hommura et al. (2008) found that these membranes could be degraded by H_2O_2, and that membrane degradation comprised two modes, as shown in Figure 4.7: (1) an unzipping reaction at unstable polymer end groups and (2) scission of main chains and formation of new unstable polymer end groups at the severed points. Unzipping of the main chains started from the carboxyl groups and formed new carboxyl groups. Scission of main chains from new carboxyl groups at the severed points of the polymer main chains could initiate new unzipping reactions. The number of chain scissions was associated with how long the membranes were exposed to H_2O_2. There were about 14 severed points per PFSI molecule at the point of 700 h exposure time in H_2O_2 according to the authors' calculations (Hommura et al., 2008), but no further scission process was discussed.

Based on both experimental and computed thermochemical and kinetic data, Coms (2008) developed two chain scission mechanisms: the sulfonyl radical mechanism and the hydrogen radical mechanism.

4.3.1.1.2.1 Sulfonyl Radical Mechanism   The C–S bonds to the sulfonic acid group are the weakest bonds in the hydrocarbon and fluorocarbon families. This lability may play a significant role in the overall fuel cell chemical stability of perfluorosulfonic acid (PFSA) membranes. According to calculated results, Coms (2008) proposed that the acidic proton can be abstracted by the hydroxyl radical to form the $-SO_3\bullet$ radical if it resides on the $-SO_3$ group as $-SO_3H$ at a low membrane hydration level. Simultaneously, $-SO_3H$ (not the sulfonate anion, because it is chemically rather inert) can react with H_2O_2 to form perfluorosulfonyl peroxide, as shown in Figure 4.8. The perfluorosulfonyl peroxide then

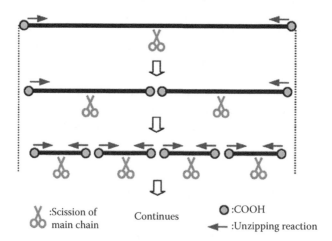

FIGURE 4.7   Degradation mechanism of PFSI membrane, proposed by Hommura et al. (Reprinted from Hommura, S. et al. 2008. *Journal of the Electrochemical Society* 155: A29–A33. With permission from Electrochemical Society.)

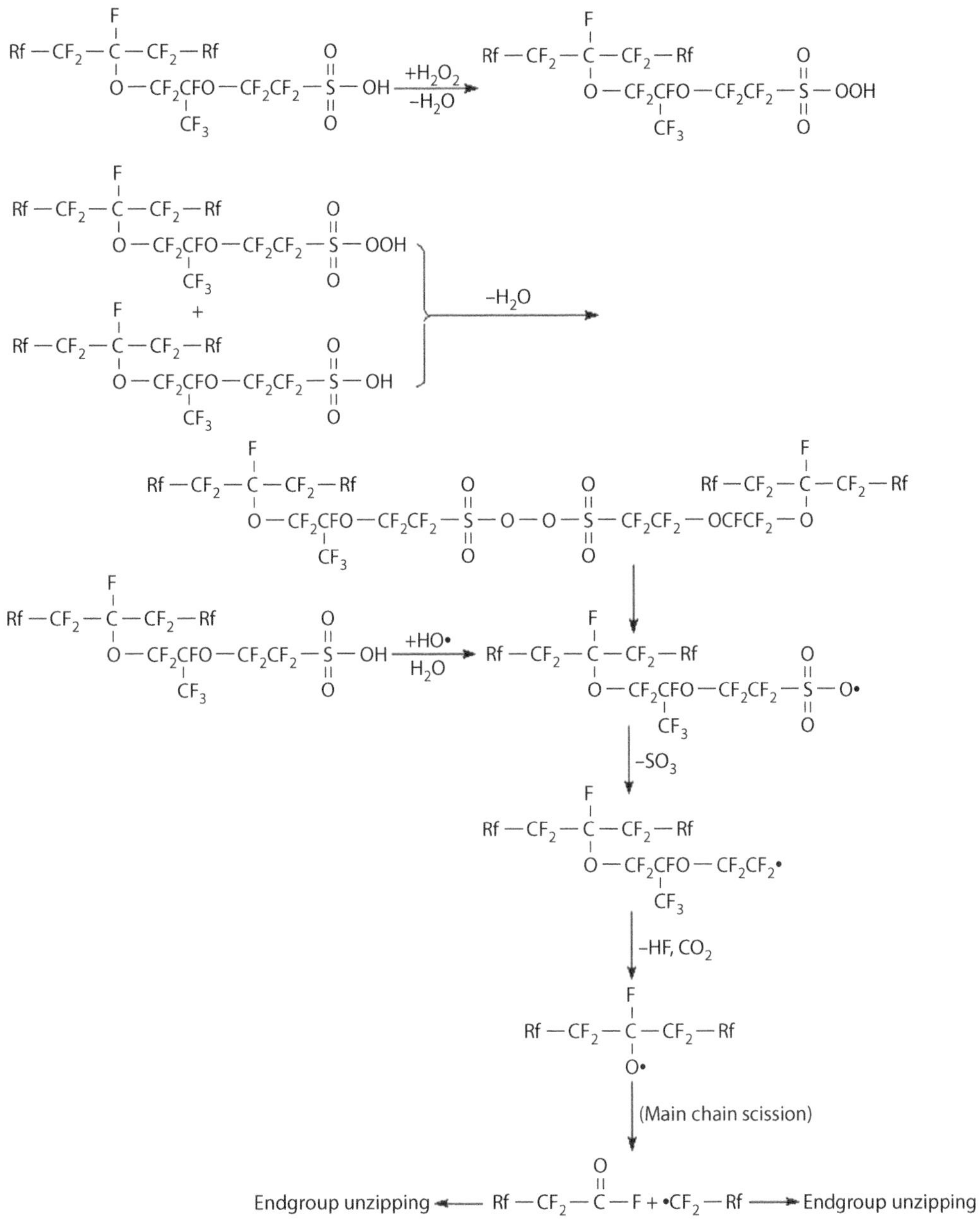
+H2O2
-H2O
-H2O
+HO•
H2O
-SO3
-HF, CO2
(Main chain scission)
Endgroup unzipping
Endgroup unzipping

FIGURE 4.8 Sulfonyl radical initiating chain scission mechanism, proposed by Coms. (Redrawn from Coms, F. D. 2008. *ECS Transactions* 16: 235–255.)

reacts with perfluorosulfonic acid to form bissulfonyl peroxide. The bissulfonyl peroxide decomposes via O–O bond homolysis at a mild temperature to perfluorosulfonyl radicals, which fragment to form perfluororadicals and sulfur trioxide. Once the perfluororadicals are formed they will continue to degrade, ultimately forming oxyradicals at the junction with the main chain of the PFSI backbone and resulting in scission of the main chain.

In this mechanism, the location of the proton has a significant impact on the chemistry in the dry membrane of a fuel cell. Sulfonic acid groups are far more reactive than sulfonate anions, and so this mechanism provides a plausible explanation for why membranes degrade faster at lower relative humidity (RH).

4.3.1.1.2.2 Hydrogen Radical Mechanism In addition to the sulfonyl radical mechanism, Coms (2008) also entertained another plausible mechanism of initiation and main-chain scission, named the hydrogen radical mechanism, in which hydrogen gas is considered a hydrogen atom donor present in an operating fuel cell. The reaction of hydroxyl radical and hydrogen generates a hydrogen radical that can be very aggressive toward the C–F bonds of PFSIs. The calculation conducted by Coms indicates that abstractions of secondary and tertiary C–F bonds could be kinetically feasible. The scission process reaction is shown in Figure 4.9.

4.3.1.1.2.3 Scission of Side Chains Free radicals generated in PEMFCs can attack not only the main chains of the ionomer molecules but also the side groups, especially in the presence of some metal counterions such as Cu(II) and Fe(III). Kadirov et al. (2005) used ESR resonance to investigate perfluorosulfonic Nafion and Dow PFSI membranes neutralized by Fe(III) and found $ROCF_2CF_2$ radicals generated from the ionomer side groups. They concluded that cations are involved in PFSI membrane degradation processes in PEMFCs, with the point of attack appearing to be at or near the side groups of the ionomer (Kadirov et al., 2005; Danilczuk et al., 2007).

Chen and Fuller (2009a) investigated the effect of humidity on the degradation of Nafion membrane and found the main-chain unzipping process to be dominant under high humidity, while a decrease

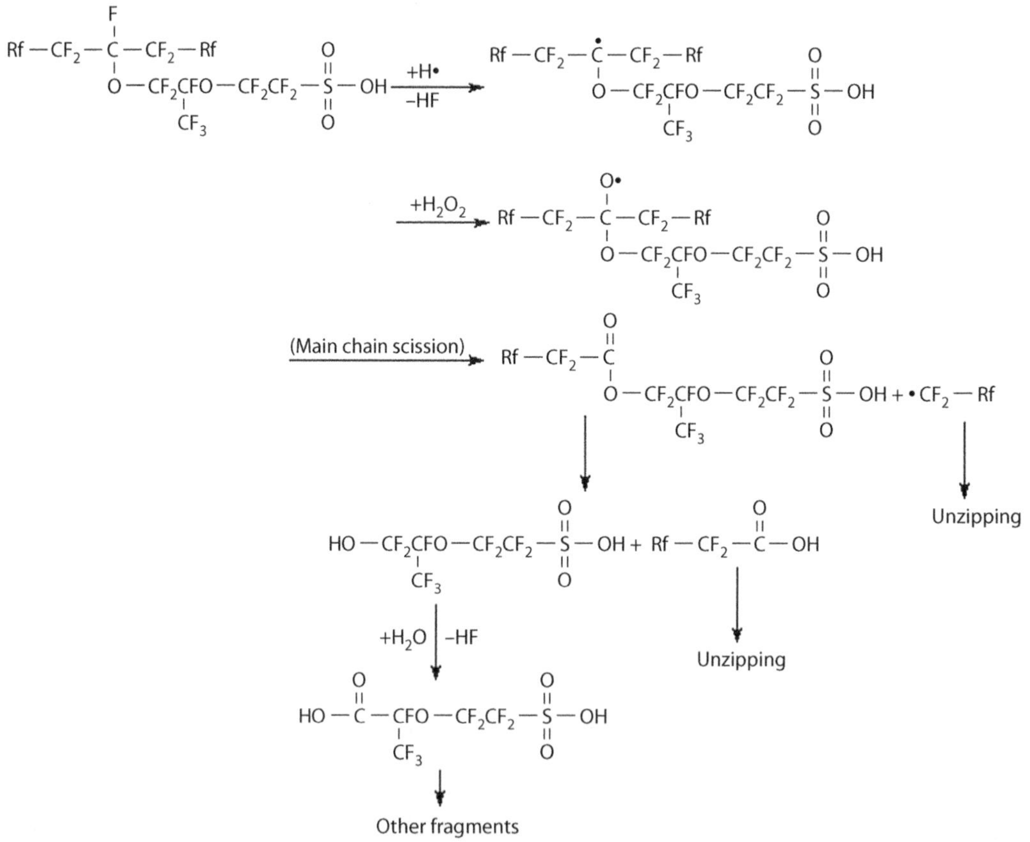

FIGURE 4.9 Main-chain scission process initiated by fluorine atom abstraction by hydrogen radical, proposed by Coms. (Redrawn from Coms, F. D. 2008. *ECS Transactions* 16: 235–255.)

in humidity enhanced the side-chain scission process. At low RH, their investigations with Fourier transform infrared spectroscopy (FTIR) showed loss of the C–O–C structure, corresponding well with the detection of trifluoroacetic acid product. Accordingly, the authors proposed that the hydroxyl radicals can attack the ether group in the middle of the side chain, as shown in Figure 4.10, resulting in further ionomer degradation.

Cleavage of ester linkages has also been found by other research groups. After studies on the chemical durability of PFSI and model compounds under simulated fuel cell conditions, Zhou et al. (2007) found that the relative reactivity of carboxylic chain ends and ether linkages is approximately 500. Commercial perfluorosulfonic acid products contain minimal carboxylic acid end groups, and the side-chain concentrations are 2–3 orders of magnitude higher than the carboxylic acid end groups; based on this comparison, the authors proposed that cleavage of side chains from the polymer may be the dominant pathway in commercial PFSI membranes with low carboxyl content. Kabasawa et al. (2009) investigated the venting water of an operating fuel cell using ion chromatography combined with mass spectrometry (IC/MS) and found the drainwater contained, as decomposition products, bisulfate and fluorine ions derived from the side chains of the Nafion membrane. Their results suggest that ether linkages, which connect the ionic side-chain groups to the polymer backbones, are susceptible to attack by radicals. Almeida and Kawano (1998) investigated the effects of x-ray radiation on Nafion 117 membrane and demonstrated the presence of peroxyl-free radicals in the membrane. Their Raman and infrared spectra of the irradiated membrane samples showed the appearance of new bands at 1458, 1700, and 1773 cm^{-1}, which could be associated with the presence of C = O and C = C groups. In addition, the Raman and IR spectra intensity changes were associated with the stretching modes of C–O, C–S, and SO^{3-}, indicating scissions of C–S and C–O bonds and the formation of SO_2.

4.3.1.1.3 Side-Group Attack

Collette et al. (2009) investigated the hygrothermal aging of Nafion over a long period at 80°C. The evolution of Nafion chemical structure was studied by infrared spectroscopy and nuclear magnetic resonance (NMR) imagery, which highlighted sulfonic anhydride formation, as shown in Figure 4.11; this created a crosslink between the two sulfonic groups, which led to decreased polymer hydrophilicity and a drop-off in proton conductivity. The authors also found that anhydride concentration was proportional to water concentration, implying that the reaction kinetics would be higher in wet conditions. The proposed mechanism is shown in Figure 4.11.

Chen et al. (2007) used FTIR to study the chemical structure of Nafion membranes before and after Fenton's test. After Fenton's test, a very weak absorption at 1434 cm^{-1} is evident and assigned to a

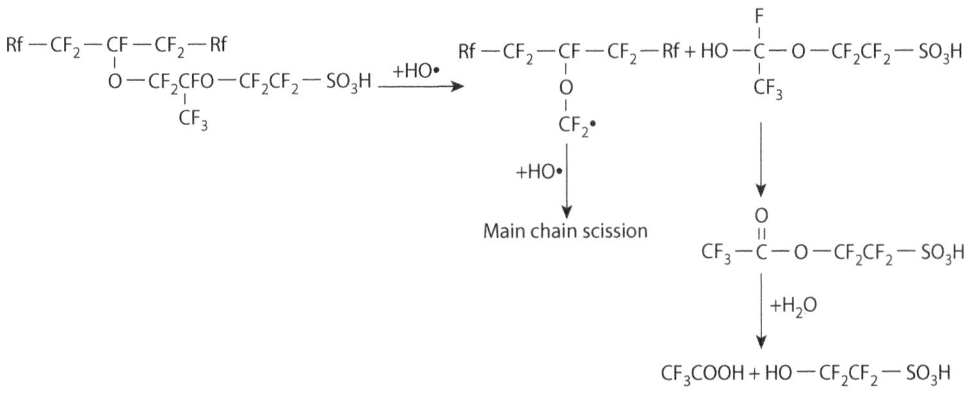

FIGURE 4.10 Side-chain scission mechanism. (Redrawn from Chen, C. and Fuller, T. F. 2009. *Polymer Degradation and Stability* 94: 1436–1447.)

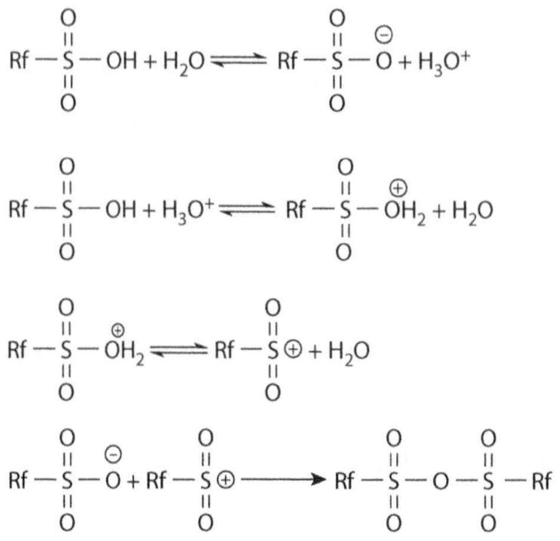

FIGURE 4.11 Anhydride formation chemical reactions of PFSIs. (Redrawn from Collette, F. M. et al. 2009. *Journal of Membrane Science* 330: 21–29.)

crosslinking S–O–S band. In the membrane degradation mechanism associated with the end group –SO_3H that the authors proposed, formation of crosslinking S–O–S bonds under the strong oxidation effect of H_2O_2 and free radicals is the first step. This anhydride may react with another sulfonic acid group to produce a sulfonate ester and release SO_2 gas. Furthermore, reaction of this ester with water produces a carboxylic acid that can be further attacked by hydroxyl radicals.

In the FTIR study on chemical aging of Nafion conducted by Alentiev et al. (2006), (–S(O)$_2$–O–) groups not observed in "native" Nafion were found after thermal treatment of Nafion films at 95°C. The same result was obtained by the authors for a Nafion film that had been stored for 14 years. Their findings indicate that sulfo-ethers and crosslinking can appear during storage, exploitation, and thermal treatment of perfluorosulfonic membranes, which should reduce the concentration of sulfo-groups and decrease the proton conductivity of Nafion membranes in the process of exploitation, such as in fuel cells.

4.3.1.2 Mitigation Strategies for Chemical Degradation

There are at least two elements in the perfluorosulfonic membrane degradation mechanism: the presence of unstable groups and linkages, and of reactive free radicals such as hydroxyl radical. Much work has been done to eliminate the unstable end groups and radicals in these PEMFC membranes.

Developing ionomer membranes that are more chemically stable is the first essential answer to improve membrane chemical durability. Curtin et al. (2004) reported that the reactive end groups could be greatly reduced by fluorinating the Nafion polymer with elemental fluorine using a traditional method. After treatment, 61% measurable reactive end groups could be removed, and after >50 h of exposure in Fenton's reagent, there was a 56% decrease in total fluoride ions generated per gram of fluorinated polymer compared with the untreated polymer. Using DuPont's newly developed proprietary methods, the fluoride ion emissions of modified membranes can be reduced 10–25-fold, thus greatly enhancing the stability of these modified membranes.

Membrane chemical degradation directly results from attack by active radicals. Hydrogen peroxide formed in fuel cells can result in radical formation. Some researchers add radical or H_2O_2 scavengers to eliminate radicals or reduce the formed H_2O_2 to harmless species. Aoki et al. (2006) confirmed that an appreciable fraction of H_2O_2 or HO• radicals was easily scavenged at Pt particles in the CL or in Pt-dispersed Nafion membranes. Danilczuk et al. (2009) found that Ce(III) in Nafion membranes with low concentrations can efficiently scavenge HO• radicals because of the Ce(III)/Ce(IV) couple redox chemistry. Ralph et al. (2008) added a H_2O_2 decomposition catalyst to the Johnson Matthey Fuel Cells (JMFC) reinforced membrane. Their test results showed that the lifetime was extended from 1200 to 1800 h with this addition, and even after 1800 h there was little sign of membrane thinning in their accelerated OCV test.

Hydrogen and oxygen crossover is the fundamental mechanism for H_2O_2 or reactive radical formation, and so reducing gas crossover might reduce membrane chemical degradation. Patil and Mauritz (2009) modified Nafion membranes via an *in situ* sol–gel polymerization of titanium isopropoxide to generate titania quasinetworks in the polar domains. During an OCV decay test they found that the voltage decay rate for the modified membrane was 3.5 times lower than that of control Nafion, and fluoride emissions for the composite were at least an order of magnitude lower than those of the control. They proposed that incorporation of the inorganic quasinetwork particles improved the mechanical and gas barrier properties of the membranes and thereby reduced both mechanical and chemical degradation.

It is generally believed that some contaminant metal ions can accelerate radical formation and membrane degradation. Kinumoto et al. (2006) found the durability of Nafion with alkali and alkaline earth metal ions as counter ions to be similar to that of H-type Nafion, while the presence of ferrous and cupric ions as counter ions significantly enhanced the Nafion decomposition rate. So, to improve membrane durability it is very helpful to decrease the concentration of ferrous and cupric ions in the membranes and the operating fuel cells.

In several membrane chemical degradation mechanisms, such as the catalytic combustion mechanism, the catalyst plays an important role in H_2O_2 or reactive radical formation. To improve the chemical durability of membranes in PEMFCs, it is thus important to develop new catalysts that produce fewer hydrogen peroxide/radical species.

4.3.2 Mechanical Degradation

4.3.2.1 Mechanism of Mechanical Degradation

Mechanical membrane degradation can lead to rapid reactant gas crossover and unwanted reactions. Sudden cell/stack failure will likely follow a mechanical breach of the weakest area of the polyelectrolyte membrane (Huang et al., 2006; Tang et al., 2007b). It has been reported that a pinhole only ~100 µm across is needed to cause failure in a full-size cell (250 cm²) (Burlatsky et al., 2006). Mechanical membrane degradation can occur in many forms, such as membrane thinning, tears, cracks, pinholes, and so on. Local stress, variations in temperature and humidity, impurities in membrane formation, chemical degradation, and other factors may cause mechanical membrane degradation in fuel cells.

4.3.2.1.1 Local Stress

In a fuel cell, the membrane is sandwiched between two bipolar plates. This structure puts it under compressive stress, which can change the membrane resistance. It has been found that the resistance of Nafion membranes increased when they were compressed, and the increase was consistent with the elastic compression of the membrane (Satterfield et al., 2006). Casciola et al. (2006) also found that membrane conductivity decay occurs only when the membrane is forced to swell anisotropically along the plane parallel to the membrane surface. In addition to the effect of compression on conductivity decay, polymer membranes in fuel cells undergo creep, which can cause membrane thinning, pinhole formation, and other failure. Stucki et al. (1998) proposed that local stress most likely triggered and/or enhanced the nonuniform thinning of the Nafion membranes in a polymer electrolyte membrane electrolyzer.

At fuel cell start-up, the dry membrane will absorb water from the feed reactants and swell. The linear expansion of Nafion N115 is 10% and 15% from 50% RH at 23°C to water-soaked at 23°C and to water-soaked at 100°C, respectively. The thickness change is 10% and 14% from 50% RH at 23°C to water-soaked at 23°C and to water-soaked at 100°C, respectively. In fuel cells, the membrane is constrained between the bipolar plates. Water absorption by the membranes is accompanied by membrane volume swelling, which creates swelling pressure against the electrode layers and bipolar plates and may subsequently induce constrained membrane creep under stress. Satterfield et al. (2006) found that the swelling pressure of a Nafion N115 membrane at 100% RH and 80°C may reach up to 0.55 MPa; the swelling pressure of N115 showed little or no dependence on temperature.

Additionally, penetration of catalyst particles or other foreign particles into the membrane can cause local high stress.

4.3.2.1.2 *Temperature and Humidity*

Temperature and humidity may change the molecular conformation and aggressive state of polymer segments in membranes, and then have great effect on the long-term mechanical properties of PFSI membranes and the performance of fuel cells. Huang et al. (2003) reported that x-ray diffraction (XRD) studies of Nafion showed the characteristic peak ($2\theta = 18°$) decaying significantly with decreased H_2/O_2 humidification temperatures; thus, insufficient water in the membrane region during fuel cell operation can destroy the membrane crystalline structure. The DuPont product data sheet for Nafion membranes reports that the tensile strength and modulus of water-soaked N115 samples decreases from 34 to 25 MPa and 114 to 64 MPa, respectively, when temperature is increased from 23°C to 100°C. At 23°C, the tensile strength and modulus of membrane samples equilibrated at 50% RH decreases from 43 to 34 MPa, and from 249 to 114 MPa for membrane samples equilibrated in water. The elongation and the break-and-tear resistance of Nafion N115 also decrease with increasing temperature and hydration. Bauer et al. (2005) found that the influence of water and temperature on the mechanical properties of H-Type Nafion membrane is rather complex: the maximum elastic modulus (E') as a function of humidity shifts to higher humidities with increasing temperature. At low temperatures, water acts mainly as a plasticizer, while at high temperatures it stiffens the membrane by the formation of hydrogen bridge bonds and oligohydrates, which act as crosslinkers between sulfonic acid groups. Plus, the glass transition of the ionic regions shifts to higher temperatures with increasing water activity. Satterfield et al. (2006) investigated the mechanical properties of Nafion and titanium dioxide/Nafion composite membranes. They found that the elastic and plastic deformation of both membranes decreased with both the temperature and water content. The composite membranes had slightly higher elastic moduli and exhibited less strain hardening than Nafion. With titanium dioxide inside, the solution-cast composite membranes also showed a 40% reduction in creep compared with Nafion N115 at 25°C and 100% RH after 3 h.

In an operating fuel cell, because the four sides of the membranes are fixed, variations in temperature and humidity during operation can induce membrane dimensional change and corresponding cyclic stresses and strains in the membrane. Tang et al. (2007b) reported that if Nafion NR111 membrane changed from a water-soaked state to a state with 25% RH at 25°C, the shrinkage stress may reach a maximum value of 2.23 MPa, which is higher than the fatigue stress safety limit of 1.5 MPa for Nafion NR111. Tang's investigation also revealed that the stress induced by temperature cycling alone was very low and would not be responsible for mechanical breach in Nafion membranes, but simultaneous temperature cycling and humidity conditions increased the membranes' mechanical degradation significantly. Investigation by Huang et al. (2006) also revealed that RH can cause the accumulation and growth of mechanical membrane damage (microcracks opening up in the direction of applied tensile stress). In their experiments, strain to failure showed a clear reduction trend as the number of RH cycles increased, and cycling from 80% to 120% RH resulted in a faster per-cycle degradation rate. They proposed that the most significant impact of RH cycling on the mechanical behavior of membranes is reduced strain-to-failure or toughness: microcracks and crazing are initially caused by RH cycling from the electrode and the electrode-electrolyte interface, and then the defects propagate into the membrane to reduce its ductility.

4.3.2.1.3 *Effect of Chemical Degradation*

Membrane failure is believed to be the result of chemical and mechanical effects acting together (Liu et al., 2001). Membrane chemical degradation can result in altered stress–strain behavior and loss of membrane mechanical strength (Huang et al., 2006; Tang et al., 2007b). Investigation shows that chemical degradation of Nafion membranes in Fenton's reagent introduces many holes and cracks in the surfaces and cross-sections of membranes (Tang et al., 2007b; Kundu et al., 2008). Such defects caused by membrane chemical degradation can act as stress concentrating points, and enlarge during fuel cell operation and changes in the membrane environment (Liu et al., 2001).

4.3.2.1.4 *Effect of Cations*

The presence of monovalent and divalent contaminant cations may affect the mechanical properties of perfluorosulfonated ionomer membranes. Kundu et al. (2005) found that the stiffness as well as the yield strength of ion-exchanged hydrated perfluorosulfonated membranes increased in order of increasing cationic radius. This mechanical effect is largely attributed to a physical crosslinking that occurs in the ionic clusters. Bigger cations have more surface area available for side chains to form crosslinkages, thereby reducing chain mobility and increasing stiffness and strength. High cation-induced crosslinking will make membranes brittle; more importantly, cation contamination of membranes can lead to fuel cell and membrane performance losses, such as cathode and anode kinetic losses, ohmic loss, and water flux and proton conductivity loss in membranes (St Pierre et al., 2000; Kelly et al., 2005a,b; Collier et al., 2006).

4.3.2.2 Mitigation Strategies for Mechanical Degradation

Obviously, care must be exercised in membrane preparation, as foreign particles introduced during fabrication can cause perforations under pressure, and some particles may decompose rapidly in operating fuel cells to form pinholes.

Reinforcement seems to be an effective way to relieve mechanical degradation of perfluorosulfonic membranes. Ralph et al. (2008) compared the physical properties of JMFC reinforced membrane with those of a 920 g/equiv equivalent weight commercial membrane, with both membranes fully hydrated. The tear propagation was some seven times higher in both the MD and the TD in the JMFC reinforced membrane compared to the commercial membrane because the reinforcement acts as an efficient crack propagation inhibitor. The dimensional stability of the reinforced membrane was also significantly improved compared to the commercial membrane. Liu et al. (2001) reported that Gore-Select membranes, which are e-PTFE reinforced membranes developed by Gore Fuel Cell Technologies, had a longer lifetime than nonreinforced membranes; in addition, membrane failure characteristics, measured by H_2 crossover rate, showed that the reinforced membranes degraded gradually whereas the nonreinforced membranes showed sudden, catastrophic failure. Increased stability in reinforced membranes was also reported by Tang et al. (2007a, 2008) using Nafion membranes. Their study found that Nafion/e-PTFE composite membranes can remain stable for over 5000 cycles, about 40% more than pure Nafion membrane, at about 3500 cycles.

Patil and Mauritz (2009) reported an *in situ*, sol–gel-derived, inorganic modification method to enhance membrane durability by reducing both physical and chemical degradation. The authors modified Nafion membranes via an *in situ* sol–gel polymerization of titanium isopropoxide to generate titania quasinetworks in the polar domains. The composite membranes showed higher modulus, lower creep, improved dimensional stability, more structural integrity with few cavities, and enhanced ability in constrained membranes to withstand contractile stresses due to humidity change. Incorporation of the inorganic quasinetwork particles improved the mechanical and gas barrier properties of the membranes and thereby reduced membrane degradation.

The mechanical properties of PFSI membranes are originally determined by the ionomer material itself. PFSI membranes have three phases: the crystalline region, the amorphous region, and the ion cluster domain. The crystalline region contributes most to mechanical membrane strength. Crystallinity of perfluorosulfonic membranes increases with increasing equivalent weight, and so the stress at break of the membranes increases with equivalent weight (Arcella et al., 2005). For long-side-chain (LSC) PFSIs such as Nafion, the proton conductivity of the ionomer decreases with increasing equivalent weight. Considering both mechanical strength and conductivity, the equivalent weight of commercial Nafion membranes is about 1100 g/equiv. However, for short-side-chain (SSC) PFSI membranes such as Hyflon ion membranes and Dow membranes, polymer main chains may interact more strongly due to shorter pendant chains. The stress at break of a Hyflon ion membrane with an equivalent weight of 860 g/equiv is about 30 MPa, which is typical of the commonly available Nafion 1100 membrane with an equivalent weight of 1100 g/equiv (Arcella et al., 2005). Additionally, SSC perfluorosulfonic membranes show high conductance with low equivalent weight. It seems that combining high ion exchange capacity and high mechanical stability is a possible and promising approach to improve membrane performance in PEMFCs, such as by changing molecular structure, increasing molecular weight, introducing stable chemical cross-link, improving crystal perfection, or introducing stable interacting particles (Kreuer et al., 2008).

4.3.3 Thermal Degradation

Several advantages accrue from operating PEMFCs at high temperatures, including enhanced electrode reactions, simplified water management and cooling systems, and increased CO tolerance (Li et al., 2003; Zhang et al., 2006). Accordingly, higher ionomer membrane thermal stability is required for high-temperature fuel cells. Additionally, catalytic combustion (Miyake et al., 2004, 2006; Inaba et al., 2006) and reactant gas starvation can cause local heat (Taniguchi et al., 2008), which can result in membrane thermal degradation.

4.3.3.1 Thermal Properties of PFSI Membranes

The thermal behaviors of Nafion membranes have been widely investigated by differential scanning calorimetry (DSC) (Moore and Martin, 1988, 1989; de Almeida and Kawano, 1999, Lage et al., 2004b; Page et al., 2005) and thermogravimetric analysis (TG) (Wilkie et al., 1991; Tiwari et al., 1998; Lage et al., 2004a,b).

DSC studies of dry Nafion PFSI membranes have revealed two endothermic peaks below 300°C, at about 120–150°C and 220–260°C. Because of the complexity of the molecular structure, there is still no consensus on the attribution of the endothermic peaks. Moore and Martin attributed the two peaks at 140°C and 260°C to the glass transition temperatures of the matrix and ionic domains, respectively (Moore and Martin, 1988). Moore and Martin also investigated the thermal properties of SSC PFSI membranes produced by Dow; the low endothermic peak (at 150–180°C, depending on the ion exchange capacity) was assigned to the matrix glass transition and the high peak to ionic cluster glass transition. In a later article, Page et al. (Page et al., 2005) attributed the high-temperature endotherm near 230°C of Nafion membrane to the melting of PTFE-like crystallites rather than to the glass transition temperature of the ionic domains as earlier suggested, and the low-temperature endotherm to the melting of small, imperfect crystallites, not to the matrix glass transition temperature or the order–disorder transition within the clusters. De Almeida and Kawano (1999) assigned the peak near 120°C to a transition into ionic clusters, the peak near 230°C to the crystalline domain's melting, and a peak near 325°C to the desulfonation process, while Moore and Martin (1989) related the endotherm at 335°C to the crystallite melting observed in their earlier study. Lage et al. (2004b) investigated the thermal behavior of Nafion membranes with various counter ions and attributed the transition in the range of 100–200°C to cluster transition.

In conclusion, H-type Nafion membranes yield three endothermic peaks below 325°C in the DSC trace. The low-temperature peak in the range of 100–200°C is considered related to the order–disorder transition within the ionic clusters, the glass transition of Nafion, or the melting of imperfect crystallites. The peak in the range of 220–260°C is related to cluster transition or the melting of PTFE-like crystallites, and the peak near 325°C is assigned to crystallite melting or membrane desulfonation.

Because of the molecular structural complexity and the diversity of segmental motions in Nafion ionomers, explanations in the limited reports on DSC data for Nafion membranes have been rather vague and inconsistent. In contrast, explanation of TG data on Nafion seems explicit. The thermal stability of Nafion membranes has been widely investigated by several groups (Wilkie et al., 1991; Tiwari et al., 1998; Lage et al., 2004a,b). Perhaps the most cited paper about Nafion thermal durability is of Wilkie et al. (1991). To study the interaction of poly(methyl methacrylate) and Nafion, Wilkie et al.

coupled TG analysis with an FTIR spectrometer to study the thermal decomposition of Nafion ionomer. They observed that the first weight loss, about 5 wt%, occurred between 35°C and 280°C; the gases were identified as H_2O, SO_2, and CO_2. The water mostly resulted from what was contained in the pristine membrane because Nafion is quite hygroscopic. SO_2 resulted from the cleavage of the C−S bond with the formation of a CF_2 radical and a SO_3H radical, as shown in Figure 4.12. The authors agreed that CO_2 was produced by the cleavage of COOH groups, as shown in Figure 4.13. In this process, sulfonic acid loses water to produce anhydride, which may react with another sulfonic acid to produce a sulfonate ester. Reaction of this ester with water will produce a carboxylic acid that can undergo decarboxylation to produce CO_2. The second weight loss was about 7 wt%, from 280°C to 355°C, generating SO_2, CO_2, SiF_4, CO, HF, and substituted carbonyl fluorides. From 355°C to 560°C, 88% of the sample volatilized. The major absorbance in this temperature region was due to HF, SiF_4, carbonyl fluorides, and C−F stretching vibrations. Additionally, the amounts of SO_2 and CO decreased dramatically at 365°C and were no longer of consequence. Based on these observations, Wilkie et al. presented a thermal decomposition mechanism for Nafion, as shown Figure 4.12. In this process, thermal decomposition is initiated with the cleavage of the C−S bond, forming a carbon-based radical and an SO_3H radical. The SO_3H radical then cleaves to generate SO_2 and a hydroxyl radical, while the carbon-based radical loses two difluorocarbenes and generates an oxygen-based radical. The oxygen-based radical further loses a substituted carbonyl fluoride and carbonyl fluoride itself. The result is a PTFE-like polymer that will degrade by a scission process combined with an end-group unzipping process.

Bas et al. (2009) investigated the thermal degradation of Nafion membranes using TG analysis coupled with mass spectroscopy (TG-MS). Besides the 4 wt% weight loss of water occurring in a broad temperature range below 200°C, there were three weight loss steps. The first step concerned 10% weight loss in the range of 280–380°C and was assigned to degradation of the pendant chains, as suggested by the detection of SO_2 and CF_3O^+ fragments. The peaks at temperatures above 380°C were both assigned to main-chain degradation.

Several groups have found that counterions have a great effect on the thermal stability of perfluorosulfonated ionomer membranes (Feldheim et al., 1993; Lage et al., 2004a,b; Bas et al., 2009). Using TG analysis, Feldheim et al. (1993) found the thermal stability of Nafion perfluorosulfonated ionomer to be

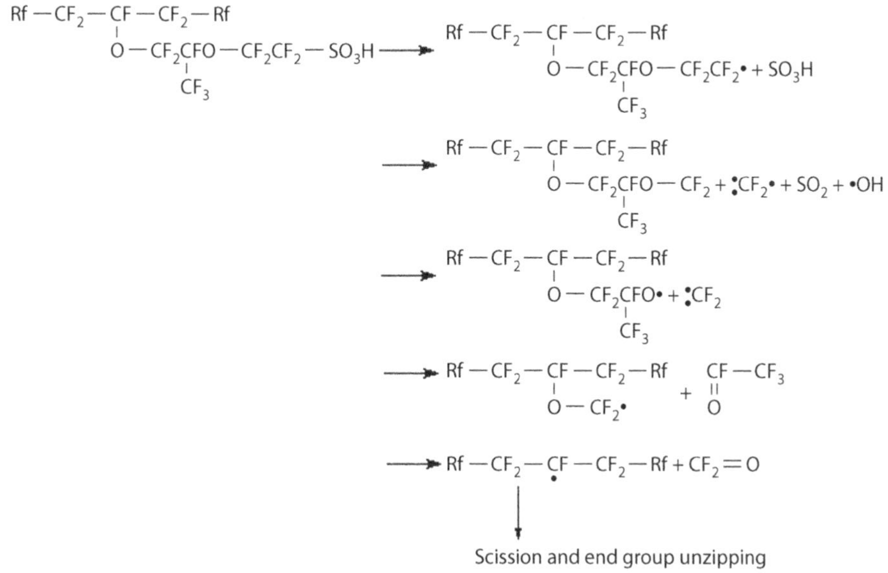

Scission and end group unzipping

FIGURE 4.12 Nafion thermal decomposition process. (Redrawn from Wilkie, C. A., Thomsen, J. R., and Mittleman, M. L. *Journal of Applied Polymer Science.* 1991, 42: 901–909.)

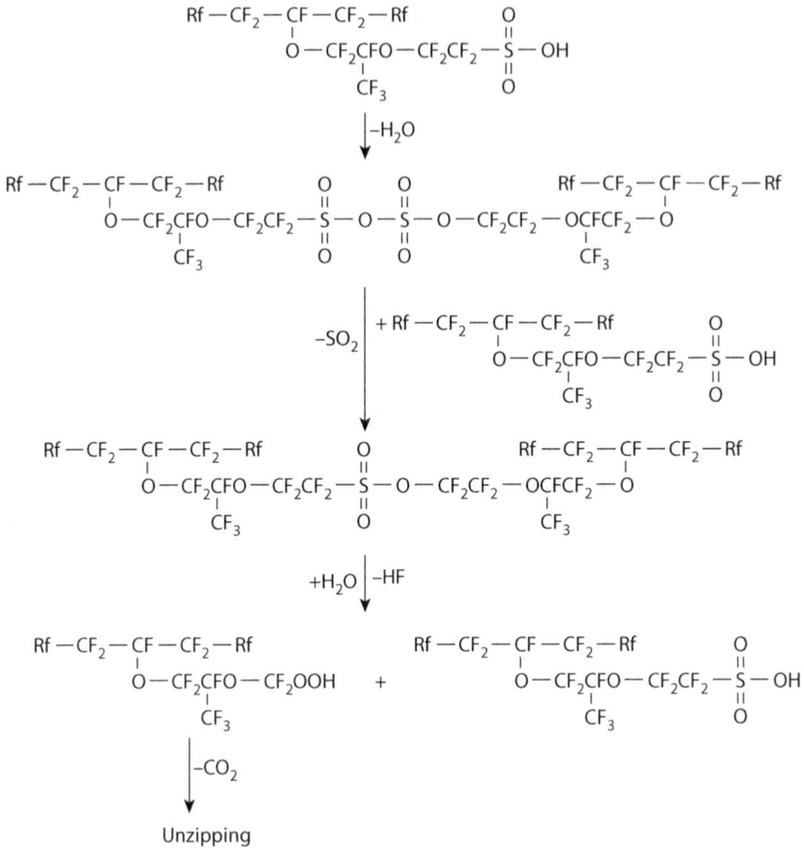

$Rf-CF_2-CF-CF_2-Rf$
$O-CF_2CFO-CF_2CF_2-\overset{O}{\underset{O}{\overset{\|}{\underset{\|}{S}}}}-OH$
CF_3

$\downarrow -H_2O$

$Rf-CF_2-CF-CF_2-Rf$
$O-CF_2CFO-CF_2CF_2-\overset{O}{\underset{O}{\overset{\|}{\underset{\|}{S}}}}-O-\overset{O}{\underset{O}{\overset{\|}{\underset{\|}{S}}}}-O-CF_2CF_2-OCFCF_2-O$
CF_3 $Rf-CF_2-CF-CF_2-Rf$ CF_3

$-SO_2 \Big|$ $+ Rf-CF_2-CF-CF_2-Rf$
$O-CF_2CFO-CF_2CF_2-\overset{O}{\underset{O}{\overset{\|}{\underset{\|}{S}}}}-OH$
CF_3

$\downarrow$

$Rf-CF_2-CF-CF_2-Rf$ $Rf-CF_2-CF-CF_2-Rf$
$O-CF_2CFO-CF_2CF_2-\overset{O}{\underset{O}{\overset{\|}{\underset{\|}{S}}}}-O-CF_2CF_2-OCFCF_2-O$
CF_3 CF_3

$+H_2O \Big| -HF$

$Rf-CF_2-CF-CF_2-Rf$ + $Rf-CF_2-CF-CF_2-Rf$
$O-CF_2CFO-CF_2OOH$ $O-CF_2CFO-CF_2CF_2-\overset{O}{\underset{O}{\overset{\|}{\underset{\|}{S}}}}-OH$
CF_3 CF_3

$\downarrow -CO_2$

Unzipping

FIGURE 4.13 Formation of CO_2 during Nafion membrane thermal decomposition. (Redrawn from Wilkie, C. A., Thomsen, J. R., and Mittleman, M. L. *Journal of Applied Polymer Science.* 1991, 42: 901–909.)

strongly dependent upon the nature of the counterions associated with the fixed sulfonate site. Nafion films showed improved thermal stability as the size of the countercations decreased, with the following thermal stability trend for alkali metals: $Li^+ \approx Cs^+ < Rb^+ < K^+ < Na^+$. The authors proposed that the initial decomposition reaction is strongly influenced by the strength of the sulfonate counterion interaction. Lage et al. (2004a) also found that the thermal stability of perfluorosulfonated ionomers with different cations follows the trend: $H^+ < Li^+ < Cs^+ < Rb^+ < K^+ < Na^+$, which is consistent with Feldheim et al.'s results. The authors proposed that the charge densities of the countercations can affect their interaction with sulfonate groups and thereby the thermal stability of perfluorosulfonated ionomers. Bas et al. (2009) investigated the thermal properties of perfluorosulfonated ionomer membranes contaminated by a series of 10 counterions and found that cations with Lewis acid strength (LAS) values lower than the Lewis basic strength $LBS_{-SO_3^-}$ of the sulfonic anion in the PFSI improved the thermal stability of the ionomer side chains, while cations with LAS values higher than the $LBS_{-SO_3^-}$ enhanced the thermal degradation rate.

4.3.3.2 Thermal Degradation of PFSI Membranes in PEMFCs

Miyake et al. (2004, 2006) believed that their OCV test results could not be reasonably explained just by a chemical degradation mechanism and therefore proposed a thermal decomposition mechanism: crossover hydrogen and oxygen react on the platinum catalyst to produce combustion heat, which can cause thermal decomposition or oxidation of the membrane. However small the combustion heat, it may still lead to gradual microscopic damage of the membrane. Inaba et al. (2006) agreed that gas crossover and catalytic combustion at the electrodes are key factors for membrane degradation: direct combustion of hydrogen and oxygen generates heat and results in thermal degradation of the membrane, and H_2O_2 formed via direct catalytic combustion can cause membrane chemical degradation.

Hydrogen starvation may cause cell reversal and result in oxygen production rather than hydrogen oxidation at the anode, through the electrolysis of water (Knights et al., 2004). Air starvation can make the cathode reduce protons to hydrogen (Taniguchi et al., 2008). It is possible for local heat to be generated due to these reactions and the subsequent direct catalytic combustion, which would lead to thermal degradation of the membrane.

Severe thermal degradation of perfluorosulfonic membranes may cause decomposition of the main chains of the molecules. It should be noted that although slight thermal degradation at low temperatures such as those in an operating PEMFC cannot decompose the PTFE-like main chains of PFSIs, it may result in decomposition of the sulfonic-end side groups. After FTIR study of Nafion membranes thermally treated at 95°C, Alentiev et al. (2006) found new bands at 1440 cm^{-1} that could be attributed to the vibrations of ether $(-S(O)_2-O-)$ —groups that are absent in "native" Nafion, indicating thermochemical dehydration of sulfonic groups at the end of the pendant groups of PFSIs. The same result was obtained for a Nafion film that had been stored for 14 years, which indicates that sulfo-ethers and cross-linking can appear at low temperatures, especially in relatively low humidity, which should reduce the concentration of sulfonic groups and decrease the proton conductivity of ionomer membranes during exploitation processes, such as in fuel cells.

4.3.3.3 Mitigation Strategies for Thermal Degradation

Mitov et al. (2006) prepared membranes based on sulfonated arylene ionomers and their crosslinked blends, and found that embedding the ionomers into textile or porous PTFE matrix and/or doping with ZrP nanoparticles improved the thermal stability, in particular increasing the onset of the SO_3H group splitting-off temperature. Although theirs were not perfluorosulfonic polymers, this is a promising mitigation method to relieve thermal degradation of perfluorosulfonic membranes.

Cation-exchanged perfluorosulfonated ionomers are more thermally stable than acid-type ionomers; in addition, cation-exchanged perfluorosulfonated ionomers will not dehydrate to form sulfo-ethers at low temperatures. An apparently good method is to use alkali metal-type perfluorosulfonated ionomer membranes to alleviate thermal degradation in fuel cells. Unfortunately, the membrane's proton conductivity will decrease with increasing concentration of alkali metal cation. But this may be a good way to change membranes to salt type for long-term storage.

Developing ionomer membranes that are more thermostable is the most essential answer to improving their thermal durability. Endoh and his coworkers at Asahi Glass (Endoh, 2006) reported that their newly developed MEA, composed of a new perfluorinated polymer composite, could be operated for more than 4000 hours at 120°C and 50% RH; compared to a conventional MEA, their new MEA could reduce the degradation rate to 1/100–1/1000. However, the detailed chemical structure of their new membrane has so far not been disclosed.

Asahi Kasei also reported their newly developed Asahi Kasei advanced membrane, which has a 20–40°C higher glass transition temperature and stronger modulus than either their conventional S1002 membranes or their thermostable membranes (Miyake et al., 2006). The detailed chemical structure of the new membranes was not released.

4.3.4 Characterization of Membrane Degradation

4.3.4.1 Fenton Test

Fenton test is almost the most widely used method in polymer electrolyte membrane degradation studies because it saves time, is well controlled, and most importantly makes it easy to obtain an undamaged sample without interference from the Pt CL during analysis (Schumb et al., 1955; Guo et al., 1999;

Kinumoto et al., 2006; Danilczuk et al., 2007; Tang et al., 2007b; Wang et al., 2008; Chen and Fuller, 2009a). Experimentally, the membranes are put into Fenton's reagent and then examinations such as FRR and scanning electron microscopy (SEM) are conducted. Membrane tests in Fenton's reagent strongly show that perfluorosulfonic membranes degrade severely because of the radicals generated in the following major steps (called the Haber–Weiss mechanism) (Schumb et al., 1955; Healy et al., 2005; Kinumoto et al., 2006; Tang et al., 2007b):

$$H_2O_2 + Fe^{2+} \longrightarrow Fe^{3+} + HO + HO^- \tag{4.5}$$

$$Fe^{2+} + HO \longrightarrow Fe^{3+} + HO \tag{4.6}$$

$$H_2O_2 + HO \longrightarrow HO_2 + H_2O \tag{4.7}$$

$$Fe^{2+} + HO_2 \longrightarrow Fe^{3+} + HO_2^- \tag{4.8}$$

$$Fe^{3+} + HO_2 \longrightarrow Fe^{2+} + H^+ + O_2 \tag{4.9}$$

It must be noted that the degradation mechanism of PFSI membranes in Fenton's reagent is probably different from what happens in an actual fuel cell, because sulfonic acid groups in the membrane react with ferrous ions when soaked in Fenton's reagent, and the ferrous-substituted membrane may have a different degradation mechanism (Hommura et al., 2008). Delaney and coworkers at Gore suggested that chain scission dominates in vapor-phase H_2O_2 tests, while end-group unzipping dominates in the Fenton's test (Delaney and Liu, 2007).

4.3.4.2 Fluoride Release Rate

Although PFSI membranes have demonstrated highly efficient and stable performance in fuel cell applications, evidence of membrane thinning and detection of fluoride ions in the product water indicate that the polymer is undergoing chemical attack. The magnitude of the FRR from the anode and cathode of a PEMFC can be considered an excellent indicator of membrane degradation and a measure of the health and life expectancy of the PFSI membrane. The FRR is useful in assessing the chemical stability of a PFSA ionomer and can identify major problems in terms of a chemical attack (Baldwin et al., 1990; Curtin et al., 2004; Healy et al., 2005; Liu et al., 2006a; Ohma et al., 2007, 2008; Hommura et al., 2008; Ralph et al., 2008).

Water from the anode and cathode vents of an operating fuel cell can be collected and analyzed with ion chromatography (Aoki et al., 2005; Inaba et al., 2006; Wang et al., 2008), an ion-selective electrode (Pozio et al., 2003; Mittal et al., 2006a), or other instruments, from which the fluoride ion emission rate is then calculated using the concentration of fluoride ions and the volume of the water samples.

4.3.4.3 Open-Circuit Voltage

Although no net electrochemical reactions take place at the anode or cathode under open-circuit conditions in a single cell, the electrolyte membranes do deteriorate significantly because of gas crossover, catalytic combustion, and other factors, resulting in increased reactant crossover plus thermal and chemical degradation of the membrane (Inaba et al., 2006). Thus, OCV tests have been an easy, *in situ*, accelerated method for testing membrane durability. However, compared to measuring the hydrogen crossover rate, the OCV method is less selective and far less quantitative. The theoretical initial OCV of H_2 fuel cells under normal operating conditions is about 1.18 V at 80°C. Generally, the actual initial OCV is about 1 V, because instead of equilibrium potentials, the electrodes show mixed potentials of hydrogen oxidation and oxygen reduction, owing to hydrogen and oxygen crossover (Inaba et al., 2006). A stable membrane shows a nearly stable OCV, while under membrane degradation conditions, the OCV declines with increasing membrane failure, which follows a negative trend that coincides with the FRR and the hydrogen crossover rate (Tang et al., 2007b).

4.3.4.4 Gas Crossover Rates

Gas crossover properties play an important role in membrane durability. Hydrogen and oxygen can permeate across perfluorosulfonic membranes, and both chemical and mechanical degradation can lead to an increase in gas crossover rate. Yoshida et al. (1998) reported that the oxygen permeability coefficient of a perfluorosulfonic membrane is 2/5 of the hydrogen permeability coefficient at 70°C. Both hydrogen and oxygen permeation rates decrease in inverse proportion to increased membrane thickness; their gas permeability coefficients are nearly constant and are independent of thickness but increase with rising temperature.

H_2 crossover rate through an electrolyte membrane can be measured using an electrochemical method, chronocoulometry (Cleghorn et al., 2003). This is used *in situ* during fuel cell operation, usually to evaluate membrane degradation. In experiments, the cathode is purged with nitrogen instead of oxygen, and an electrochemical workstation is set with a voltage about 0.4 V higher than the anode. H_2 that diffuses to the cathode can be oxidized, and the resulting protons are transported back to the fuel cell anode where they are reduced to hydrogen. The measured current is proportional to the crossover rate of hydrogen through the membrane.

4.3.4.5 High-Frequency Resistance

High-frequency resistance can be a valuable tool to elucidate any change in the electrodes' ionic conductivity, which may provide evidence for changes in ionomer structure or ionomer degradation in the electrode layer (Cleghorn et al., 2006). Cell impedance consists mainly of the proton ionic resistances of both the membrane and the recast ionomer of the electrocatalyst layers. Two main causes of increased cell impedance over time are (1) membrane degradation due to loss of sulfonic groups, resulting in increased bulk resistance or increased resistance at the interfacial zone of the membrane and CLs, and (2) degradation of the CL recast ionomer, resulting in increased resistance at the interface between catalyst clusters and the ionomer network (Saab et al., 2002, 2003; Xie et al., 2005).

4.3.4.6 *Ex situ* Characterization

FTIR spectroscopy has been widely used to characterize functional groups in polymers and organic compounds. It has also been extensively applied to characterize the chemical structure of Nafion membranes (Laporta et al., 1999; Perusich, 2000; Gruger et al., 2001; Liang et al., 2004; Hensley and Way, 2007; Hofmann et al., 2008) and probe changes in membrane chemistry due to degradation (Kinumoto et al., 2006; Delaney and Liu, 2007; Chen and Fuller, 2009a).

In studies of Nafion membranes using FTIR, several important bands have been assigned, as listed in Table 4. 2.

FTIR spectroscopy is useful for studying the chemical degradation of polyelectrolyte membranes. Alentiev et al. (2006) found ether ($-S(O)_2-O-$) groups generated in degraded Nafion membranes that were absent in "native" Nafion. From FTIR measurements of deteriorated Nafion membrane, Kinumoto et al. (2006) proposed that both main and side chains are decomposed at similar rates by radical attack because decomposition proceeds through an unzipping mechanism.

ESR spectroscopy is a useful technique to develop accelerated aging tests for free radical degradation mechanisms in membranes. Among the various analytical methods, ESR is capable of directly detecting radicals. The extraordinary sensitivity of the method permits the detection of degradation long before a mass loss is perceptible (Vogel et al., 2008). Many studies of the chemical degradation mechanism in perfluorosulfonic membranes have used ESR spectroscopy, and several ESR spectroscopy studies of membrane degradation have provided important evidence of the existence of radicals, and of membrane

TABLE 4.2   Important Infrared Active Bands and Assignments of Nafion Ionomers

Wavenumber (cm^{-1})	Assignment	Reference
500–780	$-CF_2-$	Gruger et al. (2001)
805	v (C–S)	Gruger et al. (2001)
960–978	v_s (C–O–C)	Chen and Fuller (2009a), Gruger et al. (2001), Liang et al. (2004)
983–984	v_s (C–O–C)	Chen and Fuller (2009a), Gruger et al. (2001), Perusich (2000)
1056–1060	v_s ($-SO_3^-$)	Kinumoto et al. (2006), Gruger et al. (2001), Liang et al. (2004)
1135	v_{as} ($-SO_3^-$)	Gruger et al. (2001)
1143–1148	v_{as} ($-CF_2-$)	Kinumoto et al. (2006), Gruger et al. (2001), Liang et al. (2004)
~1200	v_{as} ($-CF_2-$)	Kinumoto et al. (2006), Gruger et al. (2001), Liang et al. (2004)
1300	v (C–C)	Kinumoto et al. (2006), Gruger et al. (2001)
1450,1460	S–O–S, or S = O	Chen et al. (2009a), Kinumoto et al. (2006), Gruger et al. (2001)
1640	water	Kinumoto et al. (2006), James et al. (2001)
1695	–COOK	Delaney and Liu (2007)
1785	–COOH	Perusich (2000), Hensley and Way (2007)
2900	–OH in –COOH	Chen et al. (2009a), Perusich (2000)

fragmentation by radical attack on the main chain and side chains (Bosnjakovic and Schlick, 2004; Kadirov et al., 2005; Mitov et al., 2006; Danilczuk et al., 2007, 2009; Lund et al., 2007; Zhou et al., 2007; Ghassemzadeh et al., 2009).

NMR spectroscopy is an important technique used to identify and quantify end groups in fluoropolymers and the species extracted from degraded PFSI membranes.

Fluorinated degradation compounds of perfluorosulfonic membranes can be detected, and structural features can be deduced from NMR and/or FTIR spectrum, while mass spectrometry can help to complete the chemical structure of the segments.

4.4 Ionomer Degradation Inside the CL

4.4.1 Mechanism

In PEMFCs, the CLs are typically comprised of carbon-supported platinum catalyst and PFSIs. These CLs usually contain 20–40 wt% PFSI. The ionomer in the CLs has three roles: (1) to act as a binder between the platinum/carbon particles, (2) to provide a proton-conductive link between the bulk membrane and platinum catalyst sites for protonic current flow, and (3) to make the platinum catalyst electrochemically active by transferring protons to and from it (Young et al., 2010).

It is proposed that membrane degradation occurs because molecular H_2 and O_2 react on the surface of the platinum catalyst to form the membrane-degrading species (Aoki et al., 2005; Mittal et al., 2007), which must have an important effect on the CL ionomer.

While a significant amount of work has focused on the degradation of perfluorosulfonic membranes in PEMFCs, little has been done on ionomer degradation in CLs. Aoki et al. (2005) developed a new method to measure degradation of Nafion ionomer in CLs. They found that the ionomer showed signs of degradation in simulated PEMFC cathode and anode gas environments when platinum, hydrogen and oxygen coexisted. Greater degradation occurred under anode conditions if the partial pressure of the crossover gas was identical. Based on more experimental results after this report, the same authors proposed the decomposition mechanism of PFSI electrolyte induced by crossover of reactant gases in PEMFCs (Aoki et al., 2006). First, crossover O_2 reacts with H_2 on the surface of platinum catalyst particles in the vicinity of the interface between the CL and the membrane. In addition to the main product, H_2O, H_2O_2 is formed as a by-product. Some fraction of the H_2O_2 formed should evaporate into the gas phase, while the remaining H_2O_2 can diffuse into both the membrane and the CL, where the formation of hydroxyl radicals is enhanced by impurities such as Fe^{2+}. The hydroxyl radicals thus generated attack the PFSI to release F^-. In their experiments, the normal perfluorosulfonic membrane was decomposed by hydroxyl radicals much more easily than the ionomer coated on the catalysts. The authors proposed that many platinum catalysts can scavenge H_2O_2 and/or hydroxyl radicals, and the PFSI in the CL decomposed more slowly than the membrane in the actual cell.

Young et al. (2010) investigated the effect on PEMFC performance of using 23 and 33 wt% Nafion cathode CL designs. The authors found that the PFSI concentration in the CL had a great effect on both membrane and CL ionomer degradation. In their accelerated stress test up to 440 h, the catalyst ionomer resistance showed an overall 38% increase in the 23 wt% Nafion MEA. The overall CL ionomer resistance could not be measured accurately for the 33 wt% Nafion MEA, due to excessive hydrogen crossover, but the authors believed that it must have had a greater increase in resistance. They speculated that the initial CL ionomer fluoride loss increased the hydrophilicity and water content of the CL, which lowered the ionic resistance. Higher ionomer content in the cathode CL also provided mass-transfer pathways for contaminants such as dissolved platinum and possibly iron to diffuse into the membrane. The authors hypothesized that H_2O_2 was produced at the anode, diffused into the membrane, and then decomposed at the platinum and possibly the iron sites bound in the membrane structure. The decomposition products were transferred into the CL, where they attacked the ionomer. The majority of the PFSI degradation thus originated from the membrane rather than from the CL ionomer.

Although commercial perfluorosulfonic membranes such as Nafion 112 are relatively stable and difficult to decompose in water under conditions of ambient pressure and 100°C, Xie et al. (2005) proposed that dissolution of recast ionomer in the CL may occur in an operating fuel cell because the recast PFSI network of the CL is not as stable as its membrane counterpart. In their laboratory at Los Alamos National Laboratory they confirmed that dissolution of recast Nafion ionomer happened during a DMFC life test, by recasting an ionomer film directly from the cathode product water.

4.4.2 Mitigation Strategies

The ionomer in the CL is decomposed by radicals that attack the bulk membranes, and so methods to improve membrane stability also help to increase the durability of the CL ionomer, such as decreasing the metal ion contaminant content in fuel cells, and similar approaches. According to Young et al. (2010), suitably low ionomer concentration in the CL is good for membrane durability as well as for CL ionomer durability.

4.5 Localized Degradation

4.5.1 Platinum Precipitation

Although platinum is very stable even in high-oxidative conditions, it has been found that platinum in the CL can be dissolved into Pt^{2+}, which then diffuse into the ionomer membranes and subsequently are reduced by crossover hydrogen to form a platinum band (Ferreira et al., 2005; Ohma et al., 2008). The location of the platinum band varies with the H_2/O_2 partial pressures and is determined by the relative local flux of H_2 and O_2 within the membrane (Ferreira et al., 2005; Zhang et al., 2007). Obviously, platinum precipitation in the membranes decreases the platinum weight in the CL. In addition, some researchers have found that the platinum band in the membrane can enhance membrane degradation. On the basis of TEM and micro-Raman spectroscopy measurement results, Ohma et al. found that side chains of the PFSI in particular were decomposed around the platinum band. The magnitude of the FRR was also consistent with the location of the platinum band in the membrane during OCV testing. Their results indicate that the platinum band strongly affects membrane degradation. But, by measuring the COOH groups in membranes, Endoh et al. (2007) found that the degraded portion of the membrane was completely different from the platinum band position, which indicates that the platinum band in the membrane will not contribute to membrane degradation. The authors proposed that the surface of platinum particles in the membrane might be substantially clean, because they found that pretreated platinum black did not degrade the membranes in their OCV tests. Perhaps the morphology of the precipitated platinum in the membranes varies under different operating conditions, which would explain why their results differed from those of Ohma et al.

4.5.2 Inlets and Edges

In an operating fuel cell, the RH and gas partial pressure vary between inlet and outlet, and potential distribution exists in the vicinity of the electrode perimeter, all of which may cause local membrane degradation.

Several research groups have reported that the gas inlet area of the fuel cell is more vulnerable to membrane degradation compared to the bulk membrane (Knights et al., 2004; Yu et al., 2005; Laconti et al., 2006). All hydrogen is fed into the fuel cell via the inlet, and so hydrogen concentration in the fuel cell is highest in the inlet region and decreases gradually from inlet to outlet, which makes the hydrogen crossover through the membrane from anode to cathode higher in the inlet region than that in the outlet region. Considering that hydrogen can react with PFSI backbones thus: $-CF_2- + 2H_2 \rightarrow -CH_2- + 2HF$, which was proposed by Wilkinson and St-Pierre (2003), and the resulting $-CH_2-$ groups are very vulnerable to radical attack, Yu et al. (2005) suggested that membrane thinning in the fuel cell in hydrogen inlet region contributes to higher hydrogen concentration at localized positions in the membranes. Near the oxygen inlet, high oxygen concentration can result in high catalytic activity, and if the inlet gas streams are insufficiently humidified, the inlet region will not have enough water for heat-transfer removal, which might result in a local high-temperature region prone to membrane failure (Knights et al., 2004). Sompalli et al. (2007) developed segmented and unsegmented electrodes to examine the impact of electrode overlap on membrane failure. They found that membranes only failed in perimeter regions having just one electrode (either cathode or anode). At high RH, the cathode overlap configuration failed quickly while the anode overlap configuration remained stable. But at low RH, membrane chemical failures in the cathode overlap region accelerated slightly, while failures in the anode overlap region accelerated dramatically.

Huang et al. (2006) proposed that membrane mechanical failure can be a result of reduction in membrane ductility combined with a mechanical strain excursion caused by constrained drying in the fuel cell. RH-induced strain is not uniform in membranes. Their model prediction indicates that strain around the edge is higher than in the center region, which means that the membrane edge is more vulnerable to mechanical degradation.

4.6 Effects of Operating Conditions and other Parameters on Membrane Degradation

4.6.1 Cell Temperature

The advantages of high temperature during fuel cell operation include enhanced electrode reactions, simplified water management and cooling systems, more efficient cogeneration of heat and electricity, and reduced contamination problems (Li et al., 2003; Zhang et al., 2006; Liu et al., 2009). But it has been found that higher temperature will lead to more severe chemical degradation (Curtin et al., 2004; Healy et al., 2005). The FRR doubles after a temperature increase of 10°C for the Fenton's test (Chen et al., 2007).

Temperature effects on membrane degradation were found by Ramaswamy et al. (2008) to be linear with higher oxygen reduction reaction (ORR) activity and consequent higher H_2O_2 generation at the interface. Chen and Fuller (2009b) measured FRRs from fuel cell durability tests at different temperatures, and found that FRRs increased with increasing fuel cell temperature, indicating that high temperature increases membrane degradation due to accelerated reaction kinetics. In another paper, Chen et al. (2007) treated Nafion membranes in Fenton's reagent and also found membrane degradation rates increased with increasing temperature.

4.6.2 Catalyst

At present, platinum is the most widely used catalyst for PEMFCs owing to its excellent oxygen reduction activity. It is generally used in the form of Pt/C catalysts because of their high active surface area. Platinum alloy catalysts are also employed to reduce cost as well as enhance oxygen reduction activity. H_2, O_2, and a catalytic surface are required to generate H_2O_2 and/or radicals (Liu et al., 2006a) during membrane chemical degradation processes. Catalyst loading, particle size, and catalyst type can all affect membrane degradation. Ramaswamy et al. (2008), in experiments comparing 60% Pt/C and 30% Pt/C, reported that higher catalyst loading on the carbon support resulted in higher H_2O_2 yield and consequently higher membrane degradation, and found that Pt–Co/C alloy catalyst enriched by surface Co yielded a lower H_2O_2 current and maintained a lower level of membrane degradation. Mittal et al. (2006a) also found that Pt–Co/C catalyst led to a 50% reduction in the fuel cell FRR at 30% RH compared to Pt/C catalyst, but that the FRRs were comparable at 100% RH. Carbon alone has a higher H_2O_2 yield than a Pt/C catalyst, because the H_2O_2 produced on carbon can be reduced, or destroyed by dismutation into H_2O on platinum particles (Antoine and Durand, 2000). An RRDE study conducted by Antoine and Durand (2000) showed a weak platinum particle size effect on H_2O_2 production during the ORR; the H_2O_2 proportion increased slightly when the platinum particle size decreased. The nature of the platinum surface (alloyed or unalloyed) can also determine the ORR (Murthi et al., 2004). Mittal et al. (2006a) found that straight Pt-black catalyst could cause faster membrane degradation in fuel cells under 100% RH but that the ill effect was relevant to RH, as at 30% RH the FRR from cells with Pt-black and Pt/C catalysts dropped to the same level.

During fuel cell operation, the catalysts are exposed to extremely corrosive conditions, which may cause platinum dissolution–redeposition on the catalyst surface and platinum migration through the surface (Watanabe et al., 1994; Ferreira et al., 2005; Colón-Mercado and Popov, 2006). Additionally, the dissolved platinum species can diffuse into the ionomer membranes and form large platinum crystal particles of about 10–100 nm via reduction by crossover hydrogen (Akita et al., 2006), which severely decreases membrane stability and proton conductivity (Ohma et al., 2008). Furthermore, platinum cations originating from platinum corrosion can also induce crosslinking of sulfonic groups in the membrane, resulting in membrane stiffening (Iojoiu et al., 2007).

4.6.3 Relative Humidity

Controlling the activity of water in the reactant streams is critical to the durability of ionomer membrane separators in PEMFCs.

Generally, lower humidity leads to higher membrane degradation (Sompalli et al., 2007) and a higher FRR (Liu et al., 2006b), although gas permeation through a perfluorosulfonic membrane decreases with decreasing gas humidification (Inaba et al., 2006; Inaba, 2009). In their research work, Chen and Fuller (2009a) also found that RH greatly affected the main-chain unzipping process and side-chain scission process of the PFSI molecules. The main-chain unzipping process was dominant under high humidity, whereas a decrease in humidity enhanced the side-chain scission process, creating a large number of weak end groups and accelerating degradation. Although the ion exchange capacity and conductivity of the membranes both decreased in the fuel cells under high and low RH, after treatment with sulfuric acid the ion exchange capacity and conductivity of membrane samples tested under conditions of higher humidity recovered to near that of the pristine membranes, which indicated that the decreased conductivity was mainly caused by cation contamination. Ion exchange capacity measurements showed an unrecoverable 8.3% decrease in capacity, and membrane conductivity showed a 10% drop for a sample tested at low humidity due to the loss of sulfonic acid groups in the side-chain scission process, even after sulfuric acid treatment of the membranes. Simultaneously, it was found that the maximum force, ultimate tensile strength, and breaking strength decreased with humidity, showing the strong relationship between chemical and mechanical degradation.

Yu et al. (2005) found that inadequate water content can accelerate membrane physical degradation and ultimately result in membrane holes and reactant gas crossover. Knights and coworkers (2004) observed this effect through a series of tests in which the same type of nonoptimized cell design was run under varying humidification conditions. As conditions became drier, fuel cell failure due to significant reactant gas crossover occurred after shorter lifetimes. Huang et al. (2003) used XRD to investigate the effect of gas humidification temperature on degradation of Nafion polymer electrolyte. They found that the intensity of the characteristic XRD peak of Nafion membrane ($2\theta = 16.5°$) decays steeply with decreasing H_2/O_2 humidification temperatures, falling to even lower than that of the fresh sample, which indicated that insufficient water in the membrane region during fuel cell operation can destroy the membrane crystalline structure and hence lower the fuel cell efficiency. X-ray photoelectron spectroscopy analysis of these membranes showed that membrane degradation resulted from the interaction between fluorine and carbon as well as hydrogen, yielding the $-HCF-$ or $-CCF-$ configurations.

For SSC ionomers such as Hyflon ion membranes, Merlo et al. (2007) arrived at the same conclusion as for LSC ionomer membranes such as Nafion. In the case of H_2/O_2 PEMFCs the degradation rate is largely influenced by the reactant humidification level, increasing when the humidification level is low.

One proposed mechanism to explain the finding that low RH enhances membrane degradation is that the concentration of H_2O_2 is higher under low RH conditions. The boiling temperature of H_2O_2 is 150°C, which is higher than that of water, and hence they accumulate with ferrous ions in the membrane under low humidification and result in high membrane degradation (Inaba, 2009). *Ex situ* model experiments with trifluoromethane sulfonic acid and Pt/C catalyst conducted by Zhang and Mukerjee (2006) revealed that when the H_2O / SO_3^- ratio is lowered the H_2O_2 yield increases. So, it is possible that the H_2O_2 concentrates inside the membrane at low RH. Considering the results from Hommura et al. (2008) that PFSI membranes can be decomposed by H_2O_2 alone, the higher H_2O_2 concentration mechanism seems reasonable. But some researchers have proposed that H_2O_2 alone would account at best for a small fraction of the PFSI membrane degradation rate in a fuel cell (Mittal et al., 2007; Vogel et al., 2008). Considering that the hydroxyl radical is the only oxidant capable of abstracting a hydrogen atom from a carboxylic acid intermediate and thereby propagating the perfluorosulfonic main-chain unzipping process in the highly oxidizing environment of an operating PEMFC (Coms, 2008), that metal ions such as Fe^{2+} and Cu^{2+} greatly enhance the membrane decomposition rate (Kadirov et al., 2005; Kinumoto et al., 2006; Liu et al., 2006a), and that the hydroxyl radical can be stabilized by water (Autrey et al., 2004), it may be that rapid membrane degradation at low RH results from the synergistic effect of increased hydroxyl radical production due to the concentration of H_2O_2 and contaminant metal ions and the loss of water stabilization for hydroxyl radicals.

4.6.4 Gas Pressure

Gas crossover is believed to be a fundamental mechanism leading to chemical degradation of perfluorosulfonic membranes in PEMFCs (Liu et al., 2006a,b). The gas crossover rate is directly related to the partial pressure of the reactant gases. Hydrogen crossover drastically increases with increasing gas pressure. In Inaba et al.'s experiment (2006), the hydrogen crossover current density at 0.2 MPa was 5 times higher than at atmospheric pressure.

Gas partial pressure has an important effect on membrane chemical degradation in an operating fuel cell. Liu et al. (2006b) found that in an operating fuel cell the FRR showed a strong dependence on H_2 pressure in the high partial pressure region, from 20 to 200 kPa_{abs}. The FRR decay with H_2 pressure was close to second-order behavior, while there was no significant change in FRR when the oxygen partial pressure was lowered from 200 to 40 kPa, which indicated that using pure O_2 under these accelerated conditions will not lead to higher chemical degradation rates than using air. The authors' results indicate that fuel cell systems employing high-pressure pure H_2 will sustain more chemical degradation compared to systems with low pressure, appropriately diluted H_2 without fuel starvation. When changing the partial pressure of H_2 by a factor of 5 by mixing 20% H_2 with 80% N_2, Mittal et al. (2006b) found that the FRR fivefold.

4.6.5 Membrane Thickness

Since it is believed that membrane degradation occurs because molecular H_2 and O_2 react to form membrane-degrading species such as H_2O_2 or reactive radicals, and because membrane thickness is related to reactant gas crossover, some groups have investigated the effect of membrane thickness on membrane degradation.

Liu et al. (2005) concluded that H_2O_2 concentration appeared to depend primarily on membrane thickness: the thinner the membrane, the higher the H_2O_2 concentration due to greater H_2 or O_2 crossover. However, a greater H_2 crossover also creates a lower cathode mixed potential, reducing the OCV. Therefore, membrane thickness impacts H_2O_2 production and decomposition (Young et al., 2010). Chen and Fuller (2009b) also found that the rate of membrane degradation slows in fuel cells with thicker membranes.

Mittal et al. (2007) evaluated the membrane degradation rate as a function of membrane thickness at a fixed inlet reactant concentration for an operating fuel cell. They found that the FRRs from cells with Nafion membranes 50, 150, 350, and 525 µm thick were 2.02, 2.96, 2.32, and 1.67 µmol/h cm², respectively, indicating that the membrane degradation rate increased with increasing membrane thickness from 50 to 150 µm, then decreased from 150 to 525 µm. The authors speculated that the concentration of certain species responsible for membrane degradation may not be the only rate-limiting factor in the overall degradation reaction, and an increase in membrane thickness or in the reaction volume leads to an increase in FRR when membrane thickness increases from 50 to 150 µm. After a certain thickness, the degradation rate is limited by the concentration of reactive species due to the decrease in H_2 and O_2 crossover rates, which causes a decrease in FRR as the membrane thickness is further increased from 150 to 525 µm.

4.6.6 Feed Starvation

Hydrogen starvation can cause cell reversal, which can make the anode potential rise to what is required to oxidize water, resulting in oxygen production through the electrolysis of water instead of hydrogen oxidation at the anode (Knights et al., 2004). It is possible that oxygen thus generated might cause local heat generation and result in membrane breakthrough (Taniguchi et al., 2004).

For cell reversal during air starvation, the cathode is forced to reduce protons to hydrogen. Taniguchi et al. (2008) the possibility that the presence of hydrogen in the air electrode could chemically generate heat on the platinum particles and result in local hot spots in the MEA, which would lead to membrane failure.

4.6.7 Contaminant Ions

It is believed that metal ions can catalyze the formation of hydroxyl and peroxyl radicals from H_2O_2, and these radicals are thought to be at the origin of membrane degradation (Stucki et al., 1998; Pozio et al., 2003; Panchenko et al., 2004; Healy et al., 2005; Kinumoto et al., 2006; Xie and Hayden, 2007). Kinumoto et al. (2006) investigated the effects of impurity cations on Nafion membrane degradation in H_2O_2 and found that the presence of Fe^{2+} and Cu^{2+} greatly enhanced the membrane decomposition rate, while alkali and alkaline earth metal ions did not have any specific catalytic effect on the decomposition of H_2O_2 and thus on the degradation of perfluorosulfonic membranes.

After observation of membrane degradation using ESR, Kadirov et al. (2005) proposed that the presence of Fe(II), Fe(III), or Cu(II) ions in combination with H_2O_2 and UV irradiation doubtlessly leads to extensive radical formation on PFSI membranes. With concomitant UV irradiation, these ions may play a crucial role in radical formation and membrane degradation.

4.7 Summary

Great efforts have been made by many groups to investigate the degradation mechanisms for electrolyte membranes, in an attempt to enhance fuel cell durability. However, a great deal of further investigation is still needed. With respect to membrane durability, major achievements have been made by stabilizing the ionomer to improve its chemical stability, and by using reinforcement methods to enhance its mechanical stability. The answer lies in developing new ionomer materials and corresponding membranes that will possess higher thermal stability, better mechanical properties to combat gas crossover, and greater chemical stability against reactive radicals. It is also important to develop new catalysts that produce less harmful species.

References

Akita, T., Taniguchi, A., Maekawa, J. et al. 2006. Analytical TEM study of Pt particle deposition in the proton-exchange membrane of a membrane-electrode-assembly. *Journal of Power Sources* 159: 461–467.

Alentiev, A., Kostina, J., and Bondarenko, G. 2006. Chemical aging of Nafion: FTIR study. *Desalination* 200: 32–33.

Almeida, S. H. and Kawano, Y. 1998. Effects of X-ray radiation on Nafion membrane. *Polymer Degradation and Stability* 62: 291–297.

Antoine, O. Bultel, Y., and Durand, R. 2001. Oxygen reduction reaction kinetics and mechanism on platinum nanoparticles inside Nafion. *Journal of Electroanalytical Chemistry* 499: 85–94.

Antoine, O. and Durand, R. 2000. RRDE study of oxygen reduction on Pt nanoparticles inside Nafion®: H_2O_2 production in PEMFC cathode conditions. *Journal of Applied Electrochemistry* 30: 839–844.

Antoine, O. and Durand, R. 2001. *In situ* electrochemical deposition of Pt nanoparticles on carbon and inside Nafion. *Electrochemical and Solid State Letters* 4: A55–A58.

Aoki, M., Uchida, H., and Watanabe, M. 2005. Novel evaluation method for degradation rate of polymer electrolytes in fuel cells. *Electrochemistry Communications* 7: 1434–1438.

Aoki, M., Uchida, H., and Watanabe, M. 2006. Decomposition mechanism of perfluorosulfonic acid electrolyte in polymer electrolyte fuel cells. *Electrochemistry Communications* 8: 1509–1513.

Arcella, V., Troglia, C., and Ghielmi, A. 2005. Hyflon ion membranes for fuel cells. *Industrial & Engineering Chemistry Research* 44: 7646–7651.

Autrey, T., Brown, A. K., Camaioni, D. M., Dupuis, M., Foster, N. S., and Getty, A. 2004. Thermochemistry of aqueous hydroxyl radical from advances in photoacoustic calorimetry and ab initio continuum solvation theory. *Journal of the American Chemical Society* 126: 3680–3681.

Baldwin, R., Pham, M., Leonida, A., McElroy, J., and Nalette, T. 1990. Hydrogen-oxygen proton-exchange membrane fuel cells and electrolyzers. *Journal of Power Sources* 29: 399–412.

Bas, C., Reymond, L., Danerol, A.-S., Alberola, N. D., Rossinot, E., and Flandin, L. 2009. Key counter ion parameters governing polluted Nafion membrane properties. *Journal of Polymer Science Part B: Polymer Physics* 47: 1381–1392.

Bauer, F., Denneler, S., and Willert-Porada, M. 2005. Influence of temperature and humidity on the mechanical properties of Nafion 117 polymer electrolyte membrane. *Journal of Polymer Science Part B: Polymer Physics* 43: 786–795.

Bosnjakovic, A. and Schlick, S. 2004. Nafion perfluorinated membranes treated in Fenton media: Radical species detected by ESR spectroscopy. *The Journal of Physical Chemistry B* 108: 4332–4337.

Bro, M. I. and Sperati, C. A. 1959. Endgroups in tetrafluoroethylene polymers. *Journal of Polymer Science* 38: 289–295.

Burlatsky, S., Cipollini, N., Condit, D., Madden, T., and Atrazhev, V. 2006. Multi-scale modeling considerations for PEM fuel cell durability. *ECS Meeting Abstracts* 502: 1189.

Casciola, M., Alberti, G., Sganappa, M., and Narducci, R. 2006. On the decay of Nafion proton conductivity at high temperature and relative humidity. *Journal of Power Sources* 162: 141–145.

Chen, C. and Fuller, T. F. 2009a. The effect of humidity on the degradation of Nafion membrane. *Polymer Degradation and Stability* 94: 1436–1447.

Chen, C. and Fuller, T. F. 2009b. XPS analysis of polymer membrane degradation in PEMFCs. *Journal of The Electrochemical Society* 156: B1218–B1224.

Chen, C., Levitin, G., Hess, D. W., and Fuller, T. F. 2007. XPS investigation of Nafion® membrane degradation. *Journal of Power Sources* 169: 288–295.

Chlistunoff, J. and Pivovar, B. 2007. Hydrogen peroxide generation at the Pt/Recast-Nafion film interface at different temperatures and relative humidities. *ECS Transactions* 11: 1115–1125.

Cirkel, P. A. and Okada, T. 2000. A comparison of mechanical and electrical percolation during the gelling of Nafion solutions. *Macromolecules* 33: 4921–4925.

Cirkel, P. A., Okada, T., and Kinugasa, S. 1999. Equilibrium aggregation in perfluorinated ionomer solutions. *Macromolecules* 32: 531–533.

Cleghorn, S., Kolde, J., and Liu, W. 2003. Catalyst-coated composite membranes. In *Handbook of Fuel Cells: Fundamentals, Technology and Applications*, eds. W. Vielstich, A. Lamm and H. A. Gasteige, pp. 566–565. Chichester: JohnWiley & Sons Ltd.

Cleghorn, S. J. C., Mayfield, D. K., Moore, D. A. et al. 2006. A polymer electrolyte fuel cell life test: 3 years of continuous operation. *Journal of Power Sources* 158: 446–454.

Collette, F. M., Lorentz, C., Gebel, G., and Thominette, F. 2009. Hygrothermal aging of Nafion®. *Journal of Membrane Science* 330: 21–29.

Collier, A., Wang, H., Zi Yuan, X., Zhang, J., and Wilkinson, D. P. 2006. Degradation of polymer electrolyte membranes. *International Journal of Hydrogen Energy* 31: 1838–1854.

Colón-Mercado, H. R., and Popov, B. N. 2006. Stability of platinum based alloy cathode catalysts in PEM fuel cells. *Journal of Power Sources* 155: 253–263.

Coms, F. D. 2008. The chemistry of fuel cell membrane chemical degradation. *ECS Transactions* 16: 235–255.

Connolly, D. J. and Gresham, W. F. 1966. *Fluorocarbon vinyl ether polymers*, US 3,282,875.

Curtin, D. E., Lousenberg, R. D., Henry, T. J., Tangeman, P. C., and Tisack, M. E. 2004. Advanced materials for improved PEMFC performance and life. *Journal of Power Sources* 131: 41–48.

Danilczuk, M., Bosnjakovic, A., Kadirov, M. K., and Schlick, S. 2007. Direct ESR and spin trapping methods for the detection and identification of radical fragments in Nafion membranes and model compounds exposed to oxygen radicals. *Journal of Power Sources* 172: 78–82.

Danilczuk, M., Schlick, S., and Coms, F. D. 2009. Cerium(III) as a stabilizer of perfluorinated membranes used in fuel cells: *In situ* detection of early events in the ESR resonator. *Macromolecules* 42: 8943–8949.

de Almeida, S. H. and Kawano, Y. 1999. Thermal behavior of Nafion membranes. *Journal of Thermal Analysis and Calorimetry* 58: 569–577.

Delaney, W. E. and Liu, W. 2007. Use of FTIR to analyze *ex-situ* and *in-situ* degradation of perfluorinated fuel cell ionomers. *ECS Transactions* 11: 1093–1104.

Endoh, E. 2006. Highly durable MEA for PEMFC under high temperature and low humidity conditions. *ECS Transactions* 3: 9–18.

Endoh, E., Hommura, S., Terazono, S., Widjaja, H., and Anzai, J. 2007. Degradation mechanism of the PFSA membrane and influence of deposited Pt in the membrane. *ECS Transactions* 11: 1083–1091.

Endoh, E., Terazono, S., Widjaja, H., and Takimoto, Y. 2004. Degradation study of MEA for PEMFCs under low humidity conditions. *Electrochemical and Solid-State Letters* 7: A209–A211.

Feldheim, D. L., Lawson, D. R., and Martin, C. R. 1993. Influence of the sulfonate countercation on the thermal stability of Nafion perfluorosulfonate membranes. *Journal of Polymer Science Part B: Polymer Physics* 31: 953–957.

Ferreira, P. J., la O', G. J., Shao-Horn, Y. et al. 2005. Instability of Pt/C electrocatalysts in proton exchange membrane fuel cells. *Journal of the Electrochemical Society* 152: A2256–A2271.

Gebel, G. and Loppinet, B. 1996. Colloidal structure of ionomer solutions in polar solvents. *Journal of Molecular Structure* 383: 43–49.

Ghassemzadeh, L., Marrony, M., Barrera, R., Kreuer, K. D., Maier, J., and Müller, K. 2009. Chemical degradation of proton conducting perfluorosulfonic acid ionomer membranes studied by solid-state nuclear magnetic resonance spectroscopy. *Journal of Power Sources* 186: 334–338.

Grot, W. 2007. *Fluorinated Ionomers*, 9, Norwich: William Andrew Publishing.

Gruger, A., Régis, A., Schmatko, T., and Colomban, P. 2001. Nanostructure of Nafion® membranes at different states of hydration: An IR and Raman study. *Vibrational Spectroscopy* 26: 215–225.

Guo, Q., Pintauro, P. N., Tang, H., and O'Connor, S. 1999. Sulfonated and crosslinked polyphosphazene-based proton-exchange membranes. *Journal of Membrane Science* 154: 175–181.

Healy, J., Hayden, C., Xie, T. et al. 2005. Aspects of the chemical degradation of PFSA ionomers used in PEM fuel cells. *Fuel Cells* 5: 302–308.

Hensley, J. E. and Way, J. D. 2007. Synthesis and characterization of perfluorinated carboxylate/sulfonate ionomer membranes for separation and solid electrolyte applications. *Chemistry of Materials* 19: 4576–45784.

Hofmann, D. W. M., Kuleshova, L., D'Aguanno, B. et al. 2008. Investigation of water structure in Nafion membranes by infrared spectroscopy and molecular dynamics simulation. *The Journal of Physical Chemistry B* 113: 632–639.

Hommura, S., Kawahara, K., Shimohira, T., and Teraoka, Y. 2008. Development of a method for clarifying the perfluorosulfonated membrane degradation mechanism in a fuel cell environment. *Journal of the Electrochemical Society* 155: A29–A33.

http://www2.dupont.com/FuelCells/en_US/assets/downloads/dfc101.pdf [Online]. [Accessed].

http://www2.dupont.com/FuelCells/en_US/assets/downloads/dfc201.pdf [Online]. [Accessed].

Huang, C., Tan, K. S., Lin, J., and Tan, K. L. 2003. XRD and XPS analysis of the degradation of the polymer electrolyte in H_2–O_2 fuel cell. *Chemical Physics Letters* 371: 80–85.

Huang, X., Solasi, R., Zou, Y. et al. 2006. Mechanical endurance of polymer electrolyte membrane and PEM fuel cell durability. *Journal of Polymer Science Part B: Polymer Physics* 44: 2346–2357.

Imbalzano, J. F. and Kerbow, D. L. 1988. *Stable tetrafluoroethylene copolymers* US4743658.

Inaba, M. 2004. Degradation mechanism of polymer electrolyte fuel cells. In *14th International Conference on the Properties of Water and Steam*, Kyoto, pp. 395–402.

Inaba, M. 2009. Chemical degradation of perfluorinated sulfonic acid membranes. In *Polymer Electrolyte Fuel Cell Durability*, eds. F. N. Büchi, M. Inaba and T. J. Schmidt, pp. 57–69. New York, NY: Springer.

Inaba, M., Kinumoto, T., Kiriake, M., Umebayashi, R., Tasaka, A., and Ogumi, Z. 2006. Gas crossover and membrane degradation in polymer electrolyte fuel cells. *Electrochimica Acta* 51: 5746–5753.

Inaba, M., Yamada, H., Tokunaga, J., and Tasaka, A. 2004. Effect of agglomeration of Pt/C catalyst on hydrogen peroxide formation. *Electrochemical and Solid-State Letters* 7: A474–A476.

Iojoiu, C., Guilminot, E., Maillard, F. et al. 2007. Membrane and active layer degradation following PEMFC steady-state operation. *Journal of the Electrochemical Society* 154: B1115–B1120.

James, P. J., Antognozzi, M., Tamayo, J., McMaster, T. J., Newton, J. M., and Miles, M. J. 2001. Interpretation of contrast in tapping mode AFM and shear force microscopy. A study of Nafion. *Langmuir* 17: 349–360.

Jiang, S. H., Xia, K. Q., and Xu, G. 2001. Effect of additives on self-assembling behavior of Nafion in aqueous media. *Macromolecules* 34: 7783–7788.

Kabasawa, A., Saito, J., Miyatake, K., Uchida, H., and Watanabe, M. 2009. Effects of the decomposition products of sulfonated polyimide and Nafion membranes on the degradation and recovery of electrode performance in PEFCs. *Electrochimica Acta* 54: 2754–2760.

Kadirov, M. K., Bosnjakovic, A., and Schlick, S. 2005. Membrane-derived fluorinated radicals detected by electron spin resonance in UV-irradiated Nafion and Dow ionomers: Effect of counterions and H_2O_2. *The Journal of Physical Chemistry B* 109: 7664–7670.

Kelly, M. J., Egger, B., Fafilek, G., Besenhard, J. O., Kronberger, H., and Nauer, G. E. 2005a. Conductivity of polymer electrolyte membranes by impedance spectroscopy with microelectrodes. *Solid State Ionics* 176: 2111–2114.

Kelly, M. J., Fafilek, G., Besenhard, J. O., Kronberger, H., and Nauer, G. E. 2005b. Contaminant absorption and conductivity in polymer electrolyte membranes. *Journal of Power Sources* 145: 249–252.

Kinoshita, K. 1990. Particle size effects for oxygen reduction on highly dispersed platinum in acid electrolytes. *Journal of The Electrochemical Society* 137: 845–848.

Kinumoto, T., Inaba, M., Nakayama, Y. et al. 2006. Durability of perfluorinated ionomer membrane against hydrogen peroxide. *Journal of Power Sources* 158: 1222–1228.

Knights, S. D., Colbow, K. M., St-Pierre, J., and Wilkinson, D. P. 2004. Aging mechanisms and lifetime of PEFC and DMFC. *Journal of Power Sources* 127: 127–134.

Kreuer, K. D., Schuster, M., Obliers, B. et al. 2008. Short-side-chain proton conducting perfluorosulfonic acid ionomers: Why they perform better in PEM fuel cells. *Journal of Power Sources* 178: 499–509.

Kundu, S., Simon, L. C., and Fowler, M. W. 2008. Comparison of two accelerated Nafion degradation experiments. *Polymer Degradation and Stability* 93: 214–224.

Kundu, S., Simon, L. C., Fowler, M., and Grot, S. 2005. Mechanical properties of Nafion™ electrolyte membranes under hydrated conditions. *Polymer* 46: 11707–1115.

Laconti, A., Liu, H., Mittelsteadt, C., and McDonald, R. 2006. Polymer electrolyte membrane degradation mechanisms in fuel cells—Findings over the past 30 years and comparison with electrolyzers. *ECS Transactions* 1: 199–219.

Laconti, A. B., Hamdan, M., and McDonald, R. C. 2003. Mechanisms of chemical degradation. In *Handbook of Fuel Cells—Fundamentals, Technology and Applications*, eds. W. Vielstich, H. A. Gasteiger and A. Lamn, 647pp. Chichester: JohnWiley & Sons Ltd.

Lage, L. G., Delgado, P. G., and Kawano, Y. 2004a. Thermal stability and decomposition of Nafion membranes with different cations—Using high-resolution thermogravimetry. *Journal of Thermal Analysis and Calorimetry* 75: 521–530.

Lage, L. G., Delgado, P. G., and Kawano, Y. 2004b. Vibrational and thermal characterization of Nafion membranes substituted by alkaline earth cations. *European Polymer Journal* 40: 1309–1316.

Laporta, M., Pegoraro, M., and Zanderighi, L. 1999. Perfluorosulfonated membrane (Nafion): FT-IR study of the state of water with increasing humidity. *Physical Chemistry Chemical Physics* 1: 4619–4628.

Lee, J. S., Quan, N. D., Hwang, J. M. et al. 2006. Polymer electrolyte membranes for fuel cells. *Journal of Industrial and Engineering Chemistry* 12: 175–183.

Lee, S. J., Yu, T. L., Lin, H. L., Liu, W. H., and Lai, C. L. 2004. Solution properties of Nafion in methanol/water mixture solvent. *Polymer* 45: 2853–2862.

Li, Q. F., He, R. H., Jensen, J. O., and Bjerrum, N. J. 2003. Approaches and recent development of polymer electrolyte membranes for fuel cells operating above 100°C. *Chemistry of Materials* 15: 4896–915.

Liang, Z. X., Chen, W. M., Liu, J. G. et al. 2004. FT-IR study of the microstructure of Nafion® membrane. *Journal of Membrane Science* 233: 39–44.

Liu, H., Coms, F. D., Zhang, J., Gasteiger, H. A., and LaConti, A. B. 2009. Chemical degradation: Correlations between electrolyzer and fuel cell findings. In *Polymer Electrolyte Fuel Cell Durability*, eds. F. N. Büchi, M. Inaba and T. J. Schmidt, 71–118. New York, NY: Springer.

Liu, H., Gasteiger, H. A., Laconti, A., and Zhang, J. 2006a. Factors impacting chemical degradation of perfluorinated sulfonic acid ionomers. *ECS Transactions* 1: 283–293.

Liu, W., Ruth, K., and Rusch, G. 2001. Membrane durability in PEM fuel cells. *Journal of New Materials for Electrochemical Systems* 4: 227–231.

Liu, H., Zhang, J., Coms, F., Gu, W., Litteer, B., and Gasteiger, H. A. 2006b. Impact of gas partial pressure on PEMFC chemical degradation. *ECS Transactions* 3: 493–505.

Liu, W. and Zuckerbrod, D. 2005. *In situ* detection of hydrogen peroxide in PEM fuel cells. *Journal of the Electrochemical Society* 152: A1165–A1170.

Lousenberg, R. D. 2005. Molar mass distributions and viscosity behavior of perfluorinated sulfonic acid polyelectrolyte aqueous dispersions. *Journal of Polymer Science Part B—Polymer Physics* 43: 421–428.

Lund, A., Macomber, L. D., Danilczuk, M., Stevens, J. E., and Schlick, S. 2007. Determining the geometry and magnetic parameters of fluorinated radicals by simulation of powder ESR spectra and DFT calculations: the case of the radical RCF2CF2• in Nafion perfluorinated ionomers. *The Journal of Physical Chemistry B* 111: 9484–9491.

Merlo, L., Ghielmi, A., Cirillo, L., Gebert, M., and Arcella, V. 2007. Resistance to peroxide degradation of Hyflon® ion membranes. *Journal of Power Sources* 171: 140–147.

Mitov, S., Vogel, B., Roduner, E. et al. 2006. Preparation and characterization of stable ionomers and ionomer membranes for fuel cells. *Fuel Cells* 6: 413–424.

Mittal, V. O., Kunz, H. R., and Fenton, J. M. 2006a. Effect of catalyst properties on membrane degradation rate and the underlying degradation mechanism in PEMFCs. *Journal of the Electrochemical Society* 153: A1755–A1759.

Mittal, V. O., Kunz, H. R., and Fenton, J. M. 2006b. Is H_2O_2 involved in the membrane degradation mechanism in PEMFC? *Electrochemical and Solid-State Letters* 9: A299–A302.

Mittal, V. O., Kunz, H. R., and Fenton, J. M. 2007. Membrane degradation mechanisms in PEMFCs. *Journal of The Electrochemical Society* 154: B652–B656.

Miyake, N., Wakizoe, M., Honda, E., and Ohta, T. 2006. High durability of Asahi Kasei Aciplex membrane. *ECS Transactions* 1: 249–261.

Miyake, N., Wakizoe, M., Honda, E., and Ohta, T. 2004. Durability of Asahi Kasei Aciplex membrane for PEM fuel cell application, Abstract 1880. In *206th ECS meeting*, October 3–8, Honolulu, HA.

Moore, R. B. and Martin, C. R. 1988. Chemical and morphological properties of solution-cast perfluorosulfonate ionomers. *Macromolecules* 21: 1334–1339.

Moore, R. B. and Martin, C. R. 1989. Morphology and chemical properties of the Dow perfluorosulfonate ionomers. *Macromolecules* 22: 3594–3599.

Morgan, R. A. and Sloan, W. H. 1986. *Extrusion finishing of perfluorinated copolymers*, US 4626587.

Murthi, V. S., Urian, R. C., and Mukerjee, S. 2004. Oxygen reduction kinetics in low and medium temperature acid environment: Correlation of water activation and surface properties in supported Pt and Pt alloy electrocatalysts. *The Journal of Physical Chemistry B* 108: 11011–11023.

Ohma, A., Suga, S., Yamamoto, S., and Shinohara, K. 2007. Membrane degradation behavior during open-circuit voltage hold test. *Journal of the Electrochemical Society* 154: B757–B760.

Ohma, A., Yamamoto, S., and Shinohara, K. 2008. Membrane degradation mechanism during open-circuit voltage hold test. *Journal of Power Sources* 182: 39–47.

Page, K. A., Cable, K. M., and Moore, R. B. 2005. Molecular origins of the thermal transitions and dynamic mechanical relaxations in perfluorosulfonate ionomers. *Macromolecules* 38: 6472–6484.

Panchenko, A., Dilger, H., Kerres, J. et al. 2004. In-situ spin trap electron paramagnetic resonance study of fuel cell processes. *Physical Chemistry Chemical Physics* 6: 2891–2894.

Patil, Y. and Mauritz, K. A. 2009. Durability enhancement of Nafion fuel cell membranes via *in situ* sol-gel-derived titanium dioxide reinforcement. *Journal of Applied Polymer Science* 113: 3269–3278.

Paulus, U. A., Wokaun, A., Scherer, G. G. et al. 2002. Oxygen reduction on carbon-supported Pt – Ni and Pt–Co alloy catalysts. *The Journal of Physical Chemistry B* 106: 4181–4191.

Perusich, S. A. 2000. Fourier transform infrared spectroscopy of perfluorocarboxylate polymers. *Macromolecules* 33: 3431–3440.

Pianca, M., Barchiesi, E., Esposto, G., and Radice, S. 1999. End groups in fluoropolymers. *Journal of Fluorine Chemistry* 95: 71–84.

Pozio, A., Silva, R. F., De Francesco, M., and Giorgi, L. 2003. Nafion degradation in PEFCs from end plate iron contamination. *Electrochimica Acta* 48: 1543–1549.

Ralph, T. R., Barnwell, D. E., Bouwman, P. J., Hodgkinson, A. J., Petch, M. I., and Pollington, M. 2008. Reinforced membrane durability in proton exchange membrane fuel cell stacks for automotive applications. *Journal of the Electrochemical Society* 155: B411–B422.

Ramaswamy, N., Hakim, N., and Mukerjee, S. 2008. Degradation mechanism study of perfluorinated proton exchange membrane under fuel cell operating conditions. *Electrochimica Acta* 53: 3279–3295.

Rozière, J. and Jones, D. J. 2003. Non-fluorinated polymer materials for proton exchange membrane fuel cells. *Annual Review of Materials Research* 33: 503–555.

Saab, A. P., Garzon, F. H., and Zawodzinski, T. A. 2002. Determination of ionic and electronic resistivities in carbon/polyelectrolyte fuel-cell composite electrodes. *Journal of the Electrochemical Society* 149: A1541–A1546.

Saab, A. P., Garzon, F. H., and Zawoszinski, T. A. 2003. The effects of processing conditions and chemical composition on electronic and ionic resistivities of fuel cell electrode composites. *Journal of the Electrochemical Society* 150: A214–A218.

Satterfield, M. B., Majsztrik, P. W., Ota, H., Benziger, J. B., and Bocarsly, A. B. 2006. Mechanical properties of Nafion and titania/Nafion composite membranes for polymer electrolyte membrane fuel cells. *Journal of Polymer Science Part B: Polymer Physics* 44: 2327–2345.

Schumb, W. C., Satterfield, C. N., and Wentworth, R. L. 1955. *Hydrogen Peroxide*, p. 492, New York, NY: Reinhold Pub. Co.

Sethuraman, V. A., Weidner, J. W., Haug, A. T., and Protsailo, L. V. 2008. Durability of perfluorosulfonic acid and hydrocarbon membranes: Effect of humidity and temperature. *Journal of the Electrochemical Society* 155: B119–B124.

Sompalli, B., Litteer, B. A., Gu, W., and Gasteiger, H. A. 2007. Membrane degradation at catalyst layer edges in PEMFC MEAs. *Journal of the Electrochemical Society* 154: B1349–B1357.

St Pierre, J., Wilkinson, D. P., Knights, S., and Bos, M. L. 2000. Relationships between water management, contamination and lifetime degradation in PEFC. *Journal of New Materials for Electrochemical Systems* 3: 99–106.

Stucki, S., Scherer, G. G., Schlagowski, S., and Fischer, E. 1998. PEM water electrolysers: Evidence for membrane failure in 100 kW demonstration plants. *Journal of Applied Electrochemistry* 28: 1041–1049.

Szajdzinska-Pietek, E., Schlick, S., and Plonka, A. 1994. Self-assembling of perfluorinated polymeric surfactants in nonaqueous solvents. electron spin resonance spectra of nitroxide spin probes in Nafion solutions and swollen membranes. *Langmuir* 10: 2188–2196.

Tang, H., Pan, M., Wang, F., Shen, P. K., and Jiang, S. P. 2007a. Highly durable proton exchange membranes for low temperature fuel cells. *The Journal of Physical Chemistry B* 111: 8684–8690.

Tang, H., Peikang, S., Jiang, S. P., Wang, F., and Pan, M. 2007b. A degradation study of Nafion proton exchange membrane of PEM fuel cells. *Journal of Power Sources* 170: 85–92.

Tang, H. L., Pan, M., and Wang, F. 2008. A mechanical durability comparison of various perfluocarbon proton exchange membranes. *Journal of Applied Polymer Science* 109: 2671–2678.

Taniguchi, A., Akita, T., Yasuda, K., and Miyazaki, Y. 2004. Analysis of electrocatalyst degradation in PEMFC caused by cell reversal during fuel starvation. *Journal of Power Sources* 130: 42–49.

Taniguchi, A., Akita, T., Yasuda, K., and Miyazaki, Y. 2008. Analysis of degradation in PEMFC caused by cell reversal during air starvation. *International Journal of Hydrogen Energy* 33: 2323–2329.

Teranishi, K., Kawata, K., Tsushima, S., and Hirai, S. 2006. Degradation mechanism of PEMFC under open circuit operation. *Electrochemical and Solid-State Letters* 9: A475–A477.

Tiwari, S. K., Nema, S. K., and Agarwal, Y. K. 1998. Thermolytic degradation behavior of inorganic ion-exchanger incorporated Nafion-117. *Thermochimica Acta* 317: 175–182.

Vogel, B., Aleksandrova, E., Mitov, S. et al. 2008. Observation of fuel cell membrane degradation by ex situ and *in situ* electron paramagnetic resonance. *Journal of the Electrochemical Society* 155: B570–B574.

Wang, F., Tang, H., Pan, M., and Li, D. 2008. *Ex situ* investigation of the proton exchange membrane chemical decomposition. *International Journal of Hydrogen Energy* 33: 2283–2288.

Watanabe, M., Tsurumi, K., Mizukami, T., Nakamura, T., and Stonehart, P. 1994. Activity and stability of ordered and disordered Co–Pt alloys for phosphoric acid fuel cells. *Journal of the Electrochemical Society* 141: 2659–2668.

Wilkie, C. A., Thomsen, J. R., and Mittleman, M. L. 1991. Interaction of poly(methyl methacrylate) and Nafions. *Journal of Applied Polymer Science* 42: 901–909.

Wilkinson, D. P. and St-Pierre, J. 2003. Durability. In *Handbook of Fuel Cells: Fundamentals Technology, and Applications*, eds. W. Vielstich, H. A. Gasteiger and A. Lamn, pp. 611–626. Chichester: John Wiley & Sons Ltd.

Xie, J., Wood III, D. L., Wayne, D. M., Zawodzinski, T. A., Atanassov, P., and Borup, R. L. 2005. Durability of PEFCs at high humidity conditions. *Journal of the Electrochemical Society* 152: A104–A113.

Xie, T. and Hayden, C. A. 2007. A kinetic model for the chemical degradation of perfluorinated sulfonic acid ionomers: Weak end groups versus side chain cleavage. *Polymer* 48: 5497–5506.

Yoshida, N., Ishisaki, T., Watakabe, A., and Yoshitake, M. 1998. Characterization of Flemion® membranes for PEFC. *Electrochimica Acta* 43: 3749–3754.

Young, A. P., Stumper, J., Knights, S., and Gyenge, E. 2010. Ionomer degradation in polymer electrolyte membrane fuel cells. *Journal of the Electrochemical Society* 157: B425–B436.

Yu, J., Matsuura, T., Yoshikawa, Y., Islam, M. N., and Hori, M. 2005. Lifetime behavior of a PEM fuel cell with low humidification of feed stream. *Physical Chemistry Chemical Physics* 7: 373–378.

Zhang, J., Litteer, B. A., Gu, W., Liu, H., and Gasteiger, H. A. 2007. Effect of hydrogen and oxygen partial pressure on Pt precipitation within the membrane of PEMFCs. *Journal of the Electrochemical Society* 154: B1006–B1011.

Zhang, J., Xie, Z., Zhang, J. et al. 2006. High temperature PEM fuel cells. *Journal of Power Sources* 160: 872–891.

Zhang, L. and Mukerjee, S. 2006. Investigation of durability issues of selected nonfluorinated proton exchange membranes for fuel cell application. *Journal of the Electrochemical Society* 153: A1062–A1072.

Zhou, C., Guerra, M. A., Qiu, Z.-M., Zawodzinski, T. A., and Schiraldi, D. A. 2007. Chemical durability studies of perfluorinated sulfonic acid polymers and model compounds under mimic fuel cell conditions. *Macromolecules* 40: 8695–8707.

5

Porous Transport Layer Degradation

Walter Mérida
*University of British
Columbia*

5.1 Introduction

The transport of energy, mass, and charge is at the heart of proton exchange membrane fuel cell (PEMFC) operation. The porous layers in modern membrane electrode assemblies (MEAs) lie at the interface between the macroscopic phenomena occurring in the flow channels and the micro- and nanoscopic processes in the catalyst layers (see Figure 5.1). These layers must deliver the reactants and remove the products from the electrochemical reactions at the fuel cell electrodes. They must also provide connections to the current collecting plates with minimal thermal and electrical resistances.

Despite recent progress, the transport and degradation phenomena within the MEA layers are still poorly understood, and there are no reliable predictive models to guide design optimization. The knowledge gaps are partly due to the difficulties in spanning and linking processes that occur over wide length- and timescales. The phenomena of interest range from convective transport along the flow channels in commercial stacks (10^3 m), to the transport within the catalyst agglomerates and the hydrophilic pathways in ionic conductors (10^{-10} m). The gradual and cumulative degradation processes affecting product lifetimes (measured in thousands of hours) are compounded with effects due to short-lived deviations or sudden changes in the normal operating conditions (e.g., start-up from frozen conditions).

The commercialization efforts to characterize and alleviate degradation have been focused on priorities related to membrane and catalyst materials. However, recent work has demonstrated that the long-term losses due to porous layer degradation can be comparable or larger than those associated with the conventional priorities (Schulze et al., 2007). The following sections review the theoretical and experimental progress reported recently. The cited bibliography includes representative contributions, and preference has been given to articles that lead to more comprehensive overviews in each relevant area.

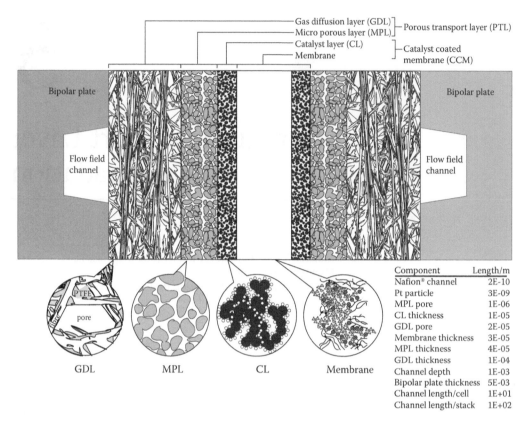

Component	Length/m
Nafion® channel	2E-10
Pt particle	3E-09
MPL pore	1E-06
CL thickness	1E-05
GDL pore	2E-05
Membrane thickness	3E-05
MPL thickness	4E-05
GDL thickness	1E-04
Channel depth	1E-03
Bipolar plate thickness	5E-03
Channel length/cell	1E+01
Channel length/stack	1E+02

FIGURE 5.1 A schematic representation of the functional layers in a MEA. The relevant length scales are typical for current PEMFC designs, and span more than ten orders of magnitude.

5.2 Material Types, Functions, and Requirements

The PEMFC terminology has evolved to accommodate the improvements in MEA design. Initially, the terms gas diffusion media, and more commonly, gas diffusion layer (GDL) referred to thin, macroscopically porous structures made of carbonaceous fibers. The fibers were bonded, woven, or pressed into textiles, papers, felts, and mats. However, the incremental design improvements unveiled the importance of several concurrent processes that include: heat and electron transport through solid phases (carbon and fluoropolymer binders), convective and diffusive species transport in the gas phase (reactants, water vapor), and liquid water transport (product and condensed water). The direct alcohol fuel cells must also consider the recirculation, transport, and phase separation of the fuels in aqueous solution and the CO_2 evolved during fuel oxidation (see Figure 5.2). In light of such complexity, some researchers (Pharoah et al., 2006) refer to porous transport layers (PTL) required for the multiphase transport of mass, charge, and heat through sequential or concurrent mechanisms (diffusion, conduction, convection, etc.). To balance the importance of this observation while maintaining consistency with published work, the nomenclature in Figure 5.1 will be used to describe the PTL sublayers, structures, and functions.

The GDL provides an electrical and thermal connection to the flow-field plates and facilitates species transport, through the macroscopic pores. It also improves materials utilization by delivering reactants not only to the exposed MEA areas in the flow-field channels, but also to the areas under the bipolar plate contacts (albeit at reduced rates due to porosity reduction upon compression). Conversely, the

FIGURE 5.2 A scanning electron microphotograph of a MEA (left), and a PTL/MPL assembly (right). The thick membrane and electrodes in the MEA are typical of direct alcohol applications. The structural differences between the GDL and PTL are clearly visible on the right. (From Fraunhofer Institute for Solar Energy Systems. With permission.)

electronic current collection from MEA areas not in contact the plates is enabled by charge transport along the GDL planes.

The primary functions of the microporous layer (MPL) are to control the water fluxes near the catalyst layer, and to reduce the contact resistance. The MPL is denser, more conductive, and more hydrophobic* than the GDL. These characteristics improve the electrical contact to the CL, and create a gradient of hydrophobicity (Wood et al., 2002) that makes it possible to maintain adequate hydration in the membrane and ionomer. The more hydrophilic and open structures in the GDL are designed to facilitate the transport of excess liquid water from the electrodes to the flow-field channels. Recent measurements indicate that the MPL reduces the liquid saturation in the GDL by limiting the number of dendritic breakthrough pathways (Gostick et al., 2009a). See Sections 5.3.4 through 5.3.6 for additional detail.

Early MEA fabrication relied on catalyst deposition directly onto the GDL materials (via spraying, screen printing, or blade application), and subsequent bonding to the membrane. Other MEA designs incorporate mass-produced, catalyst-coated membranes (CCMs) that are separated from the GDL by one or more sublayers. These sublayers can be fabricated from multiwalled carbon nanotubes (Kannan et al., 2009), doped polyaniline (PANI) (Cindrella and Kannan, 2009), or more commonly, carbonaceous particles with polymeric binders. The pore diameters in the resulting MPL can be two orders of magnitude smaller than the corresponding pores in the GDL (e.g., 10^{-5} m and 10^{-7} m, respectively).

Low-humidity operation and high-temperature membranes may render humidification systems obsolete in the future. However, current PEMFC designs rely on humidified reactants and operating temperatures below the normal boiling point of water. Under typical operating conditions, PEMFCs are exposed to pressure, temperature, and humidity gradients and potential cycling (Berning and Djilali, 2003; Burheim et al., 2010). Hence, the transport of humidified reactants and products must consider concurrent condensation, evaporation, and possibly freezing. These considerations are crucial at the cathodes, where gaseous oxygen delivery must be balanced with liquid water extraction during the electrochemical oxygen reduction reaction (ORR). Hydrophobic PTLs are required at the anodes also,

---

* The difference in hydrophobicity is due to microstructural factors (e.g., pore and particle sizes) and composition (e.g., PTFE loadings), not due to significant changes in materials.

to avoid flooding during sudden changes in temperature or operating conditions (e.g., cold cell start-up that could lead to condensation from saturated reactants).

The effective water management at the PTL is achieved by balancing the hydrophilic and hydrophobic properties of the composite structures. The GDL and MPL contain poly-tetrafluoroethylene (PTFE) or other fluorocarbons that bind and partially coat the conductive (carbonaceous) networks. The fluorocarbon films improve mechanical strength and structural integrity, provide corrosion protection, and create hydrophobic regions within the nominally hydrophilic carbon fibers or particles.[*] The overall porosity, pore size distributions, electronic conductivity, and net hydrophobicity are influenced by the amount of fluorocarbon present.

The hydrophobic and hydrophilic pathways created by the fluoropolymer impregnation enable concurrent gaseous and liquid phase transport. Liquid water condensation or penetration (flooding) into the PTL reduces the number of hydrophobic pathways available for gaseous transport (the gaseous transport rates through liquid films are several orders of magnitude smaller). To accommodate for flooding failures, larger catalyst loadings are required to achieve large (>1 A cm^{-2}) current densities. Other negative effects include fuel efficiency reductions due to larger concentration overpotentials, and catalyst degradation due to localized reactant starvation. As a result of all these challenges, significant efforts have been devoted to the optimization of liquid water transport through the GDL and MPL.

5.2.1 Manufacturing Processes

The processes relevant to GDL manufacture are based on technologies from the paper and textile industries. They usually include several processing steps at high temperatures and pressures, binder impregnation, and pressing, weaving, or moulding fibers of small diameter (e.g., 10^{-5} m). The most common topologies in commercial GDL materials are papers, cloths, and felts. Although metals have been considered (Zhang et al., 2008), the vast majority of products use carbonaceous fibers due to their low cost, light weight, and adequate chemical, electrical, thermal, and mechanical properties. Detailed reviews on the material characteristics and common processing routes have been reported previously (Mathias et al., 2003). Only a brief overview of the most common process is described here.

The carbon content in continuous, previously stabilized (~500 K in air) carbon fibers is increased progressively by pyrolysis in an inert atmosphere. The high-temperature carbonization process (1400–1650 K) reduces the molecular weight of the resulting fibers, and includes complex reactions (dehydrogenation, hydrogen transfer and isomerization, etc.) that are required to remove the volatile species that may be present (H_2, N_2, O_2, etc.). The pyrolysis temperature determines the degree of carbonization (e.g., the carbon content can exceed 90 and 99 wt% for processes at 1200 and 1600 K, respectively) (UIPAC, 1997).

After carbonization, small fiber segments are cut and dispersed with additional binders (e.g., polyvinyl alcohols) in a water solution. The mixture is then transferred to paper-making equipment and the resulting rolls are impregnated with a resin which is then cured at moderate temperature (430 K) in air. Once all the solvents have been removed, the paper can be cut and pressed or moulded into the desired thicknesses or shapes. This batch process requires intermediate temperatures and pressures (450 K at 500 kPa). Thicker GDL samples can be obtained by pressing two or more individual sheets together. However, recent work on heterogeneous porosity distributions indicates that the ply-molding processes can result in large variation between the surface and core regions (Fishman et al., 2010).

The GDL sheets are then heated to higher temperatures (2500–3300 K) that are required for graphitization (i.e., the phase transition from amorphous carbon to crystalline graphite). This transformation is crucial to enhance the electrical, thermal, and mechanical material properties. It also changes the surface characteristics and chemical response to the fuel cell operating environments (e.g., hydrophobicity and corrosion resistance). Differences in the process conditions (temperature, prevailing atmosphere or vacuum, etc.) can result in large differences in material properties. Hence, care must be

---

[*] Despite the nominal hydrophilic nature of individual particles or fibers, the untreated PTL structures are hydrophobic.

exercised when comparing graphitized materials without information on the extent of transformation or the degree of crystallinity achieved. Wood and Borup note that the effects of this step on GDL durability may be just as significant as the fluoropolymer impregnation required for adequate water management (Wood and Borup, 2009).

PTFE is the most common fluoropolymer additive for GDL materials. It is applied by immersion into aqueous emulsions or suspensions, subsequent drying, and sintering at 423 K. The amount of PTFE remaining in the GDL material influences the overall hydrophobicity and porosity. According to Mathias et al. the PTFE distribution across the GDL thickness is sensitive to the drying procedures (Mathias et al., 2003). Rapid drying leads to high PTFE concentrations at the surfaces, while slower processes result on more homogeneous concentration profiles.

The typical MPL materials start as an emulsion of carbon black particles mixed with fluoropolymers in aqueous solution. The miscibility and partial surface area coverage are controlled by adding surfactants and light alcohols to the mixture. The resulting slurry consists of up to 20 wt% solids, and it can be applied to the GDL substrate by spraying, screen-printing, and manual deposition.

5.3 Properties and Characterization Protocols

The materials in the PTL sublayers are porous and anisotropic. Hence, the related mechanical thermal and electrical properties are not homogeneous. Several experimental and modeling efforts have investigated the periodic anisotropy introduced by the operational topologies (e.g., the changes in porosity and thermal conductivity due to the compression under the bipolar plates) (Nitta et al., 2008). However, the bulk of the PEMFC literature reports measurements along the directions that are normal and parallel to the PTL planes. Further differentiation is required for materials produced via lamination, binding, or rolling because the material properties measured in the longitudinal and transverse machine directions are different.

5.3.1 Mechanical Properties

The TPL materials must be able to accommodate and distribute the forces required to seal the cell perimeter and minimize thermal and electrical contact resistances. These related loads can be significant, and depending on the flow-field topologies, the forces are applied on areas that are much smaller than the (geometric) electrode areas. Hence, elastic or plastic deformation of the PTL materials is common, and GDL intrusion into the flow channels has been reported. These deformations increase the pressure drop across the flow-field channels, and generate inhomogeneous porosities and pore size distributions. The resulting thermal and electrical flow maldistributions can have detrimental impacts both on performance and durability.

The data on the mechanical properties of PTL materials are sparse. Mathias et al. reported on the compressive deflection as a function of compressive force for carbon fiber paper and carbon cloth materials (no specifications on material composition or manufacturer were provided) (Mathias et al., 2003). The maximum applied load was 2.75 MPa and the maximum compressive strain was approximately 23% and 53% for the fiber paper and cloth samples, respectively. Significant but decreasing hysteresis was observed on both materials after 10 cycles (from zero to maximum load).

The stress–strain curves for PTL materials over a wide range of compressive loads (0–5 MPa) have been measured (Escribano et al., 2006; Nitta et al., 2008). Nitta et al. attributed the variations at small loads (<0.2 MPa) to small differences in surface roughness across the samples. They also discussed two linear regions on the curves that were attributed to the sequential deformation due to the collapse of hydrophilic and hydrophobic pores. More recently, Sadeghi et al. (2008, 2010a,b) reported measurements on circular samples of 78% porous Toray® carbon papers under static and cyclic loads of up to 1.5 MPa. These measurements revealed significant hysteresis in the measured parameters (porosity, thickness, etc.) with subsequent stabilization within a few cycles. A summary of representative data sets is plotted in

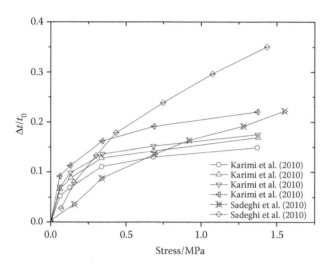

FIGURE 5.3 Approximate stress–strain curves for GDL materials calculated from representative, published data. (Reprinted from the *Electrochimica Acta*, 55, Karimi, G., Li, X., and Teertstra, P. Measurement of through-plane effective thermal conductivity and contact resistance in PEM fuel cell diffusion media, 1619–1625, Copyright (2010); *Journal of Power Sources*, 196, Corrected proof, Sadeghi, E., Djilali, N., and Bahrami, M. Effective thermal conductivity and thermal contact resistance of gas diffusion layers in PEM fuel cells. Part 1: Effect of compressive load, 246–254, Copyright (2011), with permission from Elsevier.)

Figure 5.3. These results are consistent with the results from Mathias et al. (2003), and they illustrate the variations on the mechanical properties across the GDL materials available commercially.

5.3.2 Electrical Conductivity

Owing to the preferential fiber orientation, the measured, in-plane PTL electrical conductivity ($\sigma_=$) is larger than the corresponding value for conduction in the direction perpendicular to the plane ($\sigma_\perp$). For materials fabricated using lamination, binding, or rolling processes, the values for $\sigma_=$ can differ significantly (25%) upon measurement in the longitudinal and transverse machine directions (e.g., 2×10^4 and 1.5×10^4 S m^{-1} for TGP-H-060) (Kleemann et al., 2009). Upon assembly, the PTL materials will deform under compression, and the effective properties will depend on the applied loads, the material composition, and the microscopic surface roughness (Nitta et al., 2008; Chi et al., 2010).

The contact between two (nominally flat) surfaces is known to occur as varying numbers of random, clustered microcontacts (Dyson and Hirst, 1954). A thermal or electrical contact resistance can result from quantum effects and from constrictions to the localized electronic flow near asperities (i.e., the microscopic, nonquantum contacts of arbitrary size and varying cluster distributions). The type of electronic transport and the size dependence of the constriction resistance depend on the electronic charge transport at microscopic scales.

The well-known result reported by Holm (1954) for circular contacts can only be applied when the contact diameter is much larger than the mean free path of electrons. In contrast, the expression proposed by Sharvin (1965) can only be used when the contact dimensions are much smaller. Mikrajuddin et al. (1999) have reconciled these two extremes with a single expression derived from a general theory for size-dependent contacts. The quantification and treatment of the surface roughness have been improved since Greenwood and Williamson developed expressions relevant to elastic, curved surfaces (Greenwood and Williamson, 1966). The effects of adhesion and plastic deformation have been incorporated into the relevant models (Jeng and Wang, 2003) and the issues related to scaling and instrument resolution have been addressed via fractal and multiscale quantification techniques (Jackson and Streator, 2006; Laua and Tang, 2009).

Through-plane measurements of total GDL conductivity with varying compressive loads have been measured with four-probe arrangements in axially symmetric configurations (see Section 5.3.3). Several efforts with multiple GDL samples in a stack have been reported. However, the separation of interfacial and effective bulk conductivities ($\sigma_\perp$) in thin, highly conductive samples can be challenging.

To elucidate this problem, Kleeman et al. reported a modified four-probe arrangement using wires of small diameter (30 μm) embedded into copper measuring electrodes (Kleemann et al., 2009). The small probe dimensions reduced the potential field distortion, and enabled the calculation of the average voltages at constant current. Further refinements included the separation of the combined resistance due to the MPL and the contact resistance at the catalyst layer. The combined resistance was set to zero for compressive loads larger or equal to 2 MPa, and the incremental changes were recorded as the loads were reduced from this arbitrary threshold (see Figure 5.4).

Another approach relies on making the measurements with and without the GDL samples, and then subtracting the interfacial contact resistances from electrode materials with known properties. Ismail et al. applied this technique to separate the contact resistance using copper and 316 stainless-steel electrodes. The through-plane or tridimensional measurements must also consider the averaged contact resistances at the interfaces between layers, and the anisotropy within the porous structures. These differences manifest themselves as large variations in the measured resistivity across materials and topologies.

Based on modeling efforts reported previously (Zhou and Liu, 2006; Wu et al., 2008; Zhang et al., 2009), Ismail et al. have reported on the effects of PTFE treatment on the GDL electrical conductivity (Ismail et al., 2010). These authors emphasized the importance of in-plane measurements, and characterized GDL and PTL samples from SGL Technologies. The experiments implemented the four-probe method and the correction factors proposed by Smits (1958). These factors were based on earlier work by Uhlir (1955), and Valdes (1954) who used the method of images to solve an electrostatic system with conducting and non conducting boundaries. Smits developed suitable geometry-dependent correction factors for measurements on thin sheets. This method has been relevant to the semiconductor industry for more than fifty years, and similar correction factors for the proximity of sample boundaries in a

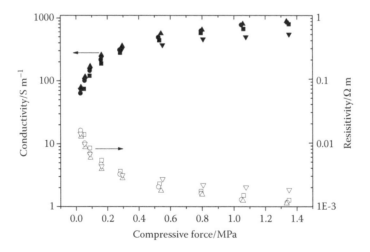

FIGURE 5.4 Calculated electrical conductivity as a function of applied compressive load. The values were calculated based on the data reported by Kleemann et al. (2009). (Reprinted from the *Journal of Power Sources*, 190, Kleemann, J., Finsterwalder, F., and Tillmetz, W. Characterisation of mechanical behaviour and coupled electrical properties of polymer electrolyte membrane fuel cell gas diffusion layers. *Selected Papers presented at the 11th ULM ElectroChemical Days*, 92–102, Copyright (2009), with permission from Elsevier.)

variety of geometries have been developed. Despite these efforts, and until recently, the four-probe correction factors had required relatively large samples and accurate probe positioning. With the advent of microfabricated four-point probes (Petersen et al., 2002), accurate measurements have become possible on small areas. However, and as reported by Kleemann et al. (2009), small-scale measurements are affected by the proximity of insulating sample boundaries. The correction factors discussed before can be applied, but they can become complex, and require accurate knowledge of the sample geometries or probe positions. Such requirements are difficult to fulfill in realistic PTL environments. Thorstensson et al. (2009) have reported on a methodology that enables accurate and correction-free conductivity measurements on thin sheets. These authors used conformal mapping to demonstrate that accurate measurements are possible on small samples with shapes having one or more planes of symmetry (squares, rectangles, circles, etc.). They concluded that the preferred sample geometry would be a rectangle or stripe with the probes aligned normal to the long edges.

The availability of microscopic four probe sensors and techniques could enable accurate *ex situ* and *in situ* measurements on PTL materials. However, the highly conductive sublayers pose a challenge even with such arrangements. Due to the porosity in the MEA layers, in-plane measurements yield effective conductivities that incorporate the contact resistances between the microstructural layer subunits (e.g., the carbon fibers or carbon particles).

5.3.3 Thermal Conductivity

The heat dissipation at the fuel cell cathodes relies on the GDL and MPL effective thermal conductivity, which strongly depends on the anisotropic packing and thermal contact of the carbon particles, fibers, and binder. The heating rate of a PEMFC producing liquid water can be calculated from the electrical power P_e, and the operating cell voltage, V_c:

$$\dot{Q} = P_e \left(\frac{1.48V}{V_c} - 1 \right)_{\text{(in Watts)}}$$

(5.1)

Figure 5.5 shows the heat-to-power ratio calculated from published PEMFC performance data. This heat must be rejected at low temperatures (<100°C) and the type of cooling system that can be used will

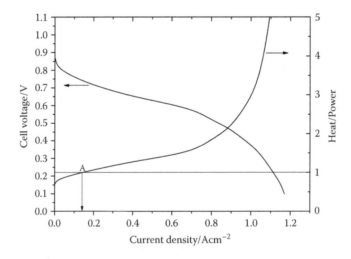

FIGURE 5.5 Typical polarization data and the corresponding heat-to-electrical power ratios. At high current densities, adequate heat transfer is required to accommodate the heat loads. Even at the moderate current density marked (A), the heat-to-power ratio is larger than unity.

be determined by the power level and the operating point. As illustrated, the heat loads increase significantly at the high current densities required in, for example, automotive applications. Even small temperature variations can significantly affect the water phase-change rates, thereby affecting the overall mass transport (Djilali and Lu, 2002).

Humidified air is less dense than dry air and the difference in densities can be used effectively to remove heat and water from the cathodes (i.e., the oxidant and coolant streams can be the same). Small fans and blowers can be used to provide more compact cell designs. The associated parasitic losses are small (e.g., <1%) even with large excess flows. For low power levels the cell cooling can be provided by convective and radiative heat transfer. The cell designs based on this approach must balance relatively open structures with the penalty of lower volumetric power densities. Applications requiring more electrical power (>1 kW$_e$) must consider active cooling schemes almost without exception.

Analytical (Sadeghi et al., 2008) and experimental efforts have been reported recently (Khandelwal and Mench, 2006; Nitta et al., 2008; Burheim et al., 2010; Karimi et al., 2010; Sadeghi et al., 2010, 2011; Teertstra et al., 2011). Most of the reported work has relied on very similar implementations of steady-state, heat flux methods (Khandelwal et al., 2006; Nitta et al., 2008; Karimi et al., 2010; Sadeghi et al., 2010, 2011; Teertstra et al., 2011). These methods impose unidirectional and insulated heat fluxes across material samples under compression. The temperature is measured in the direction parallel to the heat flux and the total thermal resistance is calculated by extrapolation of the thermal profiles. These methods have been used to make *ex situ* measurements on modern GDL materials with estimated accuracies of 5–10% (Khandelwal et al., 2006; Ramousse et al., 2008; Burheim et al., 2010; Karimi et al., 2010). The interfacial thermal contact resistances and their dependence on surface conditions have been addressed by controlling the applied contact pressure, and by coating the contacting surfaces with conductive films (Huang et al., 2006). Deviations from adiabatic conditions (e.g., radial conduction, radiation losses, etc.) can be addressed by using vacuum insulation and materials with low surface emissivity.

Figure 5.6 summarizes representative data sets for the measured thermal conductivity as a function of temperature and compressive load (Khandelwal et al., 2006; Burheim et al., 2010; Karimi et al., 2010; Sadeghi et al., 2010, 2011; Teertstra et al., 2011). In all the reported work, the total thermal resistance is expressed as the sum of effective (bulk) and interfacial components that depend on the total number of samples, n:

$$R_t(n) = nR_{GDL} + (n+1)R'_C + 2R''_C \qquad (5.2)$$

The calculated thermal contact resistance between samples R' is small (Burheim et al., 2010) and can be neglected (Karimi et al., 2010). The effective GDL conductivity R_{GDL} is isolated by carrying out experiments with multiple samples (Karimi et al., 2010) or identical samples of different thickness. Sadeghi et al. (2008, 2011) calculated and measured the effective, bulk thermal conductivity and the thermal contact resistance R'' separately. A related analytical model was developed to predict the thermal conductivity and thermal contact resistance of GDL materials.

The steady-state, unidirectional heat flux methods have provided sparse but reliable data for thermal conduction across the PTL planes. However, the tridimensional variations and the effects of humidity or partial flooding on the thermal conductivities across different materials of interfaces are not known. As illustrated in Figure 5.6, the differences in material composition, topology, and conditions result in large variations in the thermal conductivity.

Transient methods such as the flash laser technique (FLT) enable fast, accurate, and localized measurements, and they eliminate the contact-resistance effects. However, these measurements are difficult or impossible on semitransparent samples such as the porous GDL materials. Special sample preparation such as thin surface coatings with well-characterized materials can be used (Kehoe et al., 1998). If such approaches prove successful, they will enable thermal diffusivity $a(T)$ and specific heat $C_p(T)$

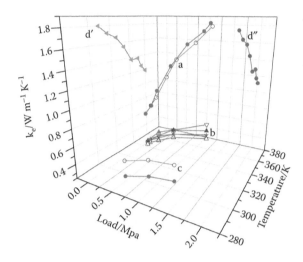

FIGURE 5.6 The calculated effective thermal conductivity based on the data sets reported by (a) Sadeghi et al. (2010), (b) Karimi et al. (2010), (c) Burheim et al. (2010), and (d) Khandelwal et al. (2006). For the last data set, the two curves (d′ and d″) correspond to the high and low ends of the applied compressive load. (Reprinted from the *Journal of Power Sources,* 196, Corrected proof, Sadeghi, E., Djilali, N., and Bahrami, M. Effective thermal conductivity and thermal contact resistance of gas diffusion layers in proton exchange membrane fuel cells. Part 2: Hysteresis effect under cyclic compressive load, 246–254, Copyright (2011); *Electrochimica Acta,* 55, Karimi, G., Li, X., and Teertstra, P. Measurement of through-plane effective thermal conductivity and contact resistance in PEM fuel cell diffusion media, 1619–1625, Copyright (2010); *Journal of Power Sources,* 195, Burheim, O. et al. *Ex situ* measurements of through-plane thermal conductivities in a polymer electrolyte fuel cell, 249–256, Copyright (2010); and *Journal of Power Sources,* 161, Khandelwal, M. and Mench, M. M. Direct measurement of through-plane thermal conductivity and contact resistance in fuel cell materials, 1106–1115, Copyright (2006), all with permissions from Elsevier.)

measurements. If the bulk density of the material is known $\rho(T)$ a direct thermal conductivity measurement is possible due to the interdependence of these three properties (Min et al., 2007):

$$k(T) = \rho(T)C_p(T)a(T) \tag{5.3}$$

Since the method was first proposed by Parker et al. (1961), improvements to eliminate artifacts arising from heat loss finite pulse duration (Takahashi, 1984), and uneven beam profiles (Baba and Ono, 2001) have been reported. Further improvements include the development of fast infrared thermometers, and new data analysis algorithms. The trend toward *in situ* characterization via microsensors may be coupled with these refinements to enable the full characterization of PTL materials under transient conditions.

5.3.4 Hydrophobicity

The low temperatures of operation ($T < 373$ K), the reliance on humidified reactants, and the production of water imply that a liquid phase will be present at the PEMFC cathodes. Moreover, the product water must be removed at sufficiently high rates to avoid accumulation in the PTL pores, and maintain open pathways for the gaseous reactant delivery to the catalyst layers. Due to the slow kinetics for oxygen reduction, the transport of oxygen and water can limit the cell performance.

Despite its importance, the capillary transport through the irregular pores in the GDL and MPL are still poorly understood. Most of the reported work has applied the theoretical foundations developed at the beginning of the nineteenth century, and more recent empirical formulations adapted from oil extraction engineering, hydrology, and catalysis (Dullien, 1988).

In 1805, Thomas Young introduced two fundamental concepts: the mean curvature H for the surface S separating a liquid and a gas, and a contact angle θ_c between this surface and a surface supporting the liquid. He concluded that θ_c depends only on the materials involved, and not on geometric, topological, or gravitational effects (Young, 1805).

These concepts were formalized by Laplace who derived mathematical expressions for the mean curvature, and developed a differential equation that must be satisfied on S (Laplace, 1805). In modern notation (Gostick et al., 2010) the resulting Young–Laplace equation relates the pressure differential across the surface Δp to the mean curvature H and provides a definition for the surface tension σ:

$$\Delta p = 2\sigma H \tag{5.4}$$

Variations of this expression have been widely used in the fuel cell literature pertaining to pore size distributions in the GDL. However, the original work by Young was concerned with cylindrical tubes of circular and constant cross sections immersed vertically in a large liquid reservoir (see Figure 5.7). For this special case, the principal curvatures of S are both equal to the inverse of a single radius of curvature R. This radius can in turn be written in terms of the radius a of the cylindrical tube:

$$\Delta p = 2\sigma\left(\frac{\cos\theta_c}{a}\right) \tag{5.5}$$

The theory was based on mechanical and hydrostatic equilibrium concepts, which were unified by C. F. Gauss using the Principle of Virtual Work. The unification introduced expressions for gravitational,

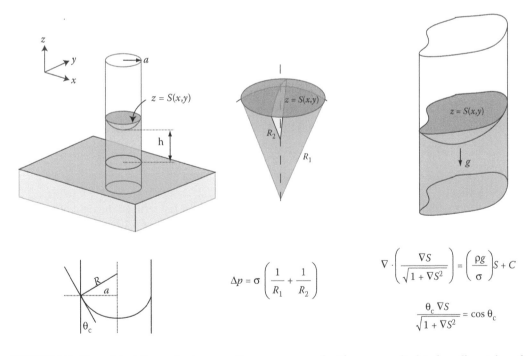

FIGURE 5.7 The original Young–Laplace equation was concerned with narrow cylindrical capillary tubes of circular and constant cross section (left). For noncircular cross sections, the capillary pressure depends on the two principal radii of curvature (middle). For irregular cross sections under the influence of a vertical gravitational field, a second-order differential equation and the related boundary conditions must be satisfied (right).

wetting, and free-surface energy terms. At the time, the required thermodynamic and hydrodynamic theories were not developed sufficiently to provide precise definitions, and some of these limitations are still present today. Moreover, a great deal of confusion and controversy has permeated the published literature with respect to the relevant conventions and terminology. See, for example, Shikhmurzaev (2008), Ip and Toguri (1994), and Dussan (1979).

Despite its restricted applicability, Equation 5.5 has been used to quantify the pore size distributions in modern PTL materials (see the next section). An abbreviated version of the general mathematical description is included here to illustrate the limitations of the reported expressions, and the care needed to apply them. The full mathematical analysis is available from intermediate texts on differential geometry (Struik, 1961; Patrikalakis et al., 2010) and summaries are available from specialized monographs (Langbein, 2002; Finn, 2002b, 2006), and reports in the public domain (Concus and Finn, 1991, 1995).

Formally, the normal curvature k_n at any point on a surface, is given by the ratio

$$k_n = \frac{L + 2M\lambda + N\lambda^2}{E + 2F\lambda + G\lambda^2} \tag{5.6}$$

The terms E, F, G and L, M, N are the coefficients of the first and second fundamental forms, respectively (Patrikalakis et al., 2010). The first fundamental form enables the calculation of the lengths of the curves and the angles between the curves on a surface. The second fundamental form quantifies the surface's deviation from a plane (i.e., for a plane $L = M = N = 0$, and the curvature is zero corresponding to infinite radii of curvature). The normal curvature depends only on a single derivative $\lambda = dv/du$, and its extreme values can be obtained by setting $dk_n/d\lambda = 0$. The resulting values k_{min} and k_{max} correspond to the minimum and maximum (i.e., the principal) curvatures at p. The usual convention[*] is to define the normal curvature as positive when the centre of curvature of the normal section curve is on the same side as the surface normal. This definition is consistent with a positive sign for Δp in Equation 5.4 when the larger pressure is in the concave side of the surface.

In three-dimensional space, a surface can be described explicitly ($z = S(x, y)$), implicitly ($S(x, y, z) = 0$), and in parametric form ($\mathbf{S}(u, v) = [x(u, v), y(u, v), z(u, v)]$). In most cases, it is convenient to express the surface explicitly, convert it to a parametric form called a Monge patch (Patrikalakis et al., 2010), and relate the mean curvature to the unit normal vector $\mathbf{N}$ at p:

$$H \equiv \frac{1}{2}\nabla \cdot \mathbf{N} = \frac{1}{2}\nabla \cdot \left(\frac{\mathbf{r}_u \times \mathbf{r}_v}{|\mathbf{r}_u \times \mathbf{r}_v|}\right) = \frac{1}{2}\nabla \cdot \left(\frac{\nabla S}{|\nabla S|}\right) \tag{5.7}$$

The vectors $\mathbf{r}_u$ and $\mathbf{r}_v$ are defined on the plane tangent to S, and the last term on the right of equation is only valid for $|\nabla S| \neq 0$. For an explicit surface $z = S(x, y)$, Equations 5.6 or 5.7 can be used to obtain (Finn, 2002b, 2006; Patrikalakis et al., 2010):

$$H = \frac{EN + GL - 2FM}{2(EG - F^2)} = \frac{(1 + S_x^2)S_{yy} - 2S_x S_y S_{xy} + (1 + S_y^2)S_{xx}}{2[1 + S_x^2 + S_y^2]^{3/2}} = \frac{k_{min} + k_{max}}{2} \tag{5.8}$$

The subscripts on the third term indicate partial differentiation, and therefore, the general problem is reduced to finding a capillary surface S whose mean curvature satisfies Equation 5.8.

---

[*] The opposite convention is used in numerically controlled machining applications.

In the absence of an external (e.g., gravitational) field, H is constant. Under a constant and vertical gravitational acceleration g, the general differential equation yields

$$\nabla \cdot \left(\frac{\nabla S}{\sqrt{1 + \nabla S^2}} \right) = \left(\frac{\rho g}{\sigma} \right) S + C \qquad (5.9)$$

In this expression C is a constant and, and the capillary constant $\rho g/\sigma$ depends on the (incompressible) fluid's density (Finn, 2002b, 2006). The relevant boundary condition depends on the constant angle θ_c at which S meets a rigid material boundary:

$$\frac{\theta_c \nabla S}{\sqrt{1 + \nabla S^2}} = \cos \theta_c \qquad (5.10)$$

More complex geometries including convex surfaces, wedges, (Concus and Finn, 1993), and exotic containers (Concus et al., 1995) have been investigated. The techniques required to describe capillary transport in these geometries may be more relevant to PTL materials than the corresponding approaches for cylindrical tubes or sand packs (see Section 5.3.6). The implications for the existence of nonunique surface solutions, liquid behavior at corners, and surface symmetry are important (Finn, 2002b, 2006) but the theory is still incomplete. Moreover, in their current form, the recent advances cannot be applied to PTL topologies directly, and developers must rely on empirical models or data.

The following sections summarize the reported experimental techniques for porous media characterization. The limitations inherent to the use of the original Young–Laplace equation to PTL materials will become apparent in light of the preceding discussion.

5.3.5 Porosity and Pore Size Distributions

The porosity, ε, is quantified by the ratio of void volume to total volume in PTL materials. This ratio can be calculated if the sample composition and dimensions are known accurately (e.g., PEFT content, sample thickness, densities, etc.).

The reported experimental approaches include pore intrusion or immersion using suitable liquids. The most common technique is mercury intrusion porosimetry (MIP) that involves pressurizing the liquid metal in a porous matrix incrementally in order to displace air or a wetting fluid. Mercury is used because it does not wet most surfaces due to its extremely high surface tension that also enables measurements under vacuum.

Water, decane, and other liquids have been used in attempts to measure the total, hydrophobic, and hydrophilic pore volumes (Dohle et al., 2002; Williams et al., 2004). Figure 5.8 shows published data on the pore size distributions for PTL materials. However, the discussion in the preceding section makes it evident that these data can provide limited qualitative information only.

5.3.6 Contact Angles

Given the importance of contact angles for capillary flow calculations, the relevant measurements on PTL materials have been the focus of intense efforts. The initial approaches implemented sessile drop measurements of contact angles on GDL surfaces. However, these measurements do not provide adequate quantification of the wetting and transport characteristics within the porous structures.

Owing to the surface roughness, the contacts between the GDL material and the liquid water are determined by the microscopic asperities described in Section 5.3.2. The finite interaction volumes contain air, and the measured contact angles are larger than the corresponding values for a flat surface. Moreover, the surface phenomena are not sufficient to describe the transport processes as the fluid wets and penetrates the GDL surface.

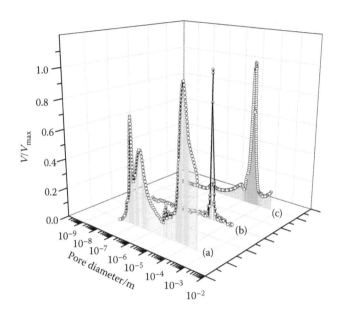

FIGURE 5.8 Calculated pore size distributions from data sets reported by (a) Cindrella et al. (2009), (b) Cheung et al. (2009), and (c) Lee et al. (2007). These data were obtained via intrusion porosimetry, and they have been normalized to the largest gravimetric void volume (in m³kg⁻¹) in each data set. (Reprinted from the *Journal of Power Sources*, 194, Cindrella, L. et al. Gas diffusion layer for proton exchange membrane fuel cells—A review. XIth Polish Conference on Fast Ionic Conductors, 146–160, Copyright (2009); *Journal of Power Sources*, 187, Cheung, P., Fairweather, J. D., and Schwartz, D. T. Characterization of internal wetting in polymer electrolyte membrane gas diffusion layers, 487–492, Copyright (2009); and *Journal of Power Sources*, 164, Lee, C. and Mérida, W. Gas diffusion layer durability under steady-state and freezing conditions, 141–153, Copyright (2007), all with permission from Elsevier.)

To improve the current models, the contact angles in the interior of PTL samples have been calculated via sorption experiments (Gurau et al., 2006; Gurau and Mann Jr., 2010; Parry et al., 2010) that were based on modified versions of the Washburn method (Washburn, 1921).

Unlike the Young–Laplace analysis, which was based on mechanical equilibrium, this method describes the dynamic sorption of a liquid into a porous medium. The analysis is based on balancing viscous and capillary forces, and it neglects inertia or gravitational effects. The mass of absorbed liquid m in time t is related to the fluid density ρ, dynamic viscoscity μ, and surface tension σ_{LV} at the liquid–vapor interface:

$$m^2 = t \frac{\mu}{C_W \rho^2 \sigma_{LV}} \cos\theta_c \tag{5.11}$$

The Washburn constant C_W can be obtained from measurements with liquids of known properties and contact angles. The unknown contact angle with a different liquid can be determined by subsequent measurements using the calculated C_W which is assumed to depend on the porous topology only. This method provides average values, but does not consider local interactions.

The Owens–Wendt theory (Good and Girifalco, 1960; Owens and Wendt, 1969) accounts for local interactions and separates the surface tension into dispersive and polar components (σ^p and σ^d) for the solid–vapor, and liquid–vapor interfaces (σ_{SV} and σ_{LV}) (Parry et al., 2010; Wood et al., 2010b):

$$(1 + \cos\theta_c)\frac{\sigma_{LV}^p + \sigma_{LV}^p}{2\sqrt{\sigma_{LV}^d}} = \sqrt{\sigma_{SV}^d}\sqrt{\frac{\sigma_{LV}^p}{\sigma_{LV}^d}} + \sqrt{\sigma_{SV}^d} \tag{5.12}$$

This expression makes it possible to obtain solid surface information by making independent contact angle measurements using liquids with known dispersive and polar contributions to σ. The surface tension of the solid–vapor interface can be calculated from the slope and the ordinate intercept in plots following Equation 5.12. Alternative approaches based on the Wilhelmy plate method (Wilhelmy, 1863) and variations thereof have been reported recently.

In the original method, a thin plate is immersed partially in a test fluid in a direction parallel to one of the larger dimensions. The vertical component of the force acting on the plate is measured gravimetrically. The force balance yields a relationship between the contact angle and the surface tension. Refinements to this method have been suggested to incorporate viscous effects (Ramé, 1997) and variations have been used to measure the contact angles for individual PTL fibers immersed in water droplets (Borup et al., 2007). A related work (Wood, 2007) reported calculations for heterogeneous sessile drop contact angles based on improvements to the Cassie equation (Cassie, 1948).

Despite these efforts the quantification of the water transport characteristics via contact angle measurements remains elusive, and further refinements are required to describe the dynamic water transport in the PTL. Moreover, the capillary transport of liquid water must be coupled to the concurrent transport of water vapor and reactant gases.

Recent efforts have focused attention on the relationship between capillary pressure and PTL saturation. Extensive and thorough reviews on the relevant literature, theory, and experimental techniques are available (Wood, 2007; Gostick et al., 2010; Sole, 2010). In all cases, the PTL materials are characterized by the relationship between capillary pressure and the water saturation in the porous media.

The early work on capillary flow in PTL materials (Wang et al., 2001) adapted expressions developed for petroleum engineering applications in the 1940s, and later modifications applied to sand packs (Leverett, 1941; Udell, 1985; Ho and Udell, 1995). The discussion in Section 5.3.4 makes it abundantly clear that the capillary flows through PTL topologies require different treatment, and this observation is corroborated by recent empirical results.

Sole reviewed the conventions used to define liquid water saturation in the PTL (Sole, 2010). These definitions considered the total dry pore volume in the PTL, and the possibility of isolated volumes of trapped fluid that are not in hydraulic contact (and hence not in thermodynamic equilibrium) with the invading or retreating fluid. Sole made the observation that after the first full or partial saturation a new reference point may be required to account for immobilized fluid that cannot be pneumatically displaced from isolated pores.

Gostick et al. published a review of the advantages and disadvantages associated with the available experimental techniques (Gostick et al., 2010). These authors identified two general methods that can be used to measure capillary pressure curves in the PTL. One method injects water into the porous materials at a constant rate while the pressure is monitored (Harkness et al., 2009). The other method controls the capillary pressure and measures the water uptake as a function of time gravimetrically (Gostick et al., 2009b). The reported capillary pressure curves present the features illustrated in Figure 5.9.

The published data demonstrate that positive capillary pressure* is required to inject water into the PTL independently of the amount of fluoropolymer present (i.e., independently of the presence of nominally hydrophilic pore networks). Conversely, negative capillary pressure is required to remove water from the PTL media that have once been saturated partially or fully. Hence, liquid water in the PTL cannot be removed via passive PEMFC design (see Section 5.5). According to Gostick et al. the shapes of the capillary surfaces in PTL media do not depend on the contact angle only (i.e., the irregular and heterogeneous microstructure also plays a role). This observation is supported by recent advances in the general theory for capillary surfaces (Finn, 2006), and the combined effects of materials and structure may be difficult to isolate.

---

* The capillary pressure is defined as the difference between the liquid (water) and gas (air) pressures: $P_W - P_A$. See Section 5.3.4 for additional details on sign conventions and surface curvature.

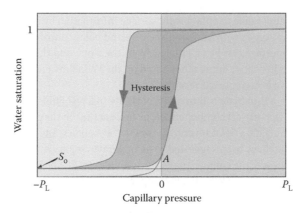

FIGURE 5.9 A typical capillary pressure curve for GDL materials. Positive or negative pressures are required to inject or remove water. An initial saturation path 0-A-1 may result in a drainage curve ending in S_o, and subsequent saturation curves S_o-1-S_o (resulting from isolated volumes that are inaccessible to pneumatic displacement). Typical values for $|P_L|$ are < 3 kPa.

Recent attempts have investigated capillary flows in the GDL through similarity model experiments (Kang et al., 2010). In this approach, the GDL and catalyst layer materials were substituted by a woven glass fiber mat, and a layer of hydrogel spheres. Such efforts may provide insight into the underlying phenomena (e.g., by nondimensional analysis), but their application to PTL materials requires further refinements.

Unlike the phenomena and characterization techniques described in Sections 5.3.1 through 5.3.3, the water transport in the PTL lacks adequate theoretical foundations. The current literature questions the assumptions in the original Young–Laplace formulations, and it emphasizes the lack of reliable models. The clarification and consolidation of the relevant terminology, theory, and experimental techniques has been aided by the recent efforts, but finding predictive relationships to PTL performance, degradation, and durability remains a challenge.

5.4 Degradation Mechanisms

A reduction in the output voltage is the most common quantifier for PEMFC degradation. As illustrated in Figure 5.10, several authors have reported voltage reductions in PEMFC stacks at rates between 2 and 60 μV h^{-1} (Wu et al., 2008). The corresponding cell voltage reduction can be significant over the product lifetime (e.g., 100 mV in 1600 h).

In practice, stationary fuel cell applications must provide 8000 h or uninterrupted service at 80% of the rated power, 40,000 h of system durability, and 90% availability (Dicks and Siddle, 2000). The corresponding figures for mobile applications are 5000 h lifetime with less than 5% power degradation. Portable applications have been reported to require approximately 1000 h (Cleghorn et al., 2003).

The durability targets make it necessary to use accelerated testing (AT) methods. These methods quantify product reliability and degradation by subjecting materials, components, and subsystems to higher than normal stress conditions (e.g., by increasing temperature, pressure, etc.). The basic indicators under normal operation (time to failure, product life, degradation rates, etc.) are then estimated by statistical data fitting to aging models. Reviews on the statistical formalism of these models and their application to AT methodologies are available (Nelson, 2005a,b; Escobar and Meeker, 1999).

In all cases, the main AT objectives are (1) to measure product reliability and degradation in a short time (by exaggerating the effects of the stress conditions), and (2) to unveil failure modes that could occur under extreme conditions (e.g., by forcing product failure during testing). Most of the reported work assumes that the reliability can be quantified by symmetric product life distributions that are fully

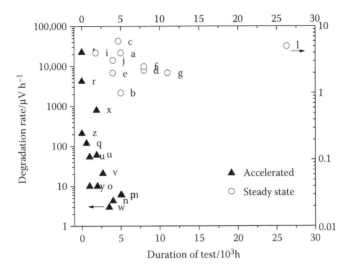

FIGURE 5.10 A summary of the reported degradation rates under steady-state and accelerated degradation conditions. The labels a through l, and m through z correspond to the references given in Tables 1 and 2 of Wu et al. (2008), respectively. (Reproduced from the *Journal of Power Sources*, 184, Wu, J. et al. A review of PEM fuel cell durability: Degradation mechanisms and mitigation strategies, 104–119, Copyright (2008), with permission from Elsevier.)

characterized by two parameters (e.g., mean and standard deviation). Typically, these parameters are modeled as simple functions of the imposed accelerating variable s (e.g., mean product life $\tau = k_0 + k_1 s + \cdots$). The relationships between s and the unknown, constant coefficients (k_i, $i = 0, 1, 2, 3, \ldots$) are usually based on Arrhenius relationships, exponential decay, and other physical models. AT methods estimate these coefficients from experimental data, and extrapolate the product life under normal operation (see Figure 5.11). Despite their logical simplicity, AT methods suffer from various limitations (Meeker and Escobar, 1998).

The reported PEMFC reliability data have used these methods to generate interpolations (temporal or otherwise) of product life within extended lifecycle test widows. However, there has been no rigorous

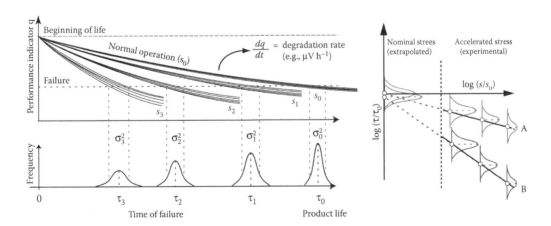

FIGURE 5.11 Product life has been estimated from the time to failure under accelerated testing conditions. The time to failure distributions under two different stress factors A and B (at stress levels s_1, s_2, etc.) will be dissimilar. The extrapolated values under normal operation ($s = s_0$) will cluster around the product life τ_0.

application of these methods to the PTL materials, and the body of published work amounts to data clusters across a large and mostly unpopulated parameter space. More importantly, these results have been used to extrapolate product life outside the testing regimes (i.e., to make estimates beyond the end of testing or outside the range of stress levels considered).

As indicated in Section 5.1, the central PTL function is the effective transport of mass, heat, and electronic charge. Hence, degradation must be interpreted and quantified with respect to deviations from initial or optimal performance baselines. However, the quantification of important properties is still inadequate, and it is challenging to find appropriate quantities to use for the ordinate axis in Figure 5.11. For example, the usual quantifiers for hydrophobicity (contact angles, PTFE content, etc.) cannot be interpreted without specific reference to local (microscopic) environments and conditions. The changes in mass transport effectiveness are usually measured as increases in the corresponding overpotential (i.e., not as intrinsic material properties), and these are coupled to other regions or operating parameters outside or independent of the PTL (flooding or inadequate ionomer hydration in the catalyst layer, etc.). These and other limitations are evident in the continuing efforts to translate these measurements into specific constrains for PTL material designs. Nevertheless, the degradation mechanisms affecting the PTL can be classified in three categories: physical, chemical, and electrochemical. The following sections summarize recent progress in applying limited AT methods, and connecting *ex situ* and *in situ* measurements (on structure, composition, history, etc.) to PTL functions.

5.4.1 Physical Degradation: Mechanical and Thermal Effects

The current understanding of mechanical PTL degradation is limited (Lee et al., 1999; and Mérida, 2007; Wilde et al., 2004) and has not received the attention devoted to other degradation mechanisms (e.g., hydrophobicity changes).

Lee and Mérida reported on PTL durability under steady-state, cycling, and freezing conditions (Lee et al., 2007). Their study focused on freezing effects (50 cycles from 20 to –35°C) and GDL compressive strain over 1500 h under simulated aging conditions. The maximum stress applied to the TPL samples was similar to those reported in Section 5.3.1. However, the reported compressive strain for the PTL samples (<1%) was lower than the values reported recently for similar applied loads and materials (10–15%). The effects on electrical resistivity, bending stiffness, air permeability, porosity, and water vapor diffusion were measured. A more recent effort by Kitahara et al. (2010) reported on permeability changes and focused on the MPL design parameters and humidity conditions. This effort made use of the Young–Laplace equation and it illustrates the limitations discussed in Sections 5.3.4, 5.3.5, and 5.3.6.

Wu et al. reported on the effects of compression and large flow rates on three sample sets corresponding to different preaging conditions: (a) no stress, (b) hot-pressing (135°C and 3 MPa for 300 s), and (c) assembly simulation (1.38 MPa for 30 min) (Wu et al., 2010). The PTL for the anode and cathode used the same GDL (of 6% PTFE-treated TORAY® TGPH-060 carbon fiber paper, with original thickness = 190 μm). The MPL was screen printed onto the GDL substrate with identical carbon loadings, but different PTFE contents (10 and 20 wt% for the anode and cathode sides, respectively). These authors reported that the through-plane electrical resistivity was larger on the samples with the larger PTFE content (cathode), and increased in sets b and c after the preaging procedures.

The aging procedure was common to all samples and it consisted of exposing them to constant gas flow rates at high temperature (120°C) and constant flows of hydrogen and air previously saturated at 80°C. The flows were 7.88 and 31 SLPM, corresponding to stoichiometries* of 7.5 and 11.8 for the anode and cathode at 1 A cm^{-2}. Notably, the electrical resistivity in all samples was close to the original values after the 200 h aging process. The authors attributed this change to the loss of nonconductive (PTFE) material. They also measured the changes in porosity via mercury intrusion porosimetry (MIP). For the reasons discussed in Section 5.3 the information on the corresponding pore size distributions can only

---

* The term stoichiometry refers to the ratio of the actual reactant flow to the flow required for a given current load.

be used for qualitative descriptions. However, the authors discussed the changes in total porosity which can be measured via MIP.

The preceding discussion illustrates the effects of overall compression and temperature. Additional effects have also been attributed to the inhomogeneous (and usually periodic) compression exerted by the bipolar plate topologies. In a recent effort, Chi and coworkers reported a numerical model that considers the nonuniform compression and the resulting variations in species concentration, temperature, and current distributions (Chi et al., 2010). Such models can become effective design tools for new PTL materials, but they require careful validation against experimental data.

Additional degradation mechanisms may be related to water freezing inside the GDL and MPL pores, and interfacial delamination after 100 freeze/thaw cycles from –40°C to 70°C has been reported (Kim et al., 2008). The changes were attributed to ice formation and subsequent GDL deformation. Other efforts (Mukundan et al., 2006) focused on freeze/thaw cycles down to –80°C using dry ice. The experimental results proved that multiple cycles could lead to interfacial delamination and GDL failure of the fuel cell, and that the preparation procedure for the MEA and the type of GDL could have a significant impact on freeze/thaw durability. The water expansion upon freezing can also result in fluoropolymer loss or structural changes due to fiber detachment, but the reported effects on GDL materials upon freezing and thawing cycles range from significant to negligible (Knights et al., 2004; Oszcipok et al., 2006; Alink et al., 2008). The treatment and elucidation of the associated transport phenomena and possible microstructural changes will require new fundamental research similar to the efforts required to explore liquid capillary flow. See, for example, He and Mench (2006a).

All these efforts illustrate that the direct thermal and mechanical effects can be significant, but may not be sufficient to explain the overall degradation mechanisms. Secondary mechanical degradation mechanisms may be present, and they could be similar to those suggested by Wood et al. in the context of hydrophobicity changes (see Section 5.4.3).

Cleghorn and coworkers reported that composite membranes show improved mechanical stability, lower hydrogen permeation rates, and improved resistance to tear and perforation propagation (mostly due to strong reinforcement materials like PTFE). (Cleghorn et al., 2003). The measured dimensional changes upon hydration (i.e., swelling) were smaller than those found in nonreinforced membranes like Nafion®. More importantly, the dimensional changes were similar in both directions in the membrane plane (i.e., the manufacturing and the transverse directions) while the corresponding changes in traditional membranes can be different.

The original data set was normalized to the changes corresponding to 20% relative humidity (RH) at ambient temperature, and it also contained points corresponding to immersion in liquid and boiling water. The variations were larger for the latter conditions, but the results at intermediate relative humidities are more representative of *in situ* RH during normal operation. Under these conditions, the effects on hydration are smaller than 10%, but preferential or nonisotropic dimensional changes could become important for monolithic PEMFC designs with only small tolerances for expansion/contraction (e.g., applications complying with the strict packaging and form factor requirements in portable applications).

5.4.2 Chemical and Electrochemical Degradation: Mass Transport Effects

The most important PTL function is the balance between adequate water management, and the delivery of reactants at sufficiently high rates (to produce the desired current). Due to the slow kinetics for oxygen reduction, this balance is critical at the fuel cell cathodes. Hence, the most important mass transport degradation processes are related to water and oxygen transport.

There have been several reports on significant changes to the overall PTL hydrophobicity. In general, the GDL and MPL surfaces become more hydrophilic after prolonged use (Liu and Case, 2006). Wood has suggested primary and secondary hydrophobicity degradation causes (Wood, 2007). In this analysis, the local environment and operating conditions have a direct (primary) effect on PTL degradation, while the degradation processes in other regions or cell components (e.g., the MEA) can have secondary effects.

Most of the published literature on accelerated methods to simulate chemical and electrochemical degradation references the work of Frisk and coworkers at 3M (Frisk et al., 2004). These authors developed a method to accelerate GDL oxidation by aging samples in a 15% hydrogen peroxide solution at 82°C. The changes were quantified by MPL weight loss, and reductions in hydrophobicity.

Up to 60% MPL weight loss was reported under 500 h, and the contact angles in conventional GDL materials were reduced significantly (from approximately 141° to 100°) after approximately 300 h in the aging solution.

These results agree with more recent data on single fiber contact angles using a variation of the Wilhelmy method (Borup et al., 2007; Wood et al., 2009). These measurements agreed with prior work that reported hydrophobicity changes with operating temperature and prevailing environment. Accelerated testing was carried out by exposing PTL materials to aging environments and temperatures (sparging nitrogen or oxygen in water at 60°C or 80°C for 480–680 h). These authors concluded that the presence of oxygen was more important than the temperature effects for the observed reduction in the single-fiber contact angles. In this case, the corresponding reduction in the static contact angle was approximately 40°.

Lee and Mérida reported MPL material loss under aging conditions, but no significant increase in the through plane dry gas permeability (Lee et al., 2007). However, the PTL was not exposed to the saturation levels typical of PEMFC operation. Other degradation effects associated with structural changes (e.g., mass loss) have been reported as changes in the pore size distributions, but the significance of these data is unclear in light of the inability of intrusion porosimetry to yield structural information. Nevertheless, such changes illustrate that chemical as well as structural changes can have an impact on the PTL interaction with liquid water.

Wilkinson and St Pierre have described the effects of long-term operation or exposure to excess liquid water (Wilkinson and St.-Pierre, 2003). GDL surfaces suffered significant hydrophobicity reduction in the exposed (flow channel) areas after 11,000 h of operation, These authors also described two identical MEAs tested with helox, oxygen, and air at the cathodes. One of the MEAs was previously soaked for 1000 h in deionized water at 80°C. The soaked MEA showed larger mass-transport losses at high current densities (the differences at low current densities were small). These results are consistent with changes in the water-management characteristics in the GDL and in the catalyst layer (e.g., changes in porosity, loss of PTFE binder, accumulation of hydrophilic impurities, etc.). Postmortem visual analysis revealed that the flow field channels clearly delineated by the different hydrophobicity characteristics in the channel and land areas. Degradation can also be quantified by measuring the current density at constant potential, and the ohmic resistance. The latter can be measured via current interrupt (Wu et al., 2010) or high-frequency resistance (HFR)* measurements (Wood et al., 2009). The variation of these two quantities under RH scans can be measured following the method described by Mathias and coworkers.

The data sets reported by Mathias et al. (2003) and more recently by Wu et al. (2010) showed a current density maximum at intermediate reactant RH (i.e., RH < 100%). The cell setup used by Mathias et al. did not incorporate an MPL and a drastic reduction in the current density was observed after reaching full reactant saturation and the resulting flooding. Wu et al. and Wood et al. used MPLs and their results illustrate the improvements that are possible with composite PTLs at high humidity. Beyond the performance improvements, systematic combinations of GDL and MPL characteristics may unveil important information on the relevant degradation mechanisms.

Wood et al. reported a revealing experiment by combining GDL and MPL materials that defined PTFE content gradients in opposite directions (Wood et al., 2009). Hydrophilic and hydrophobic MPL interfaces were prepared with GDL/MPL assemblies with varying PTFE contents (20 and 10 wt, and 5 and 23 wt%, at the GDL and MPL of hydrophilic and hydrophobic interfaces, respectively).

---

* The HFR corresponds to the extrapolated real-axis intercept for impedance spectra plotted on the Argand plane. The reported frequency ranges vary, but they usually include measurements at 1 kHz or larger frequencies.

The hydrophobic assembly provided effective water management at high humidity but poor performance under dry reactant conditions. The hydrophilic assembly provided acceptable performance over the entire range of RH conditions. Moreover, and more importantly in the context of durability, these authors used the RH sensitivity method to characterize the effect of *ex situ* aging on the hydrophobic assembly.

The aging simulation consisted of immersion in deionized water at 80°C and in the presence of sparging air for 1600 h. Notably, the current density at 0.7 V, and the HFR on the aged, hydrophobic sample very similar to those obtained with a hydrophilic assembly. These results are consistent with carbon oxidation, and they agree with the independent measurements on single fibers subjected to similar aging processes.

Additional experiments have demonstrated that the PTFE concentration, changes during fuel cell operation. Schulze et al. reported PTFE decomposition associated with a decrease in electrode hydrophobicity after 1600 h of operation with nonhumidified reactant gases (Schulze et al., 2007). The same authors also reported that PTFE decomposition, which results in critical water imbalance and partial loss of electrochemical performance, can be determined by x-ray photoelectron spectroscopy (XPS).

The reported work on electrochemical degradation is more sparse, but it includes AT methods that consisted of placing GDL samples in a 0.5 M sulfuric acid solution, applying large potentials (>1 V), and measuring the corrosion current or maintaining the samples at high potentials for predetermined time intervals (Frisk et al., 2004; Hicks, 2005). The samples aged at higher potentials (e.g., 2.6 V) showed decreased permeability under humidified conditions (suggesting possible pore infiltration and flooding), and the dry gas permeability decreased with the applied aging potential (Hicks, 2005). Both the chemical and electrochemical aging methods have demonstrated the drastic effects on hydrophobicity, which in turn, impacts the rate determining electrochemical step at the cathodes.

The changes with respect to oxygen transport have been quantified by increases in the associated mass transport overpotential, whose separation from the total voltage losses can be accomplished by Tafel slope analysis (Gasteiger et al., 2003; Neyerlin et al., 2005; Williams et al., 2005; Schmidt, 2006; Schmidt and Baurmeister, 2006). Gasteiger et al. reported on an expression for combined mass transport overpotentials (anodic and cathodic) and the application of multicomponent gas analysis (e.g., using air, oxygen, and helox: a 21–79% mixture of helium in oxygen). With such oxidant mixtures, the differences in diffusivity can be used to discern the losses associated with topological differences (e.g., pore sizes in the MPL and GDL). Williams et al. identified six overpotentials including a separate expression for the PTL overpotential.

Beuscher used experimental data to obtain extrapolated values of the limiting current density in PEMFCs. He also used an electrical model to separate the mass-transport losses associated with the catalyst layer and the PTL. He concluded that the PTL losses represent approximately 25% of the total mass-transport resistance (Beuscher, 2006).

More recent attempts have been made to link transport properties (e.g., diffusivity) to electrochemical properties (impedance) (Kramer et al., 2008), and long-term degradation (Wood and Borup, 2010a). The individual mass transport losses have also been isolated via current-interrupt methods (Stumper et al., 2005) and reference electrodes (Herrera et al., 2009). In all cases, the changes in mass transport overpotential provide a reliable quantifier for mass transport degradation, but the challenge remains to link these measurements with the local phenomena at each layer (see Figure 5.12). The direct measurement of species concentration *in situ* could elucidate the relevant mechanisms and the regions or layers most susceptible to degradation.

Previous efforts relevant to oxygen transport have used two-electrode amperometric sensors such as the Clark and Mackereth cells (Clark, 1959). These sensors are inexpensive, accurate, and small, but they require frequent calibration, consume oxygen, and suffer from long-term drift. Other options include cyclic voltammetric sensors with remote three-electrode cells, and variations on galvanic techniques to monitor limiting currents at steady state (Fan et al., 1991; Haug and White, 2000; Utaka et al., 2009).

Several efforts incorporating microsensors (He et al., 2006b) and other *in situ* techniques have been reported. Wilkinson and coworkers implemented microthermocouples (Wilkinson et al., 2006) to map

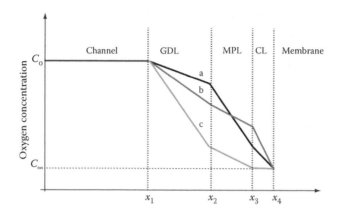

FIGURE 5.12 Three hypothetical concentration profiles across MEA layers. The limiting current densities achievable with each profile a, b, and c are indistinguishable.

the current distribution in an operating cell. He et al. used a thin film gold thermistor for *in situ* temperature measurements in the membrane. The sensor was placed between two Nafion membranes, and protected by a thin parylene film (He et al., 2006b). Lee et al. have also reported a thin-film, microtemperature sensor (2200 Å Au film on a Cr/AlN substrate) that could be used *in situ* (Lee et al., 2011).

More recently, solid-state and optical sensors for humidity and oxygen detection have been reported (Khijwania et al., 2005; Steele et al., 2006, 2009; Lee et al., 2009; Inman et al., 2010). These sensors are small, chemically stable, and minimally intrusive in typical PEMFC environments. They also enable transient measurements that approach the temporal resolution provided by available electrochemical and data-acquisition hardware (<1 s).

Basu et al. measured the RH and temperature in a PEMFC cathode using a tunable diode laser (Basu et al., 2006). More recently, David et al. have reported on *in situ* Bragg grating sensors fabricated from small-diameter (125 μm) optical fibers. These sensors have been used to measure temperature at multiple points within an operating PEM fuel cell (David et al., 2009, 2010). These efforts exemplify a trend toward direct, *in situ* noninvasive measurements.

5.4.3 Mitigation Strategies

The durability of stacks and systems has been improved by focusing on the durability of expensive materials and components (catalysts, membranes, etc.). In contrast, the durability of PTL materials has received attention only recently.

The importance of changes in physical properties (mechanical strength, thermal and electrical conductivity, etc.) has been identified and characterized, but the degradation with respect to water management is the current priority. The reported experimental evidence suggests that chemical and electrochemical degradation mechanisms are the most important, and that they are enabled by the presence of liquid water. As a result, several strategies have been implemented to mitigate the operational and (potential) lifetime effects of water within the PEMFC or stack. Passive mitigation strategies include pressure drop optimization in the flow-field plates, temperature gradients along the flow fields, and counterflow operation (Mathias et al., 2003). Although these do not target PTL durability directly, their implementation can have beneficial effects on the GDL and MPL materials.

The mitigation measures for PTL degradation include optimizing the composition (e.g., by varying PTEF loadings) and manufacturing processes (e.g., by producing GDL materials at higher temperatures to increase the crystallinity in graphite) (Borup et al., 2007). In light of the large effects due to the porous microstructures, changes in the final structures may have equally beneficial effects (e.g., from random to periodic or regular networks). Similarly, and although the MPL was introduced to improve performance, the recent work on hydrophobicity gradients (Wood et al., 2002) may extend durability by extracting the liquid water more effectively, or by directing the water fluxes to areas of lower susceptibility.

Such an approach has been suggested by Gerteisen et al. (2008). Their work was based on the premise that positive pressure is needed for through-plane transport of liquid water, and that this is higher than that for in-plane transport. This premise has been corroborated experimentally by the work reported on saturation versus capillary pressure (Gostick et al., 2010). Based on this premise, Gerteisen et al. created water transport channels to support the through-plane transport, and lower the liquid water saturation (thereby increasing the overall oxygen diffusion to the catalyst layer). The hypotheses were tested experimentally with a small single cell (1 cm^2). The cathode GDL under the channel exposed to the reactant was perforated with a laser. The regular perforations had a diameter of 80 μm and a gap distance of 1 mm. The performance of the modified GDLs was compared via potentials sweeps and chronoamperometry. These authors concluded that the perforated GDL provided enhances performance against flooding. However, the connection to improved durability would require validation via AT or long-term monitoring.

Other approaches have attempted to decouple the water removal from the oxidant delivery (Litster et al., 2007). In this work, a porous carbon flow-field plate was used as a wick to redistribute water passively within the cell.

In a related work, Buie et al. proposed an active water removal scheme via electroosmotic pumping at the PEMFC cathodes (Buie et al., 2006). The pump operation required approximately 10% of the available power, and it illustrates the drawbacks associated with active control: lower efficiency, and increases in complexity and costs.

While the effects of freezing on PTL materials require further clarification, the effects on the overall fuel cell system have been the focus of intense efforts. In 2005, Pesaran and coworkers published a report on the PEMFC freeze and rapid start (Pesaran et al., 2005). They found that the number of relevant patents exceeded the number of scientific articles published at the time of publication. They also identified the three most important mitigation strategies. They consisted of preventing freezing events by maintaining the stacks warm, by "thaw-at-start" methods, and by using effective insulation (which could delay the onset of stack freezing for up to several days following system turn-off). While these strategies were developed for the overall system, they would have a beneficial impact on the PTL.

5.5 Summary

The transport of energy, mass, and charge is at the heart of PEMFC operation. The PTL provides a crucial link between macroscopic phenomena in the flow channels and the micro- and nanoscopic processes in the catalyst layers. Typical PTL configurations consists of a diffusion layer (GDL), and a microporous layer (MPL). The GDL provides an electrical and thermal connection to the flow-field plates and facilitates species transport through macroscopic pores. The MPL controls the water fluxes near the catalyst layer, and reduces the contact thermal and electrical resistances.

Despite their importance, the transport and degradation phenomena within the PTL are still poorly understood, and there are no reliable predictive models to guide design optimization. As a result, accelerated testing methods are required to link physical material properties and functions to degradation rates and durability. Despite recent progress, further refinements are required to provide an operational context for the property variations. Moreover, the quantification of PTL degradation is challenging and remains unresolved because some of the relevant theories (e.g., capillary flow in porous media) are inadequate, and the required characterization methods (e.g., *in situ* species concentration measurement) are still under development.

The most important degradation is related to changes in the mass transport. Of these, the changes associated with liquid water, oxygen, and water vapor have the greatest impact on performance and durability. The most important degradation mechanisms include chemical and electrochemical hydrophobicity changes on the surfaces and porous layers.

The proposed active mitigation strategies (e.g., pneumatic or electroosmotic water removal, drying steps on shutdown, etc.) are complex and increase costs. The passive mitigation strategies to prolong stack and system durability (e.g., temperature and pressure gradients, counterflow operation, etc.), and those introduced as changes in manufacturing are insufficient to meet the durability requirements in many applications.

References

Alink, R., Gerteisen, D., and Oszcipok, M. 2008. Degradation effects in polymer electrolyte membrane fuel cell stacks by sub-zero operation—An in situ and ex situ analysis. *Journal of Power Sources* 182: 175–187.

Baba, T. and Ono, A. 2001. Improvement of the laser flash method to reduce uncertainty in thermal diffusivity measurements. *Measuring Science Technology* 12: 2046–2057.

Basu, S., Renfro, M. W., and Cetegen, B. M. 2006. Spatially resolved optical measurements of water partial pressure and temperature in a PEM fuel cell under dynamic operating conditions. *Journal of Power Sources* 162: 286–293.

Berning, T. and Djilali, N. 2003. *Journal of the Electrochemical Society* 150: A1589.

Beuscher, U. 2006. Experimental method to determine the mass transport resistance of a polymer electrolyte fuel cell *Journal of the Electrochemical Society* 153: A1788.

Borup, R., Meyers, J., Pivovar, B. et al. 2007. Scientific aspects of polymer electrolyte fuel cell durability and degradation. *Chemical Reviews* 107: 3904–3951.

Buie, C. R., Posner, J. D., Fabian, T. et al. 2006. Water management in proton exchange membrane fuel cells using integrated electroosmotic pumping. *Journal of Power Sources* 161: 191–202.

Burheim, O., Vie, P. J. S., Pharoah, J. G., and Kjelstrup, S. 2010. *Ex situ* measurements of through-plane thermal conductivities in a polymer electrolyte fuel cell. *Journal of Power Sources* 195: 249–256.

Cassie, A. B. D. 1948. Contact angles. *Transactions of the Faraday Society* 44: 11–16.

Cheung, P., Fairweather, J. D., and Schwartz, D. T. 2009. Characterization of internal wetting in polymer electrolyte membrane gas diffusion layers. *Journal of Power Sources* 187: 487–492.

Chi, P. H., Chan, S. H., Weng, F. B. et al. 2010. On the effects of non-uniform property distribution due to compression in the gas diffusion layer of a PEMFC. *International Journal of Hydrogen Energy* 35: 2936–2948.

Cindrella, L. and Kannan, A. M. 2009. Membrane electrode assembly with doped polyaniline interlayer for proton exchange membrane fuel cells under low relative humidity conditions. *Journal of Power Sources* 193: 447–453.

Cindrella, L., Kannan, A. M., Lin, J. F. et al. 2009. Gas diffusion layer for proton exchange membrane fuel cells—A review. XIth Polish Conference on Fast Ionic Conductors. *Journal of Power Sources* 194: 146–160.

Clark, L. C. 1959. *Electrochemical device for chemical analysis*. US Patent No. 2,913,386.

Cleghorn, S., Kolde, J., and Liu, W. 2003. Chapter 44: Catalyst coated composite membranes. In *Handbook of Fuel Cells*, eds. A. L. W. Vielstich, H. A. Gasteiger, pp. 566–575. Etobicoke, Canada: John Wiley & Sons.

Concus, P. and Finn, R. 1991. Capillary surfaces in exotic containers. In *AIAA Aerospace Sciences Meeting*, January 9–12, 1991, Reno, NV. National Aeronautics and Space Administration (NASA), pp. 1–4.

Concus, P. and Finn, R. 1993. Capillary surfaces in a wedge: Differing contact angles. *Contract Number DE-AC03-76SF00098*. Applied Mathematical Sciences Subprogram of the Office of Energy Research, U.S. Department of Energy. National Aeronautics and Space Administration (Grant NAG3–1143), and National Science Foundation (Grant DMS89–02831).

Concus, P., Finn, R., and Weislogel, M. 1995. Proboscis container shapes for the USML-2 interface configuration experiment. In *Ninth European Symposium Gravity Dependent Phenomena in Physical Sciences*, May 2–5, 1995, Berlin, Germany. US National Aeronautics and Space Administration, pp. 1–8.

David, N. A., Wild, P. M., Hu, J., and Djilali, N. 2009. In-fibre Bragg grating sensors for distributed temperature measurement in a polymer electrolyte membrane fuel cell. *Journal of Power Sources* 192: 376–380.

David, N. A., Wild, P. M., Jensen, J., Navessin, T., and Djilali, N. 2010. Simultaneous *in situ* measurement of temperature and relative humidity in a PEMFC using optical fiber sensors. *Journal of the Electrochemical Society* 157: B1173–B1179.

Dicks, A. and Siddle, A. 2000. Assesment of commercial prospects of molten carbonate fuel cells. *Journal of Power Sources* 86: 316–323.

Djilali, N. and Lu, D. 2002. Influence of heat transfer on gas and water transport in fuel cells. *International Journal of Thermal Sciences* 41: 29–40.

Dohle, H., Schmitz, H., Bewer, T., Mergel, J., and Stolten, D. 2002. Development of a compact 500 W class direct methanol fuel cell stack. *Journal of Power Sources* 106: 313–322.

Dullien, F. A. L. 1988. Two-phase flow in porous media. *Chemical Engineering & Technology* 11: 407–424.

Dussan, E. B. 1979. On the spreading of liquids on solid surfaces: static and dynamic contact lines. *Annual Review of Fluid Mechanics* 11: 371–400.

Dyson, J. and Hirst, W. 1954. The true contact area between solids. *Proceedings of the Physical Society* B67: 309–312.

Escobar, L. A. and Meeker, W. Q. 1999. *Statistical Methods for Reliability Data*. New York, NY: John Wiley & Sons.

Escribano, S., Blachot, J.-F., Ethève, J., Morin, A., and Mosdale, R. 2006. Characterization of PEMFCs gas diffusion layers properties. *Journal of Power Sources* 156: 8–13.

Fan, D., White, R. E., and Gruberger, N. 1991. Diffusion of a gas through a membrane. *Journal of Applied Electrochemistry* 22: 770–772.

Finn, R. 2002a. Eight remarkable properties of capillary surfaces. *The Mathematical Intelligencer* 24: 21–33.

Finn, R. 2002b. Some properties of capillary surfaces. *Milan Journal of Mathematics* 70: 1–23.

Finn, R. 2006. Capillary surfaces. In *Encyclopedia of Mathematical Physics*, eds. F. Jean-Pierre, L. N. Gregory and T. Tsou Sheung, pp. 431–445. Oxford: Academic Press.

Fishman, J. Z., Hinebaugh, J., and Bazylak, A. 2010. Microscale tomography investigations of heterogeneous porosity distributions of PEMFC GDLs. *Journal of the Electrochemical Society* 157(11): B1643–B1650.

Frisk, J. W., Hicks, M. T., Atanasoski, R. T. et al. 2004. MEA component durability. In *Fuel Cell Seminar*, November 1–5. San Antonio, TX, USA.

Gasteiger, H. A., Gu, W., Makharia, R., Mathias, M. F., and Sompalli, B. 2003. Beginning of life MEA performance—Efficiency loss contributions. In *Handbook of Fuel Cells: Fundamentals, Technology, and Applications*, Chapter 46, eds. W. Vielstich, A. Lamm, and H. A. Gasteiger, pp. 593–610. New York, NY: John Wiley & Sons.

Gerteisen, D., Heilmann, T., and Ziegler, C. 2008. Enhancing liquid water transport by laser perforation of a GDL in a PEM fuel cell. *Journal of Power Sources* 177: 348–354.

Good, R. J. and Girifalco, L. A. 1960. A theory for estimation of surface and interfacial energies. III. Estimation of surface energies of solids from contact angle data. *The Journal of Physical Chemistry* 64: 561–565.

Gostick, J. T., Ioannidis, M. A., Fowler, M. W., and Pritzker, M. D. 2009a. On the role of the microporous layer in PEMFC operation. *Electrochemistry Communications* 11: 576–579.

Gostick, J. T., Ioannidis, M. A., Fowler, M. W., and Pritzker, M. D. 2009b. Wettability and capillary behavior of fibrous gas diffusion media for polymer electrolyte membrane fuel cells. *Journal of Power Sources* 194: 433–444.

Gostick, J. T., Ioannidis, M. A., Fowler, M. W., and Pritzker, M. D. 2010. Characterization of the capillary properties of gas diffusion media. In *Modeling and Diagnostics of Polymer Electrolyte Fuel Cells*, eds. C.-Y. Wang and U. Pasaogullari, pp. 225–254. New York, NY: Springer.

Greenwood, J. A. and Williamson, J. B. P. 1966. Contact of nominally flat surfaces. *Proceedings of the Royal Society* A295: 300–319.

Gurau, V., Bluemle, M. J., De Castro, E. S. et al. 2006. Characterization of transport properties in gas diffusion layers for proton exchange membrane fuel cells: 1. Wettability (internal contact angle to water and surface energy of GDL fibers). *Journal of Power Sources* 160: 1156–1162.

Gurau, V. and Mann Jr, J. A. 2010. Technique for characterization of the wettability properties of gas diffusion media for proton exchange membrane fuel cells. *Journal of Colloid and Interface Science* 350: 577–580.

Harkness, I. R., Hussain, N., Smith, L., and Sharman, J. D. B. 2009. The use of a novel water porosimeter to predict the water handling behaviour of gas diffusion media used in polymer electrolyte fuel cells. *Journal of Power Sources* 193: 122–129.

Haug, A. T. and White, R. E. 2000. Oxygen diffusion coefficient and solubility in a new proton exchange membrane. *Journal of the Electrochemical Society* 147: 980–983.

He, S. and Mench, M. 2006a. One-dimensional transient model for frost heave in polymer electrolyte fuel cells. I. Physical model. *Journal of the Electrochemical Society* 153: A1724–A1731.

He, S., Mench, M. M., and Tadigadapa, S. 2006b. Thin film temperature sensor for real-time measurement of electrolyte temperature in a polymer electrolyte fuel cell. *Sensors and Actuators A: Physical* 125: 170–177.

Herrera, O., Mérida, W., and Wilkinson, D. P. 2009. Sensing electrodes for failure diagnostics in fuel cells. *Journal of Power Sources* 190: 103–109.

Hicks, M. 2005. MEA and stack durability for PEM fuel cells. *DOE Hydrogen Program Review 3M/DOE Cooperative Agreement No. DE-FC36-03GO13098.* 3M Company.

Ho, C. K. and Udell, K. S. 1995. Mass transfer limited drying of porous media containing an immobile binary liquid mixture. *International Journal of Heat and Mass Transfer* 38: 339–350.

Holm, R. 1954. *Electric Contacts Handbook.* Berlin: Springer-Verlag.

Huang, K.-L., Lai, Y.-C., and Tsai, C.-H. 2006. Effects of sputtering parameters on the performance of electrodes fabricated for proton exchange membrane fuel cells. *Journal of Power Sources* 156: 224–231.

Inman, K., Wang, X., and Sangeorzan, B. 2010. Design of an optical thermal sensor for proton exchange membrane fuel cell temperature measurement using phosphor thermometry. *Journal of Power Sources* 195: 4753–4757.

Ip, S. W. and Toguri, J. M. 1994. The equivalence of surface tension, surface energy and surface free energy. *Journal of Materials Science* 29: 688–692.

Ismail, M. S., Damjanovic, T., Ingham, D. B., Pourkashanian, M., and Westwood, A. 2010. Effect of polytetrafluoroethylene-treatment and microporous layer-coating on the electrical conductivity of gas diffusion layers used in proton exchange membrane fuel cells. *Journal of Power Sources* 195: 2700–2708.

Jackson, R. L. and Streator, J. L. 2006. A multi-scale model for contact between rough surfaces. *Wear* 261: 1337–1347.

Jeng, Y. R. and Wang, P. Y. 2003. An elliptical microcontact model considering elastic, elastoplastic, and plastic deformation. *Transactions of the ASME* 125: 232–240.

Kang, J. H., Lee, K.-J., Nam, J. H. et al. 2010. Visualization of invasion-percolation drainage process in porous media using density-matched immiscible fluids and refractive-index-matched solid structures. *Journal of Power Sources* 195: 2608–2612.

Kannan, A. M., Kanagala, P., and Veedu, V. 2009. Development of carbon nanotubes based gas diffusion layers by *in situ* chemical vapor deposition process for proton exchange membrane fuel cells. *Journal of Power Sources* 192: 297–303.

Karimi, G., Li, X., and Teertstra, P. 2010. Measurement of through-plane effective thermal conductivity and contact resistance in PEM fuel cell diffusion media. *Electrochimica Acta* 55: 1619–1625.

Kehoe, L., Kelly, P. V., and Crean, G. M. 1998. Application of the laser flash diffusivity method to thin high thermal conductivity materials. *Microsystem Technologies* 5: 18–21.

Khandelwal, M. and Mench, M. M. 2006. Direct measurement of through-plane thermal conductivity and contact resistance in fuel cell materials. *Journal of Power Sources* 161: 1106–1115.

Khijwania, S. K., Srinivasan, K. L., and Singh, J. P. 2005. An evanescent-wave optical fiber relative humidity sensor with enhanced sensitivity. *Sensors and Actuators B: Chemical* 104: 217–222.

Kim, S., Ahn, B. K., and Mench, M. M. 2008. Physical degradation of membrane electrode assemblies undergoing freeze/thaw cycling: Diffusion media effects. *Journal of Power Sources* 179: 140–146.

Kitahara, T., Konomi, T., and Nakajima, H. 2010. Microporous layer coated gas diffusion layers for enhanced performance of polymer electrolyte fuel cells. *Journal of Power Sources* 195: 2202–2211.

Kleemann, J., Finsterwalder, F., and Tillmetz, W. 2009. Characterisation of mechanical behaviour and coupled electrical properties of polymer electrolyte membrane fuel cell gas diffusion layers. *Journal of Power Sources, Selected Papers presented at the 11th ULM ElectroChemical Days* 190: 92–102.

Knights, S. D., Colbow, K. M., St-Pierre, J., and Wilkinson, D. P. 2004. Aging mechanisms and lifetime of PEFC and DMFC. *Journal of Power Sources* 127: 127–134.

Kramer, D., Freunberger, S. A., Flückiger, R. et al. 2008. Electrochemical diffusimetry of fuel cell gas diffusion layers. *Journal of Electroanalytical Chemistry* 612: 63–77.

Langbein, D. 2002. *Capillary Surfaces. Shape–Stability–Dynamics, in Particular Under Weightlessness.* Heidelberg: Springer-Verlag.

Laplace, P. S. 1805. Traite de Mecanique Celeste: supplement au Livre X. Theorie de l'action capillaire In *Oeuvres completes.* Paris: Gauthier-Villars, pp. 358–498.

Laua, Y. Y. and Tang, W. 2009. A higher dimensional theory of electrical contact resistance. *Journal of Applied Physics* 105: 124902-1–124902-10.

Lee, C. and Mérida, W. 2007. Gas diffusion layer durability under steady-state and freezing conditions. *Journal of Power Sources* 164: 141–153.

Lee, B., Roh, S., and Park, J. 2009. Current status of micro- and nano-structured optical fiber sensors. *Optical Fiber Technology* 15: 209–221.

Lee, C.-Y., Weng, F.-B., Cheng, C.-H. et al. 2011. Use of flexible micro-temperature sensor to determine temperature *in situ* and to simulate a proton exchange membrane fuel cell. *Journal of Power Sources* 196: 228–234.

Lee, W.-K., Ho, C.-H., Van Zee, J. W., and Murthy, M. 1999. The effects of compression and gas diffusion layers on the performance of a PEM fuel cell. *Journal of Power Sources* 84: 45–51.

Leverett, M. C. 1941. Capillary behaviour in porous solids. *Transactions of the AIME* 142: 159–172.

Litster, S., Buie, C. R., Fabian, T., Eaton, J. K., and Santiago, J. G. 2007. Active water management for PEM fuel cells. *Journal of the Electrochemical Society* 154: B1049–B1058.

Liu, D. and Case, S. 2006. Durability study of proton exchange membrane fuel cells under dynamic testing conditions with cyclic current profile. *Journal of Power Sources* 162: 521–531.

Mathias, M. F., Roth, J., Fleming, J., and Lehnert, W. 2003. Chapter 42: Difusion media materials and characterisation. In *Handbook of Fuel Cells—Fundamentals, Technology and Applications.* Chichester: John Wiley & Sons, pp. 517–537.

McNaught, A. D. and Wilkinson, A. 1997. *The IUPAC Compendium of Chemical Terminology.* The International Union of Pure and Applied Chemistry. Cambridge, UK: Royal Society of Chemistry.

Meeker, W. Q. and Escobar, L. A. 1998. Pitfalls of accelerated testing. *IEEE Transactions on Reliability* 47: 114–118.

Mikrajuddin, A., Shi, F. G., Kim, H. K., and Okuyama, K. 1999. Size-dependent electrical constriction resistance for contacts of arbitrary size: from Sharvin to Holm limits. *Materials Science in Semiconductor Processing* 2: 321–327.

Min, S., Blumm, J., and Lindemann, A. 2007. A new laser flash system for measurement of the thermophysical properties. *Thermochimica Acta* 455: 46–49.

Mukundan, R., Kim, Y., Garzon, F., and Pivovar, B. 2006. Freeze/thaw effects in PEM fuel cells. *Transactions of the Electrochemical Society* 1: 403–413.

Nelson, W. B. 2005a. A bibliography of accelerated test plans. *IEEE Transactions on Reliability* 54: 194–197.

Nelson, W. B. 2005b. A bibliography of accelerated test plans. Part II—References. *IEEE Transactions on Reliability* 54: 370–373.

Neyerlin, K. C., Gasteiger, H. A., Mittelsteadt, C. K., Jorne, J., and Gu, W. 2005. Effect of relative humidity on oxygen reduction kinetics in a PEMFC. *Journal of the Electrochemical Society* 152: A1073–A1080.

Nitta, I., Himanen, O., and Mikkola, M. 2008. Thermal conductivity and contact resistance of compressed gas diffusion layer of PEM fuel cell. *Fuel Cells* 8: 111–119.

Oszcipok, M., Zedda, M., Riemann, D., and Geckeler, D. 2006. Low temperature operation and influence parameters on the cold start ability of portable PEMFCs. *Journal of Power Sources* 154: 404–411.

Owens, D. K. and Wendt, R. C. 1969. Estimation of the surface free energy of polymers. *Journal of Applied Polymer Science* 13: 1741–1747.

Parker, W. J., Jenkins, R. J., Butler, C. P., and Abbott, G. L. 1961. Flash method of determining thermal diffusivity, heat capacity, and thermal conductivity. *Journal of Applied Physics* 32: 1679–1684.

Parry, V., Appert, E., and Joud, J. C. 2010. Characterisation of wettability in gas diffusion layer in proton exchange membrane fuel cells. *Applied Surface Science* 256: 2474–2478.

Patrikalakis, N. M., Maekawa, T., Patrikalakis, N. M., and Maekawa, T. 2010. Differential geometry of surfaces. In *Shape Interrogation for Computer Aided Design and Manufacturing*. Berlin: Springer, pp. 49–72.

Pesaran, A. A., Kim, G.-H., and Gonder, J. D. 2005. *PEM Fuel Cell Freeze and Rapid Startup Investigation*. Colorado, US: National Renewable Energy Laboratory.

Petersen, C. L., Hansen, T. M., Boggild, P. et al. 2002. Scanning microscopic four point conductivity probes. *Sensors and Actuators A* 96: 53–58.

Pharoah, J. G., Karan, K., and Sun, W. 2006. On effective transport coefficients in PEM fuel cell electrodes: Anisotropy of the porous transport layers. *Journal of Power Sources* 161: 214–224.

Ramé, E. 1997. The interpretation of dynamic contact angles measured by the Wilhelmy plate method. *Journal of Colloid and Interface Science* 185: 245–251.

Ramousse, J., Didierjean, S., Lottin, O., and Maillet, D. 2008. Estimation of the effective thermal conductivity of carbon felts used as PEMFC gas diffusion layers. *International Journal of Thermal Sciences* 47: 1–6.

Sadeghi, E., Bahrami, M., and Djilali, N. 2008. Analytic determination of the effective thermal conductivity of PEM fuel cell gas diffusion layers. *Journal of Power Sources* 179: 200–208.

Sadeghi, E., Djilali, N., and Bahrami, M. 2010. Effective thermal conductivity and thermal contact resistance of gas diffusion layers in proton exchange membrane fuel cells. Part 2: Hysteresis effect under cyclic compressive load. *Journal of Power Sources* 195(24): 8104–8109.

Sadeghi, E., Djilali, N., and Bahrami, M. 2011. Effective thermal conductivity and thermal contact resistance of gas diffusion layers in PEM fuel cells. Part 1: Effect of compressive load. *Journal of Power Sources* 196: 246–254.

Schmidt, T. J. 2006. Durability and degradation in high-temperature polymer electrolyte fuel cells. *ECS Transactions* 1: 19–31.

Schmidt, T. J. and Baurmeister, J. 2006. Durability and reliability in high-temperature reformed hydrogen PEFCs. *ECS Transactions* 3: 861–869.

Schulze, M., Wagner, N., Kaz, T., and Friedrich, K. A. 2007. Combined electrochemical and surface analysis investigation of degradation processes in polymer electrolyte membrane fuel cells. *Electrochimica Acta* 52: 2328–2336.

Sharvin, Y. V. 1965. *Soviet Physics JEPT* 21: 655.

Shikhmurzaev, Y. D. 2008. On Young's (1805) equation and Finn's (2006) [']counterexample'. *Physics Letters A* 372: 704–707.

Smits, F. M. 1958. Measurement of sheet resistivities with the fourßpoint probe. *Bell Systems Technical Journal* 37: 711–718.

Sole, J. D. 2010. *Investigation of water transport parameters and processes in the gas diffusion laser of PEM fuel cells*. Doctoral Dissertation, Virginia Polytechnic Institute and State University.

Steele, J. J., Taschuk, M. T., and Brett, M. J. 2009. Response time of nanostructured relative humidity sensors. *Sensors and Actuators B: Chemical* 140: 610–615.

Steele, J. J., Van Popta, A. C., Hawkeye, M. M., Sit, J. C., and Brett, M. J. 2006. Nanostructured gradient index optical filter for high-speed humidity sensing. *Sensors and Actuators B: Chemical* 120: 213–219.

Struik, D. J. 1961. *Lectures on Classical Differential Geometry*, ed. Dirk J. Struik. Reading, MA: Addison-Wesley.

Stumper, J., Haas, H., and Granados, A. 2005. *In situ* determination of mea resistance and electrode diffusivity of a fuel cell. *Journal of the Electrochemical Society* 152: A837–A844.

Takahashi, Y. 1984. Measurement of thermophysical properties of metals and ceramics by the laser-flash method. *International Journal of Thermophysics* 5: 41–52.

Teertstra, P., Karimi, G., and Li, X. 2011. Measurement of in-plane effective thermal conductivity in PEM fuel cell diffusion media. *Electrochimica Acta* 56(3): 1670–1675.

Thorstensson, S., Wang, F., Petersen, D. H. et al. 2009. Accurate microfour-point probe sheet resistance measurements on small samples. *Review of Scientific Instruments* 80: 053902-1–10.

Udell, K. S. 1985. Heat transfer in porous media considering phase change and capillarity—The heat pipe effect. *International Journal of Heat and Mass Transfer* 28: 485–495.

Uhlir, A. J. 1955. The potentials of infinite systems of sources and numerical solutions of problems in semiconductor engineering. *Bell Systems Technical Journal* 34: 105–111.

Utaka, Y., Tasaki, Y., Wang, S., Ishiji, T., and Uchikoshi, S. 2009. Method of measuring oxygen diffusivity in microporous media. *International Journal of Heat and Mass Transfer* 52: 3685–3692.

Valdes, L. 1954. Resistivity measurements on germanium for transistors. *Proceedings of the I.R.E.* 42: 420–427.

Wang, Z. H., Wang, C. Y., and Chen, K. S. 2001. Two-phase flow and transport in the air cathode of proton exchange membrane fuel cells. *Journal of Power Sources* 94: 40–50.

Washburn, E. W. 1921. The dynamics of capillary flow. *Physical Review* 17: 273.

Wilde, P., Mändle, M., Murata, M., and Berg, N. 2004. Structural and physical properties of GDL and GDL/BPP combinations and their influence on PEMFC performance. *Fuel Cells* 4: 180–184.

Wilhelmy, L. 1863. Ueber die Abhängigkeit der Capillaritäts-Constanten des Alkohols von Substanz und Gestalt des benetzen festen Körpers. *Annalen der Physik und Chemie* 119: 177–217.

Wilkinson, D. P. and St.-Pierre, J. 2003. Durability. In *Handbook of Fuel Cells—Fundamentals, Technology and Applications*, Chapter 47. Chichester: John Wiley & Sons, pp. 611–626.

Wilkinson, M., Blanco, M., Gu, E. et al. 2006. *In situ* experimental technique for measurement of temperature and current distribution in proton exchange membrane fuel cells. *Electrochemical and Solid-State Letters* 9: A507–A511.

Williams, M. V., Begg, E., Bonville, L., Kunz, H. R., and Fenton, J. M. 2004. Characterization of gas diffusion layers for PEMFC. *Journal of the Electrochemical Society* 151: A1173–A1180.

Williams, M. V., Kunz, R. H., and Fenton, J. M. 2005. Analysis of polarization curves to evaluate polarization sources in hydrogen/air PEM fuel cells. *Journal of the Electrochemical Society* 152: A635.

Wood, D. L. 2007. *Fundamental material degradation studies during long-term operation of hydrogen/air PEMFCs*. Doctoral Dissertation, University of New Mexico.

Wood, D. L. and Borup, R. L. 2009. Durability aspects of gas-diffusion and microporous layers. In *Polymer Electrolyte Fuel Cell Durability*, eds. F. N. Büchi, M. Inaba, and T. J. Schmidt, eds. New York, NY: Springer.

Wood, D. L. and Borup, I. R. L. 2010a. Estimation of mass-transport overpotentials during long-term PEMFC operation. *Journal of the Electrochemical Society* 157: B1251–B1262.

Wood, D. L., Iii, Rulison, C., and Borup, R. L. 2010b. Surface properties of PEMFC gas diffusion layers. *Journal of the Electrochemical Society* 157: B195–B206.

Wood, D. L., Grot, S. A., and Fly, G. 2002. *Composite Gas Distribution Structure for Fuel Cell*. US Patent No. 6,350,539.

Wu, J., Martin, J. J., Orfino, F. P. et al. 2010. *In situ* accelerated degradation of gas diffusion layer in proton exchange membrane fuel cell: Part I: Effect of elevated temperature and flow rate. *Journal of Power Sources* 195: 1888–1894.

Wu, J., Yuan, X. Z., Martin, J. J. et al. 2008. A review of PEM fuel cell durability: Degradation mechanisms and mitigation strategies. *Journal of Power Sources* 184: 104–119.

Young, T. 1805. An essay on the cohesion of fluids. *Philosophical Transactions of the Royal Society London* 95: 65–87.

Zhang, F.-Y., Advani, S. G., and Prasad, A. K. 2008. Performance of a metallic gas diffusion layer for PEM fuel cells. *Journal of Power Sources* 176: 293–298.

Zhang, S., Yuan, X., Wang, H. et al. 2009. A review of accelerated stress tests of MEA durability in PEM fuel cells. *International Journal of Hydrogen Energy* 34: 388–404.

Zhou, T. and Liu, H. 2006. Effects of the electrical resistances of the GDL in a PEM fuel cell. *Journal of Power Sources* 161: 444–453.

6

Degradation of Bipolar Plates and Its Effect on PEM Fuel Cells

Lars Kühnemann
Center for Fuel Cell Technology

Thorsten Derieth
Center for Fuel Cell Technology

Peter Beckhaus
Center for Fuel Cell Technology

Angelika Heinzel
Center for Fuel Cell Technology

6.1 Introduction

Bipolar plates as main part of fuel cell stacks are responsible for the transport and separation of all media within the stack, the transport of electrons and heat, and have to realize a smooth mechanical environment for the cells, respectively, membrane electrode assembly (MEA) and gaskets. Therefore, the requirements for bipolar plates derived from various publications regarding stack development do focus on ohmic resistance, thermal stability, mechanical properties, and—with respect to the complex electrochemical environment within the cells—corrosion targets. As the long-term stability of materials is highly dependent especially on fuel cell concepts and operational parameters, numerous bipolar plate technologies exist based on a wide range of materials and production methods. Due to the variety of applications for fuel cells ranging from high current and highly dynamic automobile stacks, via highly efficient stationary systems in constant and long-term operation to small power units with random operation modes, this bandwidth of materials is necessary to supply the best suitable bipolar plate technology for the application and its operating parameters.

Within this chapter, first the requirements for bipolar plates are summarized, followed by a discourse on the appropriate qualification procedures to certify the quality of bipolar plates. For the two main technology chains, metallic and graphite-based bipolar plates, a summary on documented degradation mechanisms is given. Finally, a review on different approaches for materials and coatings used, production methods, and resulting stabilities of the realized bipolar plates is done.

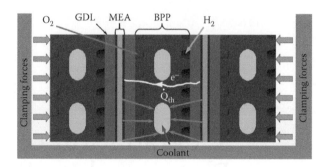

FIGURE 6.1   Schematic view of a PEM fuel cell cross-section. (Adapted from Kreuz, C. 2008. *Dissertation*: PEM-Brennstoffzellen mit spritzgegossenen Bipolarplatten aus hochgefülltem Graphit-Composite, Gerhard-Mercator-University in Germany, 04/2008.)

6.1.1   Requirements on Bipolar Plates

Bipolar plates, the essential hardware device in fuel cells, primarily need to have adequate mechanical strength. Therefore, bipolar plates must withstand high clamping pressures up to 3 MPa in general (Xing, 2010). They also have to be chemically stable, even at electric potentials resulting from open-circuit voltage (OCV) of the fuel cell. As illustrated in Figure 6.1, bipolar plates within a fuel cell have to conduct electric current, distribute and separate fuel, oxidant, and coolant, and enable water management and thermal management (Cooper, 2004). Bipolar plates also have to endure alternating loads caused by thermal dilation of fuel cell components in operation cycles. The major technical challenges for commercialization of PEM fuel cells are to reduce costs, weight, and volume. The same requirements apply for bipolar plates as an integral part of PEM fuel cells constituting the backbone of a stack. Hence, the development of alternative bipolar plates using cheap raw materials and mass production techniques is an important development goal. Furthermore, an essential point for bipolar plate materials is the chemical and stress environment at operating temperatures inside a PEM fuel cell stack. Bipolar plates for use in low-temperature PEM fuel cells are exposed to an operating environment with a pH of 2–3 (Nikam, 2005; Lafront, 2007) at temperatures around 80°C, whereas high-temperature bipolar plates have to withstand temperatures up to 200°C at similar pH value.

In order to assess the technical suitability of bipolar plates and bipolar plate materials for PEMFC the alignment with generally accepted main properties is necessary. In the long history of fuel cell research and efforts of commercialization, these general requirements changed and were subjected to diversification. As the improvement of one component of a fuel cell stack is a less complicated task than the improvement of the whole stack, up to now the discourse of technical targets for fuel cell commercialization has arrived at the conclusion to further diversify technical and cost problems regarding components such as membrane, gaskets, and not least bipolar plates.

Hence, technical and bipolar plate specific targets in the form of measurable values can be obtained from the current international literature today. The change of bipolar plate criteria for PEMFC can be shown by the well-established and documented targets of the United States Department of Energy (DOE).

Table 6.1 shows a lack of homogeneity in the definition of bipolar plate requirements. Nevertheless, there has been a differentiation between graphitic and metallic bipolar plates for PEMFC since 2006, as the ways of fabrication and feasible geometries are widely different and stationary, portable and automotive applications are distinct.

TABLE 6.1 Change of Technical Targets Concerning PEMFC Bipolar Plates (2003–2009)

Year	DOE Targets

(a)

2003

Design Criteria for Bipolar Plate Materials		
S. no.	Material Selection Criteria	Limit
1	Chemical compatibility	Anode face must not produce disruptive hydride layer; cathode face must not passivate and become nonconductive
2	Corrosion	Corrosion rate <0.016 mA/cm^2
3	Cost	Material + fabrication <US\$ 0.0045 cm^{-2}
4	Density	Density <5 g/cm^3
5	Dissolution	Minimization of dissolution (for metallic plates)
6	Electronic Conductivity	Plate resistance <0.01 Ω cm^2
7	Gas diffusivity/impermeability	Maximum average gas permeability <1.0 × 10^{-4}cm^3/s cm^2
8	Manufacturability	Cost of fabrication (see 3) should be low with high yield
9	Recyclable	Material can be recycled during vehicle service, following a vehicle accident, or when the vehicle is retired
10	Recycled	Made from recycled material
11	Stack Volume/kW	Volume <1 l/kW
12	Strength	Compressive strength >22 lb/in.^{2}
13	Surface Finish	>50 μm
14	Thermal Conductivity	Material should be able to remove heat effectively
15	Tolerance	>0.05 mm

(b)*

2006

Porvair Progress Toward Meeting DOE Bipolar Plate Property and Cost Targets			
	Units	2010/2015 Targets	Current Status
Cost	$/kW	6/4	<150
Weight	kg/kW	<1	<0.5
H$_2$ Permeation Rate	c/cm^2/sec (80°C, 3 atm)	2 × 10^{-6}	2 × 10^{-5}
Corrosion Resistance	mA/cm^2	<1	
Electrical Conductivity	S/cm	>100	>500
Resistivity	Ohm/cm^2	<0.01	
Flexural Strength	MPa	>4 (crush)	>34 (flexural)
Flexibility	% deflection at Mid-span	3–5	>10

(c)*

- Cost: $6/kg
- Weight: <1 kg/kW
- Corrosion: <1 mA/cm^2
- Conductivity: >100 S/cm
- H$_2$ permeation rate: <2 × 10^{-6} cm^3/sec•cm^2 at 80°C and 3 atm
- Resistivity: <0.01 Ω•cm^2
- Flexural strength: >4 MPa
- Flexibility: 3–5% at mid-span

continued

TABLE 6.1   (continued) Change of Technical Targets Concerning PEMFC Bipolar Plates (2003–2009)

Year	DOE Targets

(e)

Performance Requirements for PEM Fuel Cell Bipolar Plates		
Property	Units	Value
Tensile Strength— ASTM D638	MPa	>41
Flexural Strength— ASTM D790	MPa	>59
Electrical Conductivity	S cm^{-1}	>100
Corrosion Rate	μA cm^{-2}	<1
Contact Resistance	mΩ cm^2	<20
Hydrogen Permeability	cm^3 (cm^2 s)$^{-1}$	<2.10^{-6}
Mass	kg/kW	<1
Density—ASTM D792	g cm^{-2}	<5
Thermal Conductivity	W (m K)$^{-1}$	>10
Impact Resistance Unnotched) ASTM D-256	J m^{-1}	>40,5

(d)

2009

Characteristic	Units	2010/2015	Program 2009 Status
Cost^a	S/kW	5/3	TBD
Weight	kg/kW	<0.4	0.57
H$_2$ Permeation Flux	cm^3 sec^{-1} cm^{-2} @ 80°C, 3 atm (equivalent to <0.1 mA/cm^2)	<2 × 10^6	TBD
Corrosion	μA/cm^2	<1^b	<1^b
Electrical Conductivity	S/cm	>100	>1,000
Resistivity^c	Ohm-cm	0.01	<0.01
Flexural Strength	MPa	>25	>55
Flexibility	% deflection at mid-span	3 to 5	TBD

* Corrosion rate of 1 mA/cm^2 must be a typing error. Target is 1 μA/cm^2.
^{a} Bipolar plates for PEMFC in general (Metha, 2003).
^{b} Carbon/carbon composite bipolar plates for PEMFC (DOE, 2006).
^{c} Clad metal bipolar plates for PEMFC (DOE, 2006).
^{d} Graphite/polymer composite PEMFC bipolar plates for higher temperatures (120°C) and automotive applications (DOE, 2009).
^{e} Metal PEMFC bipolar plates for transportation applications (Antunes, 2010).

As can be seen from the illustrations (b) and (d) in Table 6.1 many targets like corrosion resistance,* electrical conductivity respectively resistivity[†] and flexural strength are easily reached by current composite bipolar plates. Consequently, these requirements are not addressed in composite bipolar plate development (DOE, 2006). However, current composite bipolar plates are too heavy compared to the DOE targets for transportation applications. Therefore, composite bipolar plates in most instances are developed for stationary applications where there is a fewer issue of mass reduction but a target of 40,000 h of lifetime (DOE, 2009) going with composite bipolar plates being insusceptible for corrosion.

Hence, metallic bipolar plates, promising a significant mass and volume reduction, are preferable for portable and transportation applications where a target of 5000 operation hours of lifetime (DOE, 2009) is in agreement with the corrosion susceptibility of metals and stainless steel.

---

* Corrosion resistance is illustrated by the value of current density obtained at specific voltage potentials in a fuel cell simulating environment (e.g., H$_2$SO$_4$ (pH 3) or 1.0 M H$_2$SO$_4$ (accelerated)) (see Section 6.2.1).
† Electrical conductivity respectively the resistivity is often used for material characterization (composite) and is given volume specific in S cm^{-1}, respectively in Ω cm^{-1}. As metals show unlikely higher bulk conductivity but a high contact resistance for the characterization of metallic bipolar plates the area specific contact resistance in mΩ cm^2 is used (see Section 6.2.1).

6.1.2 Materials for Graphite Composite Bipolar Plates

Owing to the chemical inertness of some selected polymers and the corrosion stability of carbon filler materials, composite-based bipolar plates are the first choice for applications demanding long-term stability.

To achieve adequate conductivities, the composite bipolar plates generally consist of a polymer, which functions as a binding matrix, and a high content of conductive filler materials. Composite plates offer an economic route for producing bipolar plates. For example, the composite material can be produced in an extruder and subsequently injection molded or compression molded to bipolar plates.

Despite cost savings and fast manufacturing, composites have some obstacles to overcome. To obtain the desired electrical and thermal conductivities the filling contents are usually much higher than the percolation threshold concentration. Such high filler content may result in poor mechanical properties and eventually poor gas tightness. Thus, an optimal compromise between the contradictory requirements has to be found. As a consequence, a range of different materials for composite based bipolar plates is under investigation or already commercially available. Table 6.2 depicts the properties of some composite bipolar plates under development.

It can be seen that different filling materials as well as spectrum of thermoplastics or thermosetting matrices have been investigated. As mentioned before, the requirements with respect to material properties contradict each other and vigorous research activities worldwide concentrate on the development

TABLE 6.2   Properties of Some Composite-Based Bipolar Plates (Extended)

Manufacturers/ Patents	Polymer	Filler (wt %)	σ(S/cm) In-plane	σ(S/cm) Through-plane	Strength Flexural (MPa)	Strength Tensile (MPa)	Strength Impact (J/m)
US 6,248,467 (LANL)	Vinyl Ester	68 GP	60		30	23	
Commercial			105		21	19	
Premix Inc.	Vinyl Ester	68 GP	85		28	24	
BMC 940	Vinyl Ester		100	50	40	30	
Plug Power	Vinyl Ester	68 GP	55		40	26	
US 4,214,969 (GE)	PVDF	74 GP	119		35–37		
US 5,942,347 (GTI)	Phenolic	77.5 GP	53				
US 6,171,720 (ORNL)	Phenolic	CF	200–300				
DuPont				25–33	53	25	
SGL			100	20	40		
H2 Economy			67				
Virginia Tech	PPS	76 GP&CF	271	19	96	58	81
U Akron	Epoxy	50 EG&CB	300–500	77–79	72	45	144
ZBT GmbH	PP	>80 GP&CB		70–120	77		
Schunk	Thermosetting plastics	80–90 GP&CB	111	53	40		
GrafTech	Thermosetting plastics	GP	1429	33	50–70	30–50	
Bac2 Ltd (EP 1105)	Patented conductive binder	GP	207	35	30		13.56
DOE target value			100		59	41	40.5

Source: Reproduced from Cunningham, B. and Baird, D. G. 2006. *Journal of Materials Chemistry* 16: 4385–4388.
Note: GP = graphite, CB= carbon black, CF= carbon fiber, EG= expanded graphite.

of composite bipolar plates to reach certain technical targets. Therefore, a large number of components and different production methods have been investigated or are still under research. This chapter reviews different approaches of composite bipolar plate production with the focus on the components.

6.1.2.1 Polymer Matrices for Composite Bipolar Plates

In principle, there are two types of polymers that can be used in composite bipolar plates; thermoplastics and thermosets. Thermoplastics include polymers such as polypropylene (PP) and polyethylene terephthalate (PET). These generally have lower glass transition temperatures (T_g)—in case of semicrystalline thermoplastics melting temperatures (T_m)—that restricts their use in high-temperature fuel cells. However, there are some high-temperature thermoplastics that have been investigated by (Derieth et al., 2008) for use in high-temperature fuel cells as shown in Table 6.3.

Semi crystalline thermoplastic polymers were identified as being preferable as a matrix for high-temperature bipolar plates due to their sharp T_m and low melt viscosities that enable good dispersion of carbon fillers. PPS was found to be the most suitable due to its higher chemical stability and higher heat deflection temperature. Liquid crystal polymer (LCP) composites have also been identified for bipolar plates by Wolf and Willer (2006) and King et al. (2008) and have potential for use in high-temperature PEM-fuel cells. The second group of polymers that have been used in low as well as high-temperature composites bipolar plates are thermosets. These include vinyl esters, epoxy, and phenolic resins. They are widely investigated and show a high state of research.

Furthermore, intrinsically conducting polymers (ICPs) such as polyanilines (PANI), polypyrroles, and polyphenylenes have been used as part of conductive composites on account of their much higher conductivity than other polymers. They are often mixed with other polymers as investigated by Taipalus (2001) and Totsra and Friedrich (2004). On the other hand Dweiri (2007) have found the addition of PANI to a composite of PP/carbon black/graphite for bipolar plates to be unsuitable for processing due to the poor thermal stability of PANI. The company Bac2 is currently the only manufacturer to use a patented electrically conductive polymer called ElectroPhen® in their composite bipolar plates.

6.1.2.2 Metal-Filled Composite Bipolar Plates

First, a few studies on metal-filled composite bipolar plates are briefly described. At Los Alamos National Laboratory (LANL) composite bipolar plates filled with porous graphite and stainless steel and bonded with polycarbonate (Hermann, 2005) has been developed. Kuo (2006) investigated in composite bipolar plates based on austenitic chromium–nickel–steel (SS316L) incorporated in a matrix of PA 6. Their results showed that these bipolar plates are chemically stable. Furthermore, Bin et al. (2006) reported a metal-filled bipolar plate using polyvinylidene fluorid (PVDF) as the matrix and titanium silicon carbide (Ti_3SiC_2) as the conductive filler and obtained an electrical conductivity of 29 S cm^{-1} with 80 wt% filling content.

TABLE 6.3 Thermoplastic Polymers Tested for High-Temperature Composites

Thermoplastic	T_m/T_g (°C)	Structure
PPO	280–310	100% amorphous
PES	225	100% amorphous
PSU	180	100% amorphous
PEI	370–410	100% amorphous
PA 12	220	30–35% semi crystalline
PPS	280	28–35% semi crystalline

Source: Reproduced from Affolter, S. 1999. Long-term behaviour of thermoplastic materials. Internet: https://institute.ntb.ch/fileadmin/Institute/MNT/NTB_MNT_Polymerics_Ageing_SA.pdf

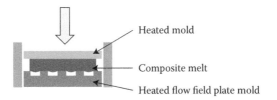

FIGURE 6.2 Schematic view for bipolar plate production via compression molding. (Adapted from Hamilton P. J. and Pollet B. G. 2010. *Fuel Cells* 10(4): 489–509.)

6.1.2.3 Carbon-Filled Composite Bipolar Plates

Owing to the chemically nearly inert nature, low costs and sustainable availability of the substance class of carbon-based fillers are preferred as filler materials for bipolar plates and carbon composite bipolar plates have been extensively investigated. The following section is structured in the different ways of composite bipolar plate production and the resulting eligible materials.

6.1.2.3.1 Compression Molding of Composite Bipolar Plates

A common production method for composite bipolar plates is compression molding. Normally, the production of the composites takes place in a kneader or twin screw extruder in which the filling particles get homogeneously dispersed within the polymeric matrix. The composite is then fed into a heated mold cavity and subsequently compression molded into bipolar plates either with gas channels or with further machining steps. In case of thermosetting polymers a chemical reaction has to occur before the bipolar plate can be removed from the mold; whereas for a thermoplastic matrix the mold has to be cooled below the melting temperature of the polymer. Both effects, the chemical reaction time or the cool-down and heat-up sequence dictate and increase the cycle time for the process of compression molding which in turn contributes to cost-intensive bipolar plates. The principle of compression molding is shown in Figure 6.2.

Wolf and Willer (2006) reported an investigation with liquid crystal polymer-based bipolar plates. As the conductive filler, a combination of carbon fibers and carbon blacks at a loading level below 40 wt% was used and a conductivity-level up to 6 S/cm was achieved. Furthermore, they suggested a model of synergetic interaction effects of different fillers used in this study as depicted in Figure 6.3.

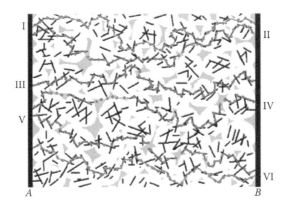

FIGURE 6.3 Model of conductive paths in a composite. Carbon fibers (black) and carbon black (gray) form six continuous paths (gray dots) from A to B. Without the carbon black, four of the paths are cut, only paths I and VI remain for conductivity. (Adapted from Wolf, H. and Willer, M. 2006. *Journal of Power Sources* 153: 41–46.)

Dweiri and Sahari (2007) studied a combination of graphite and carbon black in polypropylene for compression-molded bipolar plates. The obtained conductivity was 7 S/cm, even at a carbon content as high as 80 wt%. An adequate study was done by Yin et al. (2007) and a very high loading of 85 wt% graphite was necessary to achieve an electrical conductivity of more than 100 S/cm. The development of polyester, graphite, and glass fiber composites processed by compression molding to bipolar plates with a conductive layer on top of the plate surface has been reported by Huang et al. (2005). This bipolar plate achieved high electrical bulk conductivity of more than 200 S/cm at 65 wt% filling content. Blunk et al. (2003) investigated in PP/Graphite/CF and PVE/graphite/CF composite bipolar plates via compression molding and the highest conductivity could be achieved for the PVE/CF composite of approximately 250 S/cm (in-plane) and 6 S/cm through-plane.

Barbir et al. (1999) developed an epoxy/graphite composite bipolar plate with an active area of approximately 300 cm^2 and bulk conductivities of 345 S/cm. Even though the cost of the material was $8.5/kW the manufacturing costs amounted to $145/kW. Heo et al. (2007) investigated in composite bipolar plates with expanded graphite and a phenolic resin. They achieved a high conductivity of 250 S/cm by concurrent high flexural strength of 50 MPa. Both of them meet the DOE targets.

The primary advantage of compression molding is that high proportions of carbon fillers can be used in the composite, as low viscosities are not essential to the compression process. Hence, high conductivities can be achieved. One disadvantage of compression molding is that the mechanical compression forces during compression causes the fillers predominantly to align in plane rather than through plane according to Blunk et al. (2003) which is perpendicular to the direction of electrical flow in a fuel cell. Moreover, the long cycle time represents a market-entry barrier if larger quantities and lower costs of bipolar plates are required.

6.1.2.3.2 Injection Molding of Composite Bipolar Plates

Compared to compression molding, injection molding of composite bipolar plates has advantages, such as automated production, short cycle time, good reproducibility, and accurate size and shape. The principle of injection molding is shown in Figure 6.4.

On the other hand, it also has disadvantages, such as expensive mold costs and limitations in feasible plate size and thickness. A suitable melt viscosity is rather helpful to realize a satisfying mold filling during injection operation. But the high-carbon filler content results in high melt viscosity and high thermal conductivities. During injection molding operation the hot and molten composite flows into a colder injection molding tool and the good thermal conductivity leads to a rapid solidification of the molten composite inside the mold cavity. Therefore, high injection pressures are required, which give rise to complexities during the mold filling operation. Nevertheless, due to above-mentioned advantages, injection molding of composite bipolar plates is a very promising technology and has been successfully approved. Heinzel et al. (2004) developed low-cost composite bipolar plates by injection molding in cycle times below 20 s, yet in 2004. In addition, and among others, Kreuz (2008) deeply

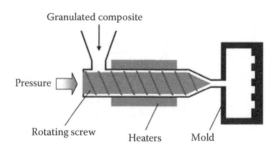

FIGURE 6.4   Schematic view for bipolar plate production via injection molding. (Adapted from Hamilton P. J. and Pollet B. G. 2010. *Fuel Cells* 10(4): 489–509.)

investigated the rheological process behavior of these composites and described the process and the properties of the bipolar plates in detail. In this study, the achieved electrical conductivity ranged from 5 to 150 S/cm. Mighri et al. (2004) prepared composite bipolar plates by injection molding, filled with up to 60 wt% of graphite and a mixture of it with carbon black or carbon fibers in a PP- and a PP-MAh-grafted matrix as well as in PPS. Additionally, they developed prototype plates using a two-shot injection overmolding procedure over aluminum plates. In the overmolded design, it was found that contact resistance between the aluminum and the polymer blend increases the resistivity significantly. The best conductivity for the bipolar plates (17 S/cm), were obtained by direct injection molding of the PPS-based blends. Kuo (2006) investigated in carbon-black-filled composites, bonded with PA 6 and produced bipolar plates by injection molding. At a loading level of 35 wt% they achieved a conductivity of 12 S/cm, which has been found to decrease dramatically when the composites were heated. Derieth et al. (2008) investigated in different graphite fillers and combinations with carbon blacks in PP. The focus in this study was the optimization of the process for both production steps, the compounding at a twin screw extruder and the injection molding operation. The process ability as well as the conductivity could be significantly improved by using a combination of spherical graphite and carbon black and the best-achieved conductivity of injection-molded bipolar plates was 23 S cm^{-1}.

Furthermore, it has been shown by Mighri et al. (2004) and Derieth et al. (2008) that there are two distinct orientations of carbon filler at the core and the surface for injection-molded bipolar plates (Figure 6.5). The core is the larger domain of the two and is where graphite particles are primarily orientated along the constant velocity flow lines during injection operation. This orientation also happens to be the direction of current flow in a fuel cell. The smaller surface layer domain contains graphite particles which are mainly orientated parallel to the surface. The orientation inside the plate core may be an important benefit of the injection process as graphite fillers are often electrically anisotropic showing good conductivity in-plane but poorer conductivity through-plane.

6.1.2.3.3 Innovative Approaches of Composite Bipolar Plates Manufacturing

An innovative approach of producing composite bipolar plates was presented by Xiao et al. (2006). They prepared a poly(arylene disulfide)/graphite nanosheet composite by *in situ* ring opening polymerization of cyclic arylene disulfide oligomers with dispersed nanographite and achieved an electrical conductivity higher than 100 S/cm at filler loadings as low as 50 wt%.

A fiber alignment method and conductive tie layer (CTL) have been proposed by Blunk et al. (2003) as a means of improving conductivity of composite bipolar plates (Figure 6.6). The authors have suggested a CTL which seeks to reduce the high contact resistance between the bipolar plate and the gas diffusion layer (GDL) interface. Even as the through-plane conductivity was improved by these processes, the

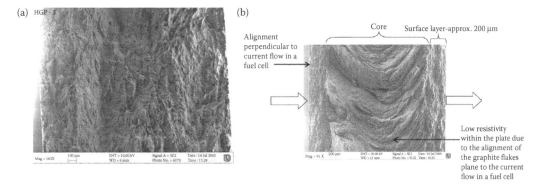

FIGURE 6.5 Cross-sectional SEM of fracture surface of compression molded (a) and injection-molded bipolar plate (b) showing graphite orientation along velocity flow lines (Adapted from Derieth, T. et al. 2008. *Journal of New Materials for Electrochemical Systems* 11: 21–29.).

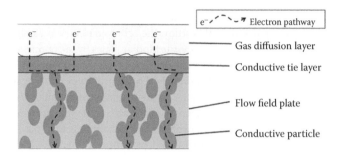

FIGURE 6.6 Conductive tie layer mechanism. (Adapted from Blunk, R. H. J. et al. 2003. *AIChE Journal* 49: 18–29.)

amount of contact resistance reduction was still not sufficient. A further disadvantage of this approach is the extra processing, either sanding or machining process needed to remove additional land height or the addition of a CTL. This CTL approach has been used by Oh et al. (2004) who electroplated Ni or Pd–Ni onto graphite composite plates.

Another approach is slurry molding and this technique was firstly used to manufacture "carbon–carbon" composites by Besmann et al. (2003). This process mixes phenolic resin with carbon fillers in water to create slurry which is fed out and vacuum molded into a preform. A second process called carbon chemical vapor infiltration (CVI) is then used to seal the plate for gas impermeability and for improvement of electrical conductivity. This process has been further developed by Huang et al. (2005) to reduce the cost caused by the CVI process and by Cunningham et al. (2007) to improve the properties of the bipolar plates. However, the mechanical properties of the bipolar plates were not found to be as high as the solely wet-lay-based plates (see Figures 6.7 and 6.8).

6.1.2.3.4 Global Manufacturers

In 1999, Hentall et al. (1999) preferred expanded graphite as material for bipolar plates and they revealed that this kind of graphite offers significant cell performance gains compared to standard graphitic

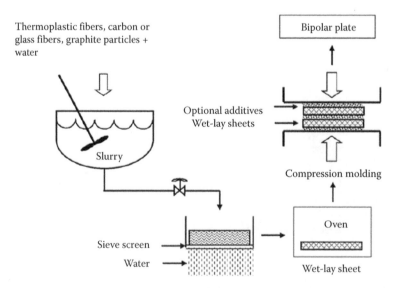

FIGURE 6.7 Slurry molding process by Huang (2005). (Adapted from Huang, J., Baird, D. G. and McGrath, J. E. 2005. *Journal of Power Sources* 150: 110–119.)

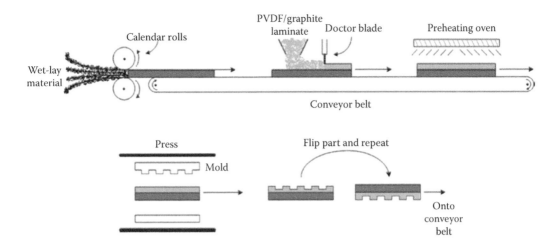

FIGURE 6.8 Wet-lay laminate approach by Cunningham et al. (2007). (Adapted from Cunningham, B. D., Huang, J., and Baird, D. G. 2007. *Journal of Power Sources* 165: 764–773.)

material. The reason is probably the compressibility of this material that enables intimate contact with the electrodes and/or GDL which leads to minimized interfacial contact resistance.

Flexible or expanded graphite was first developed for bipolar plates by GrafTech International Ltd. (previously UCAR) in 2000 (Mercuri, 2000). Expanded graphite can be impregnated with a resin and then compressed to bipolar plates including the required structures as depicted in Figure 6.9. GrafTech has claimed that their bipolar plates were found in 85% of fuel cell vehicles and 12 out of 14 bus programs worldwide (GrafTech, 2009). The reason therefore has been primarily their extensive use by Ballard Power Systems Inc. Current research (DOE, 2009) is focused on optimization of such flexible composites for high-temperature fuel cell operation.

But besides GrafTech, many more bipolar plate manufacturers offer composite bipolar plates (produced by different methods and various materials) for use in PEM fuel cells as shown in Table 6.4.

6.1.2.3.5 Interim Conclusion: Materials for Composite Bipolar Plates

In the majority of all these studies, the aspect of using bipolar plates for the different PEM fuel cell technologies has not been so far considered as it should be. In consideration of degradation mechanism it is

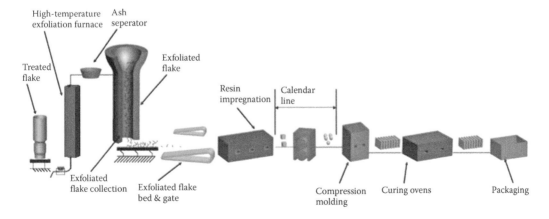

FIGURE 6.9 Production process of bipolar plates with expanded graphite (Grafcell®). (Adapted from Hamilton P. J. and Pollet B. G. 2010. *Fuel Cells* 10(4): 489–509.)

TABLE 6.4 Global Bipolar Plate Manufacturers

Bipolar Plate Manufacturers	Company Information
GrafTech International	One of the world's largest manufacturers of carbon-based materials. They claim that their GrafCell bipolar plates are found in 85% of fuel cell vehicles and 12 out of 14 bus programmes worldwide.
Poco Graphite (Entegris)	POCO was purchased by Entegris in 2008. They manufacture graphite and composite FFPs for fuel cells.
Wilhem Eisenhut GmbH	Until SGL Group. They produce composite Sigracet bipolar plate.
DuPont	Operating worldwide, DuPont offers a wide range of innovative products and services for markets. DuPont primarily manufactures the membrane for fuel cells but now offers thermoplastic composite bipolar plates.
Schunk GmbH	Manufactures high graphite content bipolar plates using a compression molding process. They also offer stacks with modular assembly ability.
ZBT GmbH	A public research institution in Germany for fuel cell technology and further energy technologies. They focused on injection molding bipolar plates.
Bac2 Ltd	Formed in 2002 to develop commercial-scale electrically conductive polymer composites. Manufactures composite carbon bipolar plates using their patented Electrophen polymer.
Bulk Molding Compounds Inc.	The largest producer of thermoset bulk molding compounds in North America. BMC vinyl ester composites have been developed for bipolar plates.
Pacific Fuel Cell Corp.	Manufactures composite bipolar plates using a compression molding process.
Dana Holding Corporation	Produce composite graphite-based and metallic bipolar plates. Also developing advanced coatings that enhance the properties of the bipolar plates.

Source: Adapted from Hamilton P. J. and Pollet B. G. 2010. *Fuel Cells* 10(4): 489–509.

of particular importance to divide the polymeric matrices according to the different PEM fuel cell technologies.

First, composite bipolar plates are used in low-temperature PEM fuel cells with an operating temperature of approximately 80°C and a sulfuric acid- and water environment. Second, there is the high-temperature PEM fuel cell technology with working temperatures of more than 160°C in a phosphoric acid- and water-sphere and third, the direct methanol PEM fuel cell which operates at 60°C and methanol as fuel. Furthermore, different cooling concepts with various cooling media, for example, glycol is used that requires corresponding properties of the used polymers.

It can be concluded that questions regarding the fuel cell technology, the process ability, long-term stability, heat resistance, chemical inertness, resistance against hydrothermal stress, and the cost play a major role for the right choice of a polymer matrix.

Nonetheless, it can be seen that a range of fillers and polymeric matrices were studied and different ways were used for the production of composite bipolar plates. Graphite and carbon black as fillers are to be found in nearly every study, and accordingly, some polymers are preferred to be used for composite bipolar plates. The reasons for both the fillers and the matrices are obvious. Graphite and carbon black have outstanding chemical stability against corrosion when compared with metallic fillers; they achieve an adequate conductivity and are obtainable at a reasonable price. In case of the matrices the chemical corrosion resistance is also a main criterion, and polyolefin materials, fluoropolymers, polyphenylene sulfide (PPS), and phenolic resins are particularly favored.

6.1.3 Materials for Metal Bipolar Plates

Thin metal sheets (0.1–0.5 mm) as base material for bipolar plates offer a variety of advantages. Metals in general can be processed with common conversion technologies, such as stamping and high-pressure metal forming, which leads to cost-efficient production.

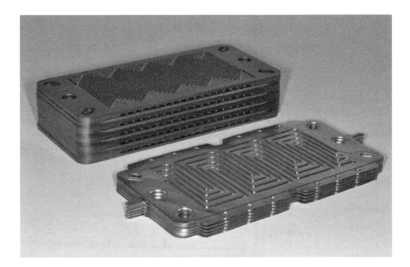

FIGURE 6.10 Comparison of a stack of five bipolar plates made from compound (behind) and metal (in front). (Adapted from Kühnemann, L., Beckhaus, P., and Bekeschus, G. 2010. Photographed at ZBT GmbH, Duisburg, Germany.)

Furthermore, metals combine important physical characteristics for fuel cell applications like high electric conductivity, high thermal conductivity, and gas impermeability. Bipolar plates made of metal sheets also exhibit reduced construction height and weight (Figure 6.10). Hence, metallic bipolar plates combine optimal properties for mass production and automotive applications.

Some disadvantages still exist: During operation, anode and cathode side are forced to pass crucial electrochemical potentials leading to corrosion for most stainless steels.

As shown in Figure 6.11, during OCV operation the anode side is forced to a potential of 0 V versus the reversible hydrogen electrode, which is for most stainless steels below potentials for active corrosion, whereas the cathode side is pushed to 1 V which represents the beginning of the trans passive corrosion region for stainless steels. Another case is operation under load at which the cathode potential declines to potentials of intrinsic passivity and the anode potential can take values about 100 mV meaning active corrosion (Scherer, 2009). Thus, under certain operation conditions and in the aggressive acidic environment* of PEM fuel cells, stainless-steel bipolar plates may suffer dissolution. Leached ions can poison the membrane which may lead to decreased power density of each cell. In addition, passive layers growing on metal surfaces increase contact resistance which offsets the positive conductivity properties of metals and again reduces fuel cell performance (Wang, 2003; Wu, 2008). Key aspects for the evaluation of metallic material for PEM fuel cell bipolar plates are the understanding of corrosion mechanisms in a fuel cell environment and the development of conductivity and contact resistance during fuel cell operation. Thus, a large quantity of research work was involved in finding suitable metals for fuel cell application until today. In the following, the different approaches of testing and valuating as well as the selection of material will be discussed.

Although it is a matter of common knowledge that stainless steel is quite prone to corrosion in fuel cells, bare substrates of different alloys were tested in past material investigations. In 1998, Hornung and Kappelt (1998) selected different iron-based materials for Solid Polymer Fuel Cell bipolar plates by using the pitting resistance equivalent (PRE = %Cr + 3.3%Mo + 30%N) as corrosion resistance criterion. The authors exposed that some iron-based materials with PRE ≥25 (the material compositions are not given

---

* Gaseous and condensed water, pH values of up to 2–4 and temperatures in a range of 40–80°C.

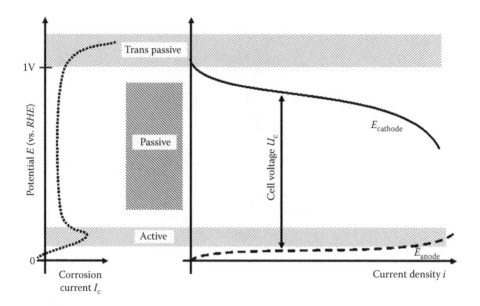

FIGURE 6.11 Relation between the current density-potential curve against the *RHE* (<u>r</u>eversible <u>h</u>ydrogen <u>e</u>lectrode) of stainless steel (left) and anode and cathode polarization curve of a polymer electrolyte fuel cell (right). (With kind permission from Springer Science+Business Media, *Polymer Electrolyte Fuel Cell Durability. Influence of metallic bipolar plates on the durability of polymer electrolyte fuel cells.* 2009. pp. 243–256. Scherer, J., Münter, D., and Ströbel, R.)

in the text) show similar corrosion behavior compared to a nickel-based material in a PEMFC model electrolyte. Regarding contact resistance, only a gold-coated sample performed well enough.

Makkus et al. (2000) tested an array of stainless-steel materials, namely 1.4439 (317 LN), 1.4404 (SS 316L), 1.4541 (SS 321), 1.4529 (926), and 1.3974 in PEMFC single cells. The samples were selected on the basis of their commercial availability and elemental composition with respect to the content of chromium, nickel, and molybdenum. This affects passivation and susceptibility to different types of corrosion. All materials released 2–200 times more contaminating ions to the MEA than the illustrated target of less than 5 mg metal concentration per 1 kg sample weight after 100 h of cell operation. The increase of contact resistance is shown to be dependent on compaction pressure but leads to unacceptable performance decay for all materials. For 1.4439 and 1.4541 indications for pitting corrosion were observed. (Davies, 2000a) found uncoated stainless steel 1.4401 (SS 316), 1.4841 (SS 310), and 1.4539 (SS 904L) to be suitable for the use in polymer electrolyte fuel cells.[*]

However, this conclusion is opposed to most findings in research literature (Hentall, 1999; Hodgson, 2001; Wind, 2002; Pozio, 2008), somehow attesting an increase in contact resistance and corrosion susceptibility causing fuel cell performance decay for uncoated stainless steels.

Today, uncoated stainless-steel materials are still in focus of research, as bare substrates promise to be the most cost-effective solution for metal bipolar plates. Nevertheless, it becomes obvious that a surface modification method or protective coating applied to metal bipolar plates is essential to improve their corrosion resistance.

Following the uncoated stainless steel, uncoated and gold-coated 316L, which is the most reviewed stainless steel, often represents the lower and upper constraints of corrosion prevention evaluation.

---

[*] Long-term tests (3000 h) were carried out at 50°C cell temperature. Corrosion is strongly temperature dependent.

6.2 Degradation Testing Protocols

6.2.1 State-of-the-Art and Standards

In order to certify the suitability of bipolar plate base material or coatings with respect to degradation in PEM fuel cells, two main characteristic criteria, interfacial contact resistance between bipolar plate surface and GDL and the corrosion affinity, emerged from the pool of criteria shown in Table 6.1 as standard characteristics. Since some requirements such as gas permeability and mechanical strength are not expected to be affected by fuel cell operation, electric properties are known to be subjected to surface alterations due to, for example, corrosion.

For the characterization of electrical surface properties the measurement of the interfacial contact resistance (ICR) between the surface of a plate or a material sample and a carbon paper representing the gas diffusion media in a fuel cell has been established (Wang, 2003; Nam, 2007; Han, 2009; Joseph, 2005). Since the ICR value is dependent on compaction force, this measurement is often carried out in a range of 0–30 kg cm^{-2} regarding compaction force for fuel cell mounting.

For the characterization of corrosion affinity potentiostatic and potentiodynamic voltammetry measurements are established. Typically, those tests are conducted with a three-electrode setup* in a fuel cell simulating solution, where the current density is measured for different potentials. In accordance with pretreatment of the PFSA membrane in sulfuric acid (Chu, 1999), H_2SO_4 solution (up to 1.0 molarity) is proposed as the electrolyte. Using dynamic voltammetry, the potential is swept from about -0.3 V_{SCE} to 1.2 V_{SCE} (scanning rate 1 mV min^{-1} to 1 mV s^{-1}) as fuel cell operating potential is in this range. Using this method, corrosion affinity can be ranked in the order of magnitude. For long-term conclusions regarding the suitability of a material, potentiostatic voltammetry is reasonable. Therefore, the same setup is used to keep the voltage at a constant value for a long period up to some thousand hours. For corrosion-resistant material no increasing current density should be obtained. The ICR value should neither be increased by this long-term immersion test.

The alignment of material research for bipolar plate application with these two main characteristics is already depicted by the Annual Progress Report 2006 of the DOE Hydrogen Program (DOE, 2006) in which the achievements in material research for (metallic) bipolar plates are valued by ICR, current density, and specific costs (Figure 6.12).

A good overview over achievements in past and current research is given by Table 6.5 representing quantitative results extracted from the literature cited in Section 6.4.2.

6.2.2 *Ex Situ* Test Equipment and Procedures

6.2.2.1 ICR Measurement

A method for quick evaluation of interfacial contact resistance was carried out first by Davies et al. (2000b), the established measurement procedure is sometimes called "Davies method" (Wang, 2003; Nam, 2007; Han, 2009). When the measurement was presented, the authors arranged a setup (Figure 6.13a) consisting of two pieces of gas diffusion media Carbel®CL sandwiched between the samples of interest. Those were clamped between two copper plates. As a constant electrical current of 5 A was fed through the arrangement by a power supply unit, the potential difference between the two samples was measured while a gradually increasing compaction force was applied to the copper plates. The interfacial resistance then was calculated according to the following equation:

$$R = \frac{V \cdot A_S}{I} \tag{6.1}$$

---

* Working electrode, counter electrode, and reference electrode (SCE).

DOE Targets for Bipolar Plates and the Status for the Materials Investigated

Goal/Alloy	ICR@140 N/ cm² mΩ·cm²	Current @ −0.1 V_{SCE} (H₂ purge), μA/cm²	Current at 0.6 V_{SCE} (Air purge), μA/cm²	Cost* $/kW
DOE 2010 Goal	*10*	*<1*	*<7*	*6*
316L	154	+2.5 ~ +12	0.7 ~ 11	3.41
349™	110	−4.5 ~ −2.0	0.5 ~ 0.8	3.61
AISI446	190	−2.0 ~ −1.0	0.3 ~ 1.0	4.08
2205	130	−0.5 ~ +0.5	0.3 ~ 1.2	3.53
201	158	−0.5 ~ +8.5	0.8 ~ 2.0	2.18
AL219	730	−3.3 ~ −1.5	1.0 ~ 3.0	2.65
Nitrided 446	6.0	−1.7~ −0.2	0.7 ~ 1.5	N/A
Modified 446	4.8	−9.0 ~ −0.2	1.5 ~ 4.5	N/A
Nitrided AL29-4C™	6.0	−6.5 ~ −3.0	0.3 ~ 0.5	N/A

Note*: Cost data were based on the average 2005 trading price of cold rolled coil 316 steel at London Metals Exchange and the market prices of the metals. The assumed stack was 6 cells/kW for a PEMFC and the dimensions of a bipolar plate are 24 cm × 24 cm × 0.0254 cm (which gives a 400 cm² utilization surface area in a 0.01 inch thick sheet).

FIGURE 6.12 DOE Status of investigated materials for metallic bipolar plates by 2006 (electrolyte: 1.0 M H₂SO₄ + 2 ppm F- at 70°C). (Adapted from DOE Progress Report 2006, http://www.hydrogen.energy.gov/annual_progress06_fuelcells.html)

where R is the interfacial contact resistance, V is the potential difference across the samples, I is the electrical current, and A_S is the surface contact area. The authors accented that this method is valid for the use as a comparative guide and "no strict resistive data should be quoted" as the measured resistive losses contain three interfacial components: twice the interface between sample and Carbel®CL and once the interface between Carbel®CL and Carbel®CL.

Until today, this method has undergone slight change in setup and a diversification of measured and calculated resistances. Mostly, this measurement is conducted as illustrated in Figure 6.13b where one sample (equal properties on both surfaces assumed, here: nitrided chromium on 316L stainless steel) is sandwiched between two carbon papers which are clamped between two gold-plated current collectors which are adjustable in compaction force. In order to calculate the contact resistance between carbon paper and coating, two different measurements have to be performed: R_{total} is the measurement of the complete chain of all resistive components. R_{extra} is the measurement of one carbon paper placed between the gold-plated current collectors. Hence, the interfacial contact resistance can be calculated via the following equations and presumptions:

$$R_{total} = 2 \cdot R_{coating-carbon} + 2 \cdot R_{carbon-gold} + 2 \cdot R_{carbon} + R_{sample} + R_{circuit} \tag{6.2}$$

$$R_{extra} = 2 \cdot R_{carbon-gold} + 2 \cdot R_{carbon} + R_{circuit} \tag{6.3}$$

$$R_{coating-carbon} = \frac{R_{total} - R_{extra}}{2} \tag{6.4}$$

where $R_{coating-carbon}$ is the interfacial contact resistance between coating and carbon paper.

TABLE 6.5 Corrosion Research Results Extracted from Cited Literature

Material	Coating/Surface Modification	Electrolyte	Corrosion Current Density [$\mu A\ cm^{-2}$]	Potentiostatic Results	ICR @ Pressure [$m\Omega\ cm^2$/ $kg\ cm^{-2}$]	Ref.
304SS	—	$10^{-3}M\ H_2SO_4 +$ $1.5 \times 10^{-4}M$	0.48	—	—	
304SS	Cr	HCl + 15 ppm HF	0.10	—	30/15	
316LSS	—		0.46	—	—	
316LSS	Cr		0.20	—	21/15	
316LSS	—	0.5 M H_2SO_4 at room temperature	60	—	15	
316LSS	Electrochemical treatment not specified		15	—	11	
316LSS	—	0.5 M H_2SO_4 at 80°C	23	Unstable	—	
316LSS	Cr pack cementation		0.2	~10 $\mu A\ cm^{-2}$ at +0.6 V	—	
316LSS	—	1 M H_2SO_4 at 80°C	~100	—	35.3/15	
316LSS	Cr pack cementation (2.5 h at 1050°C)		~1	—	17/15	
316LSS	Cr pack cementation (23 h at 1050°C)		~1	—	233.5/15	
316LSS	—	0.5 M H_2SO_4 + 2 ppm F^- at room temperature	2000[a]	—	45/20	
316LSS	Cr pack cementation (3 h at 1100°C)		0.3[a]	0.075 $\mu A\ cm^{-2}$ at −0.1 V	17/20	
316LSS	Cr pack cementation (3 h at 900°C)		0.3[a]	0.35 $\mu A\ cm^{-2}$ at − 0.1 V	13/20	
316SS	—	0.5 M H_2SO_4 + 2 ppm HF at 80°C	24		125/20	
316SS	Low temperature Carburization		2–4	4 $\mu A\ cm^{-2}$ at −0.1 V	100/20	
316LSS	—	0.5 M H_2SO_4 + 2 ppm HF at 80°C	152	—	~84/15	
316LSS	Nitrided Cr_2N		0.58	0.2 $\mu A\ cm^{-2}$ at −0.1 V	~66/15	
316LSS	Nitrided Cr_2N + CrN		4.99	—	~50/15	
349SS	Nitridation	1 M H_2SO_4 + 2 ppm F^- at 70°C, O_2 bubbled	~1000	20 $mA\ cm^{-2}$ at −0.1 V	9.5/18	
349SS	Nitridation	1 M H_2SO_4 + 2 ppm F^- at 70°C, H_2 bubbled	~1000	0.25 $mA\ cm^{-2}$ at +06 V		
446MSS	—	pH 3 H_2SO_4 solution at 80°C purged wit H_2	100	6 $\mu A\ cm^2$ at −0.1 V	30/15	
446MSS	Nitridation		2.0	—	7-8/15	
446MSS	Pre-oxidation + Nitridation		1.5	5–6 $\mu A\ cm^{-2}$ at −0.1 V	7-8/15	

continued

TABLE 6.5　(continued) Corrosion Research Results Extracted from Cited Literature

Material	Coating/Surface Modification	Electrolyte	Corrosion Current Density [$\mu A\ cm^{-2}$]	Potentiostatic Results	ICR @ Pressure [$m\Omega\ cm^2$/ $kg\ cm^{-2}$]	Ref.
316LSS	—	0.1 N H_2SO_4 + 2 ppm F^- at 80°C, H_2 bubbled	2.6	—	82/15	
316LSS	Cr plating		0.025	—	—	
316LSS	Cr plating + Plasma Nitridation (−300 V)		− 0.53	−0.1 $\mu A\ cm^{-2}$ at −0.1 V	5.6/15	
316LSS	—	0.5 M H_2SO_4 + 2 ppm F^- at 80°C	11.26	10 $\mu A\ cm^{-2}$ at +0.6 V	312.8/20	
316LSS	Ni ions implantation (1×10^{17} ions cm^{-2})		6.7[a]	1.3 $\mu A\ cm^{-2}$ at +0.6 V	36/20	
316LSS	—	0.5 M H_2SO_4 + 2 ppm F^- at 80°C, air bubbled	11.26	2.5–3.0 μA cm^{-2} at +0.6 V	255.4/20	
316LSS	Cr-Ni co-ion implantation (2h)		6.7	0.51	22.1/20	
316LSS	—	0.5 M H_2SO_4 at 70°C, H_2 bubbled	40.3	−10 $\mu A\ cm^{-2}$ at − 0.1 V	—	
316LSS	TiN (PVD)		1.02	−40 $\mu A\ cm^{-2}$ at −0.1 V	—	
316LSS	—	1 M H_2SO_4 + 2 ppm F^- at 70°C	57.0	—	—	
316LSS	TiN (PVD, 0.2 mTorr)		8.88	—	—	
316LSS	TiN (PVD, 0.4 mTorr)		0.28	—	—	
316LSS	TiN (PVD, 0.6 mTorr)		0.55	—	—	
316LSS	—	1 M H_2SO_4 at 70°C	76.5	—	~140/20	
316LSS	TiN/CrN (1:9)		0.19	—	~12/20	
316LSS	TiN/CrN (3:7)		0.59	—	~16/20	
316LSS	TiN/CrN (5:5)		2.79	—	~21/20	
316LSS	—	0.5 M H_2SO_4 + 2 ppm HF at 80°C, air bubbled	4.76 (at +0.6 V)	—	80/15	
316LSS	TiCrN (PVD), N_2 = 0 sccm		6.09 (at +0.6 V)	0.1 at +0.6 V	100/15	
316LSS	TiCrN (PVD), N_2 = 3 sccm		0.49 (at +0.6 V)	0.1 at +0.6 V	4.5/15	
304SS	—	0.5 M H_2SO_4 + 2 ppm HF at 80°C, H_2 purged 0.5 M H_2SO_4 + 2 ppm HF at 80°C, O_2 purged	~200	—	—	
304SS	Carbon (CVD)		~0.1	—	—	
304SS	—		~600	—	—	
304SS	Carbon (CVD)		~0.05	—	—	
404SS	—	0.5 M H_2SO_4 at room temperature	~32	—	—	
404SS	Carbon (CVD)		~8	—	—	

TABLE 6.5   (continued) Corrosion Research Results Extracted from Cited Literature

Material	Coating/Surface Modification	Electrolyte	Corrosion Current Density [$\mu A \ cm^{-2}$]	Potentiostatic Results	ICR @ Pressure [$m\Omega \ cm^2$/ kg cm^{-2}]	Ref.
316LSS	—	0.5 M H_2SO_4 + 2 ppm HF at 80°C, air bubbled	43.1	2.4 $\mu A \ cm^{-2}$ at +0.6 V	255.4/20	
316LSS	Amorpous carbon (PVD)		0.06	2.8 $\mu A \ cm^{-2}$ at +0.6 V	5.2/20	
304SS	—	1 M H_2SO_4 at room temperature	8.3	—	—	
304SS	TiC (HEMAA)		0.034	stable at 3.7 $\mu A \ cm^{-2}$ after 4h	—	
304SS	—	0.1 M H_2SO_4 (temperature not specified)	10	—	~80/20	
304SS	PANI		0.1	—	~580/20	
304SS	PPY		1	—	~500/20	
316LSS	—	0.5 M H_2SO_4 at 70°C	40	5 $\mu A \ cm^{-2}$ after 500 s at +0.6 V	—	
316LSS	PPY		2.4	10 $\mu A \ cm^{-2}$ after 2h at +0.6 V	—	
316LSS	Gold interlayer, PPY top coating	0.5 M H_2SO_4 at 70°C	5.46	−10 $\mu A \ cm^{-2}$ at −0.1 V and 7 $\mu A \ cm^{-2}$ at +0.6 V	—	
304SS	—	0.1 M H_2SO_4 at 60°C	5000	—	—	
304SS	PPY		0.2	—	—	

Source: Reproduced from Antunes, R.A., Oliveira, M. C. L., Ett, G. et al. 2010. *International Journal of Hydrogen Energy* 35: 3632–3647.

[a] Values of passive current densities.

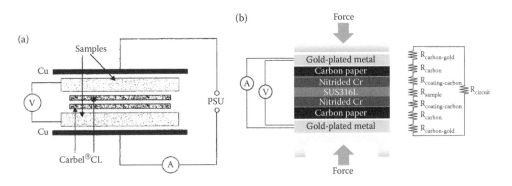

FIGURE 6.13   (a) Schematic of ICR measurement corresponding to (Davies, 2000). (b) Schematic of ICR measurement corresponding to (Han, 2009) and electrical circuit. (Han D. H. et al. 2009. *International Journal of Hydrogen Energy* 34: 2387–2395.)

$R_{carbon\text{-}gold}$ is the interfacial contact resistance between carbon paper and gold-plated current collector.

R_{carbon} is the bulk resistance of the carbon paper and is negligible.

R_{sample} is the bulk resistance of the sample and is negligible.

$R_{circuit}$ is the resistance of measurement device and circuit.

According to Equation 6.1, the interfacial contact resistance then is expressed in the area-specific unit $m\Omega\ cm^2$. As a result of this measurement, the contact resistance of materials, surface modifications, or coatings can be compared in the form of compaction force-depending curves as illustrated in Figure 6.14.

The negligence of the bulk resistance of metallic samples due to the ratio of high contact resistance to low bulk resistance is comprehensible. But for composite material development, for example, it is reasonable to take bulk properties into account. Therefore, the test setup maybe modified according to Figure 6.15. Two needles have only punctiform contact and measure the potential difference across the sample regardless of the surface properties (Kreuz, 2008). Hence, in addition to information about contact properties, information about bulk properties are provided within the same measurement due to slight intrusion of the needles into the surface and a straight-line current path assumed.

6.2.2.2 Voltammetry

As exposed in Section 6.3.1, potentiodynamic and potentiostatic voltammetry are established in current research as an *ex situ* procedure to quantify corrosion resistance of a bipolar plate material, surface modifications, and coatings and is conducted with a three-electrode setup in a fuel cell simulating environment of 1.0 M H_2SO_4 solution.

As this is a highly aggressive, accelerating test, tests with less concentrated solutions (0.1 M, 0.5 M) or solutions with additions (2–5 ppm HF, 2 ppm F[−]) can be found in the literature (Han, 2009; Yang, 2010). As dynamic polarization curves and thereof obtained corrosion potentials and corrosion current

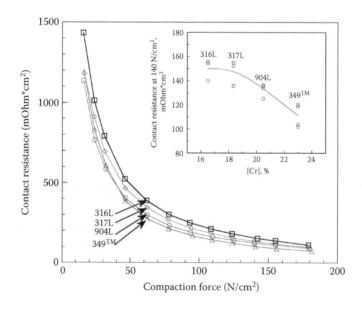

FIGURE 6.14 Interfacial contact resistance for different stainless steels. The inset shows the interfacial contact resistance of the different stainless steels at a compaction force of 140 N/cm² with respect to chromium content. (Adapted from Wang, H., Sweikart, M. A., and Turner, J. A. 2003. *Journal of Power Sources* 115: 243–251.)

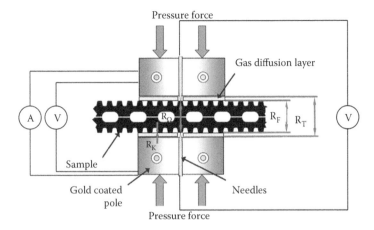

FIGURE 6.15 Enhanced setup for measuring contact resistance and bulk resistance within one measurement. Bulk properties can be obtained from a separate potential measurement across two needles electrically insulated from the main circuit. (Adapted from Kreuz, C. 2008. *Dissertation*: PEM-Brennstoffzellen mit spritzgegossenen Bipolarplatten aus hochgefülltem Graphit-Composite, Gerhard-Mercator-University in Germany, 04/2008.)

densities are affected by many test parameters,[*] certainly an alignment with widely established test parameters and targets for current densities is necessary. Therefore, test parameters established by DOE (2006) and reflected by international research literature can be aligned to the basic conditions illustrated in Figure 6.16.

By conducting potentiodynamic tests, significant characteristics such as open-circuit potential and current densities can easily be identified and brought into comparison as illustrated in Figure 6.17. Those graphs were obtained in potentiodynamic tests for bare 316L stainless-steel, chromium-plated 316L and the same with a plasma nitriding on top with respect to different bias voltages in the plasma-coating process (Han, 2009). As a result, the nitrided chromium layer lowers corrosion current densities compared to the bare substrate by one order of magnitude to less than $1~\mu A/cm^2$ for both cathodic and anodic conditions. Thereby, a quick preassessment for materials and coatings is possible. For concluding

Parameter	Anodic conditions	Cathodic conditions
Electrolyte	$1.0~M~H_2SO_4$	$1.0~M~H_2SO_4$
Temperature	60–80°C	60–80°C
Gas purge	Hydrogen	Air
Target for Current Density	$<1~\mu A/cm^2$@ $-0.1~V_{SCE}$	$<1~\mu A/cm^2$@ $0.6~V_{SCE}$

FIGURE 6.16 Basic conditions for voltammetry measurement in alignment with DOE targets. (Adapted from DOE Progress Report 2006, http://www.hydrogen.energy.gov/annual_progress06_fuelcells.html; DOE Progress Report 2009, http://www.hydrogen.energy.gov/annual_progress09_fuelcells.html)

---

[*] Type of electrolyte, electrolyte temperature, additions, gas purge, sample surface quality.

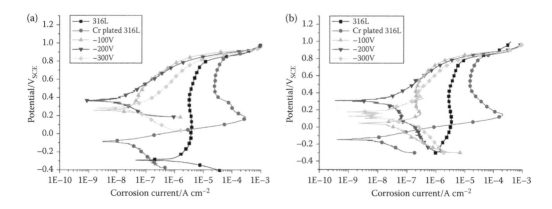

FIGURE 6.17   Potentiodynamic polarization curves of 316L, Cr plated 316L and Cr plated 316L with (−100 V, −200 V, −300 V bias voltage) plasma nitriding in 0.1 N H_2SO_4 with 2 ppm HF at 80°C with (a) air purging and (b) hydrogen purging. (Adapted from Han D. H. et al. 2009. *International Journal of Hydrogen Energy* 34: 2387–2395.)

investigations, a potentiostatic long-term measurement under the same conditions has to be performed, in which potentiodynamic results must be affirmed. In Figure 6.18, the current densities of the same samples as in Figure 6.17, measured in a potentiostatic test, are given for cathodic and anodic conditions. For both setups, the absolute values of current density stabilize at low values promising good corrosion protection (see Figures 6.17 and 6.18).

For the estimation of voltammetric measurements the knowledge of test parameters is of highest importance as obtained characteristics vary about orders of magnitude by changes in temperature, solution concentration, or gas purging. Nevertheless, voltammetry is the basic test procedure for corrosion investigation.

6.2.2.3 Immersion Tests

Especially for composite bipolar plates the long-term stability of the plastic binder materials is a very appropriate classifier for the reliability of a bipolar plate. The quantification of the period of induction of the used polymer—as explained later—is essential as after this period a rapid aging of the polymer takes place resulting in a sudden decomposition of the polymeric matrix in the composite bipolar plates.

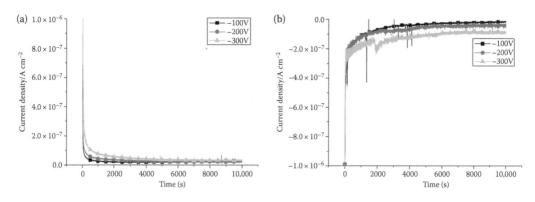

FIGURE 6.18   Potentiostatic polarization curves of the nitrided samples in 0.1 N H_2SO_4 with 2 ppm HF at 80°C with (a) air purging at 0.6 V_{SCE} and with (b) hydrogen purging at −0.1 V_{SCE}. (Adapted from Han D. H. et al. 2009. *International Journal of Hydrogen Energy* 34: 2387–2395.)

To prove such plastic behavior in advance the accelerated lifetime as the immersion test is a very helpful and commonly used test method.

Mitani and Mitsuda (2009) applied an *ex situ* accelerated aging on several thermosetting and thermoplastic-bonded composite bipolar plates. They used hot water at 150°C for 1560 h in an autoclave. This condition corresponds to 100.000 h at 90°C by the empirical rule that a 10°C rise doubles the rate of reaction. The authors determined the change in physical properties by a change in electrical resistivity, flexural strength, and weight, whereas the chemical changes were investigated by measuring the water conductivity during the immersion test. Moreover, extracted ions were analyzed by ion chromatography.

Maheshwari et al. (2007) exposed phenolic-bonded composite bipolar plates to 5% sulfuric acid solution for 5 h at 50°C. The samples were reconditioned and the differences in weight were measured. The authors also carried out water adsorption studies on several composite samples. In order to ascertain that there is no adverse effect of water adsorption on the plates, these were tested for their compressive strength, their flexural strength, and the electrical resistivity before and after water adsorption.

Wilson et al. (2000) immersed three different thermosetting composite materials in five different liquids for 1000 h at 80°C. These liquids, water, 1 M and 6 M methanol, pH 2 and pH 6 sulfuric acid cover the range from expected to unduly aggressive environments for hydrogen or direct methanol fuel cells. In addition, small squares of Nafion 112 membranes were immersed together with the samples over the 1000-h exposure period. The membrane pieces were analyzed using x-ray fluorescence spectroscopy to identify any ionic leachant species. All of the exposed mechanical test coupons plus six unexposed coupons were tested for flexural strength according to ASTM D638.

Derieth et al. (2008) immersed injection-molded polypropylene-bonded and PPS-based compression-molded composite bipolar plates. The authors exposed some sets of composite bipolar plates to different liquids (10 per investigation). Deviations in weight, thickness, surface topography, and electrical as well as mechanical properties were determined before and after exposure. Furthermore, to identify leachant ionic and organic species, the liquids were analyzed.

Bornbaum (2010) exposed epoxy-based and phenolic-bonded composite bipolar plates for 14 days in phosphoric acid ($W = 85$ wt%) at 160°C and after exposure a rockwell hardness test (HR 10/40) was applied to determine the changes in the physical and/or chemical nature of the different thermosetting matrices.

It can be seen that some immersion tests by different authors have been carried out. The chance of testing composite bipolar plates under unduly aggressive conditions in advance of time consuming in-cell-tests is a great benefit. It might be useful for following investigations to test under similar conditions but applying voltage to the test specimen. Nevertheless, for a general qualification of composite bipolar plates a long-term operation in fuel cell is necessary.

6.2.2.4 Additional Measurements

6.2.2.4.1 Water Contact Angle

A further property of bipolar plates that may be subjected to alteration during fuel cell operation is water adhesion. To enhance a sufficient drainage of water during fuel cell operation and to avoid electrode flooding by accumulated water from gas humidity and the reduction reaction of oxygen, in some literature findings bipolar plates are desired to show high hydrophobicity (Hou, 2009; Yun, 2010). Therefore, the measurement of water contact angle can be conducted (see Figure 6.19).

However, whether bipolar plates need to be desirably hydrophobic or hydrophilic to avoid water flooding, is a long-lasting and ongoing discussion and is not intended to be answered in this chapter. But to assure either one surface property or the other over the period of operation, water contact angle measurement is a quick and simple testing method and is to be conducted.

6.2.2.4.2 Chemical Analysis Regarding Contaminants

In order to determine the grade of cationic contamination of the membrane and therewith the grade of corrosion of bipolar plate material, a chemical analysis is reasonable. Either membrane material or

FIGURE 6.19   Contact angle (95°) of Cr-nitride-coated 316L with water (a) and contact angle (73°) of bare 316L substrate with water (b). (Adapted from Hou, M. et al. 2009. *International Journal of Hydrogen Energy* 34: 453–458.)

product water is analyzed after long-term fuel cell operation, or immersion solution is analyzed after a long-period potentiostatic test. In most scientific papers, in which those tests were conducted, the ICP-MS technique (Inductively coupled plasma mass spectroscopy) is used (Chung, 2008; Wang, 2008; Feng, 2010). This method combines inductively coupled plasma for ionization and a mass spectrometer to separate and detect the ions. As the membrane performance can degrade at concentration levels of 5–10 ppm of contaminants (Wang, 2008), exceeding those concentrations has to be avoided. Usually, investigated contaminating elements are Fe, Ni, Cr, and Mo as they are the principal constituents of stainless-steel alloys and, if detected, indicate dissolution of the base material.

6.2.2.4.3 Micrographs

As the functionality of coatings and surface modifications in general and especially for corrosion protection depends on the layers being homogeneous and defect free, micrographs of coated surfaces and cross-sectional views are state of the art in all fields of coating evaluation. Therefore, in coating development for fuel cell application mostly the SEM technique (scanning electron microscopy) is used (Wang, 2004; Smit, 2006; Pozio, 2008; Feng, 2010) for determining layer thickness, coating morphology, and the presence of coating defects before and after corrosion studies (Figures 6.39 through 6.42). In other works (Fukutsuka, 2007; Feng, 2008), the AFM technique (atomic force microscopy) is also employed.

6.2.2.4.4 Elemental Analysis

For knowledge of the exact composition of an alloy used as coating substrate or of the coating composition itself, more often elemental analyses have to be performed. In this context XRD (x-ray diffraction), EDX (energy-dispersive x-ray spectroscopy) and XPS (x-ray photoelectron spectroscopy) are to be named.

6.2.3 *In Situ* Test Equipment and Procedures

A powerful *in situ* diagnostic tool in terms of performance decay and degradation of fuel cells is the EIS technique (electrochemical impedance spectroscopy). A small-amplitude AC signal (frequency ranges from 10^{-1} Hz up to 10^4 Hz) is applied to an operating fuel cell via the load and the AC voltage and current response is analyzed. The resistive, capacitive, and inductive character of the impedance can be ascribed to voltage losses due to kinetic, ohmic, and mass transport losses (Scribner, 2010). As EIS is a nondestructive testing method and does not disturb the system from equilibrium, these are the main

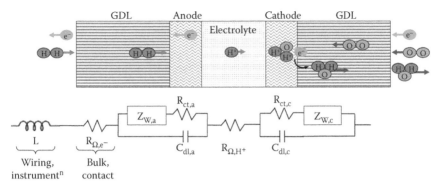

Key: GDL = gas diffusion layer, dl = double layer, ct = charge transfer, a = anode, c = cathode.

FIGURE 6.20 Example of a generalized equivalent circuit model for a single-cell fuel cell. Models regarding "real" fuel cells often contain constant phase elements (CPE). (Adapted from Scribner Associates, 2010. http://www.scribner.com/technical-papers.html)

advantages. One disadvantage is that generated information always regards the whole cell (Asghari, 2010. Local information is only available by simulating the cell.

For data interpretation an equivalent circuit model (see Figure 6.20) can be used[*] consisting of resistive, capacitive, and inductive elements associated with several sources of polarization. In order to simplify the model, often elements associated with the anode are omitted as polarization resistance of the anode is much smaller compared to the cathode (Yan, 2007; Rubio, 2008; Asghari, 2010).

There are two ways of EIS data representation: Nyquist plot and Bode plot. The most common way is the Nyquist plot; imaginary impedance is depicted versus real impedance of the cell. The advantage of Nyquist plots is that processes (circuit elements) with different time constants result in discrete impedance arcs (Figure 6.21), which give a good overview of dominating processes. The disadvantage is that corresponding frequencies are invisible in the plot.

As Bode plots represent the impedance magnitude and phase angle versus AC signal frequency, they are often shown as supplementation to Nyquist plots.

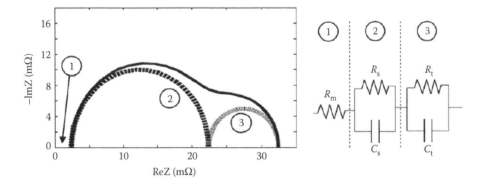

FIGURE 6.21 Nyquist plots with impedance arcs (left) corresponding to the simplified equivalent circuit model (right). (Adapted from Rubio, M. A., Urquia, A., and Domido, S. 2010. *International Journal of Hydrogen Energy* 35: 2586–2590.)

---

[*] Alternatively, the electrochemical reaction can be described by differential equations to be solved numerically (W. G. Bessler, *Solid State Ionics*, 176(11–12), March 31, 2005, pp. 997–1011).

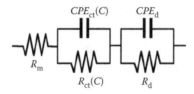

FIGURE 6.22 Simplified equivalent circuit model in (Kumagai, 2010). (Adapted from Kumagai, M. et al. 2010. *Journal of Power Sources* 195: 5501–5507.)

Rubio et al. (2010) developed two equivalent circuit models representing different reversible degradation phenomena in a fuel cell. One is valid for normal operation, cathode flooding and membrane drying and the other is valid for anode catalyst CO-poisoning. The authors turned to account that the timescale of reversible phenomena is below 10^5 s and timescale of irreversible phenomena is above 10^6 s. As PEMFC internal resistance increases with a constant rate due to irreversible degradation phenomena, the effect of reversible degradation phenomena can be decoupled once this rate is estimated. Hence, the authors conclude that the utilization of the models allows the identification of ongoing degradation phenomena.

Kumagai et al. (2010) evaluated operation performance of a fuel cell with corroded 430 stainless-steel bipolar plates by the EIS method in comparison with a fuel cell with graphite bipolar plate.

A simplified equivalent circuit model (Figure 6.22) was applied wherein R_m is associated with the sum of resistive elements such as electrolyte membrane resistance, $R_{ct}(C)$ is associated with the charge transfer resistance for cathode and R_d is associated with the diffusion resistance being affected by oxygen supply and drainage of generated water. In a long-term operation lasting 336 h electrochemical impedance spectra were obtained at the same time for both fuel cell types.

As for the fuel cell using graphite bipolar plates approximately no changes were observed for R_m, $R_{ct}(C)$, and R_d, for the fuel cell using 430 stainless-steel bipolar plates, a significant increase in R_d was determined (Figure 6.23) as documented by larger semicircles in the Nyquist plot (Figure 6.23). Thus, as

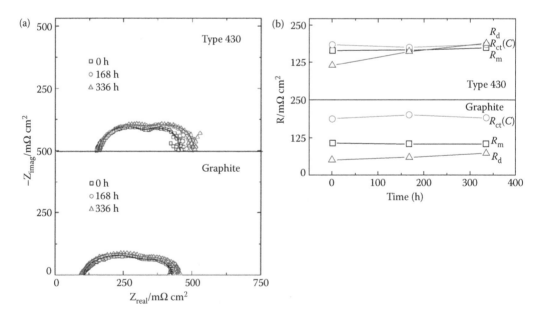

FIGURE 6.23 Nyquist plots of 430 SS and graphite bipolar plate fuel cell corresponding to increased operation time (a) and change of impedance parameters (b). (Adapted from Kumagai, M. et al. 2010. *Journal of Power Sources* 195: 5501–5507.)

R_d, which is associated with diffusion resistance, is increased due to corrosion it is assumed that corrosion products deposit on the surface of the bipolar plate or in the porous structure of GDL deteriorating oxygen supply and water drainage. The authors state that the EIS method is a suitable tool for evaluating fuel cell performance decay due to corrosion.

6.3 Degradation Mechanisms

6.3.1 Graphite Composite Bipolar Plates

Composite bipolar plates generally contain polymeric matrices as binder and filler materials such as graphite, carbon black, and others. As the degradation mechanism of these differ significantly they will be discussed separately in the following, starting with the polymeric matrices.

6.3.1.1 Polymeric Matrices

Degradation mechanisms of the polymeric matrices and the contamination effects due to migration of additives coming from the plastics into the fuel cell environment are significant concerns for the choice of bipolar plate materials. For a deeper understanding of the relevant processes a brief introduction of the chemical and physical nature of polymers is necessary.

In terms of tonnage, the bulk of plastics produced are thermoplastics, a group that includes the polyethylene, polypropylene, polystyrene, polyvinyl chloride (PVC), the nylons, polycarbonates, and cellulose acetate. In case of thermoplastics the molecules at room temperature do not have enough rotational energy to twist around the backbone chemical bonds and the polymer is rigid. At heating above a certain temperature, sufficient energy for such subchain movement is obtained, and at shearing and heating stress during processing such as extrusion, the polymer molecules partly uncoil. Chain segments slip past each other and the polymer begins to flow. On cooling the mass hardens again. If desired, the whole process of heating, shearing, and cooling may be repeated which in consequence allows bringing thermoplastic materials into any desired shape and enables the recycling of used thermoplastic resigns.

The second class of polymeric materials, the thermosetting plastics, for example, phenolic, epoxy, urea–formaldehyde, and melamine– formaldehyde resins, are formed, for example, in a mold into the desired final shape and then subjected to chemical reaction, for example, by heating in such a way that the molecules link with each another to form a cross-linked network. As the molecules are now interconnected, they can no longer slide past one another and the material is said to be set. The cross-linking process is irreversible and so, thermosetting materials can not be recycled. Figure 6.24 shows the general conditions that polymeric matrices are exposed to in processing and in fuel cell application and the possible effects to the polymer nature.

In principle, degradation of polymeric materials can occur from a variety of causes such as

- Heat (thermal degradation, thermal oxidative degradation)
- Light (photooxidation)
- Oxygen (oxidative degradation)
- Weathering (UV degradation)

In general, the ability of a polymer to resist these degradation causes is called the "stability of the plastic" and in case of PEM-fuel cell applications especially the thermal, thermal oxidative, and the oxidative degradation in interaction with the humid and acid fuel cell environment are relevant. In fact, degradation is inevitable and the resulting chain reaction will accelerate until the cycle is interrupted in some manner. The only real variable is how long it is going to take for degradation to become evident resulting in a significant loss in properties for the fuel cell performance. Degradation caused by fuel cell media can occur:

1. Physically $\rightarrow$ swelling/dissolution
2. Chemically $\rightarrow$ changes in chemical structure (minor changes in structure cause significant effects to plastic)

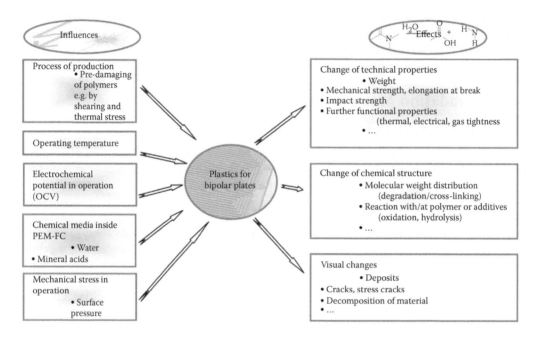

FIGURE 6.24 Influences and resulting effects on the polymeric resign in processing and in fuel cell environment. (Adapted from ZBT. 2010. Original material gained at the Center for Fuel Cell Technology.)

And the stability is depending on:

1. Swelling/dissolution/hydrolysis $\rightarrow$ differences in polarity between the polymer and the media.
2. Ability of the media to penetrate the polymer.

Furthermore, polymers degrade principally in a way shown in Figure 6.25.

During the "period of induction" the polymer is not subject to significant changes in its properties. After the consumption of additives in combination with chemical and physical effects to the polymer, it comes to an accelerated aging of the plastic and may result in a dramatic loss in properties.

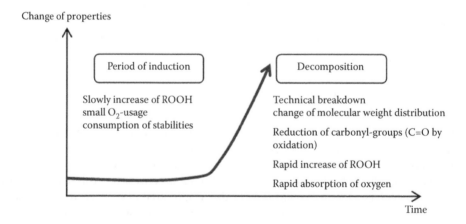

FIGURE 6.25 Degradation process of polymers. (Adapted from Krebs, C., Avondet, M. A., and Leu, K. W. 1999. Langzeitverhalten von Thermoplasten—Alterungsverhalten und Chemikalienbeständigkeit. Hanser Verlag München.)

Additionally, it has to be pointed out that for composite bipolar plates a high content of filler materials is necessary that leads to more brittle properties. Therefore, such losses in plastic properties lead to mechanical stress to the plate structure owing to swelling of the material. Furthermore, the intermediate formation of peroxides during exceptional fuel cell operation may also lead to decomposition of the polymeric matrix of bipolar plates. This leads in consequence to accelerated embrittlement of the composite material that finally would affect the stability and gas tightness of the bipolar plate.

To increase the stability and to adjust other properties, the polymer supplier or the plastic processors add a high content of additives to the polymers. But the question arises regarding whether and how additives do leach out of the polymer into fuel cell environment and what effects they have. In Bornbaum (2010), it is reported that a demolding agent may lead to Pt-degradation inside a direct methanol fuel cell environment as depicted in Figure 6.26.

The demolding agent reacts with methanol and a catalytically inactive Pt-chelate complex is formed on the surface. It can be seen, that some agents or additives—if they migrate to the topmost layer of the bipolar plate—lead to degradation of the catalyst. To emphasize, a polymer is normally adjusted with many additives, such as antiblock agents, antifog agents, antioxidants agents, lubricants, nucleating agents, UV-stabilizers, processing aids, and many more. Several of them contain di- or higher valent cations or components of wax-basis. For both classes of substances it is known, that they lead to catalyst degradation in a fuel cell.

6.3.1.2 Filler Materials

Second main material family in composite bipolar plates are the filler substances that on the one hand can contain impurities affecting the PEM fuel cell performance and on the other hand show degradation mechanisms known as carbon corrosion.

In general, impurities are known to affect fuel cell performance by various mechanisms that lead to performance loss. Such losses due to impurities can be permanent and irreversible, or temporary and reversible.

The effects of impurities of fuels have been thoroughly researched including CO, ammonia, hydrogen sulfide, hydrogen cyanide, hydrocarbons, formaldehyde, and formic acid. Extensive literature in this context is cited in Borup et al. (2007). On the cathode side, ambient air may contain impurities such as sulfur dioxide, nitrogen oxides, and particulate matter (including salts) that can affect fuel cell

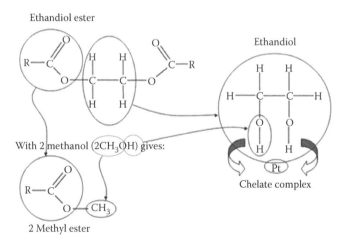

FIGURE 6.26 Degradation mechanism due to a demolding agent used in processing which affects the Pt catalyst inside a fuel cell. (Adapted from Bornbaum, S. 2010. Schunk Kohlenstofftechnik GmbH. Reproduced from Conference fee from the conference "*Kunststoffe in Brennstoffzellen*," Duisburg, ZBT.)

Ca	250	ppm
Co	<1	ppm
Cr	10	ppm
Cu	<1	ppm
Fe	100	ppm
Mo	2	ppm
Ni	8	ppm
Pb	<1	ppm
Sb	<2	ppm
Si	100	ppm
Ti	5	ppm
V	2	ppm
S	1000	ppm

FIGURE 6.27 The content of impurities in an exemplary data sheet extract of an expanded graphite. (Adapted from Timcal, 2010, http://www.timcal.com/scopi/group/timcal/timcal.nsf/pagesref/SCMM-7FGF77/$File/TIMREX_BNB90.pdf)

performance. The degradation mechanisms due to these impurities can vary, depending on the type and chemical makeup of the impurity. In all studies the degradation effect caused by impurities is clearly identified but there are still many questions regarding the impurities caused by composite bipolar plates. If impurities of the fillers do migrate out of the plate into the fuel cell environment, it may accelerate degradation phenomena of cell components similar to contamination through the fuels or cathode air. A common filler material for composite bipolar plates is expanded graphite. An extract of an exemplary data sheet of expanded graphite is shown in Figure 6.27. It clearly can be seen, that the content of impurities—especially sulfur-, silicon, calcium, and ferric-compounds—is significant.

Beside sulfur compounds, ferric traces are to be considered as particularly critical (Wu, 2008) highlighted especially trace metal ions of Fe^{2+} and Cu^{2+} which can accelerate weakening of the electrolyte membrane by catalyzing the formation of oxygen radicals due to the reaction with hydrogen peroxide (Fenton reaction).

In the 2009, DOE technical targets for bipolar plates was noted that corrosion of graphite or flexible graphite is not an issue (DOE, 2009). However, Scherer et al. (2009) claimed that the carbon content in composite bipolar plates, which is usually considered as corrosion stable, is accessible to oxidation by high electrode potentials in the presence of peroxides. The authors stated that surface oxidation of the carbon particles will result in higher contact resistance and could also change the surface properties of the bipolar plate. Furthermore, it was mentioned that introducing oxygen species into the topmost layer of a bipolar plate changed the wettability of the plate surface.

In principle, electrochemical oxidation of carbon in acid occurs by at least two anodic reaction pathways (Atanassova, 2007):

- Carbon → surface groups → CO_2
- Carbon → CO_2

And carbon corrosion is accelerated:

- During start/stop cycles
- At high voltage, OCV
- At high temperature operating condition

- At low humidity operating conditions
- Fuel starvation
- Partial coverage of hydrogen

Carbon corrosion on fuel cell electrode is thoroughly described in Fuller and Gray (2006), Tang et al. (2006), and Perry et al. (2006), whereas less has been reported regarding carbon corrosion in composite bipolar plates.

6.3.1.3 Testing Results

Mitani and Mitsuda (2009) found in their investigation that generally the physical changes in thermosetting composites tended to be greater than in thermoplastic composites. Some tested thermosetting composites tended to show a rapid increase in water conductivity, which the authors attributed to contained impurities and uncured components. Nonetheless, they obtained low-eluation materials in both thermosetting and thermoplastic composites. Finally, they found that one promising thermosetting composite and bipolar plates consisting of it were successfully operated in a 60-cell stack with hydrogen and with simulated reformate gas over 3000 h at a constant current density of 250 mA/cm^3. The authors concluded that this thermosetting-based composite bipolar plate shows over 3000 h hardly any degradation and concluded that the plate appears to be stable enough for use in a PEM fuel cell.

Maheshwari et al. (2007) found in their study that an amount of water adsorbed by almost all the samples matches the targeted value of <0.3% (Cho, 2004). In order to ascertain that there is no adverse effect of water adsorption on the plates these were tested for their compressive strength before and after water adsorption. The results show that there is no significant change in the compressive strength of the composite plate due to the small amount of water adsorbed. The values of electrical resistivity and flexural strength were also found to be similar before and after water adsorption (see Figure 6.28).

Within the investigation by Wilson et al. (2000) the XRF analysis of the Nafion membranes immersed with the composite samples indicated the presence of calcium. Calcium is present in the ash component of the graphite powder, and is able to leach out of finished plates when they are immersed in liquid. When the test liquid is not circulated or changed, noticeable amounts of calcium can accumulate. The authors also stated that in an operating fuel cell with the liquid continuously flushed, calcium does not appear to be any greater of a problem than with machined graphite plates.

The results of the flexural strength tests (Figure 6.29) show, that the flexural strengths of the exposed samples are at least equivalent to the strength of unexposed samples, within experimental scatter. Thus, the most important and encouraging result of this study is that the strengths of the various materials investigated are unaffected by exposure to various conditions including those much more aggressive

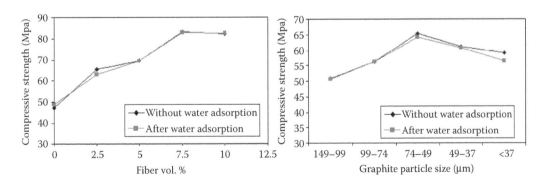

FIGURE 6.28 Variation in the compressive strength of the composite plate with increasing fiber content and variation in the compressive strength of the composite plate with decreasing graphite particle size. (Adapted from Maheshwari, P. H., Mathur R. B., and Dhami, T. L. 2007. *Journal of Power Sources* 173: 394–403.)

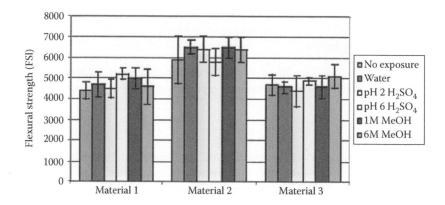

FIGURE 6.29 Flexural strengths of test samples after aggressive immersion testing. (Adapted from Wilson, M. S. et al. 2000. PEMFC stacks for power Generation. Proceedings of the DOE Hydrogen program review. NREL/CP-570–28890.)

than a fuel cell environment. Clearly, any chemical attack or hydrolysis that may be occurring is not affecting the structural component of the matrix.

Derieth et al. (2008) exposed each of the 10 PP-bonded composite bipolar plates to various liquids, in sulfuric acid ($c = 1$ mol/l) over 3000 h, at 80°C, in deionised water over 1000 h at 80°C, in methanol over 1000 h at 60°C, and in glycol over 1000 h at 80°C (all under air atmosphere, see Figure 6.30). The PPS-bonded bipolar plates were exposed to phosphoric acid ($W = 85$ wt%) over 1000 h at 160°C.

The authors concluded that in all these investigations the bipolar plates do not underlie changes in each of the determined properties. No significant differences in weight, thickness, roughness-values, and electrical and mechanical properties could be observed (Figure 6.31). On the other hand some increase in conductivity of about 0.02 µS referring to 1 g bipolar plate material could be measured

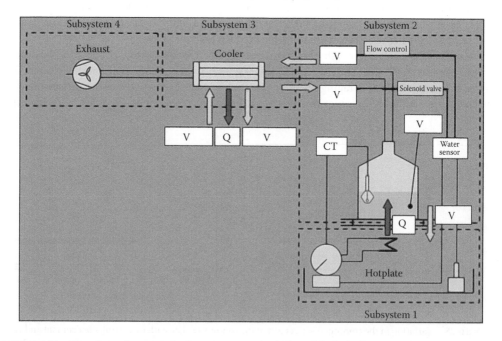

FIGURE 6.30 Flow chart of a test rig for the immersion test. (Adapted from ZBT. 2010. Original material gained at the Center for Fuel Cell Technology.)

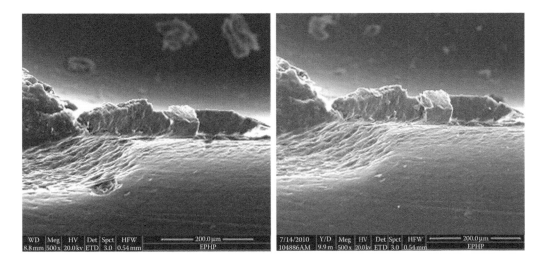

FIGURE 6.31 SEM images of a small area of a PP-bonded injection-molded bipolar plate before (left) and after (right) exposure—no changes could be observed—(time of exposure in deionised water 1000 h at 80°C). (Adapted from ZBT. 2010. Original material gained at the Center for Fuel Cell Technology.)

during the hot-water test that leads to the assumption that a small quantity of ions leaches out of the plate into the liquid.

Hence, the deionised water and the methanol in which the plates were exposed were analyzed via LC–MS and GC–MS and the result is depicted in Figure 6.32.

A few traces of different substances leach out of the plates into the deionized water. But, the ion level tends toward zero considering the test conditions and the fact that such plates have been successfully in operation in many stacks over several thousand hours (see Figure 6.33).

It can be seen that the same polypropylene matrix leaches a higher content of impurities into methanol when compared with water. This indicates that for use in a direct methanol fuel cell another type of polymer might be more appropriate, even though such polypropylene plates successfully operate in PEM fuel cells. Due to the chemically corrosion stable nature and the low additive content of PPS, it can be supposed that such plates are an appropriate choice for use in direct methanol fuel cells. At the time of writing, the authors apply a simultaneous *ex situ* test in methanol with such PPS-bonded bipolar plates.

The investigation by Bornbaum (2010) revealed significant differences depending on the chemical nature of the tested thermosetting materials as shown in Figure 6.34.

Although both polymer types are of thermosetting nature the authors operate the epoxy-based bipolar plates successfully in their low temperature fuel cell stacks, it obviously can be seen that there are significant differences between the epoxy-based and the phenolic-bonded composite and the epoxy one seems to be not suitable for use in high-temperature fuel cells.

In the investigation by Hui et al. (2010) it turned out, that the composite bipolar plates had good corrosion resistibility in the simulated solution. The aging mass loss rate was only about 0.1 wt% after 25 days. No aging mass loss after ten days could be observed. Figure 6.34 gives the curves of the conductivity and flexural strength varieties with the aging time. The results show that the conductivity of a composite plate increases first with the aging time, and then it almost keeps constant prolonging aging time. The reason therefore is a thin polymer layer on the surface of the plate and after exposure in the solution the topmost layer is dissolved. The surface resistance decreases, which increases the conductivity of composite plate. The flexural strength of composite plate decreases with the aging time rapidly as shown in Figure 6.35, but remains on the lower level afterward. The data on conductivity, mechanical properties,

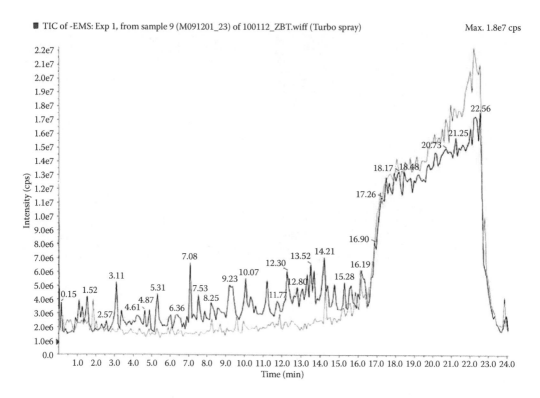

FIGURE 6.32 LC-MS-Chromatogram (negative ionisation modus) of an *ex situ* tested PP-bonded bipolar plate (time of exposure in deionised water 1000 h at 80°C)—thinner curve = reference $t = 0$ and thicker curve at $t = 1000$. (Adapted from ZBT. 2010. Original material gained at the Center for Fuel Cell Technology.)

and stability against long exposure to acid solution indicate that these composite plates will be very suitable for PEM fuel cells as the authors concluded.

6.3.2 Metal Bipolar Plates

Degradation of metallic bipolar plates can be constrained on two main mechanisms: (a) the increase of ohmic losses due to the formation of electrical passivation films on the bipolar plate's surface during cell operation; (b) the dissolution of metallic base material in the aggressive acidic environment of PEM fuel cell causing a contamination of the MEA by leached metallic ions leading to a cell performance decay.

Wang et al. (2003) found the interfacial contact resistance (ICR) between stainless steel 349™ and carbon paper to be influenced by the exposure time in a fuel cell simulating solution.[*] As can bee seen in Figure 6.36, they found a strong increase in ICR within the first 50 min and stabilization at a value 1.6 times higher than without exposition due to the formation of a passivation layer on the surface.

A similar behavior is reported by Davis et al. (2000a) for 316 and 310 stainless steels. In their work, different stainless steels were tested in long-term cell operation.[†]

The increase of ICR shown in Figure 6.37 is linked to cell performance decay after long-term operation shown in Figure 6.38.

---

[*] Potentiostatic test at +0.6 V in 1 M H_2SO_4 + 2 ppm F^- at 70°C air and hydrogen sparged.

[†] H_2 pressure 3 bar/air pressure 3 bar/H_2 utilization 70%/air utilization 35%/cell temperature 50°C/humidification 50%/ compaction force 220 N/cm² /current density 0.7 A/cm².

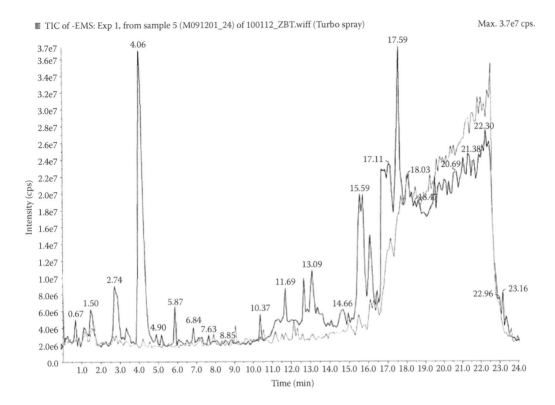

FIGURE 6.33 LC-MS-Chromatogram (negative ionisation modus) of an *ex situ* tested PP-bonded bipolar plate (time of exposure in methanol 1000 h at 60°C)—thinner curve = reference $t = 0$ and thicker curve at $t = 1000$. (Adapted from ZBT. 2010. Original material gained at the Center for Fuel Cell Technology.)

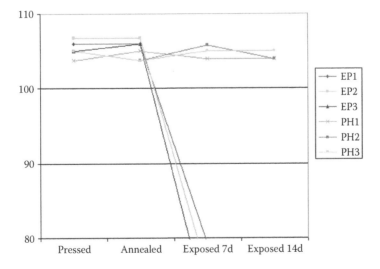

FIGURE 6.34 Comparison of epoxy-based and phenolic-bonded bipolar plates—exposed for 14 days in phosphoric acid ($W = 85$ wt%) at 160°C. (Adapted from Bornbaum, S. 2010. Schunk Kohlenstofftechnik GmbH. Reproduced from Conference fee from the conference "*Kunststoffe in Brennstoffzellen*," Duisburg, ZBT.)

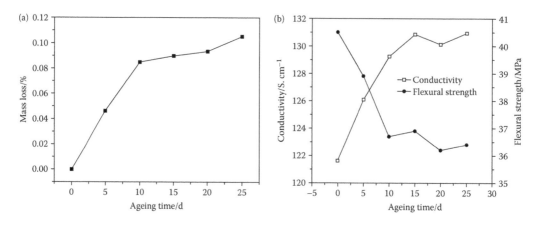

FIGURE 6.35 Mass loss of composite and properties of composite plate with different aging times. (Adapted from Hui, C. et al. 2008. *International Journal of Hydrogen Energy* 35(7): 3105–3109.)

In Makkus et al. (2000) found high concentrations of contaminants (Fe, Cr, Ni) by the inspection of MEA material used with untreated metallic bipolar plates in long-term operations. In all tests, the authors measured a cell performance decay of 30–50% during the first 300 h mainly ascribed to membrane contamination. They also assumed a link between contamination concentration and compaction pressure since they detected significantly higher contamination concentrations for a compaction pressure of 30 bar than for 4 bar.

Wu et al. (2008) highlighted three main membrane degradation mechanisms regarding the presence of cations due to corrosion of stack or system components.

1. Cations, showing stronger affinity with the sulfonic group of the PFSA (perfluorosulfonic acid) membrane than H^+, occupy active sites. Hence, membrane bulk properties, namely ionic conductivity, water content, and H^+ transference number, are changed.

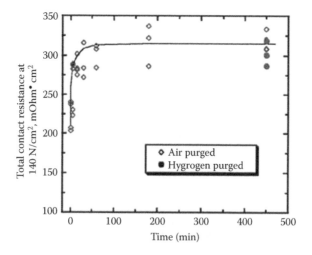

FIGURE 6.36 Increase of ICR; 349™ stainless steel after exposure in fuel cell simulating environment for different periods of time. (Adapted from Wang, H., Sweikart, M. A., and Turner, J. A. 2003. *Journal of Power Sources* 115: 243–251.)

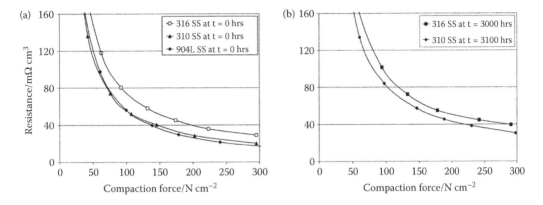

FIGURE 6.37 ICR of 316, 310, and 904 stainless steel compaction force depending before (a) and after (b) an endurance test. (Data from Davies, D. P. et al. 2000a. *Journal of Power Sources* 86: 237–242.)

2. Larger water transference numbers of foreign cations lead to an accelerated water flux, resulting in membrane dehydration (Okada, 2003).
3. Especially trace metal ions of Fe^{2+} and Cu^{2+} can accelerate depletion of the electrolyte membrane by catalyzing the formation of oxygen radicals due to the reaction with hydrogen peroxide shown in the following equations:

$$H_2O_2 + Fe^{2+} \rightarrow HO\bullet + OH^- + Fe^{3+} \tag{6.5}$$

$$Fe^{2+} + HO\bullet \rightarrow Fe^{3+} + OH^- \tag{6.6}$$

$$H_2O_2 + HO\bullet \rightarrow HO_2\bullet + H_2O \tag{6.7}$$

$$Fe^{2+} + HO_2\bullet \rightarrow Fe^{3+} + HO^{2-} \tag{6.8}$$

$$Fe^{3+} + HO_2\bullet \rightarrow Fe^{2+} + H^+ + O_2 \tag{6.9}$$

This effect may lead to membrane thinning and the formation of pinholes (Cheng, 2007; Wu, 2008).

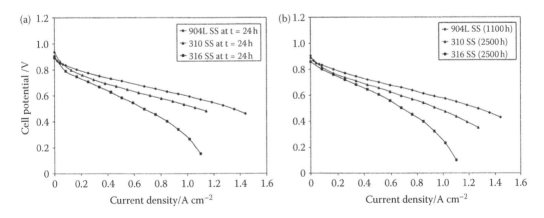

FIGURE 6.38 Polarization plot for 316, 310, and 904 stainless-steel fuel cells after 24 hours (a) and after an endurance test (b). (Data from Davies, D. P. et al. 2000a. *Journal of Power Sources* 86: 237–242.)

As increasing ohmic losses due to the formation of passivating surface films, and membrane poisoning ions that leached out as a result of corrosion, are still the most ongoing issues for metallic bipolar plates in PEM fuel cells, obviously metal has to be modified in a certain way by surface composition alteration or coating to overcome these disadvantages.

6.4 Mitigation Strategies

6.4.1 Graphite Composite Bipolar Plates

Low costs, good processability, and high performance are the general targets for composite bipolar plates and they play a major role for selection criteria of the fillers as well as for the polymeric matrices.

Besides this, for a proper long-term operation of composite bipolar plates inside a fuel cell the resistance of the polymer to chemical and physical attack is a very important issue. Furthermore, a high purity level of all source materials is required.

In case of the fillers the following criteria should be taken into account:

- Very high level of purification
- High degree of crystallinity

Besides properties such as morphology and particle size distribution the purity level differs of graphite in a wide range. In general, natural graphite has a lower purity degree compared to synthetic graphite. Hence, the first choice should be the use of synthetic graphite or eventually natural graphite with a very high level of purification (>99.8%). As mentioned in Section 6.3, some expanded or expandable graphite contain high contents of sulfur compounds and other impurities and hence it should be considered unless it is certain that no impurity traces contaminate the fuel cell environment. Generally, it can be said that the higher the purification level the better is the appropriateness of the graphite for composite bipolar plates. This thesis is valid independent of the fuel cell application and applies also for the other carbon fillers.

To minimize carbon corrosion, a high degree of crystallinity is obliging. Generally, graphite having a high degree, however, the higher the level the better the graphite resists carbon corrosion. This is particularly more important for other carbon species that naturally do not exhibit such a high degree of crystallinity than graphite.

In case of the polymeric materials several factors play a role, but generally it can be concluded that polymers that contain hydrolyzable groups, such as esters or amides, are susceptible to corrosion and decomposition in fuel cell environment. They should not be used as composite matrix material. The following properties are conducive to increase the long-term stability and should be taken into account for a targeted selection of a suitable polymeric matrix for composite bipolar plates.

- A high molecular weight
- Long length of chains and subchains
- A reduction of residue monomers
- Adding plasticizers which are harmless in a fuel cell environment
- A increase of molecule orientation
- A increase of the degree of crytallinity

However, polymers that obtain quite a few of the above-mentioned properties are usually cost intensive and challenging in processing. The best approach to achieve the partly contradicting requirements is to clarify the purpose of the bipolar plate before examining a new composite. For example, the thermoplastic resin polyphenylene sulfide (PPS) is known to be a suitable polymer for composite bipolar plates. Owing to the good heat and chemical resistance, PPS is an appropriate matrix for high-temperature composite bipolar plates as well as for use in direct methanol fuel cells due to low addivation. In this context, PPS is surely suitable for the low-temperature PEM fuel cells. On the other hand, PPS is a relatively cost-intensive thermoplastic material and challenging in processing, for example, extrusion or injection molding. Hence, clearly cheaper polymers that are usually also easier in processing, such as polypropylene, are far better for the low-temperature fuel cell purpose. Moreover, PP-bonded composite bipolar plates are thoroughly and successfully tested in low-temperature PEM fuel cells and there is no need for other materials, except that they would be better in processability and other properties, and cheaper.

Owing to their cross-linked network, thermosetting resigns are principally known to be less susceptible to corrosion than some thermoplastic materials. Nonetheless, as shown by Bornbaum (2010), a prediction of long-term stability of polymers in fuel cell operation is hardly possible. Thus, the immersion or accelerated lifetime test should be applied on every novel material which has not been operated successfully in a fuel cell over a several thousand hours. These tests may save considerable amounts of time and can be performed before a fuel cell is set up. Nevertheless and finally, only a long-term test in fuel cells at various parameters and different conditions can provide ultimate certainty as to whether a newly developed composite is suitable or not.

6.4.2 Metal Bipolar Plates

6.4.2.1 Surface Treatments

In a variety of scientific works surface treatments have been applied to metal bipolar plates to enhance durability. Most of these represent a process of chromizing, nitriding (respectively a combination of both) or carburization.

Lee et al. (2004) used an electrochemical surface treatment on stainless-steel bipolar plates to achieve a higher Cr content. Untreated and treated specimens were tested concerning electric contact resistance and corrosion current. The authors found an increased Cr content in the surface by metallurgical analyses and revealed a decreased contact resistance and corrosion current. In the text, no details about the composition of the electrolyte used for the surface treatment were given. Cho et al. (2008) carried out a chromizing pack cementation on 316L with respect to different process durations (2.5, 5, and 10 h at 1100°C). In potentiodynamic and potentiostatic tests, the lowest values of corrosion current density were obtained for a 2.5 h period of treatment. For contact resistance, the authors found slightly decreased values for 2.5 h treatment period at low compaction forces and no significant deviation from untreated 316L for all treatment periods at higher compaction forces. In a more recent work (Cho, 2009), the same authors applied two pack cementation processes with different powder mixtures and heating periods to stainless steel 316L. One process (I) was conducted with 50 wt.% Cr in the powder and 2.5 h heating time (1050°C); the other process (II) was conducted with a content of 25 wt.% Cr in the powder and 23 h heating time (1050°C). Both specimens showed the formation of a Cr-rich layer, which was strongly related to process parameters in its thickness (30 μm (I), 58 μm (II)) and in its composition. For process II a high concentration of oxygen was found according to the longer heating time. For process I, the formation of chromium carbide due to the higher Cr content in the powder was observed. Corrosion resistance was improved by both processes compared to untreated 316L stainless steel. Process II caused an increased contact resistance, whereas process I decreased contact resistance according to the presence of chromium carbide. Yang et al. (2010) suggested a lower process temperature for chromizing pack cementation applied to 316L. The authors compared a process at 900°C and a process at 1100°C to untreated stainless steel and reported a decrease of up to four orders of magnitude in the corrosion rate and one third of contact resistance for the processed material. The lower process temperature showed a slightly positive effect on contact resistance and corrosion resistance compared to the higher process temperature.

Nam et al. (2007) combined a chrome plating process with a following thermal nitriding on 316L stainless steel to improve corrosion behavior. The process parameters were chosen to form Cr_2N on the surface, which decreased corrosion current densities to acceptable values as found in potentiodynamic

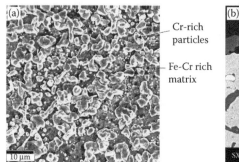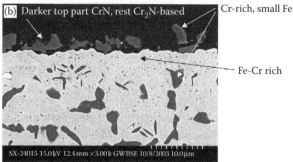

FIGURE 6.39   SEM images of thermally nitrided 349™ stainless steel; (a) top view, (b) cross-sectional view. (Adapted from Wang, H. et al. 2004. *Journal of Power Sources* 138: 86–93.)

tests. The contact resistance was also enhanced but still four times higher than that of graphite. This was ascribed to an insufficient internal layer quality. Wang et al. (2004) applied a thermal nitridation (1100°C, 2 h, pure nitrogen) to 349™ austenitic stainless steel. Although the treated material reached lower contact resistance, unacceptable high corrosion current densities were obtained in comparison to the untreated material. This was attributed to the formation of discontinuous nitride surface particles (Figure 6.39) and the modification of process parameters was suggested.

Thereupon, Lee et al. (2009) recommended a preoxidation process prior to thermal nitridation to improve both contact resistance and corrosion resistance. Applied to 446M stainless steel, the authors created better corrosion resistance relative to untreated and nitrided-only material due to the formation of a protective oxide layer. The preoxidation and nitridation as well as nitridation only resulted in decreased values for contact resistance of equal level. Han et al. (2009) implemented low-temperature plasma nitriding (ICP 2 h at 400°C) on chromium electroplated 316L stainless steel for different values of ICP bias voltages (–100 V, –200 V, –300 V). Potentiostatic and potentiodynamic tests indicated acceptable values for corrosion resistance of plasma-nitrided specimens. Contact resistance showed a dependence on bias voltage with decreasing contact resistance for increasing bias voltage.

Nikam et al. (2008) proposed a low-temperature carburization (Cao, 2003) for the improvement of corrosion resistance and electric properties of 316L specimens. In potentiostatic tests, corrosion current densities were obtained to be 4 μA cm^{-2} in anodic and 1.5 μA cm^{-2} in cathodic PEM fuel cell environment. Interfacial contact resistance of LTC 316L was lowered by approximately 24% compared to untreated material.

Feng et al. (2008, 2010) used ion implantation for modification of the passive layer of 316L stainless steel. In one study (Feng, 2008), the authors implanted nickel ions with varying implantation dosage (1, 2, 3, 4, 5 × 10^{17} Ni cm^{-2}) and found the contact resistance to be dependent on implantation dose with highest contact resistance for the highest dosage and lowest contact resistance for lowest implantation dosage, but still higher than DOE targets. In a more recent study (Feng, 2010), the same authors coimplanted chromium and nickel ions with respect to varying implantation times (0.5, 1, 2, 3 h). They found the contact resistance to be dependent on implantation time with a tendency to decrease contact resistance with longer implantation time. In both studies corrosion resistance was enhanced extensively related to bare 316L stainless-steel material.

6.4.2.2 Protective Coatings

Another way of providing higher corrosion resistance and performance stability to metallic bipolar plates is the use of conductive and corrosion-resistant coatings.

Although coating defects like pinholes and macroparticles are known to be inherent for the PVD (physical vapor deposition) technique (Chen, 2000; Andrade, 2004; Barshilia, 2004; Flores, 2007) and

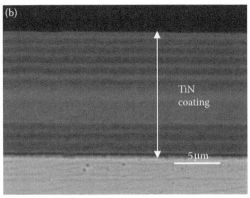

FIGURE 6.40 TiN coated 316LSS. TiN coating surface (a) and cross-sectional view of TiN coating on SS316L (b). (Adapted from Wang, Y. and Northwood, D. O. 2007. *Journal of Power Sources* 165: 293–298.)

therewith PVD coatings are susceptible to pitting corrosion, a diversity of investigations were conducted by the use of this coating process. Wang and Northwood (2007) tested a TiN PVD coating (Figure 6.40) on 316L stainless regarding corrosion resistance in the PEMFC environment.

For potentiodynamic studies, the authors observed a strongly decreased corrosion current for TiN-coated 316LSS compared to the bare 316LSS material. To simulate long-term fuel cell operation, potentiostatic tests were undertaken and revealed a threefold increase in corrosion current for cathodic operation, which was ascribed to pitting corrosion due to coating imperfects (Figure 6.41). These results confirm the studies of Cho et al. (2005) concerning TiN coating on 316 stainless-steel bipolar plates for a 1 kW-class PEMFC stack. In addition to *ex situ* tests, in which the coatings provided good electrochemical performance in a fuel cell simulating environment, 1028 h endurance operation was conducted. Drastic fuel cell performance decay was discovered for bare 316SS and TiN-coated 316SS bipolar plates. This was attributed to membrane poisoning by Fe, Cr, Ni, and Ti ions released by substrate and coating. Jeon et al. (2008) found the corrosion resistance of TiN-coated 316L stainless steel to be dependent on N_2 gas pressure during PVD process. In this work they found an N_2 partial pressure of 0.4 mTorr as optimal parameter for deposition of the corrosion-resistant TiN layer related to lowest-layer porosity.

As a possibility to provide defect-free coatings by the PVD technique, the application of multilayered coatings is often advised (Nordin, 1999; Liu, 2001). Accordingly, Nam (2010) applied multilayered

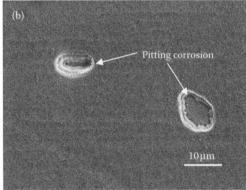

FIGURE 6.41 SEM micrographs for the coated SS316L after potentiostatic tests in simulated anode and cathode conditions; anode side (a) and cathode side (b). (Adapted from Wang, Y. and Northwood, D. O. 2007. *Journal of Power Sources* 165: 293–298.)

TiN/CrN coatings to 316L stainless steel by radio frequency magnetron sputtering for corrosion resistance improvement with respect to the ratio of TiN to CrN layer thickness. TiN/CrN layer of 1 μm thickness and ratios of 1:9, 3:7, and 5:5 were found to lower corrosion current in potentiodynamic tests and to increase charge transfer resistance (>10^5 Ω cm^2) in cathodic potentiostatic tests compared to the bare substrate. Contact resistance was reduced by one order of magnitude. For all testing results coatings with a decreasing inner TiN layer showed increasing performance.

Choi et al. (2009) investigated ternary $(Ti,Cr)N_X$ coatings created by ICP magnetron sputtering on 316L stainless steel focusing on N_2 flow rate during deposition. The results showed that the N_2 flow rate is an essential process parameter to achieve corrosion resistant coatings. All coatings with N_2 flow showed corrosion current decreased by one order of magnitude compared to the bare substrate and substrate coated without N_2 flow.

Fukutsuka et al. (2007) used PACVD (plasma-assisted chemical vapor deposition) to coat 304 stainless steel with a carbon layer and reached the DOE technical target for bipolar plate corrosion rate of less than 1 μA cm^{-2}. Chung et al. (2008) carried out a thermal CVD carbon deposition on Ni premagnetron sputtered 304 stainless steel where the Ni layer acted as a catalyst for graphitization. The deposition was conducted with respect to the ratio of C_2H_2 to H_2 in the C_2H_2/H_2 mixed gas (0.15, 0.45, 1.5) being the carbon source. Corresponding to the C_2H_2 to H_2 ratio the carbon layer morphology evolved from a filamentous layer at a ratio of 0.15 to a continuous carbon layer at a ratio of 0.45. Used as bipolar plate material in a fuel cell module, the carbon-coated material was found to have a performance similar to graphite material (Poco graphite).

An amorphous carbon coating (~3 μm thickness) was applied by Feng et al. (2009) with close field unbalanced magnetron sputter ion plating to 316L stainless steel. The coating performed promisingly in potentiodynamic and potentiostatic tests and also in ICR measurements due to a dense film protecting the base material from corrosion. The authors approved amorphous carbon coating as suitable for bipolar plates in PEM fuel cells.

Zeng and Ren (2007) used high-energy micro arc alloying (HEMAA) for TiC deposition on 304 stainless steel. The authors highlighted the superiority of this technique to PVD, for a pinhole-free layer (Figure 6.42) is formed. Potentiodynamic measurements revealed a corrosion current density of 0.034 μA cm^{-2} for the coated material as against 8.3 μA cm^{-2} for the bare 304 stainless steel. For potentiostatic

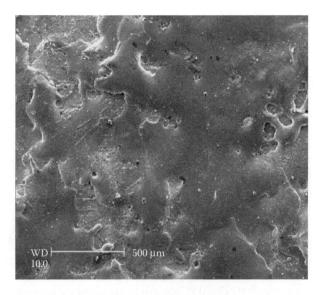

FIGURE 6.42 Surface morphology of TiC film produced by HEMAA technique on 304 stainless steel. (Adapted from Zeng, C. L. and Ren, Y. J. 2007. *Journal of Power Sources* 171: 778–782.)

measurements a stable current density of 3.7 μA cm^{-2} after 4 h and a high stability during 30 days of immersion was observed.

As most coatings represent a combination of Ti, Cr, Ni, C, and N, coating with conductive polymers is a different approach in corrosion protection of metallic bipolar plates. Joseph et al. (2005) deposited conducting polymers polyaniline (PANI) and polypyrrole (PPY) on 304 stainless steel by cyclic voltammetry polymerization. Regarding the numbers of deposition cycles (3, 8, and 15) and therewith coating thickness, they found best corrosion resistance for three cycles in potentiodynamic measurements. PANI coating was subjected to the formation of pinholes in eight cycles- and 15 cycles-deposition. Contact resistance increased for both types of coatings in comparison with uncoated 304 stainless steel. No durability prediction was given as no endurance tests were conducted in this work. Wang and Northwood (2006) investigated polypyrrole coatings on 316L stainless steel deposited by galvanostatic and cyclic voltammetric methods. They found an increasing particle size with increasing current for galvanostatic deposition and increasing particle size with increasing cycle number for cyclic voltammetry by SEM micrographs. Corrosion current density was observed to decrease by one order of magnitude for polypyrrole-coated 316L in potentiodynamic tests. No long-term tests were conducted. As the metal ions concentration in the electrolyte used for corrosion studies was found to be too high for PEM fuel cells, the same authors (Wang, 2008) carried out the deposition of a nanothick gold layer between 316L substrate and polypyrrole coating. By extrapolating ion concentration results of a 10 h potentiostatic immersion test to 5000 h target operation time for real fuel cells, the authors attested the suitability of a polypyrrole coating grown on a predeposited gold layer for concerns of ion leaching. Smit and García (2006) coated 304 stainless steel with polypyrrole layers of different thickness by cyclic voltammetry deposition. Potentiodynamic polarization curves were taken after different times of immersion (1, 3, and 24 h) in a PEM fuel cell environment simulating solution. Although corrosion current densities at first were decreased by up to four orders of magnitude after 24 h of immersion, the samples showed polarization curves similar to bare 304 stainless steel indicating that the protective properties of PPY get lost during extended immersion.

As can be seen from Sections 6.1.3 and 6.2.4, a huge variety of research work has been performed to understand degradation mechanisms metallic bipolar plates are subjected to and hence to improve durability and operation performance of metallic bipolar plates material. Currently, some results found in the extracted literature shown above fit with the technical targets of DOE, and some are close to these targets. Consequently, these materials have to be investigated in long-term fuel cell tests of some thousand hours to certify their suitability in addition to well-documented accelerated tests, which is missing in most papers. Those "full scale" endurance tests are one recommendation for future research concerning metal bipolar plates.

6.5 Summary

Bipolar plates in two ways can contribute to the limited lifetime of PEM fuel cell stacks: first is the physical decomposition of the materials or parts of the materials resulting in mechanical instability, leakage or decrease of electrical conductivity; second is the release of additives and materials to the fuel cell environment forcing degradation processes in further fuel cell parts such as gaskets, MEAs, and peripheral components. These phenomena are to be considered both for composite-based and metal-based bipolar plates.

Using composites the fuel cell contamination due to filler or plastic impurities is more essential than the overall plate's stability. The best strategy to avoid a breakdown or failures during fuel cell operation is the right choice of the polymer used. Criteria for the right choice are the costs, the processability, and, more importantly, the corrosion stability. As the stability of commercial available plates is good, most applications demanding long-term operation, higher temperatures, and other complex environments rely on composite-based plate technologies.

For metal bipolar plates, the stability of the bipolar plate, especially against corrosion, has to be brought more into focus. A suitable base material combined with a conductive and protective coating is the only way to extend the lifetime of the plates. Nevertheless, metallic bipolar plates are the preferred choice for all applications attending to small volume and weight and without high lifetime demand.

To identify and qualify materials and bipolar plates a bandwidth of testing procedures exists. But the quantification of the phenomena remains complex and is subject of recent work for both types of technologies (Decode, 2010). Also the interactions of the degradation processes of the bipolar plates with the other components like gaskets, GDL and the MEA have to be investigated with high intensity. The results from *ex situ* and *in situ* experiments have to be combined logically to determine fuel cell behavior with respect to changes in bipolar plate properties. This is especially a topic for accelerated lifetime testing procedures and their link to real fuel cell operational aspects.

Finally, a topic that has to be addressed more intensively is the link between operational aspects of fuel cells and the stability of the bipolar plates. With intelligent fuel cell control many degradation mechanisms can be avoided, but real guidelines for system-developing companies rarely exist.

References

Affolter, S. 1999. Long-term behaviour of thermoplastic materials. Internet: https://institute.ntb.ch/fileadmin/Institute/MNT/NTB_MNT_Polymerics_Ageing_SA.pdf

Andrade, E., Flores, M., Muhl, S. et al. 2004. Ion beam analysis of TiN/Ti multilayers deposited by magnetron sputtering. *Nuclear Instruments and Methods in Physics Research B* 219–220: 763–767.

Antunes, R.A., Oliveira, M. C. L., Ett, G. et al. 2010. Corrosion of metal bipolar plates for PEM fuel cells: A review. *International Journal of Hydrogen Energy* 35: 3632–3647.

Asghari, S., Mokmeli, A., and Samavati, M. 2010. Study of PEM fuel cell performance by electrochemical impedance spectroscopy. *International Journal of Hydrogen Energy* 35: 9283–9290.

Atanassova, P. 2007. Conference Fee. *Gordon Research Conference on Fuel Cells.* July 22–27, 2007, Bryant University, Smithfield, RI.

Barbir, F., Braun J., and Neutzler, J. 1999. Properties of molded graphite bipolar plates. *Journal of New Materials for Electrochemical Systems* 2: 197–200.

Barshilia, H. C., Prakash, M. S., Poojari, A. et al. 2004. Corrosion behavior of nanolayered TiN/NbN multilayer coatings prepared by reactive direct current magnetron sputtering process. *Thin Solid Films* 460: 133–142.

Besmann, T. M., Henry, J. J., Klett, J. W. et al. 2003. Carbon composite bipolar plate for PEM fuel cells. *Hydrogen and Fuel Cells Merit Review Meeting*, Berkeley, California.

Bin, Z., Bingchu, M., Chunhui, S. et al. 2006. Study on the electrical and mechanical properties of polyvinylidene fluoride/titanium silicon carbide composite bipolar plates. *Journal of Power Sources* 161: 997–1001.

Blunk, R. H. J., Lisi, D. J., Yoo, Y. E. et al. 2003. Enhanced conductivity of fuel cell plates through controlled fiber orientation. *AIChE Journal* 49: 18–29.

Bornbaum, S. 2010. Schunk Kohlenstofftechnik GmbH. Reproduced from Conference fee from the conference "*Kunststoffe in Brennstoffzellen*," Duisburg, ZBT.

Borup R., Meyers, J., Pivovar, B. et al. 2007. Scientific aspects of polymer electrolyte fuel cell durability and degradation. *Chemical Reviews* 107: 3904–3951.

Cao Y., Ernst, F., and Michal, G. M. 2003. Colossal carbon supersaturation in austenitic stainless steels carburized at low temperature. *Acta Materialia* 51: 4171–4181.

Chen, J.-Y., Yu, G.-P., and Huang, J.-H. 2000. Corrosion behavior and adhesion of ion-plated TiN films on AISI 304 steel. *Materials Chemistry and Physics* 65: 310–315.

Cheng, X., Chi, Z., Glass, N. et al. 2007. A review of PEM hydrogen fuel cell contamination: Impacts, mechanisms, and mitigation. *Journal of Power Sources* 165: 739–756.

Cho, E. A., Jeon, U. S., Ha, H. Y. et al. 2004. Characteristics of composite bipolar plates for polymer electrolyte membrane fuel cells. *Journal of Power Sources* 125: 178–182.

Cho, E. A., Jeon, U. S., Hong, S.-A. et al. 2005. Performance of a 1 kW-class PEMFC stack using TiN-coated 316 stainless steel bipolar plates. *Journal of Power Sources* 142: 177–183.

Cho, K. H., Lee, S. B., Lee, W. G. et al. 2009. Improved corrosion resistance and interfacial contact resistance of 316L stainless-steel for proton exchange membrane fuel cell bipolar plates by chromizing surface treatment. *Journal of Power Sources* 187: 318–323.

Cho, K. H., Lee, W. G., Lee, S. B. et al. 2008. Corrosion resistance of chromized 316L stainless steel for PEMFC bipolar plates. *Journal of Power Sources* 178: 671–676.

Choi, H. S., Han, D. H., Hong, W. H. et al. 2009. (Titanium, chromium) nitride coatings for bipolar plate of polymer electrolyte membrane fuel cell. *Journal of Power Sources* 189: 966–971.

Chu, D. and Jiang, R. 1999. Comparative studies of polymer electrolyte membrane fuel cell stack and single cell. *Journal of Power Sources* 80: 226–234.

Chung, C.-Y., Chen, S.-K., Chiu, P.-J. et al. 2008. Carbon film-coated 304 stainless steel as PEMFC bipolar plate. *Journal of Power Sources* 176: 276–281.

Cooper, J. S. 2004. Design analysis of PEMFC bipolar plates considering stack manufacturing and environment impact. *Journal of Power Sources* 129: 152–169.

Cunningham, B. and Baird, D. G. 2006. The development of economical bipolar plates for fuel cells. *Journal of Materials Chemistry* 16: 4385–4388.

Cunningham, B. D., Huang, J., and Baird, D. G. 2007. Development of bipolar plates for fuel cells from graphite filled wet-lay material and a thermoplastic laminate skin layer. *Journal of Power Sources* 165: 764–773.

Davies, D. P., Adcock, P. L., Turpin, M. et al. 2000. Stainless steel as a bipolar plate material for solid polymer fuel cells. *Journal of Power Sources* 86: 237–242.

Davies, D. P., Adcock, P. L., Turpin, M. et al. 2000b. Bipolar plate materials for solid polymer fuel cells. *Journal of Applied Electrochemistry* 30: 101–105.

DECODE project, http://www.decode-project.eu/about-D-3.html.

Derieth, T., Bandlamudi, G., Beckhaus, P. et al. 2008. Development of highly filled graphite composites as bipolar plate materials for low and high temperature PEM fuel cells. *Journal of New Materials for Electrochemical Systems* 11: 21–29.

DOE Progress Report 2006, http://www.hydrogen.energy.gov/annual_progress06_fuelcells.html

DOE Progress Report 2009, http://www.hydrogen.energy.gov/annual_progress09_fuelcells.html

Dweiri, R. and Sahari, J. 2007. Electrical properties of carbon-based polypropylene composites for bipolar plates in polymer electrolyte membrane fuel cell (PEMFC). *Journal of Power Sources* 171: 424.

Feng, K., Shen, Y., Liu, D. et al. 2010. Ni–Cr Co-implanted 316L stainless steel as bipolar plate in polymer electrolyte membrane fuel cells. *International Journal of Hydrogen Energy* 35: 690–700.

Feng, K., Shen, Y., Mai, J. et al. 2008. An investigation into nickel implanted 316L stainless steel as a bipolar plate for PEM fuel cell. *Journal of Power Sources* 182: 145–152.

Feng, K., Shen, Y., Sun, H. et al. 2009. Conductive amorphous carbon-coated 316L stainless steel as bipolar plates in polymer electrolyte membrane fuel cells. *International Journal of Hydrogen Energy* 34: 6771–6777.

Flores, M., Huerta, L., Escamilla, R. et al. 2007. Effect of substrate bias voltage on corrosion of TiN/Ti multilayers deposited by magnetron sputtering. *Applied Surface Science* 253: 7192–7196.

Fukutsuka, T., Yamaguchi, T., Miyano, S.-I. et al. 2007. Carbon-coated stainless steel as PEFC bipolar plate material. *Journal of Power Sources* 174: 199–205.

Fuller, T. F. and Gray, G. 2006. Carbon corrosion induced by partial hydrogen coverage. *ECS Transactions* 1: 345–352.

GrafTech. 2009. http://www.graftechaet.com/grafcell/grafcell-Home.aspx

Hamilton P. J. and Pollet B. G. 2010. Polymer electrolyte membrane fuel cell (PEMFC) flow field plate: Design, materials and characterization. *Fuel Cells* 10(4): 489–509.

Han D. H., Hong, W.-H., Choi, H. S. et al. 2009. Inductively coupled plasma nitriding of chromium electroplated AISI 316L stainless steel for PEMFC bipolar plate. *International Journal of Hydrogen Energy* 34: 2387–2395.

Heinzel, A., Mahlendorf, F., Niemzig, O. et al. 2004. Injection moulded low cost bipolar plates for PEM fuel cells. *Journal of Power Sources* 131: 35–40.

Hentall, P. L., Lakeman, J. B., Mepsted, G. O. et al. 1999. New materials for polymer electrolyte membrane fuel cell current collectors. *Journal of Power Sources* 80: 235–241.

Heo, S. I., Oh, K. S., Yun, J. C. et al. 2007. Development of preform moulding technique using expanded graphite for proton exchange membrane fuel cell bipolar plates. *Journal of Power Sources* 171: 396–403.

Hermann, A., Chaudhuri, T., and Spagnol, P. 2005. Bipolar plates for PEM fuel cells: A review. *International Journal of Hydrogen Energy* 30: 1297–1302.

Hodgson, D. R., May, B., Adcock, P. L. et al. 2001. New lightweight bipolar plate system for polymer electrotyte membrane fuel cells. *Journal of Power Sources* 96: 233–235.

Hornung, R. and Kappelt, G. 1998. Bipolar plate materials development using Fe-based alloys for solid polymer fuel cells. *Journal of Power Sources* 72: 20–21.

Hou, M., Fu, Y., Lin, G. et al. 2009. Optimized Cr-nitride film on 316L stainless steel as proton exchange membrane fuel cell bipolar plate. *International Journal of Hydrogen Energy* 34: 453–458.

Huang, J., Baird, D. G., and McGrath, J. E. 2005. Development of fuel cell bipolar plates from graphite filled wet-lay thermoplastic composite materials. *Journal of Power Sources* 150: 110–119.

Hui, C., Hong-Bo, L., Li, Y. et al. 2008. Study on the preparation and properties of novolac epoxy/graphite composite bipolar plate for PEMFC. *International Journal of Hydrogen Energy* 35(7): 3105–3109.

Jeon, W.-S., Kim, J.-G., Kim, Y.-J. et al. 2008. Electrochemical properties of TiN coatings on 316L stainless steel separator for polymer electrolyte membrane fuel cell. *Thin Solid Films* 516: 3669–3672.

Joseph, S., McClure, J. C., Chianelli, R. et al. 2005. Conducting polymer-coated stainless steel bipolar plates for proton exchange membrane fuel cells (PEMFC). *International Journal of Hydrogen Energy* 30: 1339–1344.

King, J. A., Barton, R. L., Hauser, R. A. et al. 2008. Synergistic effects of carbon fillers in electrically and thermally conductive liquid crystal polymer based resins. *Polymer Composites* 29: 421–428.

Krebs, C., Avondet, M. A., and Leu, K. W. 1999. Langzeitverhalten von Thermoplasten—Alterungsverhalten und Chemikalienbeständigkeit. Hanser Verlag München.

Kreuz, C. 2008. *Dissertation:* PEM-Brennstoffzellen mit spritzgegossenen Bipolarplatten aus hochgefülltem Graphit-Composite, Gerhard-Mercator-University in Germany, 04/2008.

Kühnemann, L., Beckhaus, P., and Bekeschus, G. 2010. Photographed at ZBT GmbH, Duisburg, Germany.

Kumagai, M., Myung, S.-T., Ichikawa, T. et al. 2010. Evaluation of polymer electrolyte membrane fuel cells by electrochemical impedance spectroscopy under different operation conditions and corrosion. *Journal of Power Sources* 195: 5501–5507.

Kuo, J. 2006. A novel Nylon-6-S316L fiber compound material for injection molded PEM fuel cell bipolar plates. *Journal of Power Sources* 162: 207–214.

Lafront, A. M., Gahli, E., and Moralis, A. T. 2007. Corrosion behavior of two bipolar plate materials in simulated PEMFC environment by electrochemical noise technique. *Electrochimica Acta* 52: 5076–5085.

Lee, K.-H., Lee, S.-H., Kim, J.-H. et al. 2009. Effects of thermal oxi-nitridation on the corrosion resistance and electrical conductivity of 446M stainless steel for PEMFC bipolar plates. *International Journal of Hydrogen Energy* 34: 1515–1521.

Lee, S.-J., Huang, C.-H., Lai, J.-J. et al. 2004. Corrosion-resistant component for PEM fuel cells. *Journal of Power Sources* 131: 162–168.

Liu, C., Leyland, A., Bi, Q. et al. 2001. Corrosion resistance of multi-layered plasma-assisted physical vapour deposition TiN and CrN coatings. *Surface and Coatings Technology* 141: 164–173.

Maheshwari, P. H., Mathur R. B., and Dhami, T. L. 2007. Fabrication of high strength and a low weight composite bipolar plate for fuel cell applications. *Journal of Power Sources* 173: 394–403.

Makkus, R. C., Janssen, A. H. H., de Bruijn, F. A. et al. 2000. Use of stainless steel for cost competitive bipolar plates in the SPFC. *Journal of Power Sources* 86: 274–282.

Mercuri, R., Angelo, R., and Gough, J. J. 2000. Flexible graphite composite for use in the form of a fuel cell flow field plate. US Patent, 6037074.

Metha, V. and Cooper, J. S. 2003. Review and analysis of PEM fuel cell design and manufacturing. *Journal of Power Sources* 114: 32–53.

Mighri, F., Huneault, M. A., and Champagne, M. F. 2004. Electrically conductive thermoplastic blends for injection and compression molding of bipolar plates in the fuel cell application. *Polymer Engineering and Science* 44: 1755–1765.

Mitani, T. and Mitsuda, K. 2009. Durability of graphite composite bipolar plates. In *Polymer Electrolyte Fuel Cell Durability*, eds. F. N. Büchi, T. J. Schmidt, and M. Inaba, pp. 257–270. Springer Science + Buisness Media, LLC, New York, NY.

Nam, D.-G. and Lee, H.-C. 2007. Thermal nitridation of chromium electroplated AISI316L stainless steel for polymer electrolyte membrane fuel cell bipolar plate. *Journal of Power Sources* 170: 268–274.

Nam, N. D., Han, J. H., Tai, P. H. et al. 2010. Electrochemical properties of TiNCrN-coated bipolar plates in polymer electrolyte membrane fuel cell environment. *Thin Solid Films*, doi:10.1016/j.tsf.2010.03.046.

Nikam, V. V. and Reddy, R. G. 2005. Corrosion studies of a copper–beryllium alloy in a simulated polymer electrolyte membrane fuel cell environment. *Journal of Power Sources* 152: 146–155.

Nikam, V. V., Reddy, R. G., Collins, S. R. et al. 2008. Corrosion resistant low temperature carburized SS 316 as bipolar plate material for PEMFC application. *Electrochimica Acta* 53(6): 2743–2750.

Nordin, M., Herranen, M., and Hogmark, S. 1999. Influence of lamellae thickness on the corrosion behaviour of multilayered PVD TiN/CrN coatings. *Thin Solid Films* 348: 202–209.

Oh, M. H., Yoon, Y. S., and Park, S. G. 2004. The electrical and physical properties of alternative material bipolar plate for PEM fuel cell system. *Electrochimica Acta* 50: 777–780.

Okada, T. 2003. Effect of ionic contaminats. In *Handbook of Fuel Cells: Fundamentals. Technology and Applications*, Vol. 3, eds. W. Vielstich, H. A. Gasteiger, and A. Lamm, pp. 628–646. John Wiley & Sons Ltd, Chichester, England.

Perry, M. L., Patterson, T. W., and Reiser, C. 2006. Systems strategies to mitigate carbon corrosion in fuel cells. *ECS Transactions* 3(1): 783–795.

Pozio, A., Zaza, F., Masci, A. et al. 2008. Bipolar plate materials for PEMFCs: A conductivity and stability study. *Journal of Power Sources* 179: 631–639.

Rubio, M. A., Urquia, A., and Domido, S. 2010. Diagnosis of performance degradation phenomena in PEM fuel cells. *International Journal of Hydrogen Energy* 35: 2586–2590.

Rubio, M. A., Urquia, A., Kuhn, R. et al. 2008. Electrochemical parameter estimation in operating proton exchange membrane fuel cells. *Journal of Power Sources* 183: 118–125.

Scherer, J., Münter, D., and Ströbel, R. 2009. Influence of metallic bipolar plates on the durability of polymer electrolyte fuel cells. In *Polymer Electrolyte Fuel Cell Durability*, eds. F. N. Büchi, T. J. Schmidt and M. Inaba, pp. 243–256. Springer Science + Buisness Media, LLC, New York, NY.

Scribner Associates, 2010. http://www.scribner.com/technical-papers.html.

Smit, M. A. and García, M. A. L. 2006. Study of electrodeposited polypyrrole coatings for the corrosion protection of stainless steel bipolar plates for the PEM fuel cell. *Journal of Power Sources* 158: 397–402.

Taipalus, R., Harmia, T., Zhang, M. Q. et al. 2001. The electrical conductivity of carbon-fibre-reinforced polypropylene/polyaniline complex-blends: Experimental characterisation and modelling. *Composite Science and Technology* 61: 801–814.

Tang, H., Qi, Z., Ramani, M. et al. 2006. PEM fuel cell cathode carbon corrosion due to the formation of air/fuel boundary at the anode. *Journal of Power Sources* 158: 1306–1312.

Timcal, 2010, http://www.timcal.com/scopi/group/timcal/timcal.nsf/pagesref/SCMM-7FGF77/$File/TIMREX_BNB90.pdf

Tsotra, P. and Friedrich, K. 2004. Short carbon fiber reinforced epoxy resin/polyaniline blends: Their electrical and mechanical properties. *Composite Science and Technology* 64: 2385–2391.

Wang, H., Brady, M. P., Teeter, G. et al. 2004. Thermally nitrided stainless steels for polymer electrolyte membrane fuel cell bipolar plates Part 1: Model Ni–50Cr and austenitic 349TM alloys. *Journal of Power Sources* 138: 86–93.

Wang, H., Sweikart, M. A., and Turner, J. A. 2003. Stainless steel as bipolar plate material for polymer electrolyte membrane fuel cells. *Journal of Power Sources* 115: 243–251.

Wang, Y. and Northwood, D. O. 2006. An investigation into polypyrrole-coated 316L stainless steel as a bipolar plate material for PEM fuel cells. *Journal of Power Sources* 163: 500–508.

Wang, Y. and Northwood, D. O. 2007. An investigation into TiN-coated 316L stainless steel as a bipolar plate material for PEM fuel cells. *Journal of Power Sources* 165: 293–298.

Wang, Y. and Northwood, D. O. 2008. An investigation into the effects of a nano-thick gold interlayer on polypyrrole coatings on 316L stainless steel for the bipolar plates of PEM fuel cells. *Journal of Power Sources* 175: 40–48.

Wilson, M. S., Zawodzinski, C., Bender, G. et al. 2000. PEMFC stacks for power Generation. Proceedings of the DOE Hydrogen program review. NREL/CP-570-28890.

Wind, J., Späh, R., Kaiser, W. et al. 2002. Metallic bipolar plates for PEM fuel cells. *Journal of Power Sources* 105: 256–260.

Wolf, H. and Willer, M. 2006. Electrically conductive LCP-carbon composite with low carbon content for bipolar plate application in polymer electrolyte membrane fuel cell. *Journal of Power Sources* 153: 41–46.

Wu, J., Yuan, X. Z., Martin, J. J. et al. 2008. A review of PEM fuel cell durability: Degradation mechanisms and mitigation strategies. *Journal of Power Sources* 184: 104–119.

Xiao, M., Lu, Y., Wang, S. J. et al. 2006. Poly(arylene disulfide)/graphite nanosheets composites as bipolar plates for polymer electrolyte membrane fuel cells. *Journal of Power Sources* 160: 165–174.

Xing, X. Q., Lum, K. W., Poh, H. J. et al. 2010. Optimization of assembly clamping pressure on performance of proton-exchange membrane fuel cells. *Journal of Power Sources* 195: 62–68.

Yan, X., Hou, M., Sun, L. et al. 2007. AC impedance characteristics of a 2 kW PEM fuel cell stack under different operating conditions and load changes. *International Journal of Hydrogen Energy* 32: 4358–4364.

Yang, L., Yu, H., Jiang, L. et al. 2010. Improved anticorrosion properties and electrical conductivity of 316L stainless steel as bipolar plate for proton exchange membrane fuel cell by lower temperature chromizing treatment. *Journal of Power Sources* 195: 2810–2814.

Yin, Q., Li, A., Wang, W. et al. 2007. Study on the electrical and mechanical properties of phenol formaldehyde resin/graphite composite for bipolar plate. *Journal of Power Sources* 165: 717–721.

Yun, Y.-H. 2010. Deposition of gold–titanium and gold–nickel coatings on electropolished 316L stainless steel bipolar plates for proton exchange membrane fuel cells. *International Journal of Hydrogen Energy* 35: 1713–1718.

ZBT. 2010. Original material gained at the center for Fuel Cell Technology.

Zeng, C. L. and Ren, Y. J. 2007. Corrosion protection of 304 stainless steel bipolar plates using TiC films produced by high-energy micro-arc alloying process. *Journal of Power Sources* 171: 778–782.

Zhu, B., Mei, B., Shen, C. et al. 2006. Study on the electrical and mechanical properties of polyvinylidenefluroide/titanium silicon carbide composite bipolar plates. *Journal of Power Sources* 161: 997–1001.

7

Degradation of Other Components

Daijun Yang
Tongji University

Junsheng Zheng
Tongji University

Bing Li
Tongji University

Jianxin Ma
Tongji University

7.1 Introduction

The proton exchange membrane fuel cell (PEMFC) is considered as one of the most promising options for future energy needs because it offers a clean, mobile power source with high-energy conversion and low pollutant emissions (Jia et al., 2001). Research on PEMFC components to increase fuel cell durability during long-term operation is one of the most urgent requirements in PEMFC studies (Bieringer et al., 2009).

In the foregoing sections, the functions, requirements, and degradation mechanisms of all the main components, including catalyst, membrane, membrane electrode assembly (MEA), bipolar plate (BPP), and gas diffusion layer (GDL), have been extensively discussed. Compared with the MEA, BPP, and GDL, the importance of the seal, endplate, and bus plate has long been ignored, and so the literature on this topic is very limited. Accordingly, studies on the degradation of and mitigation methods for the seal, endplate, and bus plate are relatively insufficient.

However, the properties of the seal, endplate, and bus plate are critical to the performance and durability of PEMFC, including single cell and stack. Seals should be very carefully designed to avoid overboard leakage and crossover leakage within a fuel cell stack. Endplates can provide uniform pressure distribution in each active area and consequently reduce contact resistance between the components inside the fuel cell. Bus plates can facilitate the output of the electric current generated by the fuel cell. So, it is essential to study the degradation mechanisms and mitigation methods for these components.

In this chapter, the functions of and current research on seal, end plate, and bus plate are introduced, and their degradation mechanisms and mitigation methods are discussed.

7.2 Functions of Seals, Endplates, and Bus Plates

7.2.1 Functions and Requirements of Seals

Seals (gaskets) are of great importance, such that material selection, structure and profile design, and processing method should be carefully considered to prevent both overboard and crossover leakage within a fuel cell stack, otherwise its performance and durability will be greatly compromised. Overboard leakage refers to fluid loss from the anode, cathode, or coolant chamber to the external environment. Crossover leakage is fluid movement across the separation of either the membrane or the BPP. Natural crossover leakage due to gas diffusion through materials, like Nafion® membrane and graphite BPP, is expected and cannot be completely avoided. However, excessive leakage is preventable through the employment of gaskets. Figure 7.1 shows the main structure of the core component—the MEA—while Figure 7.2 shows the structure of the repeat unit of a stack.

Figure 7.1 illustrates the plan view of an MEA (the cathode side is not shown), in which two pieces of GDLs, normally carbon paper, are attached to each side of the active area through hot pressing or simply gluing. The active areas, that is, catalyst layers (CLs), for both anode and cathode sides should be sealed all around by rectangular gaskets, as shown in Figure 7.1, to inhibit them from losing reactant gases. Other than the active area gasket, four circular gaskets are also shown in Figure 7.1. These are used for the manifolds of fluids, which should not be introduced into the current area. For example, H_2 gets into the anode-side GDL from the through-hole without sealant and then to the CL, but air and coolant are separated and directed into the back spaces of the anode side, that is, air and coolant flow fields, respectively. All the gaskets in Figure 7.1 are only for the anode side and all are the first type of sealing material—elastomers. Excellent elongation and durability are of great importance in elastomers used for reactant gases.

Traditional elastomers, such as polyacrylate, acrylate copolymer, butyl, neoprene, silicone, ethylene–propylene–diene-monomer (EPDM), and so on can be used in fuel cells as gaskets. For better fuel, oxygen, water, and heat resistance, fluorine-containing elastomers, such as fluorosilicone, are preferred. However, although the silicone–oxygen (siloxane) main polymer backbone is very thermally stable, it is prone to mechanical damage. To obtain a highly designable profile and shape, a cured-in-place-gasket (CIPG) sealant is also needed. The advantages of a profiled seal design as compared to a flat design have been analyzed in a review paper; it is believed that no excessive assembly force results from a profiled seal, while other components will be secured (Bieringer et al., 2009). In CIPGs, thermosetting liquid

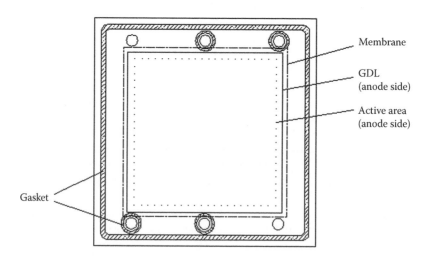

FIGURE 7.1 Plan view of an MEA with GDL and bonded with gaskets.

silicones, epoxies, acrylates, and so on are suitable. After application to the BPP, MEA, or GDL, they cure with time and temperature, and then bonded and integrate with these substrates. However, these widely used commercial sealing materials may not meet all fuel cell requirements because of drawbacks such as the leaching out of filler and component degradation in a fuel cell environment. Therefore, some professional sealant companies, such as Henkel, Dow Corning, Three Bond, Freudenbure-NOK, as well as their partners in the fuel cell field are involved in the development of new materials for PEMFCs to achieve a longer lifetime and higher reliability.

Figure 7.2 the coolant flow field lies on the opposite side of the anode flow field of plate A, which can be permanently bonded to plate B with glue since it requires almost no maintenance. For this purpose, another type of sealing material, that is, an adhesive such as epoxy, silica gel, AB glue, silicone, silicone-coated fiberglass gasket, and so on, can be used. Considering the function of this kind of material, strong bonding strength is required rather than elasticity and elongation. The material can also be used to glue the electrolyte membrane and the frames together. Cross-leakage of reactant gas depends not only on the physical robustness of the 20–50 μm-thick membrane, but also on the sealing effect at these interfaces, so in addition to losses arising from equilibrium (or Nernstian) voltage via natural reactant crossover, poor sealing will further aggravate voltage loss (Haile, 2003). Only with perfect sealing by adhesive can the interface between the two layers of MEA frames and the interface between the frames and membrane be safely sealed. For the frames, a rigid and stable polymer material such as polyethylene naphthalate (PEN), polyethylene telephthalate (PET), or polyimide (PI) are usually chosen. Hot melt adhesives, which are widely used in the garment, package, and electronic industries, can then be used because of their lower operating temperature (normally 120–160°C) and strong adhesiveness. But great care should be taken to ensure the durability of this kind of adhesive, since its compatibility with heat, water, and acid is more critical in a fuel cell environment than in the normal industries. For high-temperature PEMFCs, which have generated considerable research interest in the past decade because of their greater efficiency, simpler thermal management, and better carbon monoxide tolerance in H_2 (Zhang et al., 2006), materials with higher melting points (up to 200°C) should be chosen. Otherwise, the adhesive may move from its original place to other components of the MEA and BPP, resulting in suffocation of the GDL or abnormal adhesion between the MEA and BPP.

Since the second type of sealing material does not come into direct contact with the reactant gases or the intermediates produced along with the anode and cathode reactions (e.g., hydrogen peroxide and hydroxyl radicals) the focus here will be primarily on the requirements for the first type, elastomers.

Apart from performance and durability losses caused by malfunction or failure in the membranes, GDLs, and CLs (see previous chapters), a fuel cell stack undergoes losses from seals. The durability and efficiency of a fuel cell stack depend largely on the integrity of the various contact and sealing interfaces within individual fuel cells and between adjacent fuel cells. The durability of seals during extended operation often determines whether fuel cells can be used cost effectively. The core component, the MEA, can fail in a number of ways, but one typical failure mechanism is excessive gas crossover caused by nonideal sealing between the interfaces. Chemical and mechanical degradation of sealing materials may contribute to crossover leakage, and therefore specific requirements for these

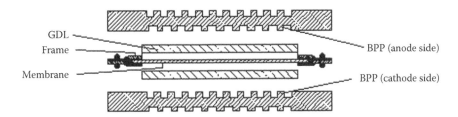

FIGURE 7.2 Sectional view of a single cell in a stack, taken along a line in the middle of the active area in Figure 7.1.

materials are necessary and have been approached in several ways. For example, for the most popular elastomer, silicon, Dow Corning advanced detailed specifications as early as 2001 (Frisch, 2001). In terms of fuel cell operating conditions, such as temperature, pressure, and chemical/electrochemical environment, as well as lifetime and cost targets for a total fuel cell stack, three key requirements exist:

1. Excellent mechanical stability. A gasket is installed in each cell to seal the fluids. The internal pressure of the fuel cell can be relatively high, and the gasket/plate or gasket/frame interfaces must resist this internal pressure during extended operation. The most important property for mechanical stability is compression set. According to the standard of the American Society for Testing and Materials (ASTM, D395), compression set is determined from the change in final thickness after removal of a force that has compressed 25% of the specimen for a specified amount of time. Compression set is one of the seal design criteria and is usually measured during durability testing in a real or simulated fuel cell environment. A large compression set value means that higher compression force is needed during fuel cell stack operation to maintain a low electrical resistance and good sealing. However, over-compression may be harmful to other components such that the gas permeability of the GDL will be changed, which in turn will influence fuel cell performance. Besides compression set, hardness, elongation, and tensile strength can also be characterized to determine the seal's mechanical stability, and all of these measurements are involved in recommended standard test methods for fuel cell gasket materials, as proposed by the United States Fuel Cell Council (USFCC).

2. Chemical and thermal stability. Sealing materials must tolerate the conditions and internal environment of a PEMFC stack, such as high humidity, temperature changes (−40–100°C), reactive chemicals, strong redox environment, and certain trace hydrocarbons and inorganic species. Case studies will be discussed in detail later in this chapter. Any gasket failure resulting in hydrogen or air leakage is highly undesirable.

3. Good machinability. One disadvantage of a PEMFC stack is that sometimes hundreds of fuel cells must be stacked on top of each other to achieve a desired current and voltage output, and these require a large number of gaskets to prevent the escape of hydrogen and air. The large number of gaskets among the various components necessitates their integration onto other components, because positioning of each independent gasket—for example, flat seals cut from sheet goods—to other components is labor-intensive, time consuming, and now outdated. Instead, gaskets can be integrated with the MEA or BPP into "one component" that is machinable, lowering the processing cost because the method is amenable to high-volume manufacture and reduces the chance of defects. Reducing the number of components also facilitates the maintenance or modification of prototype fuel cell stacks. Although few journal papers exist on this topic, numerous patents have been published for a variety of designs.

Seals are commonly formed by one of several proven methods, including the following: molding of various profiles that can be used on flat plate surfaces or assembled into a groove in the plate; dispensing an elastomer onto the plate or MEA through a CIPG process or formed-in-place-gasket (FIPG) process, then curing the elastomer into a designed profile; or liquid injection moulding (LIM), which attaches an elastomer in various profiles to the BPP or MEA. Compared to the LIM process, dispensing an elastomer using a CIPG or MIPG process is tedious, time consuming, and expensive. Therefore, the LIM process is probably the best choice, but it still requires an expensive machine and die. Whatever the method used, integrating a seal into other components is one of the key factors in achieving a successful commercial fuel cell design.

7.2.2 Functions and Requirements of Endplates

Endplate is one of the main components in PEMFCs. Apart from providing an interface bearing fluid inlet and outlet fittings, it can guarantee good performance and long lifetime for a PEMFC through optimized design and material selection. The major role of the endplate (see Figure 7.1).

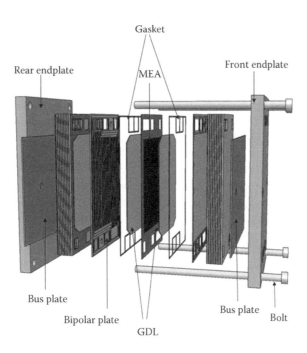

Gasket

Rear endplate

MEA

Front endplate

Bus plate

Bipolar plate

Bus plate

Bolt

GDL

FIGURE 7.3 Schematic of PEM fuel cell showing the location of the components.

Figure 7.3 is to ensure uniform pressure distribution in each active area, and thereby reduce contact resistances among all the components inside the fuel cell stack, such as BPPs, GDLs, and CLs. Various novel endplate structural designs have been advanced to fulfill these requirements. A sandwiched endplate, composed of carbon fiber epoxy faces and a foam-filled Nomex honeycomb core, was recently designed (Yu et al., 2010a). Via structural and thermal finite analysis, it was found that the sandwich-type endplate could provide uniform pressure to the stack and improve its cold-start characteristics by decreasing the thermal conductivity. Furthermore, using polyurethane foam as the filler, the flexural strength and stiffness of the sandwich structure can be increased threefold, making it promising as an endplate material. A hybrid composite endplate was designed with a proper stacking sequence to make the composite plates yield a uniform pressure on the fuel cell stack (Yu et al., 2010b). In this case, asymmetric composite endplates with a precurvature created by thermal fabrication residual stress were designed to replace heavy metallic endplates. The composite endplate had large thermal residual deformation and high flexural stiffness. Using pressure-sensitive film it was found that by virtue of this composite endplate, the actual pressure distribution was more uniform than in other symmetric endplates. Therefore, this new composite endplate is recommended to reduce weight without compromising PEMFC performance.

The components of a fuel cell stack should be held together with sufficient assembly force to prevent overboard leakage of reactants and minimize the contact resistance among adjacent unit cells (Larminie and Dicks, 2003; Barbir, 2005; Hwang et al., 2008; Yu et al., 2009), requirements accomplished by the employment of endplates. Overcompression can squeeze the GDL and change its porosity ratio, which may choke the fuel cell by impeding the flow of gases and migration of water. On the other hand, insufficient pressure may result in high contact resistance between the GDL and BPP, which will also lower fuel cell performance. Therefore, endplate materials should have high flexural stiffness to resist the high clamping force applied to the endplates by bands or bolts.

Since endplates have inlets and outlets for fuel, air, and coolant, their chemical and electrochemical stabilities are also crucial. Endplates are prone to corrosion in the presence of water, especially acid water, arising from humidification and fuel cell reactions. During corrosion, metal ions dissolved from the endplates may pollute the membrane and thus decrease the stack's power output (Kelly et al., 2005). The weight and density of endplates should be reduced as much as possible for passenger car applications, to obtain high mileage and fuel utilization. In addition, to ensure start-up in freezing conditions, glycol antifreeze is often added to the coolant water. Endplates should thus exhibit high chemical stability in glycol solution as well. Furthermore, fuel cell vehicles need high output power, and so endplates should have good electrical insulation to ensure safety. For PEMFC commercialization, the price of endplates should be as low as possible. For these reasons, an ideal endplate should possess all of the following properties: low price, high flexural stiffness, good chemical and electrochemical stability, and high electrical insulation.

Degradation of the endplate may affect PEMFC performance and durability. In some fuel cell stack assemblies, a pneumatic piston is installed adjacent to either one of the endplates. In such arrangements, the pneumatic piston uniformly applies compressive force to the stack, which permits control of the compressive force applied to the endplate. Unfortunately, the use of a pneumatic piston adds to the complexity of the fuel cell stack and can be a source of unreliability, with potentially adverse consequences if the piston-based compression system fails. For example, internal leakage may occur in the pneumatic system.

For a PEMFC stack, more attention should be paid to assembly force. On the one hand, insufficient assembly force may result in sealing problems, such as overboard leakage and internal combustion within GDLs and CLs caused by crossover leakage. On the other hand, since GDL is a porous medium and is highly sensitive to clamping force, external clamping force not only can affect the thickness but also can change the porosity/permeability of the GDL. Therefore, the endplates at both sides should provide an appropriate pressure. Besides, the contact pressure provided by endplates must be uniform, otherwise the fuel cell performance will be significantly affected. The contact states between MEAs and BPPs depend on the external pressure exerted on the endplates. If the pressure distribution among the surface of the endplates is uneven, the ohmic resistance distribution will also be uneven, resulting in nonuniform current density and heat generation distribution inside the stack. Hot spots may form on an MEA, leading to its failure, and the stack's durability will also decline. Results in a published paper showed that low clamping pressure (e.g., <5 bar) resulted in a high interfacial electric resistance between the BPPs and GDLs, and thereby reduced the PEMFC's electrochemical performance (Chang et al., 2007). In contrast, high clamping pressure levels (e.g., >10 bar) reduced the ohmic resistance, but at the same time the diffusion paths for mass transfer from the gas channels to the CL were narrowed. Therefore, no obvious performance improvement was obtained through elevated clamping pressure, while the mass transfer at a high current density was limited.

One paper (Pozio et al., 2003) investigated the possibility of stainless steel (SS316L) as a PEMFC endplate material, but found that iron cations leached out from this material led to degradation of the Nafion membrane, assessed through a massive fluoride loss. Beside of stainless steel, many materials have been used for endplates, totally classifiable into two types. The first type comprises nonmetals, including engineering plastics, polysulfone, and the like. However, the thermal stabilities of engineering plastics and polysulfone are not satisfactory, and they tend to degenerate in a fuel cell environment. The second type comprises metals, such as stainless steel, titanium, aluminum alloy, and the like. If stainless steel is used, the endplates will be not only massive but also highly heat-conductive because of a large heat capacity. However, during cold start-up, the fuel cell stack should be heated up as quickly as possible. Therefore, the heat generated inside the stack should be used to increase the internal temperature, especially that in the area near to the three-phase boundary of the MEA, which requires low thermal conductivity and low thermal inertia of the endplate material, together with high specific stiffness (i.e., stiffness divided by density). Although low-density materials such as aluminum alloy and magnesium alloy are good candidates for endplates because of their high mechanical strength, they have insufficient corrosion resistance and electrical insulation.

Because of the relatively thin dimensions and low mechanical strength of GDLs and MEAs versus the gaskets, bipolar plates, and endplates, the most important goal in stack design and assembly is to achieve a proper and uniform pressure distribution in GDLs and MEAs. The endplates that make up each fuel cell assembly are compressed and maintained in their assembled states by compressed springs, such as tie rods and bands. The tie rods extend through holes formed in the peripheral edge portion of the end-plates, and have nuts or other fastening means assembling the tie rods to the fuel cell assembly and compressing the endplates of the fuel cell assembly toward each other. The reason for employing a peripheral location for the tie rods is to avoid introducing openings or otherwise interfering with the central, electrochemically active portion of the fuel cell. Use of external tie rods requires that each of the endplates be greater in area than the stacked fuel cell assemblies interposed between them, which can significantly increase stack volume and stack weight. This is particularly undesirable in transportation applications. The associated fasteners also increase the number of component parts required to assemble a stack. To reduce the number of parts and improve volume efficiency, stack manifolds in an array can be incorporated into the compression endplates of fuel cell stacks. For example, one patent (Gorbell, et al., 1996) shows an array manifold integrated into the compression endplates of multiple fuel cell stacks.

Some resilience is generally desirable in compression endplate assemblies for various reasons, for example, to accommodate and compensate for dimensional changes and to maintain compressive force over prolonged periods of time. Examples of different resilient compression endplate assemblies are disclosed in patents (Gibb et al., 1996; Wozniczka et al., 1998). In fuel cell stack assemblies, the use of springs in conjunction with tie rods or bands is generally required to compress the stack and maintain the compressive load over time, due to the tendency of MEAs to gradually decrease in thickness while under compressive load. Optimally, the springs should impart a predetermined compressive load with minimal load variation over as large a deflection range as possible. When peripherally disposed tie rods are employed, each of the end-plates securing the fluid flow field plates and MEAs must be greater in area (and therefore create an over-hang). The amount of overhang depends upon the diameter of the springs inserted at the ends of the tie rods between the endplates and nuts securing the tie rods, since almost all of the springs' diameters should be in contact with the endplate to provide effective and uniform compressive load (Gibb et al., 2000).

The compression assembly should impart a sufficient internal compressive force to ensure good electrical contact and sealing within the stack without detrimentally deforming the stack components. The requirements defining the preferable operating range will vary depending on the characteristics of the materials from which the stack is made. Stiffness of the stack components is mainly dependent on stiffness of the endplate materials. Metallic materials have high mechanical properties and thermal stability, but their low corrosion resistance and electrical insulation are a problem. Aluminum is an ideal material for PEMFC endplates because of its low density and high rigidity. Two methods can be used to overcome its insulation problem: (1) using an insulating layer between the current collector plate and the endplate, and (2) modifying the endplate surface using methods such as applying an electrostatic coating, anodizing, and so forth (Patermarakis et al., 1993; Thornpson, 1997). In addition, to achieve the appropriate corrosion resistance to humidified reactants and cooling fluid, Teflon parts can be inserted into the endplate to act as inlet and outlet passages. Furthermore, carbon fiber polymeric composites are candidates for endplates due to their very high specific stiffness and high corrosion resistance in the acidic environment of PEMFCs that arises from the very pure water generated in the cell. However, carbon fiber-reinforced composites have rather high thermal and electrical conductivities, and so employing them in endplates requires electrical insulation to prevent short circuits and thermal insulation to reduce heat flow. For this purpose, an endplate composed of a carbon fiber polymeric composite face and a low-density polymeric foam core reinforced with Nomex honeycomb was designed in a sandwich construction because the sandwich configuration has high thermal insulation and electrical insulation, and very high specific stiffness (Zenkert, 1997a,b).

7.2.3 Functions and Requirements of Bus Plates

Typically, a PEMFC stack contains two bus plates, an anode, and a cathode, each of which has one or more current output terminals. Bus plates function as an external electric circuit, exporting the electric current from the fuel cell.

The current output terminal sends out the electric current generated by the PEMFC single cell or stack. Bus plates for all kinds of fuel cell stacks are made of noble metals such as gold or platinum, or nonnoble metals such as stainless steel, copper, or aluminum. The noble metals not only have good conductivity but also can almost avoid electrochemical corrosion and thus will not produce metallic ions that may poison the fuel cell. However, these noble metals are very expensive.

If stainless steel, copper, or aluminum is directly used to make a bus plate, electrochemical corrosion will occur if fluids pass through it, resulting in unwanted damage due to the metallic ions produced. In order to avoid this problem, nonnoble metal materials plated with gold or platinum are frequently used (Joh et al., 2008). Unfortunately, gold- or platinum-plated copper or aluminum is still relatively expensive.

In addition, the bus plate materials may influence fuel cell performance. G.-S. Kim et al. investigated PEMFC stack voltage distribution using a voltage/current distribution model (Kim et al., 2005). The bus plate materials were stainless steel, nickel, and aluminum, and their results showed that the bus plate made of stainless steel might have a better voltage distribution (Figure 7.4).

7.2.4 Assembly Modes

In a conventional PEMFC design, endplates are the two outermost components in a fuel cell assembly, providing compressive force that brings the single fuel cells together to form a stack. At present, stack assembly modes can be classified into three types. In the first type of design, the components that make up a fuel cell stack are compressed and maintained in their assembled state by tie rods; these extend through holes formed in the peripheral edge portion of the stack endplates, and have associated nuts or other fasteners for assembling the tie rods with the stack assembly and springs, or other resilient materials for pushing the front and rear endplates toward each other. One such fuel cell stack design incorporates internal tie rods extending between the endplates through openings in the fuel cell plates and MEAs (Gibb et al., 1996). The second assembly mode uses compression bands, and a typical design can be found in US Patent No. 5789091 (Wozniczka et al., 1998). In the compression band system, at least one compression band encircles the front and rear endplates, as well as all the unit cells. The resilient compression assembly brings the front endplate toward the rear endplate, thereby applying compressive force on the whole fuel cell assembly. With this method the fuel cell is compressed with the desired internal force while changes in fuel cell thickness are accommodated. In the third assembly mode, the compression fabrication comprises springs and/or a pneumatic piston, which can be employed either

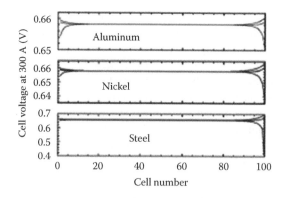

FIGURE 7.4   100-Cell stack voltage distribution model computations for one anomalous bus plate (either stainless steel, nickel, or aluminum) and one copper bus plate, at 300 A. Model curves correspond to inlet, middle, and outlet locations. (Reprinted from *Journal of Power Sources*, 152, Kim, G. S. et al. Electrical coupling in proton exchange membrane fuel cell stacks. 210–217. Copyright (2005), with permission from Elsevier.)

individually or in combination. Springs are often used as a backup to provide a compressive force if the pneumatic piston pressure is lost or inadequate for applying the desired compressive force for efficient and safe fuel cell operation.

7.3 Degradation of Seals and the Balance of the Plant

7.3.1 Degradation of Seals

Gasket degradation may have several impacts on fuel cell performance and durability over long-term operation: (1) increased compression force on the GDL, leading to decreased porosity and increased reactant transport resistance; (2) adverse effects on the hydrophilic/hydrophobic character of the GDL, which in turn influence water management within the fuel cell; (3) catalyst poisoning by the products of degradation or leaching out; and (4) combustion of air or H_2 and damage to the membrane due to cross-over leakage.

Although there is a substantial body of literature on the degradation of elastomeric materials (Bousquet and Fouassier, 1983; Vondráček and Doležel, 1984), few reports have been published on degradation and degradation mechanisms in the PEMFC environment (Tan, 2007a,b), and even these papers are all related to the degradation of silicone seals, partly because these are easily available and have been widely used in the PEMFC sector. Unfortunately, almost all of the reported results are not particularly positive.

Lifetime testing of a 23 cm² single cell for 26,300 h was performed by Cleghorn and coworkers (Cleghorn, 2006) at 800 mA cm⁻². In this cell, a silicone-coated fiberglass gasket (CHR–furon) was chosen. Severe degradation of the silicone/glass reenforced gasket was visually observable. In addition, the gasket thickness was typically reduced by approximately 25 μm, and only the glass reenforcement remained at the active area. Silicone particles were also observed on the GDL surface.

In the 7-cell stack experiment implemented by Husar (Husar et al., 2007), precision-grade silicone was used as the gasket at 60°C and 4.5 bar. However, the stack failed after just 20 h of operation, due to crossover leakage on the cathode-side gasket of cell #2, caused by an unexplainable temperature rise. Obvious discoloration and fusing to the membrane indicated the gasket's degradation.

Similar silicone degradation in real fuel cells has been observed and recorded in recent publications (Ahn et al., 2002; Schulze et al., 2004; Du et al., 2006). Compared to these test results obtained in a real fuel cell, Tan and coworkers (Tan, 2007a,b), using a simulated PEMFC environment, made more fundamental measurements of silicone rubber, such as weight loss, attenuated total reflection Fourier transform infrared (ATR-FTIR) spectroscopy, optical micrograph, elastic modulus, and atomic absorption spectrometry. All of their results indicated that over time the material's surface chemistry changed significantly via de-cross-linking and chain scission in the backbone. The ATR-FTIR spectrum revealed that temperature (varied from 60°C to 80°C) had a significant effect on degradation: the higher the temperature, the faster the material degraded.

7.3.2 Degradation of the Balance of the Plant

Endplates should be durable enough to bear the assembly pressure of a stack and protect all components, especially the BPP, from deformation. Nevertheless, to date most research has focused on the durability of the BPP.

As mentioned above, bus plates made of nonnoble materials may corrode during electrochemical reactions. Fortunately, this corrosion or deactivation process is not significant in comparison with what happens to the catalyst, membrane, and MEA. Research on the deactivation process for bus plates is limited, and is mostly focused on fuel cells that use methanol, ethanol, or formic acid (Stumper and Charles, 2008) as the fuel.

Joh et al. studied the methanol solution's corrosion effect on bus plate (Joh et al., 2008). Their results showed that effluent solution with lower pH value was likely to corrode metal bus plates. After running a stack for 17 days, they found severe corrosion where the plates were in direct contact with the methanol solution.

In Figure 7.5, to avoid bus plate corrosion, they used gold-plated brass as bus plates.

In recent years, scientists and engineers have also been looking for novel bus plate materials as well as designs. For example, Hu et al. integrated an end plate and a bus plate into an "end-bus plate" for a fuel cell (Hu et al., 2006). This novel plate included a main plate body and a conducting plate. The main plate body was made of corrosion-resistant, nonconducting material, and its inner side had a surface for flat contact with one side of the fuel cell unit attached adjacent to the main plate body. The conducting plate, made of electroconducting material, was integrally attached to the interior side of the main plate body to form an integrated body for electrically communicating with the reaction region of the unit. This novel end–bus plate may overcome conventional drawbacks, including fluid leakage between the end plate and the bus plate, and the formation of metallic ions due to electrochemical corrosion of the bus plate, which generally results in damage to the fuel cell.

7.4 Summary

Although investigations into seals, endplates, and bus plates are relatively inadequate, it is obvious that the properties of these components have a significant influence on PEMFC performance, especially durability.

As fuel cell manufacturers continue to improve system reliability and durability, the importance of durable seals is growing. At the same time, sealing technologies are advancing from the R&D stage to real applications. Satisfactory, economical sealing within a multimaterial stack is an important factor in the commercialization of fuel cells.

The endplate is a crucial component in a PEMFC stack, affecting performance and durability. This chapter has introduced the structure, function, requirements, and degradation of endplates. In addition, it has described in detail the relevant stack assembly modes and mitigation methods.

The bus plate is also important for PEMFCs, but its degradation process is not obvious and bus plate R&D is mainly focused on how to retard the corrosion process in fuel cells other than H_2/O_2 PEMFCs, such as those using methanol, ethanol, or formic acid.

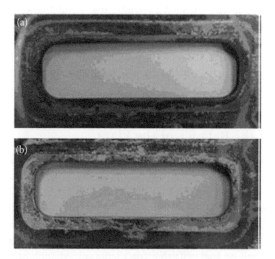

FIGURE 7.5 Photographs of corroded bus plates affected by effluent methanol solution: (a) anode bus plate and (b) cathode bus plate. (Reprinted from Joh, H. et al. 2008. *International Journal of Hydrogen Energy* 33: 7153–7162. With permission from International Association for Hydrogen Energy.)

Acknowledgments

The authors gratefully acknowledge support from the Ministry of Science and Technology of China (2007DFC61690), National Natural Science Fund of China (21006073), national "111 project" and the Henkel Professorship.

References

Ahn, S. Y., Shin, S. J., Ha, H. Y. et al. 2002. Performance and lifetime analysis of the kW-class PEMFC stack. *Journal of Power Sources* 106: 295–303.

Barbir, F. 2005. *PEM Fuel Cells: Theory and Practice.* Burlington, MA: Elsevier Academic Press; pp. 197–200.

Bieringer, R., Adler, M., Geiss, S. et al. 2009. Gaskets: Important durability issues. In Büshi F. N., Schmidt T. J., and Inaba M. (eds). *Polymer Electrolyte Fuel Cell Durability.* New York, NY: Springer Press; 271–281.

Bousquet, J. A. and Fouassier, J. P. 1983. Mechanism of photo-oxidation of an elastomer. *Polymer Degradation and Stability* 5: 113–133.

Chang, W. R., Hwang, J. J., Weng, F. B. et al. 2007. Effect of clamping pressure on the performance of a PEM fuel cell. *Journal of Power Sources* 166: 149–154.

Cleghorn, S. J. C. 2006. A polymer electrolyte fuel cell life test 3 years of continuous operation. *Journal of Power Sources* 158: 446–458.

Du, B., Guo, Q. H., Pollard, R. et al. 2006. PEM fuel cells: Status and challenges for commercial stationary power applications. *JOM Journal of the Minerals, Metals and Materials Society* 58: 45–49.

Frisch, L. 2001. PEM fuel cell stack sealing using silicone elastomers. *Sealing Technology* 93: 7–9.

Gibb, P. R. 2000. Compression assembly for an electrochemical fuel cell stack. *US Patent* No. 6057053.

Gibb, P., Voss, H., Schlosser, W. et al. 1996. Electrochemical fuel cell stack with compression mechanism extending through interior manifold headers. *US Patent* No. 5484666.

Gorbell, B. N., Wozniczka, B. M., and Chow, C. Y. 1996. Internal fluid manifold assembly for an electrochemical fuel cell stack array. *US Patent* No. 5486430.

Haile, S. M. 2003. Fuel cell materials and components. *Acta Materialia* 51: 5981–6000.

Hu, L., Zheng, L., and Guo, W. 2006. Integrated end-bus plate for fuel cell. *US Patent* No. 2006141319.

Husar, A., Serra, M., and Kunusch, C. 2007. Description of gasket failure in a 7 cell PEMFC stack. *Journal of Power Source*s 169: 85–91.

Hwang, I. K., Yu, H. N., Kim, S. S. et al. 2008. Bipolar plate made of carbon fiber epoxy composite for polymer electrolyte membrane fuel cells. *Journal of Power Sources* 184: 90–94.

Jia, N., Martin, R. B., and Qi, Z. 2001. Modification of carbon supported catalysts to improve performance in gas diffusion electrodes. *Electrochimica Acta* 46: 2863–2869.

Joh, H., Hwang, S., Cho, J. et al. 2008. Development and characteristics of a 400 W-class direct methanol fuel cell stack. *Internation Journal of Hydrogen Energy* 33: 7153–7162.

Kelly, M. J., Fafilek, G., Besenhard, J. O. et al. 2005. Contaminant absorption and conductivity in polymer electrolyte membranes. *Journal of Power Sources* 145: 249–252.

Kim, G. S., St-Pierre, J., Promislow, K. et al. 2005. Electrical coupling in proton exchange membrane fuel cell stacks. *Journal of Power Sources* 152: 210–217.

Larminie, J. and Dicks, A. 2003. *Fuel Cell Systems Explained*, 2nd ed. New York, NY: Wiley Press, pp. 67–69.

Patermarakis, G. and Papadreadis, N. 1993. Mechanism of their hydration and pore closure during hydrothermal treatment. *Electrochimic Acta* 38: 1413–1420.

Pozio, A., Silva, R. F., Francesco, M. D. et al. 2003. Nafion degradation in PEFCs from end plate iron contamination. *Electrochimica Acta* 48: 1543–1549.

Schulze, M., Knöri, T., Schneider, A. et al. 2004. Degradation of sealings for PEFC test cells during fuel cell operation. *Journal of Power Sources* 127: 222–229.

Stumper, J. and Charles, S. 2008. Recent advances in fuel cell technology at Ballard. *Journal of Power Sources* 176: 468–476.

Tan, J., Chao, Y. J., Van Zee, J. W. et al. 2007a. Degradation of elastomeric gasket materials in PEM fuel cells. *Materials Science and Engineering* A445–A446: 669–675.

Tan, J., Chao, Y. J., Li, X. D. et al. 2007b. Degradation of silicone rubber under compression in a simulated PEM fuel cell environment. *Journal of Power Sources* 172: 782–789.

Thornpson, G. E. 1997. Porous anodic alumina—Fabrication, characterization and applications. *Thin Solid Films* 297: 192–201.

Vondráček, P. and Doležel, B. 1984. Biostability of medical elastomers: A review. *Biomaterials* 5: 209–214.

Wozniczka, B., Fletcher, J., and Gibb, R. 1998. Electrochemical fuel cell stack with compression bands. *US Patent* No. 5789091.

Yu, H. N., Hwang, I. U., Kim, S. S. et al. 2009. Integrated carbon composite bipolar plate for polymer electrolyte membrane fuel cells. *Journal of Power Sources* 16: 929–934.

Yu, H. N., Kim, S. S., Su, J. D. et al. 2010a. Axiomatic design of the sandwich composite endplate for PEMFC in fuel cell vehicles. *Composite Structures* 92: 1504–1511.

Yu, H. N., Kim, S. S., Su J. D. et al. 2010b. Composite endplates with pre-curvature for PEMFC (polymer electrolyte membrane fuel cell). *Composite Structures* 92: 1498–1503.

Zenkert, D. 1997a. *The Handbook of Sandwich Construction*. Barnsley, Germany: Emas Publishing, pp. 11–50.

Zenkert, D. 1997b. *An Introduction to Sandwich Construction*. Barnsley, Germany: Emas Publishing, pp. 7–38.

Zhang, J. L., Xie, Z., Zhang, J. J. et al. 2006. High temperature PEM fuel cells. *Journal of Power Sources* 160: 872–891.

8

Contaminant-Induced Degradation

Olga A. Baturina
Naval Research Laboratory

Yannick Garsany
Naval Research Laboratory;
EXCET, Inc.

Benjamin D. Gould
Naval Research Laboratory

Karen E.
Swider-Lyons
Naval Research Laboratory

8.1 Introduction

Fuel cells (FCs) are susceptible to impurities that may affect FC performance in multiple ways: contaminate catalysts/ionomers at both the cathode and anode of the FC, change the hydrophobicity of the gas diffusion layer (GDL), or cause membrane degradation. The influence of impurities on the performance of proton exchange membrane fuel cells (PEMFC) has been a subject of several reviews and a book chapter (Borup et al., 2007; Cheng et al., 2007; Li et al., 2009; Schmittinger et al., 2008), with the most recent manuscript published in 2009 (Li et al., 2009). These reviews and book chapters comprehensively review the literature published until 2007. Considering the large number of publications on the topic, the current chapter covers mostly the literature published during the last 3–4 years with new findings in low-temperature FCs. Also, we briefly discuss contamination issues in high-temperature PEMFC and hydroxide-exchange membrane FCs.

Experimental data are discussed along with insights from modeling in order to get a fundamental understanding of the influence of impurities on the PEMFC performance. Contamination impacts are followed by a discussion on contamination mitigation, maintenance, and performance recovery strategies that are critical for life extension of the FCs operating in a real-world environment.

8.2 Contamination Sources

Contaminants are introduced into the FCs as impurities in the incoming reactant streams (fuel and oxygen) or from materials used to construct the FC. It is useful to organize contaminant species by their location in the FC: fuel contaminants, air contaminants, and system contaminants.

8.2.1 Fuel Contaminants

The origin of fuel-side contaminants comes from impurities in the feedstock from which the fuel is derived as well as from impurities introduced during the conversion process. In the near term, the hydrogen used in PEMFCs will be derived from hydrocarbon feedstocks. The majority of commercial hydrogen is produced by the steam reforming of natural gas, which produces 50 million metric ton annually worldwide (Suresh et al., 2007). The steam reforming process converts hydrocarbons into CO, CO_2, and H_2; the product of such a conversion process is commonly referred to as reformate. Natural gas contains many naturally occurring impurities, most notably H_2S and NH_3. After steam reforming, the reformate is purified to the desired H_2 grade, but the purification processes can leave hydrogen with residual quantities of CO, H_2S, and NH_3. These three impurities are of the greatest concern because they are potent catalyst poisons.

The impurities in liquid hydrocarbon feedstocks depend on the specific fuel but are largely organic sulfur species. Di-benzothiophene is often used as a model compound to study sulfur poisoning in liquid hydrocarbon reforming experiments (Palm et al., 2002). During the reforming process organic sulfur species can be oxidized to SO_2 or reduced to H_2S. CO is always a contaminant whether the liquid fuel reforming process used is steam reforming, partial oxidation, or autothermal reforming (Gould et al., 2007). The resulting traces of CO, H_2S, SO_2, and organic sulfur species (thiophenes, mercaptons, etc.) derived from hydrocarbon feedstocks can be decreased by methods such as pressure-swing adsorption or more expensive methods such as using a palladium membrane.

High-purity H_2 can be generated by the electrolysis of water, which is often done at the laboratory scale to produce a pure source of H_2, but is thermally inefficient and generally not cost effective.

8.2.2 Air Contaminants

Most PEMFCs use the ambient air as a source of O_2 and are therefore susceptible to airborne contaminants. Airborne contaminants come from many sources including both natural and man-made sources.

Volcanic exhaust (an example of natural sources of impurities) is mostly comprised of SO_2 and a small fraction of H_2S. Concentrations as high as 11% SO_2 have been reported for plumes venting at the source, but 5 ppm SO_2 is a more likely number based on air quality data for Hawaii. Natural sources of SO_2 are primarily a concern for areas with high volcanism, like Hawaii, Japan, and Iceland.

The salts in the sea spray foul the surface of the cathode catalyst through anion adsorption and damage the electrolyte membrane by cation exchange. The chloride anion is particular damaging, and may result in rapid performance loss and Pt dissolution (Matsuoka et al., 2008).

Man-made contaminants come primarily from the combustion of hydrocarbons and the aforementioned impurities in the hydrocarbons that cause fuel-side contaminants. The combustion of hydrocarbons, such as coal in power plants or gasoline in cars, produces SO_x and NO_x emissions in the atmosphere. SO_x and NO_x are two important airborne contaminants that are the subject of many studies (Gould et al., 2009; Nagahara et al., 2008). Other man-made contaminants include road de-icers, such as $MgCl_2$, which can have similar effects as sea salt, and the methanol in windshield wiper fluid. Man-made battlefield contaminants can lead to permanent performance losses in PEMFCs (Moore et al., 2000). Their sources include the combustion products of heavy fuels, explosive products, and chemical agents.

8.2.3 System Contaminants

Components that make up the FC system can be a source of contaminants entering the FC. Hydrocarbon coolants have a strong affinity for platinum catalysts and can cause fouling of the surface. Silanes used in gaskets can be unstable. Plasticizers can leach out of tubing and sealing materials. O'Leary reported some of common materials used in automotive systems that are likely impurities in PEMFCs, including antioxidants and flame retardants (O'Leary et al., 2009). A list of some of these compounds is reported by Budinski et al. (2005). Even the catalyst can contain contaminants, such as residual Cl⁻ from starting synthetic materials, and sulfur in the carbon catalyst supports (Swider et al., 1996). The impurities can be distributed via the air or hydrogen gases, or water in the humidifier can collect organic and ionic impurities.

The use of pristine water is particularly important for analytical FC and electrochemical measurements (Garsany et al., 2010) but has practical importance for the operation of humidifiers in FC systems. Figure 8.1 shows the effect of operating humidifiers filled with deionized (DI) water (400 kΩ cm nonorganic free) versus nanopure water (18 MΩ cm organic free) on FC polarization curves. The difference between the two polarization curves is significant, considering that DI water is relatively "clean" water. A 15% decrease in cathode active Pt surface area was observed during operation with DI water. The loss of performance during operation with DI water is likely caused by both cathode and anode contamination with organic and ionic species.

8.2.4 Fuel Purity Standards

Fuel purity is being standardized to a set of specifications that will be universal for all H_2 fuel for PEMFC vehicles. The guiding documents on fuel purity standards are SAE #J2719 and ISO TS-14687–2 with the final standard still being drafted. As of 2008, the acceptable levels of contaminants were determined from single cell measurements and are shown below in Table 8.1. Future specification will likely push these levels lower because of the durability requirements for automotive PEMFCs. For example, total sulfur content will likely drop to around 1 ppb.

There are many challenges facing the adoption of a set of H_2 purity standards. In particular, these standards apply to the H_2 composition at the filling station nozzle and there are difficulties assessing how the H_2 purity will change from the H_2 storage tank to filling station nozzle, as well as determining which monitoring systems and analytical methods are capable of measuring such low quantities of contaminant species in H_2.

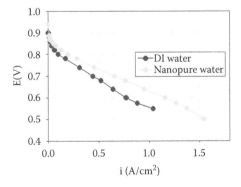

FIGURE 8.1 The influence of humidifier water purity on FC polarization curves at 80°C and 100% relative humidity. Stoichiometry: 2/2 for H_2/air.

TABLE 8.1   Threshold Levels for H_2 Fuel Impurities

Groups of Contaminants	Level
Total non-H_2 gases	100 ppm
Total sulfur compounds (H_2S, COS, CS_2, and mercaptans)	4 ppb
Total hydrocarbons (C_1 basis)	2 ppm
Total halogenated compounds	50 ppb
Individual Contaminants	**Level**
Carbon monoxide (CO)	200 ppb
Carbon dioxide (CO_2)	2 ppm
Formaldehyde (HCOH)	10 ppb
Formic acid (HCHOOH)	20 ppb
Ammonia (NH_3)	100 ppb
Oxygen (O_2)	5 ppm
Helium (He), nitrogen (N_2), argon (Ar)	100 ppm

8.3 Contamination Impacts and Mechanisms in Low-Temperature PEM FCs

8.3.1 Air-Side Contamination

Air-side contaminants that will be discussed below include neutral inorganic (SO_x, H_2S, COS, NO_x) and organic molecules (benzene, toluene, HCl, ClCN, Sarin, and Mustard), ionic species (NH_4^+, Na^+, F^-, Cl^-, NO_3^-, and SO_4^{2-}), and mixtures of neutral inorganic molecules (NO_2 + SO_2 + NO) and ionic species (Na^+ and Cl^-).

8.3.1.1 Impacts and Mechanisms

8.3.1.1.1 Sulfur Dioxide

Airborne SO_2 may decrease FC performance by adsorbing to surface sites of Pt-based cathode electro-catalysts, and making them unavailable for oxygen adsorption. The contamination effect of SO_2 in the cathode airstream is a function of its concentration (Mohtadi et al., 2004; Zhai et al., 2010) and FC oper-ating voltage (Baturina et al., 2009). Even very low concentrations of SO_2 in air are detrimental to FC performance. The voltage response of three cells exposed to three different SO_2 concentrations of 1, 2, and 10 ppm, but for the same total SO_2 dosage of 160 μmol is shown in Figure 8.2 (Zhai et al., 2010). The performance losses after exposure to 1, 2, and 10 ppm SO_2 at steady-state conditions were 216, 235, and 269 mV, respectively. Steady-state poisoning conditions were reached after approximately 24.5, 15.7, and 4.6 h of SO_2 exposure, respectively. Similar trends in changing FC performance versus SO_2 concentration for the same total SO_2 dosage of 118 μmol were reported by Mohtadi et al. (2004). In order to explain the concentration dependence of the FC performance at the same SO_2 dosage, they suggested a concentra-tion gradient effect.

The cell voltages versus time curves in Figure 8.2 are characterized by two regions with different deg-radation rates separated by an inflection point at a cell voltage of 0.63 V. The two regions are likely due to the change in the mechanism of oxygen reduction reaction (ORR) on the Pt electrocatalysts caused by a change in the SO_2 adsorption process with potential.

Baturina et al. (Baturina and Swider-Lyons, 2009) found that the poisoning effect of airborne SO_2 on PEMFC performance depends on the operating cell voltage. Figure 8.3 shows the current density versus time obtained during the contamination of a FC cathode with 1 ppm SO_2 at 0.5, 0.6, and 0.7 V.

The gradual current density decrease due to exposure to 1 ppm SO_2 in air resulted in 43%, 49%, and 54% performance drops after 3 h exposure to SO_2 for experiments conducted at 0.5, 0.6, and 0.7 V,

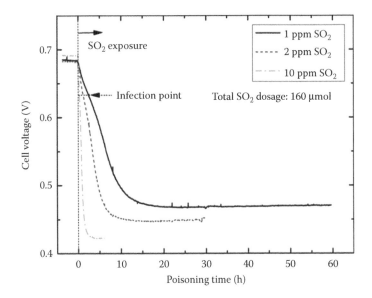

FIGURE 8.2 Cell voltage versus time during the poisoning of a PEM FC cathode to 1, 2, and 10 ppm SO₂ while operating the cell at a current density of 0.6 A cm⁻². (Reprinted from Zhai, Y. et al. 2010. *Journal of the Electrochemical Society* 157: B20–B26. With permission from The Electrochemical Society, Inc.)

respectively. Cyclic voltammetry (CV) and polarization curves were measured after exposure of the FC cathodes to SO₂ in air at 0.5–0.7 V in order to quantify the Pt surface sulfur coverages. Figure 8.4 presents a set of CV curves measured after exposure of the FC cathode to 1 ppm SO₂ at 0.5 V for 3 h. The suppression of the hydrogen region (0.08–0.30 V) in scan 1 indicates the presence of adsorbed sulfur species on the platinum surface. Sulfur oxidation in the first positive going scan was evidenced by the two broad peaks at 1.05 and 1.20 V which were assigned to the oxidation of weakly and strongly adsorbed sulfur adatoms (Contractor et al., 1978), and has recently been attributed to the oxidation of adsorbed sulfur at edge and face sites, respectively, on the Pt electrocatalyst nanoparticles (Ramaker

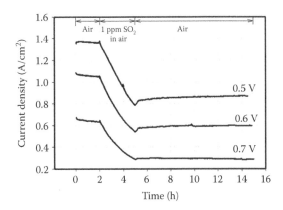

FIGURE 8.3 Current density versus time during the contamination of a PEM FC cathode with 1 ppm SO₂ at cell voltage of 0.5, 0.6 and 0.7 V. Cell running at 80°C and 100% RH. (Reprinted from Baturina, O. A. and Swider-Lyons, K. E. 2009. *Journal of the Electrochemical Society* 156: B1423–B1430. With permission from The Electrochemical Society, Inc.)

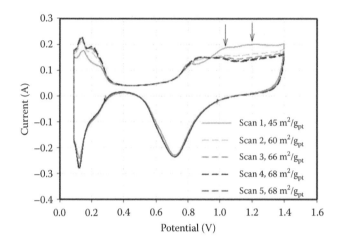

FIGURE 8.4 Cyclic voltammograms (CVs) recorded after exposure of a FC cathode to 1 ppm SO$_2$ for 3 h and purging of the cathode with neat air for 10 h while holding the cell voltage at 0.5 V. The potential scan rate is 50 mV s^{-1}, the cell temperature is 30°C and the humidifiers temperature is 50°C. (Reprinted from Baturina, O. A. and Swider-Lyons, K. E. 2009. *Journal of the Electrochemical Society* 156: B1423–B1430. With permission from The Electrochemical Society, Inc.)

et al., 2010). Overall, the reaction when going from 0.08 to 1.40 V is described by the reaction proposed by Loucka (1971):

$$Pt - S + 4H_2O \rightarrow SO_4^{2-} + 8H^+ + 6e^- + Pt \tag{8.1}$$

With successive cycles, more sulfur species were oxidized to sulfate and desorbed from the Pt catalyst surface, and the CV acquired the shape that was reminiscent of a clean Pt surface.

The coverages of the platinum surface by adsorbed sulfur species after exposure to SO$_2$ determined from the suppression of the hydrogen desorption region on the CVs shown in Figure 8.4 was found to decrease from 0.32 ± 0.02 to 0.19 ± 0.02 as the cell voltage increased from 0.50 to 0.70 V. These coverages, however, were inconsistent with coverages determined from the shift in polarization curves (plotted in Tafel coordinates) illustrated in Figure 8.5.

Sulfur coverage versus cell voltage determined from the shift in the kinetic regions of the polarization curves shown in Figure 8.5 had a minimum at 0.6 V, in agreement with maximal performance observed on the polarization curve measured after exposure to SO$_2$ at 0.6 V, as clearly seen in the inset in Figure 8.5.

Inconsistency in sulfur coverages determined from CV curves and polarization curves was assigned to: (1) different platinum sites these two methods accounted for (i.e., sites available for hydrogen adsorption/desorption and sites available for ORR, respectively) and (2) different SO$_2$ adsorption products in the hydrogen adsorption region and potential region where the ORR occurred. A cell voltage of 0.6 V was concluded to be the optimal operating FC voltage in the presence of low concentration of SO$_2$ in the cathode airstream.

Because the same Tafel slopes were observed for the poisoned cathodes versus the clean cathodes as shown in Figure 8.5, Baturina et al. (Baturina and Swider-Lyons, 2009) assumed that the potential shift due to exposure to SO$_2$ species in the kinetic region of the polarization curve could be attributed to a simple blocking effect of Pt sites by products of SO$_2$ adsorption, whereby the adsorbed sulfur species made the Pt sites unavailable for oxygen adsorption but did not inherently change the mechanism of the

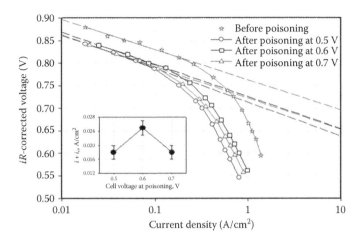

FIGURE 8.5 *iR*-corrected cell voltage versus hydrogen crossover-corrected current density in Tafel coordinates for polarization curves measured before poisoning of the FC cathode (stars) and after its exposure to 1 ppm SO₂ for 3 h at 0.5 V (circles), 0.6 V (squares), and 0.7 V (triangles). The inset shows the $i + i_x$ vs. cell voltage at poisoning at an *iR*-corrected voltage of 0.84 V. (Reprinted from Baturina, O. A. and Swider-Lyons, K. E. 2009. *Journal of the Electrochemical Society* 156: B1423–B1430. With permission from The Electrochemical Society, Inc.)

ORR. Similar conclusions were obtained by Nagahara et al. (Nagahara et al., 2008) when studying the durability of FC cathodes exposed to 0.5 ppm SO₂.

SO₂ affects performance of the FC cathode not only by blocking platinum sites and making them unavailable for oxygen adsorption, but also by changing mechanism of ORR, thus resulting in generation of hydrogen peroxide (Garsany et al., 2007b). Garsany et al. (2007b) used the rotating ring disk electrode (RRDE) methodology to probe how adsorbed sulfur species affect the catalyst activity for the ORR and the efficiency for water versus hydrogen peroxide production. RRDE permits the partial correction for diffusion limitations of the oxygen gas in solution at high potentials, allowing isolation of the ORR kinetics (Huang et al., 1979). The use of the additional ring, which surrounds the central disk with the electrocatalyst, allows quantitative detection of a possible intermediate species, hydrogen peroxide (H₂O₂) (Paulus et al., 2001). The ORR required at the PEMFC cathode is the 4-electron conversion of O₂ gas and proton (H⁺) to water (Equation 8.2). If the catalyst is inefficient, the O₂ gas is reduced to hydrogen peroxide (H₂O₂) via a 2-electron process reaction (Equation 8.3), and the H₂O₂ is detected at the ring.

$$O_2 + 4e^- + 4H^+ \rightarrow 2H_2O \qquad (8.2)$$

$$O_2 + 2e^- + 2H^+ \rightarrow H_2O_2 \qquad (8.3)$$

In these RRDE studies, the Pt/VC electrocatalysts was poisoned by S(IV) solutions of varying concentrations at 0.65 V vs. a reversible hydrogen electrode (RHE), and then brought to 0.05 V, to reduce any S^x species to S^0. CV was used to determine the amount of sulfur adsorbed on the Pt/VC electrocatalyst exposed to S(IV) solutions by correlation to the charge consumed for the irreversible 6-electron oxidation of S^0 to water-soluble sulfate that occurred when the electrodes were cycled up to about 1.50 V (see Equation 8.1). Knowing the initial sulfur coverage ($\theta_{S,i}$), the ORR activity of the poisoned electrodes was also studied. Figure 8.6 presents the family of polarization curves for the ORR of a clean Pt/VC and a contaminated Pt/VC as a function of initial sulfur species coverage ($\theta_{S,i}$), along with the fraction of hydrogen peroxide produced at the disk electrode.

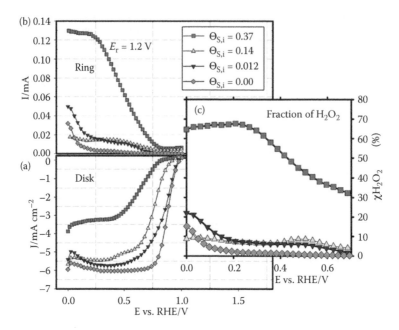

FIGURE 8.6 (a) Polarization curves for the ORR in 0.1 M HClO$_4$ solution of a clean Pt/VC and Pt/VC electrode covered with adsorbed sulfur species as a function of initial sulfur coverage ($\theta_{S,i}$). (b) Ring current following the production of hydrogen peroxide (E_{ring} = 1.2 V). (c) Fraction of hydrogen peroxide formed during the ORR of clean and Pt/VC poisoned with sulfur species as a function of initial sulfur coverage. (Reprinted from Garsany, Y., Baturina, O. A., and Swider-Lyons, K. E. 2007b. *Journal of the Electrochemical Society* 154: B670–B675. With permission from The Electrochemical Society, Inc.)

On a clean Pt/VC electrode ($\theta_{S,i}$ = 0), the ORR proceeded almost entirely through the direct 4e$^-$ reduction process to water as described by Equation 8.2. As the initial sulfur coverage on the Pt/VC electrocatalyst surface increased, the ORR activity strongly decreased with the onset potential for the ORR shifting to more negative potentials, and considerable hydrogen peroxide oxidation current on the ring electrode was observed, pointing to a change in the ORR pathway from a 4e$^-$ reduction process to water to a 2e$^-$ reduction process to hydrogen peroxide (Equation 8.3). This decrease in ORR activity due to the presence of adsorbed sulfur species on the Pt/VC electrocatalyst was accompanied by an enhanced formation of hydrogen peroxide, up to 70% for initial sulfur coverage $\theta_{S,i} \geq 0.37$. This indicates the effectiveness of the adsorbed sulfur species in reducing the number of pairs of Pt sites necessary for breaking of the oxygen–oxygen bond.

8.3.1.1.2 Other Sulfur-Containing Species (H$_2$S and COS)

Poisoning of the cathode with 1 ppm H$_2$S and 1 ppm COS in air have similar effect on a FC performance, almost undistinguishable from the effect of SO$_2$ (Gould et al., 2009), indicating that they block active sites on Pt electrocatalysts identically. Figure 8.7 illustrates the current density versus time of Pt/VC-based cathodes during their exposure to air contaminated with 1 ppm SO$_2$, H$_2$S or COS over a 12-h period while the cell voltage was held at 0.6 V (Gould et al., 2009).

As shown in Figure 8.7, all three of the sulfur impurities, SO$_2$, H$_2$S, and COS caused the same loss of current density versus time within experimental errors. Nagahara et al. (2008) also reported similar voltage decay rates for FCs exposed to SO$_2$ and H$_2$S when measured at constant current density. All three membrane electrode assemblies (MEAs) underwent a rapid loss of activity within the first 3 h of

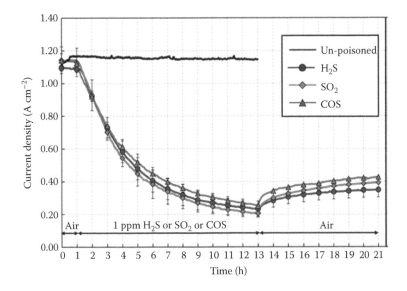

FIGURE 8.7 Current density loss during exposure of Pt/VC-based MEAs to 1 ppm H$_2$S for 12 h, 1 ppm SO$_2$ for 12 h, 1 ppm COS for 12 h while holding the cell voltage at 0.6 V. Poisoning step followed by recovery in air for 8 h while holding the cell voltage of 0.6 V. (Reprinted from Gould, B. D., Baturina, O. A., and Swider-Lyons, K. E. 2009. *Journal of Power Sources* 188: 89–95. With permission.)

poisoning, followed by an asymptotic approach to saturation current density of 0.22 A cm^{-2} presumably due to the equilibration of the Pt electrocatalysts with the contaminants. The current density improved only slightly after the MEA was operated for 8 h with neat air at 0.6 V.

8.3.1.1.3 Nitrogen Oxides (NO$_x$)

The effect of NO$_x$ species on PEMFC performance has been reported in the literature for a wide range of NO$_x$ concentrations. Figure 8.8 presents the typical voltage response of a FC exposed to different

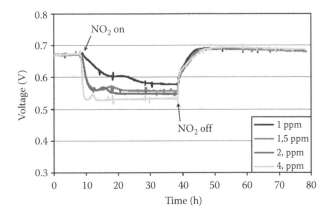

FIGURE 8.8 Cell voltage response during cathode exposure to 1, 1.5, 2, and 4 ppm NO$_2$ in air. The contamination with NO$_2$ is preceded and followed with exposure to neat air for 10 and 20 h, respectively. Cell Temp = 70°C, P_{anode} = $P_{cathode}$ = 1.5 × 10^5 Pa, Relative humidity = 60%. (Reprinted from Franco, A. A. et al. 2009. *ECS Transactions* 25: 1595–1604. With permission from The Electrochemical Society, Inc.)

concentrations of NO_2 (the thermodynamically stable species under ambient conditions), when operating at constant current mode (Franco et al., 2009).

The typical deactivation behavior is characterized by a clear voltage drop ($t = 10$–12 h) after NO_2 injection followed by a stabilization of the voltage ($t = 12$–40 h). As the NO_2 concentration in the airstream was increased from 1 to 4 ppm NO_2, the FC performance decay was increased. Increasing the PEMFC temperature and the operating current density (from 0.2 to 1 A cm^{-2}) also resulted in an increase in the cell performance loss (Franco et al., 2009). Mothadi et al. (2004) also found that the cell performance degradation was a function of NO_x concentration. Similarly Yang et al. (2006) reported significant cell performance degradation as the NO_x concentration was increased from 10 to 1480 ppm. However, the rate of poisoning of PEMFCs does not strongly depend on the NO_x bulk concentration. Both Mothadi et al. and Yang et al. observed no linear relationship between the voltage drop and the NO_x concentration. The recovery phase clearly showed that the impact of NO_2 contamination on the performance of the PEMFC was reversible. When the NO_2 was removed from the airstream, and the cell cathode side was purge with neat air, complete FC performance recovery was observed (Franco et al., 2009; Mohtadi et al., 2004; Yang et al., 2006). For Mothadi et al. (2004) the poisoning effect of NO_2 did not appear to be a catalyst surface poisoning. The comparison of cyclic voltammograms obtained with a clean cathode to a NO_2-contaminated cathode revealed the absence of any oxidation peaks corresponding to adsorbed surface species on the Pt catalyst layer. The authors concluded that the ionomer and/or the catalyst–ionomer interface could be affected by the exposure to NO_2. One of the mechanisms might be electrochemical reduction of NO_2 at the cathode to NH_4^+ which might compete with the ORR for Pt sites.

8.3.1.1.4 Ammonia (NH_3)

Ammonia is another nitrogen species that can have a deleterious effect on a PEM FC performance. NH_3 introduced at the cathode, strongly decreases the FC performance. The deactivation behavior of 48 ppm

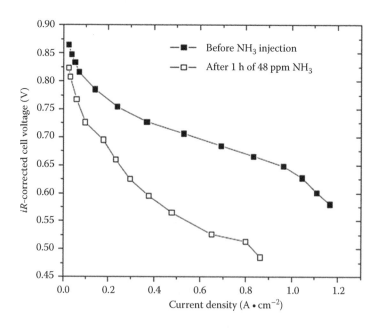

FIGURE 8.9 *iR*-corrected polarization curves measured before poisoning of the PEM FC cathode and after injection of 48 ppm of NH_3. Cell temp = 80°C, Relative humidity = 100%, 30 psi back pressure. (Reprinted from Garzon, F. H. et al. 2009. *ECS Transactions* 25: 1575–1583. With permission from The Electrochemical Society, Inc.)

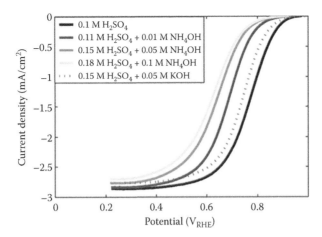

FIGURE 8.10 Polarization curves for the ORR in sulfuric acid electrolyte containing different concentrations of ammonium. The pH of the electrolyte was kept constant. Electrode rotation rate 1000 rpm, potential scan rate = 5 mV s⁻¹, room temperature. (Reprinted from *Electrochimica Acta*, 51, Halseid, R., Bystron, T., and Tunold, R. Oxygen reduction on platinum in aqueous sulphuric acid in the presence of ammonium, 2737–2742, Copyright (2006), with permission from Elsevier.)

NH_3 (injected into the humidified cathode airstream) on the polarization curves is presented in Figure 8.9 (Garzon et al., 2009).

As seen in Figure 8.9, the kinetic, ohmic, and mass transport regions of the polarization curve are negatively impacted by the presence of the ammonia in the cathode airstream. Ammonia may affect the PEM FC performance in different ways. It can reduce the ionic conductivity of the membrane, which in its ammonium form, is by a factor of 4 lower than in the protonated form (Halseid et al., 2006b). The effect of NH_4^+ on the membrane performance will be considered in details in Section 8.3.3. NH_3 can also poison the cathode catalyst in the ORR process. ORR studies in aqueous sulfuric acid solutions (Halseid et al., 2006a) showed that the effect of NH_4^+, even in moderate concentration, on the ORR was significant as shown in Figure 8.10.

The shift of onset potential for the ORR toward more positive potential as the concentration of NH_4^+ increased indicates inhibition of ORR in the presence of NH_3. CV studies showed that NH_4^+ formed both surface species and volatile species at potentials above 0.77 V versus a reversible hydrogen electrode (RHE) on the Pt electrode, and those surface species remained on the Pt surface even at low potentials. The mechanism behind the reduced ORR rate in the presence of NH_4^+ is unclear though.

Ammonium ions also caused increased H_2O_2 formation on polycrystalline Pt electrodes in sulfuric and perchloric acid aqueous solutions (Halseid et al., 2008). In FC experiments, neither the introduction of 2 ppm NH_3 or 2 ppm NO_2 in the cathode airstream, resulted in increased fluoride emission rates (Imamura et al., 2009b) indicating that no hydrogen peroxide was generated by the ammonium contaminants.

8.3.1.1.5 Inorganic Ionic Species

Matsuoka et al. (2008) investigated the effect of four anionic species (Cl^-, F^-, SO_4^{2-}, and NO_3^-) at the PEMFC cathodes using single cell tests complemented by electron probe microanalysis (EPMA) and transmission electron microscopy (TEM) conducted after the completion of the FC tests. The poisoning of the cathode airstreams by either F^-, SO_4^{2-}, or NO_3^- anions for 50 h did not affect the cell voltage. Only the presence of the Cl^- anion in the cathode airstream showed a drop in cell voltage. This drop in cell voltage could not be recovered back to its initial performance when the Cl^- anion was removed from the cathode airstream, and the cell cathode side was purged with neat air as shown in Figure 8.11.

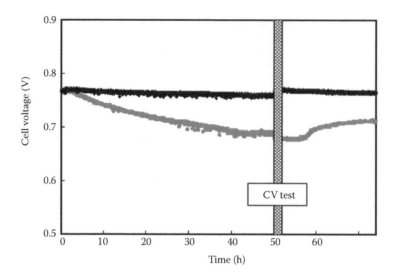

FIGURE 8.11 Effect of 3 mM solution of Cl$^-$ (gray line) and ultrapure water (black line) on the cell voltage of a PEM FC single cell. Cell temperature = 80°C, operating current density = 300 mA cm^{-2}, anode humidification = 70°C, cathode humidification = 30°C. (Reprinted from *Journal of Power Sources*, 179, Matsuoka, K. et al. Degradation of polymer electrolyte fuel cells under the existence of anion species, 560–565, Copyright (2008), with permission from Elsevier.)

CV curves measured at the FC cathodes after the 50 h of poisoning are presented in Figure 8.12. There was no change in the CV curves when F$^-$, SO$_4^{2-}$, and NO$_3^-$ were supplied to the cathode airstream, indicating no change in the Pt electrochemical surface area (ECSA). The CV curve obtained after poisoning cathode catalyst with the Cl$^-$ anion showed a significant suppression of the hydrogen adsorption/ desorption region (i.e., 0.05–0.40 V) indicating a 30% decrease in the Pt ECSA. Imamura et al. (2009a) also observed close to 30% decrease in Pt ECSA when Cl$^-$ anion was supplied to the cathode air inlet either as NaCl or HCl. Irreversible Pt surface area loss caused by Cl$^-$ has been demonstrated for Pt/C

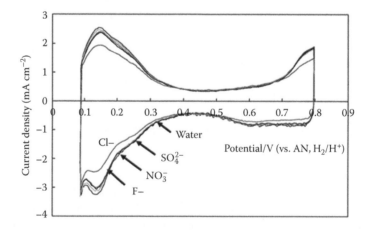

FIGURE 8.12 Cyclic voltammograms (CVs) of the cathode catalyst layer recorded after the different poisoning tests. CVs performed at room temperature and a potential scan rate of 5 mV s^{-1}. Anode gas: H$_2$ and cathode gas: N$_2$. (Reprinted from *Journal of Power Sources*, 179, Matsuoka, K. et al. Degradation of polymer electrolyte fuel cells under the existence of anion species, 560–565, Copyright (2008), with permission from Elsevier.)

MEAs (Steinbach et al., 2007). The 50 cm^2 MEAs were poisoned by 20 μM of HCl added to the deionized water used for humidification, and then their performance was partially recovered by thermal cycling including FC shutdown, flushing, and, restarting.

The decrease in the Pt ECSA after exposure to aqueous chloride solutions was suggested to be related to the accelerated Pt dissolution due to the presence of Cl$^-$ anions (Imamura and Ohno, 2009a; Matsuoka et al., 2008). Accelerated platinum dissolution below $E = 1.06$ V (vs. RHE) in 0.5 M H$_2$SO$_4$ solutions was reported by Yadav et al. (2007). In the absence of Cl$^-$ anions, the standard potential for Pt dissolution is

$$Pt \rightarrow Pt^{2+} + e^-, \quad E^0 = 1.18 \text{ V (vs. RHE)} \tag{8.4}$$

In the presence of Cl$^-$ anions Pt dissolved at 0.758 V (vs. RHE) and 0.742 V (vs. RHE) according to the following equations:

$$Pt + 4Cl^- \rightarrow [PtCl_4]^{2-} + 2e^-, \quad E^0 = 0.758 \text{ V (vs. RHE)} \tag{8.5}$$

$$Pt + 6Cl^- \rightarrow [PtCl_6]^{2-} + 4e^-, \quad E^0 = 0.742 \text{ V (vs. RHE)} \tag{8.6}$$

8.3.1.1.6 Sodium Chloride

The impact of sodium (Na$^+$) and chloride (Cl$^-$) ions (both individually and combined together) on the cell voltage and internal resistances of a PEMFC is illustrated in Figure 8.13 (Imamura and Ohno, 2009a).

The presence of NaOH in the cathode airstream caused an irreversible cell voltage drop as well as an irreversible increase in internal resistance due to replacement of protons by sodium ions both in the membrane and catalyst layer ionomers (Imamura and Ohno, 2009a). The accumulation of Na$^+$ ions in the FCs was confirmed by analysis of the effluent water during the NaOH solution supply. When the cathode was poisoned by a HCl solution, a cell voltage dropped from 0.65 to 0.50 V within the first 10 h of exposure to HCl, and then the voltage approached a constant value. The internal resistance remained mainly constant during this poisoning step. The cell voltage was partially recovered when the HCl solution was replaced by pure water suggesting blocking the platinum surface by adsorbed chloride ions. The presence of NaCl in the cathode airstream caused a voltage drop due to the combined effect of an increase in internal resistance and platinum surface area loss (due to the presence of Cl$^-$ ions). Partial recovery was observed when the NaCl solution was replaced by pure water.

Different results on the effect of NaCl at the cathode were reported by Mikkola et al. (2007). Chloride adsorption on platinum was not observed in this study, and the whole effect was assigned to the replacement of protons by sodium ions both in the membrane and catalyst layer.

8.3.1.1.7 Silicon

Silicon, from silicone-containing seals or tubing, is another poison that adsorbs on Pt. One study of a H$_2$|O$_2$ PEMFC showed the aggressive poisoning effects of silicone gasket, silicone-red® (Schulze et al., 2004). During FC operation, silicone products migrated to the cathode, and formed inactive particles of Pt, O, and Si. They concluded that the decomposition products of the silicone may contribute to the irreversible poisoning of the catalysts and may also change the hydrophilic/hydrophobic characteristic of the electrodes. Silicone contamination was also a source of degradation of a stack using silicone gaskets (Ahn et al., 2002).

8.3.1.1.8 VOCs (Benzene, Toluene, HCN, ClCN, Sarin, Mustard)

The effect of benzene contamination on the PEMFC performance is a function of operating current density (Moore et al., 2000) as the cell performance loss was more marked at higher current densities.

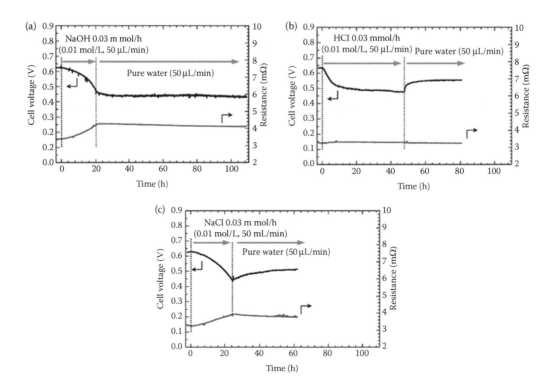

FIGURE 8.13 Changes in cell voltage and resistance caused by the introduction of (a) NaOH, (b) HCl, and (c) NaCl in the cathode airstream. Cell temperature = 80°C, operating current density = 1000 mA cm⁻². (Reprinted from Imamura, D. and Ohno, K. 2009a. *2009 Fuel Cell Seminar Extended Abstracts*: LRD 25–49. With permission from the Fuel Cell Seminar Headquaters.)

These performance drops were not fully recovered after the benzene was shut off and the FC cathode was purged with neat air over a 10-min period.

The effect of toluene in air on PEMFC performance has recently been investigated using various levels of toluene concentrations in the airstreams, under different operating conditions including different current density, relative humidity, Pt cathode loading, back pressure, and air stoichiometry (Li et al., 2008, 2009). The magnitude of toluene poisoning was a strong function of toluene concentration in the airstream and operating current density as shown in Figure 8.14.

At all current densities, the cell voltage started to decline immediately after the introduction of toluene in the cathode airstream, and then reached a plateau. These plateau voltages indicate the saturated nature of toluene contamination. Spikes in the cell voltage on the curves measured at current densities higher than 0.2 A cm⁻² were assigned by the authors to water management issues in the catalyst layer induced by the toluene contamination. Operating conditions such as back pressure, air stoichiometry, relative humidity, and Pt cathode loading had significant influences on the toluene contamination. Increasing the relative humidity, Pt cathode loading, and operating pressure led to less severe performance degradation, while increasing air stoichiometry resulted in a larger voltage drop.

Electrochemical impedance spectroscopy demonstrated that the kinetic and mass transfer resistances are significantly increased as a result of toluene contamination, while the membrane's resistance remained unchanged. Results of EIS measurement are presented as a bar chart in Figure 8.15.

Both kinetic and mass transfer resistances increase due to toluene contamination, but increases in the kinetic resistance are the more dominant contributors to the drop in cell performance. Li et al. suggested

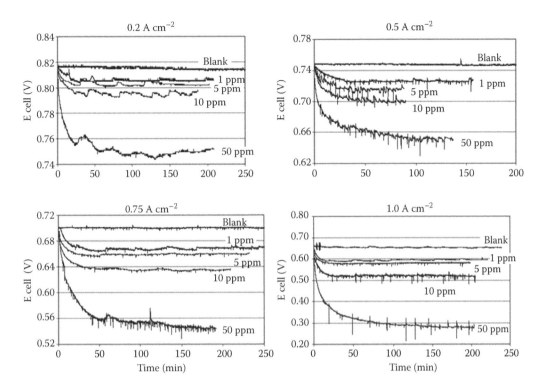

FIGURE 8.14 Voltage versus time curves with various levels of toluene at different current densities. Cell temperature = 80°C, Relative humidity = 80%, 30 psi back pressure, stoichiometry: 1.5/3.0 for H_2/air. (Reprinted from *Journal of Power Sources*, 185, Li, H. et al. Polymer electrolyte membrane fuel cell contamination: Testing and diagnosis of toluene-induced cathode degradation, 272–279, Copyright (2008), with permission from Elsevier.)

that the kinetic effect is due to the adsorption of toluene on the Pt surface that blocks Pt active sites, thus degrading the FC performance. Moreover, the adsorption of toluene may change the surface structure and hydrophobicity/hydrophilicity of the cathode's GDL and cathode catalyst layer (CCL) thus rendering the GDL/CCL more hydrophilic.

Chemical warfare agents can also seriously compromise the performance of the FC (Moore et al., 2000). The responses of the cell to poisoning by HCN and ClCN in the cathode airstream are very similar. With 1780 ppm HCN and 1560 ppm ClCN, the power output of the PEMFC was only 13% and 12% of the original value, respectively. Only partial recovery of the initial performance was obtained when the cathode airstream was purged with neat air. For Moore et al. these responses are consistent with Pt active catalyst sites being preferentially occupied by the poison gases. The decrease in cell performance by Sarin and Mustard gas to a steady value was more gradual, and subsequent to the impurity, cell recovery did not occur. Sarin and Mustard gas affected the FC performance much more slowly than HCN or ClCN. Moore et al. concluded that Sarin and Mustard gas were likely to bind irreversibly with the catalytic platinum site, but the size of the agent molecules reduced the rate of this reaction and might prevent neighboring platinum sites from being rapidly poisoned.

8.3.1.1.9 Gas Mixtures (SO_2 + NO_2 + NO in Air)

Figure 8.16 presents the cell voltage versus time obtained during the contamination of a PEM FC cathode with a gas mixture containing 1 ppm SO_2 + 0.8 ppm NO_2 + 0.2 ppm NO for 100 h, while operating the cell at a current density of 0.5 A cm^{-2} (Jing et al., 2007).

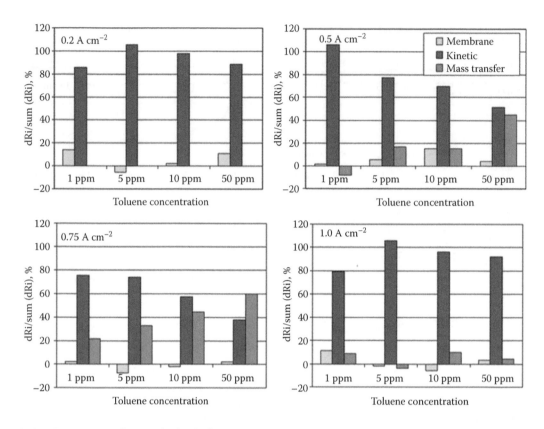

FIGURE 8.15 Contribution of individual resistance increases relative to the total resistance increase at different current densities for toluene contamination. Cell temperature = 80°C, Relative humidity = 80%, 30 psi back pressure, stoichiometry: 1.5/3.0 for H_2/air. (Reprinted from *Journal of Power Sources*, 185, Li, H. et al. Polymer electrolyte membrane fuel cell contamination: Testing and diagnosis of toluene-induced cathode degradation, 272–279, Copyright (2008), with permission from Elsevier.)

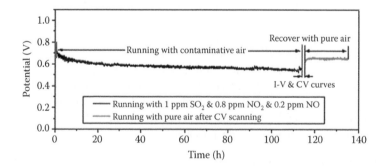

FIGURE 8.16 Cell voltage loss during exposure of a PEM fuel cell cathode to a gas mixture containing 1 ppm SO_2 + 0.8 ppm NO_2 + 0.2 ppm NO for 100 h while operating the cell at a current density of 0.5 A cm^{-2}. (Reprinted from *Journal of Power Sources*, 166, Jing, F. N. et al. The effect of ambient contamination on PEMFC performance, 172–176, Copyright (2007), with permission from Elsevier.)

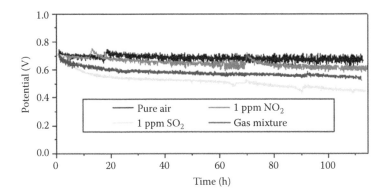

FIGURE 8.17 Cell voltage profile during exposure of a PEM fuel cell cathode to different gases for 100 h while operating the cell at a current density of 0.5 A cm^{-2}. (Reprinted from *Journal of Power Sources*, 166, Jing, F. N. et al. The effect of ambient contamination on PEMFC performance, 172–176, Copyright (2007), with permission from Elsevier.)

Figure 8.16 shows that the cell voltage started to decline immediately after the introduction of the gas mixture in the PEM FC cathode stream. The cell voltage decreased from the initial value of 0.69 to 0.53 V at the end of the 100 h poisoning experiment. CV study performed immediately after the poisoning step indicated the presence of different adsorbed species related to each component present in the gas mixture on the Pt surface.

The effect of mixed NO_2 and SO_2 on the PEM FC performance is shown in Figure 8.17.

The cell voltage loss obtained when the different gases were fed to the PEM FC cathode stream follow the sequence 1 ppm SO_2/air > gas mixture (1 ppm SO_2/air + 1 ppm NO_2) > 1 ppm NO_2/air > pure air. The effect of the gas mixture on the FC voltage loss is between that of the 1 ppm SO_2/air and 1 ppm NO_2/air. Jing et al. concluded that competitive adsorption of NO_2 and SO_2 on the Pt/VC electrocatalyst surface was the reason for this behavior. Overall, 94% of the initial cell voltage was recovered after running CV curves on the cathode poisoned with the gas mixture. This percentage recovery (94%) was the same as for the MEA poisoned with the 1 ppm NO_2/air. As the gas mixture was introduced in the cathode airstream, the adsorption of NO_2 took place first, resulting in less area remaining for SO_2 adsorption on the Pt surface as depicted in Figure 8.18.

8.3.1.2 Insights from Modeling

Developing accurate models of cathode/anode deactivation behavior is essential for predicting PEMFC performance degradation and durability, which are key to the rational design of FC systems. Modeling can also be used to provide deeper insight into the chemistry taking place at the surface by comparing different theoretical mechanisms to experimental results.

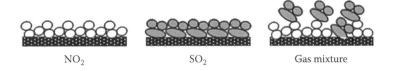

FIGURE 8.18 Schematic of air contaminants adsorbing on the Pt catalyst layer. (Reprinted from *Journal of Power Sources*, 166, Jing, F. N. et al. The effect of ambient contamination on PEMFC performance, 172–176, Copyright (2007), with permission from Elsevier.)

A kinetic model that accounts for adsorption and oxidation of contaminants at the cathode side of a PEMFC was developed by Shi et al. (2009). This model is based on five reactions adopted from the literature to describe the mechanisms of the ORR, with the limiting step given below:

$$Pt-O_2 + H^+ + e^- \leftrightarrow Pt-O_2H \tag{8.7}$$

For the adsorption and electrooxidation of the contaminant P in the cathode airstream, the following set of equations was proposed:

$$nPt + P \leftrightarrow Pt_n - P \tag{8.8}$$

$$P + nPt - O_2 \leftrightarrow Pt_n - P + nO_2 \tag{8.9}$$

$$P + nPt - O_2H + 3ne^- + 3\,nH^+ \leftrightarrow 2\,nH_2O + Pt_n - P \tag{8.10}$$

$$Pt_n - P + mH_2O \leftrightarrow Pt_n - P' + qH^+ + qe^- \tag{8.11}$$

$$Pt_n - P + lH_2O \leftrightarrow Pt_n + P''' + zH^+ + ze^- \tag{8.12}$$

Here n is the number of platinum sites occupied by a contaminant P, while n and m represent the number of water molecules involved in reactions 8.11 and 8.12. P' and P'' are the products of oxidation for contaminant P. Overall, five species such as Pt, Pt – O_2, O_2H, Pt_n – P, and Pt_n – P' are considered to occupy platinum surface.

A set of equations derived from the model assumptions complemented by Langmuir adsorption isotherm was used by Shi et al. (2009) to describe experimental steady-state and transient performance of a PEMFC in the presence of adsorbed toluene species in the cathode airstream. The model allowed the authors to qualitatively describe air-side toluene contamination observed for four different current densities (0.2, 0.5, 0.75, and 1.0 A/cm²) and three different concentrations (1, 5, and 10 ppm).

St-Pierre (2009) developed a "zero-dimensional model" that considers competitive adsorption for a contaminant with O_2 or H_2 at the cathode or anode side, respectively. This model assumes that contaminant transport through the gas flow channels, GDLs and ionomer in the catalyst layers is much faster compared to surface kinetics. The rate determining step is considered to be due to contaminant reaction or desorption of reaction product from the platinum surface. Other model assumptions include the absence of lateral interaction between adsorbates, first-order reaction kinetics, constant pressure, and constant temperature at the cathode/anode sides. Using a set of parameters, St-Pierre (2009) successfully used his model in order to describe experimental transient data obtained in the presence of SO_2, NO_2, and H_2S in the cathode airstreams.

Another approach for modeling experimental FC data in the presence of impurities at the FC cathodes was reported by Franco et al. (2009). They derived kinetic parameters for the ORR and contaminant reaction elementary steps using density functional theory analysis. Assuming the following steps for the ORR on platinum sites s:

$$O_2 + 2s \rightarrow 2Os \tag{8.13}$$

$$2Os + H^+ + e^- \rightarrow O_2Hs + s \tag{8.14}$$

$$O_2Hs + H_2O + 2s \rightarrow 3OHs \tag{8.15}$$

$$OHs + H^+ + e^- \rightarrow H_2O + s \tag{8.16}$$

and NO_2 reactions on the platinum surface

$$NO_2 + s \rightarrow NO_2s \tag{8.17}$$

$$NO_2s + s \rightarrow NOs + Os \tag{8.18}$$

$$NOs + O_2Hs \rightarrow HNO_3 + 2s \tag{8.19}$$

Franco et al. (2009) successfully interpreted unusual features ("voltage waves") observed on experimental voltage vs. time curves (see Figure 8.8) in the presence of NO_2 in the cathode airstream. A simulated voltage versus time curve is shown in Figure 8.19. The sudden voltage drop occurring after NO_2 introduction into the airstream followed by voltage stabilization was assigned to NO_2 adsorption on platinum according to Equation 8.17. Generation of adsorbed oxygen atoms according to Equation 8.18 led to an increase in the ORR rate (rising part of a "voltage wave" in Figure 8.19), while reaction of released NO with O_2H according to Equation 8.19 resulted in decrease in the ORR rate (declining part of a "voltage wave").

8.3.2 Fuel-Side Contamination

A large body of work exists on fuel-side contamination of PEMFCs. Much of it is derived from the desire to run FCs on reformate and is focused on CO and H_2S. A large fraction of the fuel-side contamination literature focuses on carbon monoxide because of its high concentration in the reformate stream (1–2%)

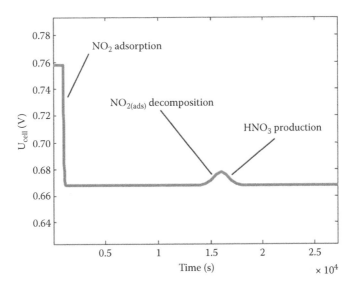

FIGURE 8.19   Simulated performance of PEMFC performance under air + NO_2. (Reprinted from Franco, A. A. et al. 2009. *ECS Transactions* 25: 1595–1604. With permission from The Electrochemical Society, Inc.)

and because of carbon monoxide's known affinity for Pt surfaces. State-of-the-art reforming processing can lower the concentration of CO to ppm-levels (Manasilp et al., 2002) but even at these low levels CO causes loss of FC performance.

8.3.2.1 Impacts

8.3.2.1.1 Carbon Monoxide (CO)

When the Pt electrocatalyst of the PEMFC anode is exposed to CO it binds to the Pt catalyst surface and occupies sites needed for the hydrogen oxidation reaction (HOR).

$$Pt - H_2 \Leftrightarrow Pt + 2H^+ + 2e^- \tag{8.20}$$

During CO adsorption to the Pt surface, the molecule stays intact and coordinates with the Pt at an atop site or as a bridged species as shown in the reaction below:

$$Pt + CO \rightarrow Pt - CO \tag{8.21}$$

$$\begin{array}{c} O \\ \| \\ C \\ / \ \backslash \\ Pt + CO \rightarrow Pt\,Pt \end{array} \tag{8.22}$$

The loss of active sites manifests itself as overpotential when contamination occurs at constant current or as a loss of current when contamination occurs at constant cell voltage. Both cases represent a loss in FC power and can occur during the exposure to a few ppm CO or less. Figure 8.20 shows the typical response of a FC when exposed to 1 ppm CO in H_2 (Bender et al., 2009). CO was injected at hour 45 and removed at hour 100. The typical deactivation behavior was observed with an initial

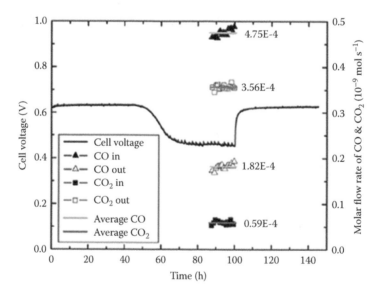

FIGURE 8.20   Cell voltage response to exposure to 1 ppm CO in H_2 at the anode at 60°C and 1 A cm^{-2}. (Reprinted from *Journal of Power Sources*, 193, Bender, G. et al., Method using gas chromatography to determine the molar flow balance for proton exchange membrane fuel cells exposed to impurities, 713–722, Copyright (2009), with permission from Elsevier.)

rapid performance loss ($t = 45$–60 h) followed by an approach to a saturation condition ($t = 60$–100 h). The cell voltage was almost entirely recoverable when carbon monoxide was removed from the H_2 stream. The work by Bender et al. (2009) is illuminating because the carbon molar flow balance for CO and CO_2 is monitored by analyzing the FC exhaust. The molar flow rates of CO and CO_2 at steady-state contamination are plotted on the second ordinate in Figure 8.20 which shows that a large fraction of the CO was converted to CO_2 in the anode at steady state. The mechanism for CO oxidation has been proposed to occur through electrochemical oxidation by water, but reaction with crossover O_2 is also a possibility.

The spatial distribution of CO contamination in the anode has been studied using the segmented cell technique, in which the anode was divided into multiple sections in the direction of H_2 flow and each segment's performance was monitored individually (Reshetenko et al., 2010). Figure 8.21b shows the response of individual cell segments when exposed to 2 ppm CO in H_2. CO contamination did not uniformly affect the performance of the FC. The current densities at the inlet segments (01–02) decreased by 20–40%, while the current densities near the exit increased by 20–40%. The increase in current density of the downstream segments balanced out for the decrease in current densities of the upstream segments to reach a constant current density of 0.8 A cm^{-2} over the entire cell. Figure 8.21a shows the overall effect of 2 ppm CO on FC performance and the typical trends in deactivation were observed when the anode became poisoned by CO.

8.3.2.1.2 Carbon Dioxide (CO₂)

Carbon dioxide is the contaminant with highest concentration in reformate (15–25%). Carbon dioxide converts into CO by the reverse water gas shift (WGS) reaction and blocks electrocatalytic sites as shown

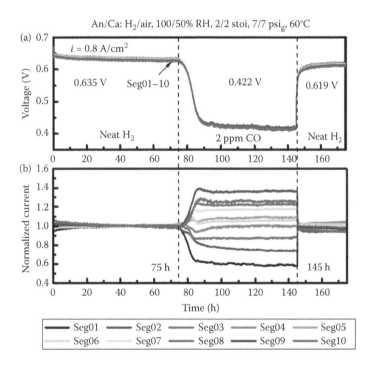

FIGURE 8.21 (a) Cell voltage response to exposure to 2 ppm CO in H_2 at the anode at 60°C and 0.8 A cm^{-2}. (b) Current response of individual cell segments to exposure to 2 ppm CO in H2. (Reprinted from Reshetenko, T., Bethune, K., and Rocheleau, R. 2010. In *217 Electrochemical Society Meeting*, Vancouver, Canada, pp. B7–B621. With permission from The Electrochemical Society, Inc.)

in reactions 8.21 and 8.22. The combination of CO and CO_2 has been shown to be deleterious with both compounds having a synergistic effect to lower FC performance.

8.3.2.1.3 Hydrogen Sulfide (H_2S)

H_2S typically exists in much lower concentrations in H_2 than CO or CO_2, but can have a deleterious effect on catalyst activity. Sulfur species are notorious catalyst poisons for dispersed metal catalysts because the unpaired electrons of sulfur strongly interact with the Pt surface, blocking sites needed for HOR. At anode potentials adsorbed H_2S is rapidly converted into sulfur ad-atoms by the reaction in

$$Pt - H_2S \rightarrow Pt - S^0 + 2H^+ + 2e^- \qquad (8.23)$$

H_2S concentrations as low as 10 ppb have been cited to cause catalyst deactivation (Garzon et al., 2006).

8.3.2.1.4 Mixture of CO and H_2S

Figure 8.22 shows the response of a PEMFC to 10 ppm H_2S in H_2 and 50 ppm CO in H_2. The cell voltage versus time plot shows rapid decrease in FC performance during the exposure to both contaminant species. Once the H_2S was removed from the H_2 the recovery of performance was minimal. The opposite was true for CO, which showed near complete recovery of performance when it was removed from the H_2. This illustrates an important difference between H_2S and CO, demonstrating that H_2S is a very potent catalyst poison in PEMFCs. When the cell voltage versus time was compared for the same concentration of CO and H_2S, H_2S caused a more rapid decrease in PEMFC performance (Shi et al., 2007a).

8.3.2.1.5 Ammonia (NH_3)

NH_3 in H_2 primarily interacts with the polymer electrolyte membrane and ionomer in the catalyst layer (Uribe et al., 2002). The influence of NH_3 on the membrane performance will be considered in more detail in Section 8.3.3 dedicated to membrane contamination. The effect of NH_3 on HOR is not completely understood. Different reports show that NH_3 may or may not strongly adsorb on the Pt catalyst and disrupt HOR (Uribe et al., 2002; Zhang et al., 2009). This disagreement is likely due to different compositions of anode catalyst layers in Uribe et al. (2002) and Zhang et al. (2009).

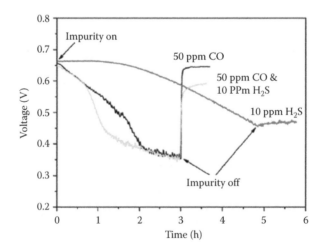

FIGURE 8.22 Cell voltage response during anode exposure to 50 ppm CO in H_2, 10 ppm H_2S in H_2, and a mixture of 50 ppm H_2S w/10 ppm CO in H_2. (Reprinted from Shi, W. et al. 2007a. *International Journal of Hydrogen Energy* 32: 4412–4417. With permission.)

8.3.2.1.6 *Hydrocarbons*

Hydrocarbons can have many different effects on FC performance depending on their structure. Methane acts as an inert species in the fuel diluting the concentration, but does not interact strongly with the catalyst surface to disrupt the HOR. Aromatic hydrocarbons interact more strongly with the catalyst surface and tend to hydrogenate to cycloparaffins (Bender et al., 2009). While this is a parasitic consumption of H_2, the effect of 20 ppm toluene on FC performance was minimal. Oxygenated hydrocarbons such as formic acid (HCOOH) and formaldehyde (HCHO) were explored in the literature because they are intermediate products of methanol reforming. The behavior of PEMFCs when exposed to formic acid or formaldehyde in H_2 is very similar to CO; it is characterized by initially rapid decreases in PEMFC performance followed by a saturation period and partial recovery when the contaminant is removed from the H_2 (Narusawa et al., 2003).

8.3.2.1.7 *Organic Sulfur Impurities*

Small amount of data exist on the impact of organic sulfur impurities in the fuel stream. This is largely because most of the organic sulfur species will be converted to H_2S and SO_2 during the reforming process. However, incomplete conversion or bypass is always a possibility in catalytic systems. Organic sulfur species will likely act like H_2S and SO_2, but more aggressively because of the added hydrocarbon component. There are data on the interaction between organic sulfur species (benzenethiol) and a Pt electrode, but they are limited to the solution phase. A suppression of the active Pt surface area was observed when exposed to organic sulfur species (Pomfret et al., 2010).

8.3.2.2 Modeling

The majority of anode deactivation modeling has focused on CO as a contaminant in the H_2 stream (Springer et al., 2001; Chan et al., 2003; Bhatia et al., 2004; Cheng et al., 2007; Shah et al., 2008) because of its importance to onboard reforming and the wealth of literature concerning CO contamination that can be used to validate the models. The models build in complexity starting with work of Springer et al. (2001) which focuses on predicting the change in polarization curve behavior during steady-state operation with H_2 containing inlet CO levels of 10–100 ppm. Chan et al. (2003) added transport equations to predict water velocities in the MEA. Bhatia et al. (Bhatia and Wang, 2004) provided a transient model of CO deactivation. All of the models are capable of fitting experimental data.

A general anode contamination model was developed by Zhang et al. (2005) and is capable of describing the effect of various contaminant species. St.-Pierre's generalized contamination model that is applicable to both the cathode and anode (St.-Pierre, 2009) has already been discussed briefly in the Section 8.3.1.2. All of the above models use similar reactions to describe the kinetics of HOR and the contamination reactions, along with Butler–Volmer expression to determine the FC's overpotential or current density. The reaction network and rate constants for a contamination process involving a general contaminant P is shown below:

$$Pt + H_2 \overset{k_{1f}}{\underset{k_{1b}}{\Longleftrightarrow}} Pt - H_2 \tag{8.24}$$

$$Pt - H_2 + Pt \overset{k_{2f}}{\underset{k_{2b}}{\Longleftrightarrow}} 2Pt - H \tag{8.25}$$

$$Pt - H \overset{k_{3f} = \exp(\alpha_3 n_3 F \eta_a / RT)}{\underset{k_{3b} = \exp(-(1-\alpha_3) n_3 F \eta_a / RT)}{\Longleftrightarrow}} Pt - P + H^+ + e^- \tag{8.26}$$

$$P + Pt \overset{k_{4f}}{\underset{k_{4b}}{\Longleftrightarrow}} Pt - P \tag{8.27}$$

$$P + Pt-H \underset{k_{5b}=\exp(-(1-\alpha_5)n_5 F\eta_a/RT)}{\overset{k_{5f}=\exp(\alpha_5 n_5 F\eta_a/RT)}{\Longleftrightarrow}} Pt-P + H^+ + e^- \tag{8.28}$$

$$P + Pt-H_2 \underset{k_{6b}}{\overset{k_{6f}}{\Longleftrightarrow}} Pt-P + H_2 \tag{8.29}$$

$$Pt-P + mH_2O \underset{k_{7b}=\exp(-(1-\alpha_7)n_7 F\eta_a/RT)}{\overset{k_{7f}=\exp(\alpha_7 n_7 F\eta_a/RT)}{\Longleftrightarrow}} Pt-P' + qH^+ + qe^- \tag{8.30}$$

$$Pt-P + mH_2O \underset{k_{8b}=\exp(-(1-\alpha_8)n_8 F\eta_a/RT)}{\overset{k_{8f}=\exp(\alpha_8 n_8 F\eta_a/RT)}{\Longleftrightarrow}} Pt + P' + qH^+ + qe^- \tag{8.31}$$

Here Equations 8.24 through 8.26 are related to electrochemical adsorption/desorption of hydrogen, while Equations 8.27 through 8.32 describe interaction of contaminant P with platinum surface in the presence of hydrogen and water. Using the above reaction network as a framework, Shi et al. (2007b) was able to develop a transient model for anode contamination with H_2S. The results of the modeling effort can be seen in Figure 8.23. Figure 8.23 shows that good agreement can be reached between the model and experimental results, predicting accurately the effect of H_2S concentration on both the initial deactivation rate and the steady-state voltage reached during saturation.

8.3.3 Membrane Contamination

Membrane contamination may affect FC performance in two different ways: (1) through replacement of protons by inorganic cations such as Na^+, Ca^{2+}, ammonium, or other metal ions (Uribe et al., 2002;

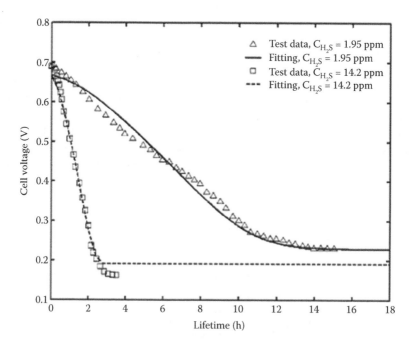

FIGURE 8.23 Comparison of experimental data and model fitting for anode deactivation with H_2S in H_2 at 0.5 A cm^{-2}. (Reprinted from Shi, Z. et al. 2007b. *Journal of the Electrochemical Society* 154: B609–B615. With permission from The Electrochemical Society, Inc.)

La Conti et al., 2003; Halseid et al., 2006b; Mikkola et al., 2007) leading to a decrease in membrane conductivity due to an increase in electroosmotic drag and reduced level of membrane humidification (Okada et al., 2002) or (2) through formation of hydroxyl or peroxyl radicals causing a decrease in membrane conductivity and eventually membrane failure (La Conti et al., 2003; Pozio et al., 2003). Inorganic cations are the most-studied impurities, as their presence is expected to strongly affect the ion-exchange membrane performance. They can be introduced into the membrane both externally (from reformate at the anode and air impurities at the cathode) and internally (from leaching catalysts (Sulek et al., 2008) or end plates (Pozio et al., 2003)). Some of these cations like Na^+ (Mikkola et al., 2007), NH_4^+ (Uribe et al., 2002), and Ni^{2+} (Sulek et al., 2008) are "benign" meaning that while replacing protons by ion-exchange reaction, they do not cause formation of radicals leading to mechanical degradation of Nafion membrane. Other cations such as Fe^{2+} cause radical formation in the presence of hydrogen peroxide through the modified Fenton's reaction (Fenton, 1894; Inaba et al., 2006):

$$Fe^{2+} + H_2O_2 \rightarrow Fe^{3+} + OH\bullet + OH^- \tag{8.32}$$

$$Fe^{2+} + OH\bullet \rightarrow Fe^{3+} + OH^- \tag{8.33}$$

$$H_2O_2 + OH\bullet \rightarrow OOH\bullet + H_2O \tag{8.34}$$

$$Fe^{2+} + OOH\bullet \rightarrow Fe^{3+} + HOO^- \tag{8.35}$$

$$Fe^{3+} + OOH\bullet \rightarrow Fe^{2+} + H^+ + O_2 \tag{8.36}$$

The $OH\bullet$ and $OOH\bullet$ radicals generated by reactions 8.32 and 8.34 attack fluoride chains of perfluorosulfonic membranes, leading to accelerated fluoride release rates and membrane degradation.

Neutral inorganic molecules introduced at the FC cathode can also affect membrane durability. Increased fluoride release rate after exposure of the FC cathode to 2 ppm SO_2 in air was reported by Imamura et al. (Imamura et al., 2007).

8.3.3.1 Impacts

8.3.3.1.1 Ammonia at the Anode

The performance losses due to introduction of ammonia at the FC anode were studied by the exposure to 1–1000 ppm NH_3 at different exposure times (Halseid et al., 2006b; Soto et al., 2003; Uribe et al., 2002). High concentrations caused dramatic decrease in FC performance that could be recovered partially after exposure to neat hydrogen. Complete recovery could be observed only after 1–3 h exposure according to Uribe et al. (2002), while Soto et al. (2003) observed full recovery of the FC performance even after exposure of the FC anode to 200 ppm HN_3 for 10 h. A FC exposed to a low concentration (1 ppm) of NH_3 for a long period of time (1 week) could not be fully recovered after several days of exposure to neat hydrogen (Halseid et al., 2006b).

The effect of NH_3 in the anode stream on the FC performance was interpreted in terms of replacement of protons by NH_4^+ ions in the proton exchange membrane and ionomer inside the anode (Uribe et al., 2002) and cathode (Soto et al., 2003) catalyst layers. An increase in the bulk membrane resistance due to replacement of H^+ by NH_4^+ is believed to account for only 5–15% of the total performance loss (Halseid et al., 2006b), the major contribution to the total performance loss comes from the losses in ionomer in the catalyst layers.

8.3.3.1.2 *Sodium Chloride*

Sodium ions introduced at the FC cathode (Mikkola et al., 2007) do not appear to be as detrimental to the FC performance as other inorganic ions. Cell performance at 0.6 V dropped only 33% after injection of 1 M NaCl solution for 96 h (Mikkola et al., 2007). Decay in the FC performance, as shown in Figure 8.24, is accompanied by a significant increase in high-frequency resistance (HFR). Both effects are thought to be due to replacement of protons by sodium ions both in membrane and iono-mers in the catalyst layers, due to much higher affinity of sodium ions in Nafion than protons (Okada et al., 2002).

Figure 8.25a shows HFR versus calculated resistance (R_{calc}) that accounts mostly for proton conduc-tivity in the membrane and catalyst layers. Correction of experimental polarization curves for R_{calc} (Figure 8.25b) leads to a perfect agreement between polarization curves shown in Figure 8.24a, suggest-ing that changing the proton conductivity of the membrane and catalyst layers is the major factor affect-ing FC performance.

8.3.3.1.3 *Sulfur Dioxide and Hydrogen Sulfide*

Membrane/ionomer performance can also be affected by neutral inorganic molecules such as SO_2 and H_2S. Membrane/ionomer degradation during and after exposure of the FC cathode to 2 ppm H_2S or SO_2 in air (Figure 8.26) was evidenced by increased fluoride emission rates (Imamura and Hashimasa, 2007). This was likely due to increased amounts of hydrogen peroxide generated on platinum catalysts when the ORR occurs in the presence of SO_2 (Garsany et al., 2007b). Recently, Imamura et al. (Imamura and Yamaguchi, 2009b) reported that an increase in fluoride emission rate was due to degradation of the ionomer in the cathode and anode catalyst layers, but not the PEM.

8.3.3.1.4 *Ammonia and Nitrogen Dioxide at the Cathode*

Neither 2 ppm ammonia nor 2 ppm nitrogen dioxide affected fluoride emission rate when introduced in the cathode airstreams, as was found by Imamura et al. (Imamura and Yamaguchi, 2009b).

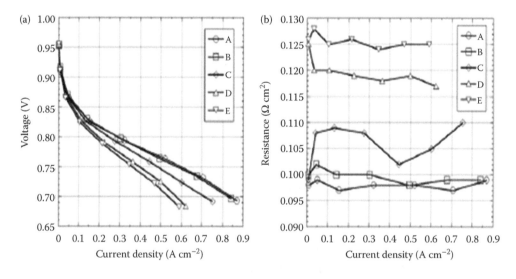

FIGURE 8.24 (a) *iR*-compensated polarization and (b) high-frequency resistance curves recorded before, during, and after the NaCl injection test. The polarization curves were recorded at (A) 1.1 h, (B) 53.4 h, (C) 71.6 h, (D) 99.2 h, (E) 167.3 h. (From Mikkola, M. S. et al. The effect of NaCl in the cathode air stream on PEMFC perfor-mance. *Fuel Cells.* 2007. 7: 153–158. Copyright Wiley-VCH Verlag GmbH & Co. KGaA. Reprinted with permission.)

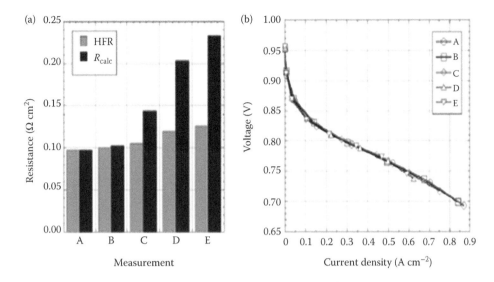

FIGURE 8.25 (a) Measured HFR and calculated resistance R_{calc} for polarization curves (A)–(E) in Figure 8.24. (b) Polarization curves from Figure 8.24, *iR*-compensated with the calculated resistance values. (From Mikkola, M. S. et al. The effect of NaCl in the cathode air stream on PEMFC performance. *Fuel Cells.* 2007. 7: 153–158. Copyright Wiley-VCH Verlag GmbH & Co. KGaA. Reprinted with permission.)

8.3.3.2 Modeling

Careful modeling of the influence of "benign" cationic contaminants that do not react with hydrogen peroxide, causing oxygen radicals formation was made by Okada in 1999 (Okada, 1999a,b). Considering water transport across the membrane in the presence of two cations (proton and foreign cation), he concluded that water management in the membrane was drastically affected by impurity ions. Replacement of protons by foreign cations in the proton exchange membrane led to increased electroosmotic

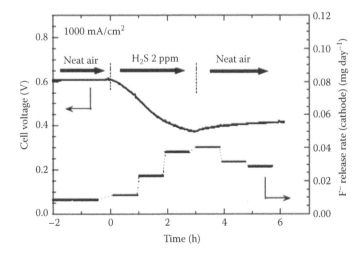

FIGURE 8.26 Effect of 2 ppm H_2S in air: change of cell voltage and F^- release rate in cathode. (Reprinted from Imamura, D. and Hashimasa, Y. 2007. *ECS Transactions* 11: 853–862. With permission from The Electrochemical Society, Inc.)

drag and reduced humidification levels in the membrane, and as a result, decreased membrane conductivity. Cations localized both at the anode/membrane and cathode/membrane interfaces caused membrane drying due to increased electroosmotic drag. However, contamination at the cathode/membrane interface caused more severe membrane dehydration.

Different approach to model the effects of cationic contaminants on the FC performance was used by Kienitz et al. (2009). Instead of focusing on modeling the water management in the membrane (Okada, 1999a,b), they developed steady-state PEM model that considered cations transport in the membrane, thus accounting for cationic movement or reorganization in the membrane. Transport phenomena included diffusion and migration of foreign cations and protons inside the membrane. Starting with the Nernst–Plank equation to describe ionic flux:

$$j_{i^+} = D_i + \frac{dC_{i^+}}{dx} + u_i + C_i + \frac{d\Phi}{dx} \tag{8.37}$$

and an equation for water transport that includes both transport by diffusion and electroosmotic drag:

$$j_W = -D_W \frac{dC_W}{dx} + \sum_{i^+} n_{\mathrm{drag},i} + j_{i^+} \tag{8.38}$$

the concentration profiles of foreign cations as a function of contamination level at the cathode and FC operating current were calculated for the system that consisted of the membrane in the presence of one impurity cation, protons, and water.

Here j_i is the molar flux of species i (mol/m²s); D_i is the diffusion coefficient (m²/s); C_i is the concentration of species i (mol/m³); x is dimensionless length of the membrane; u_i is ion mobility (m²/V s); Φ is the electric potential (V), W the subscript indicating water; and n the electroosmotic drag coefficient.

Foreign cation-concentration profiles are given in Figure 8.27a and b, respectively, as a function of current at a given contamination level and a function of contamination level at constant current. The fraction of contaminant cations at the cathode increases both with increasing contamination level and current density due to the potential gradient across the membrane leading to cationic transport by migration.

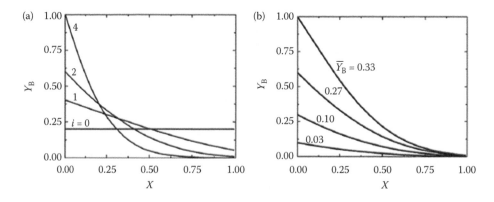

FIGURE 8.27 (a) Cation contaminant concentration profiles as a function of current for a constant average contamination of 20%. All current densities are measured in A cm⁻². (b) Cation contaminants profiles as a function of contamination level for a constant current of 2 A cm⁻². (Reprinted from *Electrochimica Acta*, 54, Kienitz, B. L., Baskaran, H., and Zawodzinski, T. A., Modeling the steady-state effects of cationic contamination on polymer electrolyte membranes, 1671–1679, Copyright (2009), with permission from Elsevier.)

Another consequence of membrane contamination by cationic impurity can be a decrease in the limiting current on polarization curve measured at high contamination levels. Due to proton deficiency at the cathode, the ORR current may become limited by diffusion of protons, but not oxygen diffusion through the CCL. This effect observed in experimental systems (Uribe et al., 2002; Halseid et al., 2006b) was qualitatively described using model assumptions proposed by Kienitz et al. (2009).

Serincan et al. (2010) also modeled PEMs and validated their model with experimental data. Na$^+$ contamination at the cathode caused a significant drop in the operating current density at the anode. Anode contamination caused dry out at the cathode.

8.4 Contamination Impacts in High-Temperature PEM FCs

Phosphoric-acid-doped polybenzimidazole (PBI) membranes are being developed as alternatives to the traditional perfluorosulfonic-acid-based (e.g., Nafion) membranes for PEMFCs. Nafion membranes cannot be used over about 80°C without loss of proton conductivity and mechanical stability (Shao et al., 2004). PBI is used at temperatures ≥160°C and thus offer easier water and thermal management than PEMFCs with Nafion (Yang et al., 2001; Li et al., 2004). PBI also has excellent thermochemical stability, mechanical properties, low gas permeability (Savadogo, 2004), and good proton conductivity (Mecerreyes et al., 2004). Another feature of PBI-based PEM FC is that their high-temperature operation (≥160°C) provides tolerance of the Pt electrode to impurities such as CO (Schmidt et al., 2006; Das et al., 2009), H$_2$S, and SO$_2$ (Schmidt and Baurmeister, 2006; Garsany et al., 2009b).

8.4.1 Carbon Monoxide at the Anode

Das et al. (2009) found that the cell voltage loss in the presence of CO in the H$_2$ stream was a function of CO content, temperature, and current density. The variation of cell voltage as a function of temperature at current densities of 0.2 and 0.5 A cm^{-2}, respectively for both 2% CO and 5% CO mixed with hydrogen is presented in Figure 8.28. The voltage of FC running on pure hydrogen mildly and linearly dropped with temperature. When the cell was operated at 0.2 A cm^{-2}, both 2% CO and 5% CO could be tolerated at 180°C without any voltage drop. As the temperature decreased from 180°C, the cell voltage drop was increased for the 5% CO compared to the 2% CO when the cell was operating at 0.2 A cm^{-2}. At a cell voltage of 0.65 V and current density of 0.2 A cm^{-2} both 2% and 5% CO could be tolerated.

The effect of 2% CO and 5% CO mixed with hydrogen on the cell voltage loss at 140°C, 160°C, and 180°C are shown in Figure 8.29. At the higher temperature of 180°C, the FC performance degradation rate was lower compared to the lower temperature of 140°C.

Das et al. explained the performance losses of the high-temperature PBI FC in the presence of a small amount of CO in the hydrogen stream by competing CO and hydrogen for adsorption sites on the Pt catalyst surface. CO binds to Pt catalyst surface by two types of bonding modes as shown in Equations 8.21 and 8.22. Because CO adsorption is strongly favored at low temperature but not at high temperature (Li et al., 2003), the more pronounced cell performance decrease at low temperature might be associated with blocking more Pt active sites by the adsorbed CO needed for hydrogen adsorption and oxidation. The authors used a detailed electrochemical analysis to determine the CO coverage on the catalyst surface as a function of temperature. The results of their CO coverage evaluation, presented in Figure 8.30, indicate that CO surface coverage of the Pt catalyst decreases significantly with the temperature.

8.4.2 H$_2$S and H$_2$S + CO at the Anode

Early studies done by Schmidt et al. indicate that PBI-based PEMFCs were tolerant to at least 10 ppm H$_2$S in the fuel stream (Schmidt and Baurmeister, 2006). More than 3000 h operation in reformate containing 5 ppm H$_2$S and 2% CO was proven.

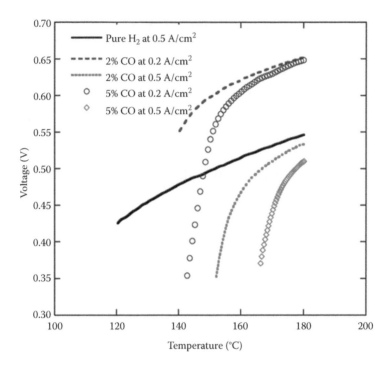

FIGURE 8.28 Variation of cell voltage as a function of temperature at current densities of 0.2 A cm⁻² and 0.5 A cm⁻², respectively, for both 2% CO and 5% CO mixed with hydrogen. (Reprinted from *Journal of Power Sources*, 193, Das, S. K., Reis, A., and Berry, K. J., Experimental evaluation of CO poisoning on the performance of a high temperature proton exchange membrane fuel cell, 691–698, Copyright (2009), with permission from Elsevier.)

8.4.3 H₂S and SO₂ at the Cathode

The tolerance of the PBI cathode catalyst to sulfur-containing species (H_2S and SO_2) in air at the cathode has also been proven. Figure 8.31 presents the performance of a PBI-based PEM FC after its cathode has been exposed to 1, 2.5, 5, and 10 ppm H_2S for a total combined time of 250 h (Garsany et al., 2009b).

Little to no deterioration of the cell performance was observed within the first 24 h of poisoning with 1 ppm H_2S. Total cell performance recovery was achieved when the contamination source was shut off and neat air was turned on for 24 h directly after the poisoning step. During exposure to 2.5 ppm H_2S for 24 h, a loss of 2.14% of the initial current density was observed but again purging the cathode side with neat air for 24 h led to complete cell performance recovery. Exposure to 5 and 10 ppm H_2S for 24 h led to 5.2% and 7.1% decreases in cell performance, respectively. Subsequent purging of the cathode inlet with neat air after the poisoning step led to complete recovery of the cell performance within 2% of the initial current density. Garsany et al. also studied the cell performance decrease versus different time intervals for PBI and Nafion-based MEAs exposed to 1 ppm H_2S and 1 ppm SO_2. The impact of the H_2S and SO_2 was found to be equivalent on the PBI-based PEM FC, with only a 1.14% loss in cell performance after 24 h exposure to 1 ppm H_2S and 1.64% loss in cell performance after 24 h exposure to 1 ppm SO_2. Complete performance recovery was obtained after purging the cathode inlet with neat air for 24 h. Contrary to PBI-based FCs, the Nafion-based FCs demonstrated 80.8% and 82.9% performance decrease after exposure to 1 ppm of H_2S and 1 ppm of SO_2 for 24 h, respectively. The cell performance recovery was only possible by potential cycling and the electrochemical conversion of adsorbed sulfur species to sulfate for the Nafion-based FCs. The mechanism for the higher tolerance of the PBI FC cathode to sulfur species is not completely understood.

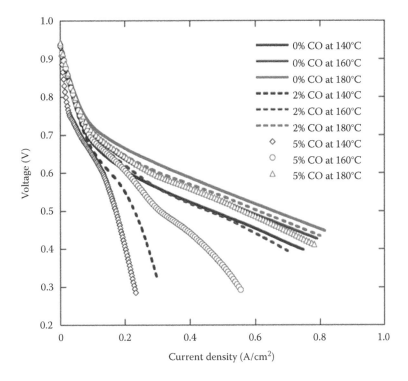

FIGURE 8.29 Polarization curves obtained with a PBI-based PEM fuel cell running with pure hydrogen and 2% CO and 5% CO mixed with pure hydrogen, respectively, at different temperatures. (Reprinted from *Journal of Power Sources*, 193, Das, S. K., Reis, A., and Berry, K. J., Experimental evaluation of CO poisoning on the performance of a high temperature proton exchange membrane fuel cell, 691–698, Copyright (2009), with permission from Elsevier.)

8.5 Contamination Impacts in Hydroxide Exchange Membrane FCs

Alkaline FCs earned notoriety with their successful deployment as both an electrical and water sources onboard the United States' Space Shuttle. These systems work effectively on pure oxygen and hydrogen, and use a liquid KOH electrolyte, with simply a separator between the cathodes and anodes. Their degradation due to impurities was kept to a minimum through the choice of materials (e.g., silicone-free components) and ultrapure gases. New types of alkaline FCs are being explored with alkaline anion exchanges membranes (AAEMs), which conduct anions (e.g., OH^-) via ammonium pendant groups on fluoropolymer, polybenzimidazole, or related backbones (Varcoe et al., 2005; Zhou et al., 2009). The use of such solid-electrolyte membranes introduces many of the benefits of Nafion-based cells, absent the corrosive, low pH environment of the Nafion. The high pH of AAEMs allows the use of nonprecious metal catalysts, as the first row-transition metals are not soluble in base, thus bringing the promise of low-cost FCs.

Contamination is a major cause of performance degradation and low durability in alkaline FCs. The main problem that faces alkaline FCs is that they are contaminated by traces of CO_2 in air. The intermediate steps in the hydroxide pathway to carbonate are written as (Vega et al., 2010):

$$CO_2 + OH^- \rightarrow HCO_3^-$$ (8.39)

$$HCO_3^- + OH^- \leftrightarrow CO_3^{2-} + H_2O$$ (8.40)

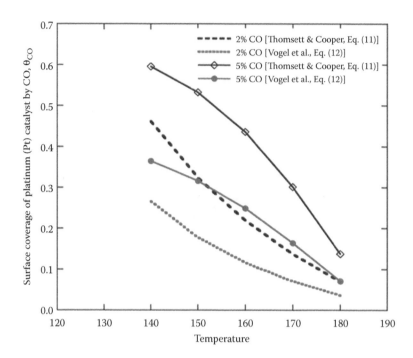

FIGURE 8.30 CO surface coverage of the Pt catalyst for 2% CO and 5% CO mixed with hydrogen as a function of temperature. (Reprinted from *Journal of Power Sources*, 193, Das, S. K., Reis, A., and Berry, K. J., Experimental evaluation of CO poisoning on the performance of a high temperature proton exchange membrane fuel cell, 691–698, Copyright (2009), with permission from Elsevier.)

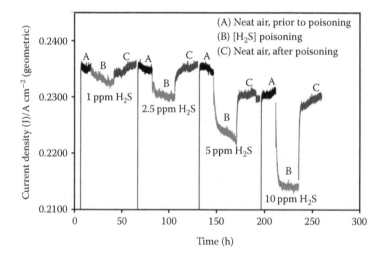

FIGURE 8.31 Transient response of a PBI MEA held at a cell voltage of 0.68 V while its cathode is exposed to air mixed with 1, 2.5, 5, and 10 ppm H_2S for 24 h. Each poisoning step was followed by recovery in neat air while the cell voltage was held at 0.68 V. (Reprinted from Garsany, Y. et al. 2009b. *Electrochemical and Solid-State Letters* 12: B138–B140. With permission from The Electrochemical Society, Inc.)

The overall reaction of CO_2 with KOH electrolyte results in precipitation of solid crystals of potassium carbonate (Tewari et al., 2006):

$$CO_2 + 2KOH \rightarrow K_2CO_3 + H_2O \tag{8.41}$$

With the formation of solid carbonate, alkaline FC performance degrades due to the depletion of the concentration of KOH in the electrolyte, lowering of the pH by the acidic carbonate, and blockage of pores and mechanical destabilization of the membrane by crystals (Al-Saleh et al., 1994; Vega and Mustain, 2010).

The impact of other poisons, such as fuel contaminants and other air contaminants, has not been extensively studied on alkaline FCs. If new nonprecious metal catalysts are developed, they are likely to adsorb contaminants less strongly than standard Pt catalysts; however, they may also be less amenable to clean-up techniques such as voltage cycling and air/fuel starvation methods, due to their narrower stability regions with respect to potential (Suntivich et al., 2010).

8.6 Performance Recovery and Contaminant Mitigation

An important part of contaminant studies is determining whether the contamination is reversible, and, if it is, how to remove the contaminant and recover the performance of the FCs. Due to the low-temperature operation of PEMFC, many of the contaminants are reversibly desorbed or removed from FC components, but some poisons can cause irreversible degradation. FC designers and practitioners create low contamination environment by using ultrahigh purity H_2, air filtration, robust seals, and components with low chemical degradation and outgassing. Contamination can be inevitable if the system is exposed to aggressive conditions (highly contaminated air) or is operated for thousands of hours. The FC can be operated in various ways to remove contaminants, and some components are less susceptive to poisoning than others and recover more easily from poisoning.

8.6.1 Platinum Catalyst Contamination and Recovery

The basic method for removing adsorbed contaminants from Pt catalysts is by potential cycling. Potential cycling of Pt is well understood, and is routinely carried out in deaerated electrolyte to understand water activation on Pt for fundamental understanding and characterization of the electrocatalyst. A representative cyclic voltammogram of nanoscale Pt on carbon in deaerated $HClO_4$ electrolyte is shown in Figure 8.32. In the anodic (positive going) sweep, OH_{ads} groups adsorb on Pt at potentials more positive than 0.7 V as water is activated on the Pt. Over about 1.0 V, O adsorbs on the Pt. The adsorbed oxygen is reduced back to water on the cathodic sweep near 0.7 V. Below 0.4 V, hydrogen adsorbs on Pt, again via the activation of water and at potentials more negative than 0.05 V H_2 evolution occurs.

The oxidation of Pt is inherent to its functionality as an oxygen reduction catalyst, as oxygen must adsorb to react, but then if it is not removed it becomes a poison. Oxygen and OH_{ads} groups from the activation of water continuously poison the Pt-based catalysts at the cathode as they adsorb on Pt and block sites for O_2 adsorption and reduction. As described in the CV above, poisoning by oxygen via water activation increases with increasing potential of the cathode. The amount of oxygen adsorbed on the Pt is higher in the presence of oxygen. The amount of OH^- poisoning can be lesser for platinum alloys, such as Pt_3Co and Pt_3Ni, because their modified electronic states decrease the adsorption and surface coverage of poisoning oxygen species (Stamenkovic et al., 2007).

Electrochemists routinely cycle Pt electrodes to "clean" them, and increase the available Pt surface area, resulting in larger areas for hydrogen adsorption/desorption. FC practitioners decrease the oxygen poisoning of Pt by air starvation methods, whereby the FC cathode is brought to low potentials (<0.1 V) for milliseconds to seconds, allowing H to adsorb and OH to be removed. When the cathode returns to its prior operating conditions, its Pt has fewer OH adsorbates and thus higher activity. The implementation

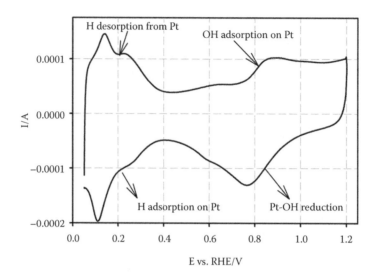

FIGURE 8.32 Cyclic voltammogram of nanoscale Pt on Vulcan carbon in deaerated 0.1 M HClO₄ showing the potential regions for OH_{ads} adsorption and desorption/reduction at potentials more positive than 0.6 V, and H adsorption and desorption at 0.05–0.4 V.

of this recovery process is most extensively discussed in the patent literature, where there are claims for dynamic operation of the FC via application of a load, with individual and combinations of steps for air starvation, applying low potentials, and/or applying an external electric field or drawing high currents, all with the same effect driving the cell voltage to low potentials to remove oxygen and other impurities from the surface of the platinum. These methods are described in general as methods to remove poisons from FC catalysts, both at the anodes and cathodes. Representative patents include: Abdou et al. (2004), Adams et al. (2003), Colbow et al. (2001), Reiser et al. (2003), and Uribe et al. (2006).

Another variation for the dynamic operation of FCs is thermal cycling (cycling between room temperature and 70°C), which has been shown to be effective in the removal of contaminants (Steinbach et al., 2007). Starting and stopping the FC is also used, during which the catalysts experience a voltage transient.

The effect of potential cycling on the oxygen reduction activity of Pt is clearly illustrated in Figure 8.33. Measurements were carried out by RDE methodology (Garsany et al., 2010), which can be used as a tool to predict performance in operational MEAs (Gasteiger et al., 2005). The magnitude of oxygen reduction activity, as compared by the current density at 0.9 V, increases significantly as the lower limit of the voltammogram is increased from 0.7 to 0.05 V, while the upper limit is held at 1.03 V. Therefore, exposing the Pt to the hydrogen adsorption region clearly improves its activity by making more sites available for oxygen reduction.

8.6.1.1 Chloride Contamination

Removal studies for Cl⁻ from Pt catalysts are still under investigation. Mitigation strategies will likely include flushing the FC with water to remove Cl⁻, and incorporating anion-exchange units (Abd Elhamid et al., 2006). Pt-alloys also have been reported to recover more completely from Cl-poisoning as reported by Steinbach et al. (2007).

8.6.1.2 Silicon Contamination

Mitigation approaches with respect to silicon contamination include avoiding the use of silicon-containing materials in FC designs, or using silicone materials that are highly stable under the aggressive chemical and electrochemical environment of the FC.

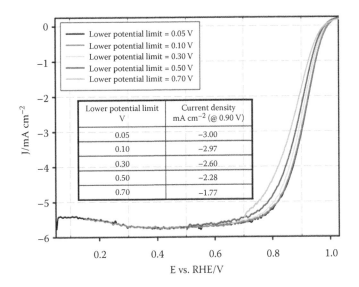

FIGURE 8.33 Oxygen reduction activity of a Pt/Vulcan carbon catalyst as measured by potential cycling by RDE methodology. The upper voltage limit is 1.03 V, and the lower voltage limit varies from 0.7 to 0.05 V. The current density for oxygen reduction at 0.9 V is highest when the electrode is cycled to a lower limit of 0.05 V, where H adsorption on the Pt electrode can displace OH_{ads} impurities from the Pt surface. Electrolyte: O_2-saturated 0.1 M $HClO_4$, 30°C, 1600 rpm, 20 mV s^{-1}. Electrode: 20 μg_{Pt} cm^{-2} of 20% Pt/Vulcan carbon (E-TEK).

8.6.1.3 Carbon Monoxide Removal and Tolerance

CO is easily oxidized on platinum, and the recovery of CO-poisoned anodes has been discussed in detail elsewhere (Gottesfeld, 1992). The CO may also be removed by fuel starvation methods discussed in the patent literature. Development of CO-tolerant electrocatalysts is also an active area of research (Huang et al., 2007). Operation of FCs at >120°C also improves their CO tolerance.

8.6.1.4 S^x Removal

The removal of sulfur species from platinum-based catalysts is a two-step process. Starting with the reaction in Equation 8.1, the S-species must first be oxidized to sulfate/bisulfate at around 1.2 V in the presence of humidified N_2. The resulting SO_4^{2-}/HSO_4^- species, although water soluble, remains on the Pt until the Pt is held at open-circuit voltage (OCV) in nitrogen (below the point-of-zero charge of the Pt), so that they can electrostatically desorb from the platinum surface (Baturina and Swider-Lyons, 2009). Figure 8.34 compares CV curves measured after "CV-cleaning" (dashed line) of the FC cathode to oxidize sulfur according to Equation 8.1 and after holding the cathode potential at OCV for 30 min (solid line) to desorb sulfate/bisulfate ions from the Pt surface. Desorption of SO_4^{2-}/HSO_4^- species is evidenced by the negative shift of the onset of OH_{ads} formation and changing the shapes of both hydrogen and oxygen anodic regions.

The most efficient way of S^x removal is potential cycling between the OCV and 1.2 V (vs. potential at the anode) when flowing hydrogen/nitrogen through the anode/cathode (Gould et al., 2010). Figure 8.35 shows that performance of the FC can be completely recovered after cycling cathode potential between the OCV and 1.2 V under hydrogen/nitrogen. Sulfur oxidation to sulfate is likely occur through the interaction with surface platinum oxide, as it was observed by Swider and Rolison (1996) for sulfur incorporated into Vulcan carbon supports.

We have also shown that S^x species are more easily desorbed from the surfaces of Pt_3Co alloy compared to Pt. We measured that Pt_3Co-based cathodes can be cleaned by running successive polarization curves in PEMFCs (Baturina et al., 2010; Garsany et al., 2007a, 2009a). Figure 8.36 presents a series of

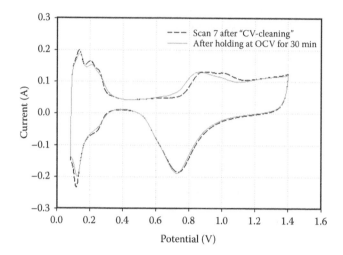

FIGURE 8.34 CV curves recorded at 50 mV/s after "CV-cleaning" of the fuel cell cathode contaminated with 1 ppm SO_2 for 3 h at 0.7 V (dashed line) and after sulfate/bisulfate removal by holding the cathode potential at OCV (62 mV) for 30 min (solid line). Cell and humidifier temperatures are 30°C and 50°C, respectively. The flow rate is 0.25/0.03 L/min (anode/cathode). (Reprinted from Baturina, O. A. and Swider-Lyons, K. E. 2009. *Journal of the Electrochemical Society* 156: B1423–B1430. With permission from The Electrochemical Society, Inc.)

successive polarization curves measured on contaminated Pt_3Co- (a) and Pt-based (b) cathodes. It is seen that more severely contaminated Pt_3Co-based cathodes are recovered more easily and more completely than Pt-based cathodes. The more facile oxidation of S^x species on Pt_3Co vs. Pt has been attributed to the difference in the adsorption potentials of oxygen species on the two materials, as the oxidation of Pt precedes S^x oxidation (Ramaker et al., 2010).

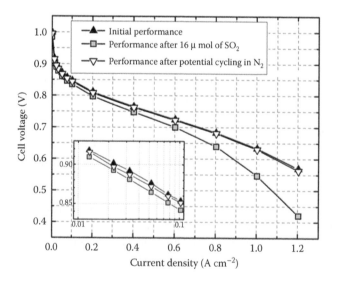

FIGURE 8.35 Polarization curves showing the PEMFC's initial performance, performance after exposure to SO_2, and after potential cycling. The inset depicts the kinetic region of the polarization curve in Tafel coordinates (cell voltage corrected for Ohmic loss vs. current density corrected for H_2 crossover current).

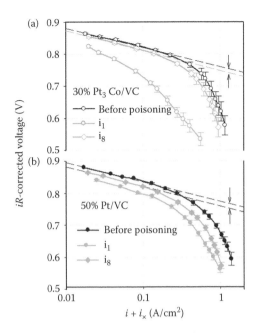

FIGURE 8.36 Polarization curves measured on 30 wt% Pt$_3$Co/VC (a, open symbols) and 50 wt% Pt/VC (b, solid symbols) before contamination with 1 ppm SO$_2$ (black circles), after contamination (gray circles) and after running eight successive polarization curves (gray diamonds). Platinum loading 0.4 mg$_{Pt}$ cm^{-2}, 80°C, 100% RH, $s = 2/2$. (Reprinted from Baturina, O. A. et al. 2010. *Electrochimica Acta* 55: 6676–6686. With permission.)

8.6.1.5 Alcohol and Hydrocarbon Removal

Alcohol oxidation from Pt is one of the most studied in FCs because it is the basis of the anode reaction in direct methanol FCs. Methanol is oxidized in a six-electron reaction to CO$_2$, H$^+$ and H$_2$O on nanoscale Pt at around 0.4 V to 0.5 vs RHE. Methanol is oxidized more efficiently at lower potentials on PtRu alloys and related materials. The mechanism of the oxidation process on alloy catalysts and the form of the catalysts has been the subject of hundreds of papers and is beyond the scope of this paper. When alcohol is present on a catalyst in low concentrations as an impurity, it can be oxidized by exposing the FC to open-circuit conditions, or implementing various air starvation and transient operation methods discussed in the patent literature cited above. Methanol oxidation to CO$_2$ and H$^+$ completes its removal from the catalyst surface. Methanol poisoning at the cathode can also be a problem at DMFC cathodes, as methanol can easily cross over the Nafion. If the FC becomes severely contaminated, even open-circuit potentials might be too low to fully oxidize the methanol, requiring aggressive procedures to oxidize the methanol from the catalysts.

Alcohols, such as ethylene glycol found in coolants, can be not only oxidized, but simply washed off the platinum surface. Ethylene glycol in 0.1 M HClO$_4$ electrolyte strongly modifies the CV characteristics of the Pt metal nanoparticles. However, when the "poisoned" electrode is transferred to a clean 0.1 M HClO$_4$ electrolyte (i.e., with no ethylene glycol), and the electrode is cycled between 0.05 V and 1.2 V (cleaning step-1 in Figure 8.37a) full recovery of the initial Pt surface area (as evidenced from the hydrogen adsorption/desorption regions) is observed. Figure 8.37b shows that surfactants and corrosion inhibitors are more difficult to remove from the Pt surface with potential cycling. When a Pt/VC electrocatalyst electrode is first cycled in a 0.1 M HClO$_4$ electrolyte containing ethylene glycol/surfactant/corrosion inhibitor mixture and then transferred to 0.1 M HClO4 electrolyte, 69% of the Pt surface area remain blocked after cleaning step-1 (cycling between 0.05 V and 1.2 V, 20 cycles). If a second cleaning step is applied (cycling between 0.05 V and 1.5 V, 20 cycles), 30% of the Pt surface area is not recovered with cycling to 1.5 V.

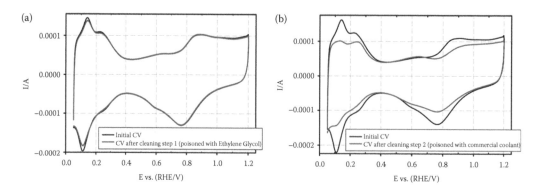

FIGURE 8.37 CV curves on Pt/VC thin-film electrodes showing cleaning of ethylene glycol and ethylene glycol containing a surfactant and a corrosion inhibitor. (a) The Pt/VC electrode is poisoned in 0.18 M ethylene glycol, and then cycled in clean 0.1 M $HClO_4$ at 20 mV s^{-1} between 0.05 and 1.2 V, 20 times. (b) The Pt/VC electrode is poisoned in 0.18 M ethylene glycol + surfactant, and then cycled in clean 0.1 M $HClO_4$ at 20 mV s^{-1} between 0 and 1.5 V, 20 times.

The adsorption of hydrocarbons and surfactants from the surface of carbon-based GDLs and bipolar plates may be irreversible because of the absence of a Pt (or related) catalyst to facilitate oxidation. More conventional clean-up methods would have to be used, such as chemical or thermal removal. While a layer of hydrocarbon or surfactant is unlikely to change conductivity properties, these adsorbates can modify hydrophilicity, and permanently affect water management, and so on, in the FC. The modification of GDL hydrophilicity has been observed for toluene, as discussed in Section 8.3.2.1.

8.6.2 Removal of Cations from Proton Exchange Membranes

It is common practice during MEA preparation to convert cation-containing PEMs to a pure proton form by boiling in concentrated sulfuric acid (Wilson et al., 1995). Imamura and Ohno (2009a) reported that flushing Na$^+$-contaminated MEAs with H_2SO_4 increased their release rate of Na$^+$; however, a sulfuric acid flush may not be practical to implement for commercially deployed PEMFCs. One mitigation strategy is to incorporate ion exchange media in porous diffusion media to prevent cation poisoning of the Nafion membranes (Abd Elhamid et al., 2006).

PEMs can be irreversibly poisoned by peroxide radicals that cause breakdown of the PFSA backbone. The issue of peroxide degradation is now being avoided through the implementation of Ce^{4+} and Mn^{3+}, which scavenge peroxide radicals (Trogadas et al., 2008).

8.7 Summary

Owing to a joint effort of different research groups during last 3–4 years some aspects of the influence of impurities on the PEMFC performance degradation are now better understood. These recent years have been marked by thorough characterization of inorganic and organic impurities by detailed electrochemical analysis and advanced characterization techniques complemented by significant progress in modeling. However, the ultimate goal of studying contaminants in PEMFCs is to develop recovery strategies or find materials that are not susceptive to poisoning. With respect to achieving this ultimate goal, more progress has been made in some areas compared to others. For example, it is certain now that membrane decomposition caused by hydrogen peroxide can be avoided by tuning Nafion properties (i.e., incorporating Ce^{3+} or Mn^{4+} compounds inside the Nafion membrane; Trogadas et al., 2008). High-performance platinum alloys were found to recover more easily from poisoning than traditional Pt catalysts (Garsany et al., 2007a, 2009a; Baturina et al., 2010). SO$_2$ can be fully removed from Pt cathodes

(Gould et al., 2010). Advanced FCs, such as high-temperature PEMFCs, using PBI membranes show promise toward being resistant to contamination (Schmidt and Baurmeister, 2006; Das et al., 2009; Garsany et al., 2009b).

However, it is not certain, for example, how to deal with impurities such as chloride in terms of recovery, even though it is well understood which particular part of catalyst-coated membrane requires attention.

Developed modeling approaches allowed the authors to describe some features of experimental curves caused by introduction of inorganic cations, organic and inorganic molecules both at the anode and the cathode of PEMFC (Cheng et al., 2007; Kienitz et al., 2009; St-Pierre, 2009; Zhang et al., 2009). However, due to complexity of the system, predictive ability of these models is still low.

In summary, as more knowledge is gained toward fundamental understanding of different aspects of the influence of impurities in PEMFCs, the degradation processes due to contaminants appear to be solvable by mitigation and recovery methods.

Acknowledgments

The authors are grateful to Office of Naval Research for financial support of their work.

References

Abd Elhamid, M. H., Mikhail, Y. M., and Blunk, R. H. 2006. *Devices incorporating porous diffusion media and bipolar plate assembly with anion exchange resin.* US Patent 2006029859.

Abdou, M., Xie, T., and Andrin, P. 2004. *Method for regeneration of performance in a fuel cell.* WO Patent 2004030118.

Adams, W. A., Gardner, C. L., Dunn, J. H., and Vered, R. 2003. *Fuel cell operational health management system.* WO Patent 2003083975.

Ahn, S. Y., Shin, S. J., Ha, H. Y. et al. 2002. Performance and lifetime analysis of the kW-class PEMFC stack. *Journal of Power Sources* 106: 295–303.

Al-Saleh, M. A., Gultekin, S., Al-Zakri, A. S., and Celiker, H. 1994. Effect of carbon dioxide on the performance of Ni/PTFE and Ag/PTFE electrodes in an alkaline fuel cell. *Journal of Applied Electrochemistry* 24: 575–580.

Baturina, O. A., Gould, B. D., Garsany, Y., and Swider-Lyons, K. E. 2010. Insights on SO_2 poisoning of Pt_3Co/VC and Pt/VC catalysts. *Electrochimica Acta* 55: 6676–6686.

Baturina, O. A. and Swider-Lyons, K. E. 2009. Effect of SO_2 on the performance of the cathode of a PEM fuel cell at 0.5–0.7 V. *Journal of the Electrochemical Society* 156: B1423–B1430.

Bender, G., Angelo, M., Bethune, K., Dorn, S., Thampan, T., and Rocheleau, R. 2009. Method using gas chromatography to determine the molar flow balance for proton exchange membrane fuel cells exposed to impurities. *Journal of Power Sources* 193: 713–722.

Bhatia, K. K. and Wang, C.-Y. 2004. Transient carbon monoxide poisoning of a polymer electrolyte fuel cell operating on diluted hydrogen feed. *Electrochimica Acta* 49: 2333–2341.

Borup, R., Meyers, J., Pivovar, B. et al. 2007. Scientific aspects of polymer electrolyte fuel cell durability and degradation. *Chemical Reviews* 107: 3904–3951.

Budinski, K. G. and Budinski, M. K. 2005. *Engineering Materials: Properties and Selection.* Upper Saddle River, NJ: Prentice-Hall.

Chan, S. H., Goh, S. K., and Jiang, S. P. 2003. A mathematical model of polymer electrolyte fuel cell with anode CO kinetics. *Electrochimica Acta* 48: 1905–1919.

Cheng, X., Shi, Z., Glass, N. et al. 2007. A review of PEM hydrogen fuel cell contamination: Impacts, mechanisms, and mitigation. *Journal of Power Sources* 165: 739–756.

Colbow, K. M., Van Der Geest, M., Longley, C. J. et al. 2001. *Method and apparatus for operating an electrochemical fuel cell with periodic reactant starvation.* WO Patent 2001001508.

Contractor, A. Q. and Lal, H. 1978. Two forms of chemisorbed sulfur on platinum and related studies. *Journal of Electroanalytical Chemistry and Interfacial Electrochemistry* 96: 175–181.

Das, S. K., Reis, A., and Berry, K. J. 2009. Experimental evaluation of CO poisoning on the performance of a high temperature proton exchange membrane fuel cell. *Journal of Power Sources* 193: 691–698.

Fenton, H. J. H. 1894. Oxidation of tartaric acid in presence of iron. *Journal of the Chemical Society, Transactions* 65: 899–910.

Franco, A. A., Barthe, B., Rouillon, L., and Lemaire, O. 2009. Mechanistic investigations of NO_2 impact on ORR in PEM fuel cells: A coupled experimental and multi-scale modeling approach. *ECS Transactions* 25: 1595–1604.

Garsany, Y., Baturina, O. A., and Swider-Lyons, K. E. 2007a. Impact of SO_2 on the kinetics of Pt_3Co/Vulcan carbon electrocatalysts for oxygen reduction. *ECS Transactions* 11: 863–875.

Garsany, Y., Baturina, O. A., and Swider-Lyons, K. E. 2007b. Impact of sulfur dioxide on the oxygen reduction reaction at Pt/Vulcan carbon electrocatalysts. *Journal of the Electrochemical Society* 154: B670–B675.

Garsany, Y., Baturina, O. A., and Swider-Lyons, K. E. 2009a. Oxygen reduction reaction kinetics of SO_2-contaminated Pt3Co and Pt/Vulcan carbon electrocatalysts. *Journal of the Electrochemical Society* 156: B848–B855.

Garsany, Y., Baturina, O. A., Swider-Lyons, K. E., and Kocha, S. S. 2010. Experimental methods for quantifying the activity of platinum electrocatalysts for the oxygen reduction reaction. *Analytical Chemistry* 82: 6321–6328.

Garsany, Y., Gould, B. D., Baturina, O. A., and Swider-Lyons, K. E. 2009b. Comparison of the sulfur poisoning of PBI and Nafion PEMFC cathodes. *Electrochemical and Solid-State Letters* 12: B138–B140.

Garzon, F. H., Lopes, T., Rockward, T. et al. 2009. The impact of impurities on long term PEMFC performance. *ECS Transactions* 25: 1575–1583.

Garzon, F. H., Rockward, T., Urdampilleta, I. G., Brosha, E. L., and Uribe, F. A. 2006. The impact of hydrogen fuel contaminates on long-term PEMFC performance. *ECS Transactions* 3: 695–703.

Gasteiger, H. A., Kocha, S. S., Sompalli, B., and Wagner, F. T. 2005. Activity benchmarks and requirements for Pt, Pt-alloy, and non-Pt oxygen reduction catalysts for PEMFCs. *Applied Catalysis B-Environmental* 56: 9–35.

Gottesfeld, S. 1992. *Prevention of carbon monoxide poisoning in fuel cells*. US Patent 279694.

Gould, B. D., Baturina, O. A., and Swider-Lyons, K. E. 2009. Deactivation of Pt/VC proton exchange membrane fuel cell cathodes by SO_2, H_2S and COS. *Journal of Power Sources* 188: 89–95.

Gould, B. D., Bender, G., Bethune, K. et al. 2010. Operational performance recovery of SO_2-contaminated proton exchange membrane fuel cells. *Journal of the Electrochemical Society* 157: B1569–B1577.

Gould, B. D., Chen, X., and Schwank, J. W. 2007. Dodecane reforming over nickel-based monolith catalysts. *Journal of Catalysis* 250: 209–221.

Halseid, R., Bystron, T., and Tunold, R. 2006a. Oxygen reduction on platinum in aqueous sulphuric acid in the presence of ammonium. *Electrochimica Acta* 51: 2737–2742.

Halseid, R., Heinen, M., Jusys, Z., and Behm, R. J. 2008. The effect of ammonium ions on oxygen reduction and hydrogen peroxide formation on polycrystalline Pt electrodes. *Journal of Power Sources* 176: 435–443.

Halseid, R., Vie, P. J. S., and Tunold, R. 2006b. Effect of ammonia on the performance of polymer electrolyte membrane fuel cells. *Journal of Power Sources* 154: 343–350.

Huang, C. Y., Chen, Y. Y., Su, C. C., and Hsu, C. F. 2007. The cleanup of CO in hydrogen for PEMFC applications using Pt, Ru, Co, and Fe in PROX reaction. *Journal of Power Sources* 174: 294–301.

Huang, J. C., Sen, R. K., and Yeager, E. 1979. Oxygen reduction on platinum in 85% orthophosphoric acid. *Journal of the Electrochemical Society* 126: 786–792.

Imamura, D. and Hashimasa, Y. 2007. Effect of sulfur-containing compounds on fuel cell performance. *ECS Transactions* 11: 853–862.

Imamura, D. and Ohno, K. 2009a. Impact of Na$^+$ and Cl$^-$ on degradation of PEMFC. *2009 Fuel Cell Seminar Extended Abstracts*: LRD 25–49.

Imamura, D. and Yamaguchi, E. 2009b. Effect of air contaminants on electrolyte degradation in polymer electrolyte membrane fuel cells. *ECS Transactions* 25: 813–819.

Inaba, M., Kinumoto, T., Kiriake, M. et al. 2006. Gas crossover and membrane degradation in polymer electrolyte fuel cells. *Electrochimica Acta* 51: 5746–5753.

Jing, F. N., Hou, M., Shi, W. Y., Fu, J., Yu, H. M., Ming, P. W., and Yi, B. L. 2007. The effect of ambient contamination on PEMFC performance. *Journal of Power Sources* 166: 172–176.

Kienitz, B. L., Baskaran, H., and Zawodzinski, T. A. 2009. Modeling the steady-state effects of cationic contamination on polymer electrolyte membranes. *Electrochimica Acta* 54: 1671–1679.

La Conti, A. B., Hamdan, M., and Mcdonald, R. C. 2003. Mechanisms of membrane degradation. *In: Handbook of Fuel Cells—Fundamentals, Technology and Application*, eds. Vielstich, V., Lamm, A. and Gasteiger, H. A., pp. 647–662. New York, NY: John Wiley & Sons.

Li, H., Zhang, J. L., Fatih, K. et al. 2008. Polymer electrolyte membrane fuel cell contamination: Testing and diagnosis of toluene-induced cathode degradation. *Journal of Power Sources* 185: 272–279.

Li, H., Zhang, J. L., Shi, Z. et al. 2009. PEM fuel cell contamination: Effects of operating conditions on toluene-induced cathode degradation. *Journal of the Electrochemical Society* 156: B252–B257.

Li, Q., He, R., Jensen, J. O., and Bjerrum, N. J. 2004. PBI-based polymer membranes for high temperature fuel cells—Preparation, characterization and fuel cell demonstration. *Fuel Cells* 4: 147–159.

Li, Q. F., He, R. H., Gao, J. A., Jensen, J. O., and Bjerrum, N. J. 2003. The CO poisoning effect in PEMFCs operational at temperatures up to 200 degrees C. *Journal of the Electrochemical Society* 150: A1599–A1605.

Loucka, T. 1971. Adsorption and oxidation of sulphur and of sulphur dioxide at platinum electrode. *Journal of Electroanalytical Chemistry* 31: 319–332.

Manasilp, A. and Gulari, E. 2002. Selective CO oxidation over Pt/alumina catalysts for fuel cell applications. *Applied Catalysis B: Environmental* 37: 17–25.

Matsuoka, K., Sakamoto, S., Nakato, K., Hamada, A., and Itoh, Y. 2008. Degradation of polymer electrolyte fuel cells under the existence of anion species. *Journal of Power Sources* 179: 560–565.

Mecerreyes, D., Grande, H., Miguel, O. et al. 2004. Porous polybenzimidazole membranes doped with phosphoric acid: Highly proton-conducting solid electrolytes. *Chemistry of Materials* 16: 604–607.

Mikkola, M. S., Rockward, T., Uribe, F. A., and Pivovar, B. S. 2007. The effect of NaCl in the cathode air stream on PEMFC performance. *Fuel Cells* 7: 153–158.

Mohtadi, R., Lee, W. K., and Van Zee, J. W. 2004. Assessing durability of cathodes exposed to common air impurities. *Journal of Power Sources* 138: 216–225.

Moore, J. M., Adcock, P. L., Lakeman, J. B., and Mepsted, G. O. 2000. The effects of battlefield contaminants on PEMFC performance. *Journal of Power Sources* 85: 254–260.

Nagahara, Y., Sugawara, S., and Shinohara, K. 2008. The impact of air contaminants on PEMFC performance and durability. *Journal of Power Sources* 182: 422–428.

Narusawa, K., Hayashida, M., Kamiya, Y. et al. 2003. Deterioration in fuel cell performance resulting from hydrogen fuel containing impurities: Poisoning effects by CO, CH4, HCHO and HCOOH. *JSAE Review* 24: 41–46.

O'leary, K., Budinski, M., and Lakshmanan, B. 2009. Methodologies for evaluating automative system contaminants. In *Canada-USA PEM Network Research Workshop*, Vancouver, Canada.

Okada, T. 1999a. Theory for water management in membranes for polymer electrolyte fuel cells—Part 1. The effect of impurity ions at the anode side on the membrane performances. *Journal of Electroanalytical Chemistry* 465: 1–17.

Okada, T. 1999b. Theory for water management in membranes for polymer electrolyte fuel cells—Part 2. The effect of impurity ions at the cathode side on the membrane performances. *Journal of Electroanalytical Chemistry* 465: 18–29.

Okada, T., Satou, H., Okuno, M., and Yuasa, M. 2002. Ion and water transport characteristics of perfluo-rosulfonated ionomer membranes with H+ and alkali metal cations. *Journal of Physical Chemistry B* 106: 1267–1273.

Palm, C., Cremer, P., Peters, R., and Stolten, D. 2002. Small-scale testing of a precious metal catalyst in the autothermal reforming of various hydrocarbon feeds. *Journal of Power Sources* 106: 231–237.

Paulus, U. A., Schmidt, T. J., Gasteiger, H. A., and Behm, R. J. 2001. Oxygen reduction on a high-surface area Pt/Vulcan carbon catalyst: A thin-film rotating ring-disk electrode study. *Journal of Electroanalytical Chemistry* 495: 134–145.

Pomfret, M. B., Pietron, J. J., and Owrutsky, J. C. 2010. Measurement of benzenethiol adsorption to nano-structured Pt, Pd, and PtPd films using Raman spectroelectrochemistry. *Langmuir* 26: 6809–6817.

Pozio, A., Silva, R. F., De Francesco, M., and Giorgi, L. 2003. Nafion degradation in PEFCs from end plate iron contamination. *Electrochimica Acta* 48: 1543–1549.

Ramaker, D. E., Gatewood, D., Korovina, A., Garsany, Y., and Swider-Lyons, K. E. 2010. Resolving sulfur oxidation and removal from Pt and Pt₃Co electrocatalysts using *in situ* x-ray absorption spectros-copy. *Journal of Physical Chemistry C* 114: 11886–11897.

Reiser, C. A. and Balliet, R. J. 2003. *Fuel cell performance recovery by cyclic oxidant starvation.* US Patent 2003224228.

Reshetenko, T., Bethune, K., and Rocheleau, R. 2010. Study of spatial PEMFC performance under CO poisoning using segmented cell approach. In *217 Electrochemical Society Meeting*, Vancouver, Canada, pp. B7–B621.

Savadogo, O. 2004. Emerging membranes for electrochemical systems—Part II. High temperature com-posite membranes for polymer electrolyte fuel cell (PEFC) applications. *Journal of Power Sources* 127: 135–161.

Schmidt, T. J. and Baurmeister, J. 2006. Durability and reliability in high-temperature reformed hydrogen PEFCs. *ECS Transactions* 3: 861–869.

Schmittinger, W. and Vahidi, A. 2008. A review of the main parameters influencing long-term perfor-mance and durability of PEM fuel cells. *Journal of Power Sources* 180: 1–14.

Schulze, M., Knori, T., Schneider, A., and Gulzow, E. 2004. Degradation of sealings for PEFC test cells during fuel cell operation. *Journal of Power Sources* 127: 222–229.

Serincan, M. F., Pasaogullari, U., and Molter, T. 2010. Modeling the cation transport in an operating poly-mer electrolyte fuel cell (PEFC). *International Journal of Hydrogen Energy* 35: 5539–5551.

Shah, A. A. and Walsh, F. C. 2008. A model for hydrogen sulfide poisoning in proton exchange membrane fuel cells. *Journal of Power Sources* 185: 287–301.

Shao, Z. G., Joghee, P., and Hsing, I. M. 2004. Preparation and characterization of hybrid Nafion-silica membrane doped with phosphotungstic acid for high temperature operation of proton exchange membrane fuel cells. *Journal of Membrane Science* 229: 43–51.

Shi, W., Yi, B., Hou, M., and Shao, Z. 2007a. The effect of H₂S and CO mixtures on PEMFC performance. *International Journal of Hydrogen Energy* 32: 4412–4417.

Shi, Z., Song, D., Zhang, J. et al. 2007b. Transient analysis of hydrogen sulfide contamination on the per-formance of a PEM fuel cell. *Journal of the Electrochemical Society* 154: B609–B615.

Shi, Z., Song, D. T., Li, H. et al. 2009. A general model for air-side proton exchange membrane fuel cell contamination. *Journal of Power Sources* 186: 435–445.

Soto, H. J., Lee, W. K., Van Zee, J. W., and Murthy, M. 2003. Effect of transient ammonia concentrations on PEMFC performance. *Electrochemical and Solid State Letters* 6: A133–A135.

Springer, T. E., Rockward, T., Zawodzinski, T. A., and Gottesfeld, S. 2001. Model for polymer electrolyte fuel cell operation on reformate feed effects of CO, H₂ dilution, and high fuel utilization. *Journal of the Electrochemical Society* 148: A11–A23.

St-Pierre, J. 2009. PEMFC contamination model: Competitive adsorption followed by an electrochemical reaction. *Journal of the Electrochemical Society* 156: B291–B300.

Stamenkovic, V. R., Mun, B. S., Arenz, M. et al. 2007. Trends in electrocatalysis on extended and nanoscale Pt-bimetallic alloy surfaces. *Nat. Mater. FIELD Full Journal Title:Nature Materials* 6: 241–247.

Steinbach, A. J., Hamilton, C. V., Jr., and Debe, M. K. 2007. Impact of micromolar concentrations of externally-provided chloride and sulfide contaminants on PEMFC reversible stability. *ECS Transactions* 11: 889–902.

Sulek, M. S., Mueller, S. A., and Paik, C. H. 2008. Impact of Pt and Pt-alloy catalysts on membrane life in PEMFCs. *Electrochemical and Solid State Letters* 11: B79–B82.

Suntivich, J., Gasteiger, H. A., Yabuuchi, N., and Yang, S.-H. 2010. Electrocatalytic measurement methodology of oxide catalysts using a thin-film rotating disk electrode. *Journal of the Electrochemical Society* 157: B1263–B1268.

Suresh, B., Yoneyama, M., and Schlag, S. 2007. Hydrogen. In *Chemical Economics Handbook.* Menlo Park: SRI Consulting.

Swider, K. E. and Rolison, D. R. 1996. The chemical state of sulfur in carbon-supported fuel cell electrodes. *Journal of the Electrochemical Society* 143: 813–819.

Tewari, A., Sambhy, V., Macdonald, M. U., and Sen, A. 2006. Quantification of carbon dioxide poisoning in air breathing alkaline fuel cells. *Journal of Power Sources* 153: 1–10.

Trogadas, P., Parrondo, J., and Ramani, V. 2008. Degradation mitigation in polymer electrolyte membranes using cerium oxide as a regenerative free-radical scavenger. *Electrochemical and Solid State Letters* 11: B113–B116.

Uribe, F. A., Gottesfeld, S., and Zawodzinski Jr., T. A. 2002. Effect of ammonia as potential fuel impurity on proton exchange membrane fuel cell performance. *Journal of the Electrochemical Society* 149: A293–A296.

Uribe, F. A. and Rockward, T. Q. T. 2006. *Cleaning (de-poisoning) PEMFC electrodes from strongly adsorbed species on the catalyst surface.* US Patent 2006249399.

Varcoe, J. R. and Slade, R. C. T. 2005. Prospects for alkaline anion-exchange membranes in low temperature fuel cells. *Fuel Cells* 5: 187–200.

Vega, J. A. and Mustain, W. E. 2010. Effect of CO_2, HCO_3^- and CO_3^{2-} on oxygen reduction in anion exchange membrane fuel cells. *Electrochimica Acta* 55: 1638–1644.

Wilson, M. S., Valerio, J. A., and Gottesfeld, S. 1995. Low platinum loading electrodes for polymer electrolyte fuel-cells fabricated using thermoplastic ionomers. *Electrochimica Acta* 40: 355–363.

Yadav, A. P., Nishikata, A., and Tsuru, T. 2007. Effect of halogen ions on platinum dissolution under potential cycling in 0.5 M H_2SO_4 solution. *Electrochimica Acta* 52: 7444–7452.

Yang, C., Costamagna, P., Srinivasan, S., Benziger, J., and Bocarsly, A. B. 2001. Approaches and technical challenges to high temperature operation of proton exchange membrane fuel cells. *Journal of Power Sources* 103: 1–9.

Yang, D. J., Ma, J. X., Xu, L., Wu, M. Z., and Wang, H. J. 2006. The effect of nitrogen oxides in air on the performance of proton exchange membrane fuel cell. *Electrochimica Acta* 51: 4039–4044.

Zhai, Y., Bender, G., Dorn, S., and Rocheleau, R. 2010. The multiprocess degradation of PEMFC performance due to sulfur dioxide contamination and its recovery. *Journal of the Electrochemical Society* 157: B20–B26.

Zhang, J., Wang, H., Wilkinson, D. P. et al. 2005. Model for the contamination of fuel cell anode catalyst in the presence of fuel stream impurities. *Journal of Power Sources* 147: 58–71.

Zhang, X., Pasaogullari, U., and Molter, T. 2009. Influence of ammonia on membrane-electrode assemblies in polymer electrolyte fuel cells. *International Journal of Hydrogen Energy* 34: 9188–9194.

Zhou, J. F., Unlu, M., Vega, J. A., and Kohl, P. A. 2009. Anionic polysulfone ionomers and membranes containing fluorenyl groups for anionic fuel cells. *Journal of Power Sources* 190: 285–292.

9

Environment-Induced Degradation

Eric Pinton
Commissariat à l'Energie
Atomique et aux Energies
Alternatives

Sébastien Rosini
Commissariat à l'Energie
Atomique et aux Energies
Alternatives

Laurent Antoni
Commissariat à l'Energie
Atomique et aux Energies
Alternatives

9.1 Introduction

Proton exchange membrane fuel cell (PEMFC) is one of the most promising technologies to produce clean electricity with only water as the exhaust product. However, water management gives rise to technical complications, especially for applications where winter operating conditions are required. The U.S. Department of Energy (DOE) defined cold-start targets for transportation fuel cell stacks: 50% of the rated power must be achieved in less than 30 s under an ambient temperature of –20°C. Unfortunately, self-start-up in these conditions with standard shutdown and start-up procedures cannot be achieved. It is commonly described in the literature that the fuel cell (FC) shutdown during start-up under subfreezing conditions is due to ice formed in the cathode catalyst layer (CCL) by the oxygen reduction reaction (ORR), thus hindering oxygen transport. This ice and the one formed during freeze–thaw cycles may entail structural damages to the membrane electrode assembly (MEA). The DOE specification in terms of durability requires a FC lifetime of 5000 h.

The next two sections of this chapter deal with the impacts of freezing and cold start on PEMFCs. For each section fundamental mechanisms, effects on performances and mitigation strategies are successively presented.

The following analyses concern, unless otherwise specified, perfluorosulfonated membranes, which represent worldwide the main current commercialized type of membrane.

9.2 Freezing

9.2.1 Fundamental Mechanisms

Comprehension of water-freezing mechanisms occurring inside a MEA is fundamental to understanding potential damages produced by a subzero environment.

Water state and water management inside a membrane during freeze–thaw process begin to be well known. Using a micro x-ray beam, the structure of a water swollen Nafion® membrane, alone or in a MEA designed for FC, was studied on cooling down to –70°C (Pineri, 2007). By scanning the membranes along their thicknesses, the water sorption–desorption process was investigated as a function of cooling/heating stages. From the scattering curves, it was deduced that the water inside the membrane at a subzero temperature is at a liquid state, commonly called "supercooled state." Ice crystals were only observed outside the membrane. In the case of the MEA, ice formation can be destructive since it is localized inside the active catalyst layers.

Experiments performed by nuclear magnetic resonance (NMR) (Guillermo, 2009) confirm observations obtained by x-ray diffraction (Pineri, 2007). The nuclear magnetic relaxation properties of liquid water and solid ice are different enough to easily discriminate between the NMR signals of these two water phases. The investigations were related to water concentration and to water mobility measurements inside a hydrated Nafion membrane emplaced in an airtight glass capillary when the temperature decreases slowly from the room temperature down to –53°C. Such a protocol was chosen to be the most closely related to the standard usage in transportation application when the PEMFC has to be turned off and stored at low temperatures. The initial water content at room temperature ranged from 8% to 20.5% in Nafion + water weight, corresponding to water content "λ" (λ = moles H_2O/mole SO_3^-) from 5.4 to 15.7. The change of water content in these conditions is displayed in Figure 9.1. The relation between water content and water concentration "C," defined as the mass of water divided by the mass of hydrated membrane, is

$$C = \frac{1}{\left(EW/18\lambda + 1\right)} \tag{9.1}$$

where EW is the molecular equivalent weight of the membrane (mass of dry membrane per mol of sulfonic acid SO_3^-: typically $EW = 1100$ g mol^{-1} for Nafion).

The continuous line in Figure 9.1 shows that a perfluorosulfonated membrane has a maximal water content "$\lambda_{max}(T)$" depending on the subzero temperature, which decreases with temperature drop. $\lambda_{max}(T)$ denotes the equilibrium swelling. It means that, if the water content at room temperature

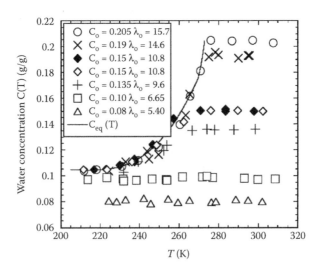

FIGURE 9.1 Water content change of Nafion 112 as a function of the ambient temperature at different initial λ_0. (Reprinted with permission from Guillermo, A. et al. 2009. NMR and pulsed field gradient NMR approach of water sorption properties in Nafion at low temperature. *J. Phys. Chem.* B 113: 6710–6717. Copyright (2009). American Chemical Society Publications.)

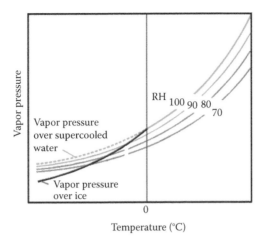

FIGURE 9.2 Schematic phase diagram of water (Data from Padfield, T. 1999, What happens to water absorbent materials below freezing? http://www.natmus.dk/cons/tp/cool/suprcool.htm).

$\lambda_0(20°C)$ is equal to $\lambda_{max}(-10°C)$, for instance, no desorption will appear between 20°C and –10°C. However, if the temperature falls below –10°C, for example, up to –20°C, water desorption occurs at the membrane interface: $\Delta\lambda_{desorb} = \lambda_{max}(-10°C) - \lambda_{max}(-20°C)$. Ice crystal formation was observed preferentially on the glass capillary surfaces because of cooler surfaces of the experimental setup. So, the simultaneous presence of supercooled water inside the membrane and of ice outside the membrane suggests an analysis in connection with water absorbent materials below freezing (Padfield, 1999). In this configuration, relative humidity (RH) is defined as the saturated water vapor pressure in atmosphere divided by the saturating vapor pressure over supercooled water.

$$RH_{<0°C}(T) = \frac{Vp_{sat,ice}(T)}{Vp_{sat,supercooled}(T)} \tag{9.2}$$

The Schematic phase diagram of water is represented in Figure 9.2. The solid line at RH = 100% is the saturated vapor pressure at positive temperature. This is extended below 0°C as a dotted line representing the vapor pressure over supercooled water at subzero temperature. The thick solid line below is the vapor pressure over ice. The diagram is distorted to exaggerate the divergence of the line for ice.

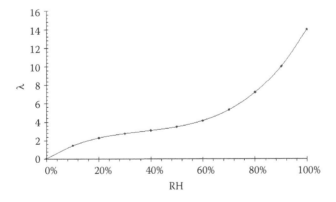

FIGURE 9.3 Evaluation of the membrane water content λ versus RH at 30°C from Springer et al. relation. (From Springer, T. E., Zawodzinski, T. A., and Gottesfeld, S. 1991. *J. Electrochem. Soc.* 138: 2334–2341. With permission from Electrochemical Society.)

It is now well known that the water uptake λ in the membrane is directly and mainly dependent on RH. As illustrated in Figure 9.3, λ decreases with RH in order to maintain the thermodynamic equilibrium at the membrane interface. Consequently, once the vapor pressure in atmosphere reaches the vapor pressure over ice as temperature decreases, the RH (Equation 9.2) also decreases with decreasing temperature. So the membrane desorbs in order to maintain the thermodynamic equilibrium at the supercooled membrane interface.

Figure 9.1 also highlights a common limit value of the water uptake around $\lambda_{min,desorp} = 6.7$, where no desorption occurs at room temperature regardless of the water concentration. This value is slightly higher than those obtained in the literature where $\lambda_{min,desorp}$ is found between 5 and 6 (Pineri, 1995; Thompson, 2006).

Finally, the desorption and sorption phenomena during cooling and heating process, respectively, have been observed as quite reversible (Thompson, 2006; Guillermo, 2009). It means that ice sublimates when temperature rises and the sublimated vapor is reabsorbed by the membrane.

9.2.2 Effects of Freeze Thaw Cycles

Although substantial research efforts have been made in the last 5 years with regard to the effects of freezing–thaw cycles on PEM fuel cell degradation, results from the literature remain still scarce and are conflicting for some of them. Conflicting and discrepant results may be ascribed to differences in the properties of the catalyst layer microstructure (such as porosity, thickness, composition, hydrophobicity, etc.), manufacturing processes of membrane electrode assemblies (MEAs) (catalyst coated on the membrane (CCM), or catalyst coated on the backing) and testing procedures. Nevertheless, most researchers converge toward relatively close observations even though some of them are still uncertain.

It is commonly accepted that freezing–thaw cycles of a fully hydrated MEA leads to performance losses at rated conditions. Results from polarization curves (Figure 9.4) indicate that voltage degradations become significant for high current densities (Hou, 2007; Alink, 2008; Pinton, 2009a; Song, 2009; Luo, 2010; Park, 2010). The order of magnitude of voltage loss is generally observed around 0.2 mV cycle^{-1} at 0.8 A cm^{-2}, namely 0.03% cycle^{-1}.

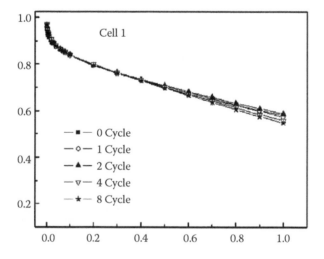

FIGURE 9.4 Performance evolution after eight freeze/thaw cycles of a fully hydrated MEA. (With kind permission from Springer Science+Business Media: *J. Applied Electrochemistry*, Sub-freezing endurance of PEM fuel cells with different catalyst coated membranes, 39, 2009, 609–615, Song, W. et al.)

FIGURE 9.5 Effects of freezing–thaw cycles on the catalyst layer surface. (a) Fresh catalyst surface, (b) catalyst surface after freeze/thaw cycles. (Reprinted from *J. Power Sources,* 160, Guo, Q. and Qi, Z., Effect of freeze–thaw cycles on the properties and performance of membrane–electrode assemblies, 1269–1274, Copyright (2006), with permission from Elsevier.)

Electrochemical impedance spectroscopy (EIS) performed at a nominal point evidenced that increases in gas diffusion limitation toward the catalytic site and activation limitation were, most of the time, preponderant to performance loss (Hou, 2007; Song, 2009; Luo, 2010; Park, 2010). However, although an active area decrease is systematically observed by cyclic voltammetry (CV) at active layers, preponderance of diffusion and activation limitations is not clearly identified. According to the literature, degradations related to freezing–thaw cycles can arise from two phenomena:

1. Structure alterations induced by ice expansion:

   As explained previously, the water inside the membrane at a subzero temperature is at a liquid state and water desorption during subzero cooling occurs at the membrane interface because of additional ice formation inside the active layers. So, it is suggested that the freezing water would compress pores of electrode layers (Hou, 2007; Song, 2009). It was also observed by scanning electron microscopy (SEM) that ice would develop preferentially in the surface defects of catalyst layers (CL) (Figure 9.5), like cracks, leading to crack network extending (Guo, 2006; Kim, 2007). Finally, some authors also observed the delamination of MEA layers (Figure 9.6) (Kim, 2007; Park, 2010). Whatever it can be, majority of the literature agreed that structure modifications

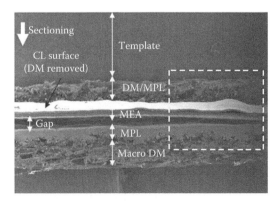

FIGURE 9.6 MEA with layer delamination after freeze–thaw cycles. (Reprinted from *J. Power Sources*, 174, Kim, S. and Mench, M. M., Physical degradation of membrane electrode assemblies undergoing freeze/thaw cycling: Micro-structure effects, 206–220, Copyright (2007), with permission from Elsevier.)

inside CL induced by ice led to irreversible performance losses, but the membrane structure remained unchanged by freeze–thaw cycles (Pineri, 2007).

2. Formation of a water film:

   Appearance of a film of water after ice melting plugged the pores of the CL (Hou, 2007). A similar interpretation is proposed that water is trapped between particles and ionomer of the CCL as a result of ice melting occurred during cold start (Ge and Wand, 2007a). Whatever it could be, both groups agreed that the residual water led to reversible performance loss.

Structure damages induced by freezing–thaw cycles inevitably have an impact on MEA durability. Unfortunately, no papers in the open literature proceeded to investigate the impact of structure damages induced by freezing–thaw cycles on fuel cell durability.

9.2.3 Mitigation Strategies

The most obvious strategy to mitigate the effects of freezing–thaw cycles on fuel cell durability is to remove water from the cells before the temperature falls to subzero. There has been practically an agreement on the fact that MEA drying prior to freezing conditions does not entail performance loss and structure damages (Guo, 2006; Hou, 2006; Song, 2009; Luo, 2010; Park, 2010). However, the drying level must be controlled to ensure that no water desorption from the membrane will occur during the cooling step, as explained in Section 9.2.1. A patent on a method for storing a fuel cell proposed a solution to control and optimize the drying level of the fuel cell (Pinton, 2009b). It included a first step of calibration from a standard membrane in order to obtain a profile of the maximal water content versus subzero storage temperatures, and a second step of calibration in order to obtain a relation between the ohmic resistance of a standard cell and the water content of the membrane at subfreezing temperatures. The method furthered a drying phase that depends on the two calibration steps.

It was nevertheless observed that drying method could possibly lead to performance drop at rated conditions (Pinton, 2009a; Luo, 2010). Consequently, if the shutdown strategy procedure includes a drying stage, it will have to be chosen carefully using the following options:

- Drying by vacuum
- Drying with dry gases with or without current
- Drying at the anode and cathode side at the same time
- Drying with nonsaturated wet gases
- Drying at ambient temperature, and so on

Another solution to mitigate freeze impact is to act on MEA structure. It is recommended to use CL with flawless surface conditions and especially with no initial cracks (Kim, 2007) in order to avoid favorable sites for ice formation. Moreover, it is suggested to use CLs rather hydrophobic than hydrophilic (Song, 2009).

9.3 Cold Start and Shutdown

9.3.1 Fundamental Mechanisms

As explained in Section 9.2.3, to avoid water desorption of the membrane at subzero temperatures which induces ice formation in the MEA layers, it is highly recommended to perform FC drying during the shutdown process. It is commonly admitted that the membrane conductivity at upzero temperatures is strongly dependent on the membrane water content and on temperature. This also applies to subzero temperatures as shown in Figure 9.7 that presents conductance of a Nafion 117 membrane during cooling stages from room to subfreezing temperatures and for different initial water contents at

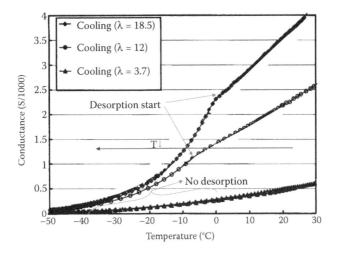

FIGURE 9.7 Ohmic conductance of Nafion 117 during cooling as a function of the temperature at different initial water contents at 30°C. (Reprinted from Thompson, E. L. et al. 2006. *J. Electrochem Soc.* 153: A2351–A2362. With permission from Electrochemical Society.)

30°C. Figure 9.7 is in accordance with Figure 9.1 where the change in slope represents the desorption phenomena during the cooling phase. This phenomenon is found to be reversible with nevertheless a weak hysteresis especially at high initial water uptake. But the hysteresis becomes less noticeable at lower water content (Thompson, 2006). Whatever it is, drying procedure prior to cold start coupled with subfreezing temperature leads to a significant rising of the membrane ohmic resistance (inverse of conductance) and consequently leads to a significant rise in the FC ohmic resistance. This implies that the FC ohmic resistance is directly linked with membrane water content. In fact, the increase of FC ohmic resistance during the drying procedure has been commonly observed in the literature. The ohmic resistance is usually determined by EIS at high frequency around 1000 Hz. This resistance is currently called the high-frequency resistance (HFR). Figure 9.8 shows the FC HFR during N_2 purge. It can be seen in parts I, III, and V that the HFR increases and reaches a plateau. During the relaxation

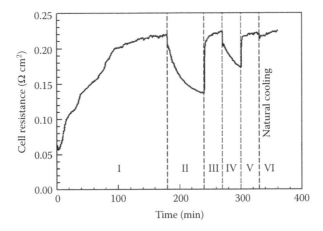

FIGURE 9.8 FC HFR during N_2 purge with 40% RH at 30°C ($\lambda \approx 3.1$ according to Figure 9.3). The cell was consecutively purged and relaxed. (Reprinted from Chacko, C. et al. 2008. *J. Electrochem. Soc.* 155: B1145–B1154. With permission from Electrochemical Society.)

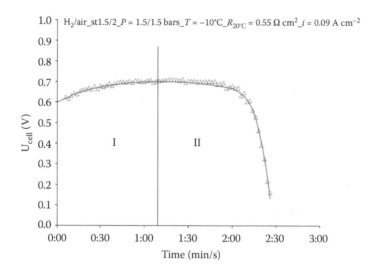

FIGURE 9.9 Typical isothermal galvanostatic cold start at –10°C for a FC initially dried prior to start-up.

periods when purge is turned off (parts II, IV) HFR decreases due to water redistribution through the MEA layers. While the exact mechanism for HFR relaxation remains elusive, the most plausible explanation is that drying entails gradient of water concentration inside the membrane and the ionomer of the CLs owing to internal diffusion of water. Thus, during relaxation period, MEA layers seek to obtain water concentration equilibrium.

Figure 9.9 illustrates a typical galvanostatic isothermal cold start for an FC initially dried prior to start-up. The maximal survival time at –10°C for a low current density of 0.1 A cm^{-2} is quite short, around 2.5 min. Two characteristic periods and their corresponding transitions are distinguished:

- Period I: the FC voltage increases.
- Transition: the FC voltage reaches its maximal value.
- Period II (called "starvation phase"): the FC voltage decreases reaching the zero value.

It can be noted that the general FC behavior during a potentiostatic cold start is similar to the one under galvanostatic mode (Figure 9.10). However, the starvation stage is longer for potentiostatic start-up as the current density and so the kinetic of water production depend on local environment

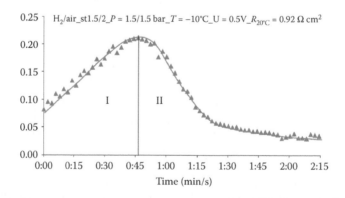

FIGURE 9.10 Typical isothermal potentiostatic cold start at –10°C for a FC initially dried prior to start-up.

conditions in MEA layers which are continually changing due to ice formation, whereas the amount of water produced by unit time remains constant for the galvanostatic mode.

Water management analysis from the membrane coupled with successive EIS measurements performed during isothermal galvanostatic cold start at low current densities (Figure 9.11a) enable us to formulate a phenomenological interpretation of the overall FC electrochemical response. The following interpretation is mainly supported by the work in references (Oszcipok, 2005; Ge and Wang, 2007a,b; Tajiri, 2007; Chacko, 2008; Thompson, 2008a,b; Pinton, 2009c).

9.3.1.1 Period I

When the electronic load is connected ($t = 0$ s), part of the water produced from the ORR at the cathode side is absorbed by the dried membrane. This leads to an increase in the membrane conductance as the water content grows (Figure 9.7), and therefore to a decease in the ohmic resistance (R_Ω in Figure 9.11b and Figure 9.12). At the same time, the activation (R_{ct}) and diffusion (R_{dif}) resistances remain quite stable (R_{ct} and R_{dif} in Figure 9.11b and Figure 9.12). This implies that no ice is formed in the CCL and that the major part of the produced water is absorbed by the membrane. Moreover, because of the very low vapor pressure at subfreezing temperatures, the water vapor transported into the exiting reactant can be considered as negligible (Thompson, 2008a). It is highly believed that a substantial water gradient exists inside the membrane because of the competition between the electroosmotic drag and the water diffusion (Thompson, 2008b). This gradient can be significant as the diffusion coefficient of water in the membrane decreases substantially with decreasing water content and temperature (see Figure 9.13).

It should be pointed out that the effects of subfreezing temperatures and of water content on the electroosmotic drag coefficient found in the literature are not very consistent. Some researchers indicate that the electroosmotic drag coefficient is proportional to λ, others claim that it is constant and independent of λ, or that it depends on λ but with a minimal value for $\lambda = 0$. According to Meier and Eigenberger (Meier, 2004) the electroosmotic drag coefficient ranges from 1 to 1.9 when λ changes from 0 to 14.

9.3.1.2 Transition

The current density reaches a maximal value when the ohmic or HFR resistance is minimal. At this time, the membrane at the cathode side is almost saturated and additionally produced water cannot be totally absorbed and thus begins to flow and to freeze in the CCL. However, no ice is formed at the anode catalyst layer (ACL). This has been observed in the literature by different means: scanning electron microscopy (SEM) (Thompson, 2008a), MEA layers hydrophobicity (Oszcipok, 2005) and visual observations (Ge and Wang, 2007b).

9.3.1.3 Period II

Activation and diffusion resistances (R_{ct} and R_{dif} in Figure 9.11c and Figure 9.12) rise dramatically due to pores of CCL being plugged by ice/frost, thus hindering oxygen transport and reducing the active electrochemical surface area (ECSA). This obviously leads to the FC starving and to the FC shutdown. R_Ω increases as well, but R_{ct} and R_{dif} evolution is clearly preponderant compared to R_Ω increase. The increase of R_Ω can be explained by the presence of ice inducing additional electrical contact resistance whereas the membrane conductivity remains almost stable or decreases very slightly. It is suggested that all the ice formed inside the CCL does not locate necessarily at the same place during the cooling step and during cold-start step. Actually, during the cooling step of a fully hydrated MEA (no drying) that has previously operated at rated conditions, no cell resistance jump is observed (Pinton, 2009c), whereas cell resistance increases sharply during the cold-start starvation stage once the water content in the membrane at the cathode side is saturated. Thus, ice formed in a wet MEA due to FC cooling without operation locates only in the hydrophilic pores. During the cold-start starvation phase, ice does not only locate in the pores of the CCL, but also at the interfaces of CCL and diffusion layers and/or at the

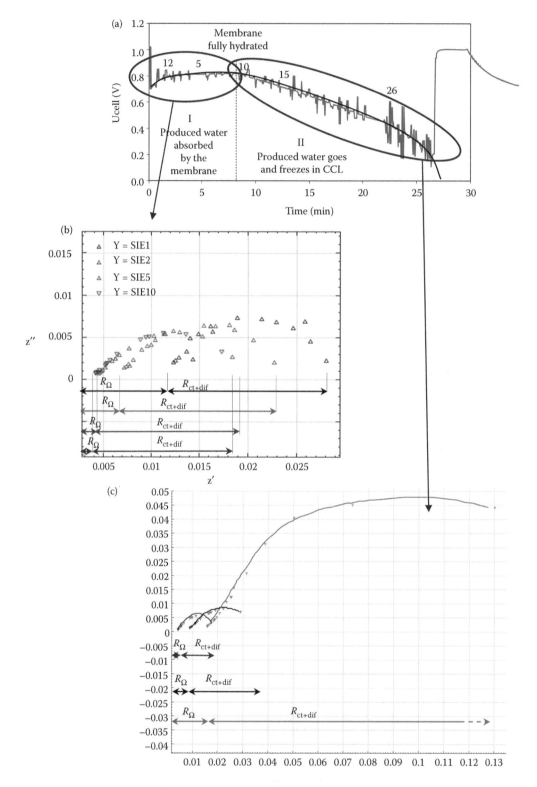

FIGURE 9.11 Successive EIS during isothermal galvnostatic cold start at 0.023 A cm^{-2} and at −10°C for an initially drying FC prior to start-up (a). Resistance changes during the period I (b). Resistance changes during the period II (c).

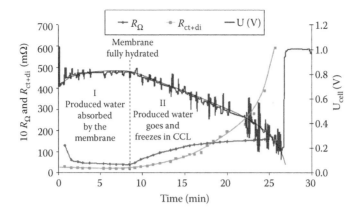

FIGURE 9.12 Changes of R_Ω, R_{ct}, and R_{dif} versus time from successive EIS in Figure 9.11.

interface of particle agglomerates of ionic and electronic conductors inside the CCL. This interpretation almost agrees with what Ge and Wang have proposed (Ge and Wang, 2007a) that ice also locates between Pt particles and the ionomer (Figure 9.14). Oszcipok et al. (Oszcipok, 2007) observed experimentally a small step of contact resistance increase when cold-start starvation occurs. They suggest that this step could be explained by contact resistance increase between cell layers, while membrane resistance is continuously decreasing.

Once the load is disconnected and the gas flows cut off, the global cell resistance decreases reaching a constant value. The membrane at the cathode side is saturated because it is in equilibrium with frozen water and water vapor inside the pores and the interstice of CCL. This frozen water sublimates to maintain the thermodynamic equilibrium at the membrane interface because of water content gradient

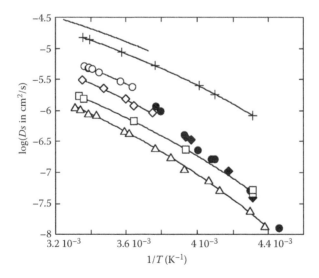

FIGURE 9.13 Self-diffusion coefficients $Ds(T)$ of water in Nafion. Open symbols are for temperature domains where water concentration is constant, full symbols are for temperature domains where concentrations decrease: (○,●) $\lambda = 14.6$, (◇) $\lambda = 9.6$, (□) $\lambda = 6.65$, and (△) $\lambda = 5.4$. $Ds(T)$ is also shown for water in sulfuric acid (+). (Reprinted with permission from Guillermo, A. et al. 2009. NMR and pulsed field gradient NMR approach of water sorption properties in Nafion at low temperature. *J. Phys. Chem. B* 113: 6710–6717. Copyright (2009). American Chemical Society Publications.)

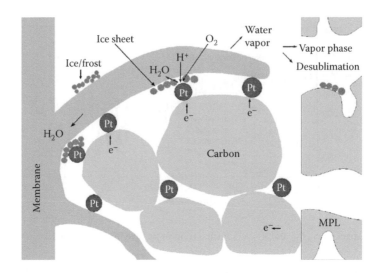

FIGURE 9.14 Schematic of ice formation and microscale distribution in the CCL. (Reprinted from Ge, S. and Wang, C. Y. 2007a. *J. Electrochemical Soc.* 154: B1399–B1406. With permission from Electrochemical Society.)

remaining along the thickness of the membrane. The sublimated vapor is thereby absorbed by the membrane through water diffusion from the cathode to the anode. Subsequently, ice content and therefore the contact resistances decrease. This process continues until homogeneous water saturation is obtained inside the membrane thickness (Pinton, 2009c).

Figure 9.15 summarizes the supposed water content "λ" evolving with time in the membrane thickness, and the time evolution of the membrane resistance, contact resistance, activation, and diffusion resistances. It is worth noting that R_Ω increase during the cold-start starvation phase is not systematically observed. As already mentioned in Section 9.2 about freezing, this may probably be ascribed to differences in the properties of the CLs, manufacturing process of MEAs, and testing procedures.

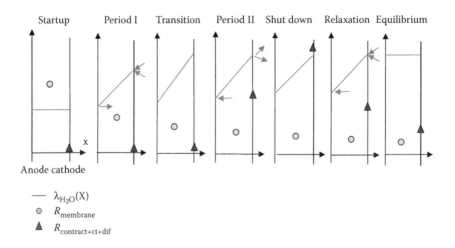

FIGURE 9.15 Schematic change in water content and resistances during an isothermal galvanostatic cold start and shutdown for a FC initially dried prior to start-up.

9.3.2 Effects of Operating Parameters on Cold-Start Capability

This section presents the effects of the main operating parameters on cold-start capability, which includes:

- The initial water content λ_{ini}
- The imposed current or imposed voltage
- The temperature

9.3.2.1 Effect of the Initial Water Content λ_{ini} on Cold-Start Capability

Figure 9.16 presents the effect of the FC drying level on potentiostatic cold start. For a cell without drying, cold-start starvation is immediate (major part of the produced water moves and freezes in the CCL), whereas a dried cell can operate during few tens of seconds before cold-start starvation. The higher is the initial level of drying, the longer is the starvation priming. Actually, with an initial drying level up to $R_{20°C} = 0.55\ \Omega\ cm^2$, more water can be absorbed by the membrane before water saturation is reached. Then, for drying levels higher than $R_{20°C} = 0.55\ \Omega\ cm^2$, current density is significantly reduced mainly because of a lower proton conductivity of the membrane. Thus, more time is needed to saturate the membrane. This analysis can also be applied to a galvanostatic cold start (Figure 9.17). Moreover, Figure 9.17 shows that, at the transition point, voltage is maximal, that is to say the membrane water uptake at the cathode side/interface is maximal. Also, the maximal voltage V_{max} decreases with λ_{ini}. This means that the FC ohmic resistance is higher at higher λ_{ini}. Actually, an increase in the water gradient between the anode and the cathode sides of the membrane is observed mainly due to the decrease of the diffusion coefficient of water with lower λ. Thus, the maximal water storage capacity of the membrane defined by Equation 9.3 is not fully utilized as initial water uptake λ_{ini} decreases ($\lambda_{max}(T)$ is the maximal water content of the membrane depending on the subzero temperature as explained in Section 9.2.1). The literature also suggested that for a successful cold self-start the membrane should not be too dry (Oszcipok, 2006; Ge, 2007b; Pinton, 2009c).

$$\Delta\lambda_{max} = \Delta\lambda_{max}(T) - \lambda_{ini} \tag{9.3}$$

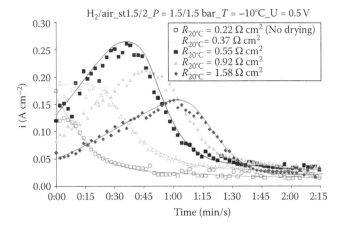

FIGURE 9.16 Change in FC current density for isothermal potentiostatic cold start at 0.5 V and –10°C for different dried levels obtained at 20°C prior to start-up. (Reprinted from *J. Power Sources*, 186, Pinton, E. et al., Experimental and theoretical investigations on a proton exchange membrane fuel cell starting up at subzero temperatures, 80–88, Copyright (2009), with permission from Elsevier.)

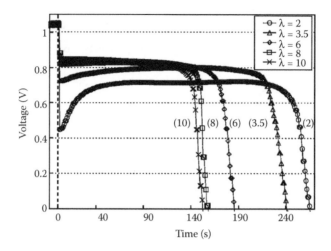

FIGURE 9.17   Change in FC voltage for isothermal galvanostatic cold start at 0.05 A.cm^{-2} and –20°C for different dried levels obtained at 20°C prior to start-up. (Reprinted from Thompson, E. L. et al. 2008a. *J. Electrochem Soc* 155: B625–B634. With permission from Electrochemical Society.)

9.3.2.2 Effect of the Imposed Current or Imposed Voltage on Cold-Start Capability

Figure 9.18 displays the effect of voltage on isothermal potentiostatic cold starts. A decrease in voltage leads to shorter FC operation before starvation as more water is produced at higher current densities, and thus, less time is needed to saturate the membrane with a voltage decrease. Therefore, to take the maximum benefit from the reaction heat produced in order to answer to automotive applications (start-up time < 30 s), starting at low voltage (0.3 < cell votage < 0.5 V) and high current density seems preferable (Pinton, 2009c; Schießwohl, 2009). This analysis can also be applied to galvanostatic cold start (Figure 9.19). The current integration is directly proportional to the cumulated water produced during galvanostatic cold start. Figure 9.20 shows the cumulated charge from results shown in

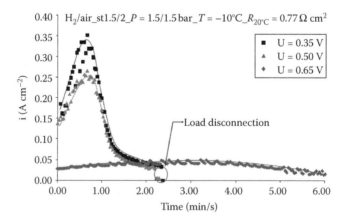

FIGURE 9.18   Change in FC current density for different isothermal potentiostatic cold starts at –10°C and at an identical dried level at 20°C prior to start-up. λ_{ini} corresponds to $R_{20°C} = 0.5\ \Omega\ \text{cm}^{-2}$. (Reprinted from *J. Power Sources*, 186, Pinton, E. et al., Experimental and theoretical investigations on a proton exchange membrane fuel cell starting up at subzero temperatures, 80–88, Copyright (2009), with permission from Elsevier.)

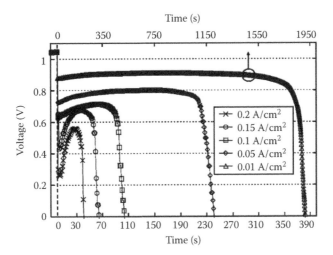

FIGURE 9.19 Change in FC voltage for different isothermal galvanostatic cold starts at –20°C and at an identical initial dried level $\lambda_{ini} = 3.5$. (Reprinted from Thompson, E. L. et al. 2008a. *J. Electrochem Soc* 155: B625–B634. With permission from Electrochemical Society.)

Figure 9.19. It reveals that the cumulated amount of water produced decreases with the increase of current density, which can be explained in part by an increase in the water gradient between the anode and the cathode sides of the membrane. The increased water gradient is mainly due to the rise of water transport by electroosmotic phenomena from anode to cathode side. So, the maximal water storage capacity of the membrane $\Delta\lambda_{max}$ (Equation 9.3) is not fully utilized and decreases as current density increases.

9.3.2.3 Effect of the Imposed FC Temperature on Cold-Start Capability

As described in the literature (Ge, 2007b; Tajiri, 2007; Pinton, 2009c) in Figure 9.21 and Figure 9.22, the FC is able to operate indefinitely without shutdown for subfreezing temperatures, called "$T_{minStart}$," ranging from –1 to –3°C. "$T_{minStart}$" value depends on MEA structure. It was observed by Ge and Wang (Ge, 2007b) with a visualization method that, for this temperature range, only liquid water emerges on the surface of CCL once the membrane is fully hydrated. No ice appears. It implies that either the CL temperature is

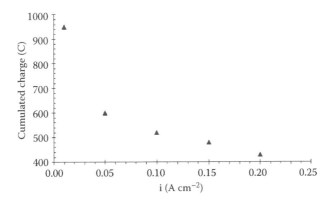

FIGURE 9.20 Cumulated charge versus current density from results in Figure 9.19.

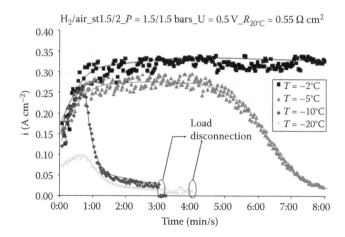

FIGURE 9.21 Change in FC current density for isothermal potentiostatic cold starts at different imposed temperatures at 0.5 V and at an identical dried level at 20°C prior to start-up. λ_{ini} corresponds to $R_{20°C} = 0.5\ \Omega\ cm^{-2}$. (Reprinted from *J. Power Sources*, 186, Pinton, E. et al., Experimental and theoretical investigations on a proton exchange membrane fuel cell starting up at subzero temperatures, 80–88, Copyright (2009), with permission from Elsevier.)

slightly higher than the ambient temperature because of the thermal gradient in GDL resulting from heat generation by the electrochemical reaction or a minor degree of freezing point of water within CL pores which is lowered in confined spaces as the size decreases (Ge, 2007b). For temperatures slightly below $T_{minStart}$, like –5°C in Figure 9.21 and –3°C in Figure 9.22, even if FC shutdown occurs, the amount of produced water is clearly more than the membrane and CCL storage capacity. Thus, it is likely that the water produced in the CCL remains in the liquid state for the reason described previously, and is

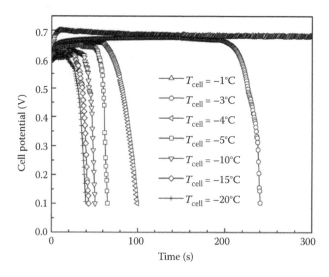

FIGURE 9.22 Change in FC voltage for isothermal galvanostatic cold starts at different imposed temperatures at 0.1 A cm^{-2} and at an identical initial dried level. (Reprinted from *Electrochimica Acta*, 52, Ge, S. and Wang, C.-Y., Characteristics of subzero start up and water/ice formation on the catalyst layer in a polymer electrolyte fuel cell, 4825–4835, Copyright (2007), with permission from Elsevier.)

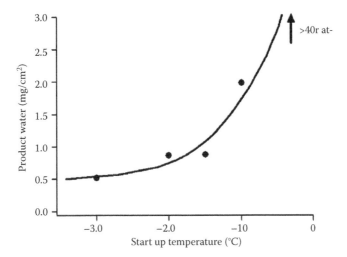

FIGURE 9.23 Relation between product water and isothermal galvanostatic start-up temperature at 0.04 A cm^{-2} and at an identical initial FC dried level $\lambda_{ini} = 6.2$. (Reprinted from *J. Power Sources*, 165, Tajiri, K. et al., Effects of operating and design parameters on PEFC cold start, 279–286, Copyright (2007), with permission from Elsevier.)

transported to the GDL in liquid phase by capillary forces and may freeze in the GDL pores or be further transported to the gas channels. For temperatures noticeably below $T_{minStart}$, like –10°C in Figure 9.21 and –4°C in Figure 9.22, performances and the operating time drop quickly and sharply. No water droplets, frost, or ice have been observed by Ge and Wang (Ge, 2007b) on the CL surface at any time, namely no water emerges out of the CCL. The relation between the produced water and the subzero start-up temperature is highly nonlinear (Figure 9.23). This is mainly due to the decrease in the maximal membrane water uptake with temperature drop (Figure 9.1) and to a lesser extent because of a lower water diffusivity (higher water gradient) and lower vapor pressure (lower transport of water vapor).

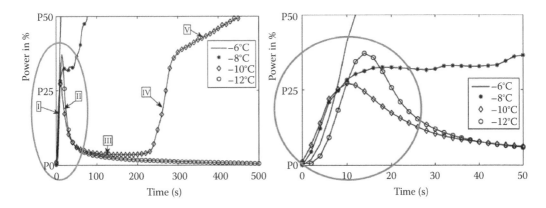

FIGURE 9.24 Time evolution of power for freeze start with long purge prior to start-up with an average starting voltage of 0.45 V at freeze climatic chamber temperatures: $T_{Start} = -6°C$ (–), $T_{Start} = -8°C$ (*), $T_{Start} = -10°C$ ($\Diamond$) and $T_{Start} = -12°C$ (o). (Reprinted from *J. Power Sources*, 193, Schießwohl, E. et al., Experimental investigation of parameters influencing the freeze start ability of a fuel cell system, 107–115, Copyright (2009), with permission from Elsevier.)

9.3.2.4 Effect of the Free FC Stack Temperature on Cold-Start Capability

Imposed FC temperature is usually performed on single cell level in order to evidence the physical phenomena occurring from start-up to freezing shutdown. However, this is not representative of real conditions of a FC stack evolving in subzero ambient temperature. So, in order to investigate FC stack on cold-start capability, experiments are usually carried out on a scale 1 or on a short FC stack located in a climatic chamber (Oszcipok, 2006; Alink, 2008; Bégot, 2008, 2010; Schießwohl, 2009). Data from the open literature are rather scarce, but those available never show a successful cold self-start (FC warming is ensured by the thermal losses of the electrochemical reaction) according to DOE specifications (50% of the rated power after 30 s at –20°C). The best performances have been obtained by Schießwohl et al. (Schießwohl, 2009) with PEMFC system made of a 60-cells stack with graphite bipolar plates (2.3 kW). It succeeded to satisfy DOE requirement but at a minimal ambient temperature of –6°C with an optimal specific shutdown procedure (FC drying level) and optimal specific start-up procedure (average cell voltages = 0.45 V with a shortened cooling loop) (Figure 9.24). For an ambient temperature of –10°C, Figure 9.24 shows that the FC begins to starve and successfully starts once the temperature is above 0°C.

Begot et al. (Bégot, 2010) also succeeded but only after 40 min of self-start at –9°C with a PEMFC system made of a 52 cells stack with 156 cm² of active area and with graphite compound bipolar plates (3.5 kW). The FC stack was dried at room temperature prior to cold start and the current was progressively increased as the cell voltages remain stable around 0.76 V.

Schießwohl et al. (Schießwohl, 2009) highlight that the drying procedure must be carried out carefully in order to avoid high heterogeneity of the cell voltages at the start-up beginning. Even if the cell voltages are quiet homogeneous at the start-up beginning, an important disparity appears during the starvation stage (see Figure 9.25) probably due to the heterogeneity of the cell temperatures. The cell temperatures are expected to be lower for the end cells that are in contact with the massive end plates and higher for the middle cells of the stack.

Note that, if the FC operates at its rated temperature, when the ambient temperature suddenly drops down to subzero temperature, practically no effect is observed on FC stack performances. Actually, the thermal losses occurring inside the stack are usually sufficient to maintain its rated temperature (Bégot, 2010).

9.3.3 Effects of Cold-Start Cycles

Data from the literature about impacts of cold-start cycles on FC degradations are also scarce (Oszcipok, 2005; Yan, 2006; Ge and Wang, 2007a; Alink, 2008; Pinton, 2009a,c; Schießwohl, 2009). Nevertheless, performance losses at rated conditions observed after cold start until a complete freezing starvation (that is to say when performances are near zero, cell voltage <0.3 V with galvanostatic cold start) are clearly higher than the ones observed after freezing–thaw cycles of a non dried cell. The order of magnitude of voltage loss is around 6 mV cycles^{-1} at 0.8 A cm^{-2} for cold starts at –10°C (or about 1% cycles^{-1}). This is namely more than 30 times higher than the one observed after freezing–thaw cycles of a fully hydrated cell. It confirms the interpretation in Section 9.3.1 about the fundamental mechanism where ice formed inside the CCL does not totally locate at the same place during the cooling step (weak performance degradation) and during the cold-start step (high-performance degradation). For recall, it is suggested that during the starvation phase of a cold start, ice would not only locate in the pores of the CCL, but also at the interfaces of CCL and diffusion layers and/or at the interface of particle agglomerates of ionic and electronic conductors inside the CCL.

Regardless, structural alterations and water trapped in CCL described in Section 9.2.2 about effects of freezing–thaw cycles are also causes of performance degradations during cold start. However, they are emphasized during cold start as ice is produced in the core of the CCL and near the catalytic sites. Although an active area decrease is systematically observed by CV at the CCL, preponderance between ohmic, diffusion, and activation limitations is not clearly identified.

Structural damages caused by cold start must doubtlessly have an impact on MEA durability. Unfortunately, no papers in the open literature proceeded to durability investigation impact (for remind DOE specification = 5000 h).

It can be noted that, it was experimentally evidenced as expected (Pinton, 2009a) that, if a dried FC prior to cold start operates only in part I of the start-up (Figures 9.9 and 9.10) where the produced water by the electrochemical reaction is absorbed by the membrane, no damages occur.

9.3.4 Mitigation Strategies

Mitigation strategies can be classified in four topics (Peseran, 2005):

1. Water removal at FC shutdown
2. Keeping FC warm to prevent freezing
3. Heating FC using waste heat from FC operation
4. Heating FC using an external energy source

9.3.4.1 Water Removal at FC ShutDown

Water is removed from the fuel cell stack before it is shut down in order to avoid frozen water (ice) and therefore ice damages. For more detail, see Section 9.2.3 where this part has already been presented in connection with mitigation strategies to prevent freezing–thaw cycle impacts.

9.3.4.2 Keeping FC Warm to Prevent from Freezing

FC is kept warm by adding heat and efficient thermal insulation in order to keep the FC at a temperature above zero while the system is stored in subfreezing environment. The main advantage is that no ice damage can occur. However, this solution requires:

- Complex, and therefore costly, energy demanding control
- Great fuel consumption which limits the storage protection time available
- High and efficient insulation that increases the stack volume and weight

9.3.4.3 Heating FC Using Waste Heat from FC Operation

Supplying H_2 and oxidant directly to the fuel cell while the electrical power feeds the load and the thermal losses due to the electrochemical reaction enables to warm up the stack. The main advantage is that it is a relatively simple design that typically requires no system changes. However, this solution:

- Requires a specific shutdown procedure to remove water (see Section 9.3.4.1).
- Can take several minutes to reach operational temperature because of the high thermal capacity of the stack and of the coolant circuit. FC stack materials with low heat capacity are recommended, for instance, some thin metallic bipolar plates instead of thick graphite bipolar plates.

9.3.4.4 Heating FC Using an External Energy Source

FC is warmed up by adding an external energy source sized to respond to the required start-up delay or by using the power train battery in electrical vehicles (EV). The main advantage is that the FC stack can rapidly reach a positive temperature. However, this solution requires:

- A specific shutdown procedure to remove water (see Section 9.3.4.1).
- A more complex system in components and in control which has an impact on system reliability and cost.
- An increase of the FC system volume and weight affecting vehicle performance, and increases in the installed power requirements and cost. In the case of EVs, this will impact the driving range.

9.4 Summary

This chapter deals with impacts of PEM fuel cell degradations ascribed to a subzero environment.

The state of water and change in water content of a perfluorosulfonated membrane as a function of temperature is a key point in understanding the PEMFC behavior during freezing–thaw cycles:

1. Water inside the membrane at subzero temperatures remains at a liquid state called "supercooled state."
2. Membrane has a maximal water content depending on subzero temperature "$\lambda_{max}(T_{<0°C})$" that decreases with temperature drop once the temperature is below 0°C. Thus, subzero cooling entails water desorption of a fully hydrated membrane leading to additional ice formation inside the active layers of an MEA.

Water desorption and sorption phenomena during cooling and heating processes at subzero temperatures, respectively, are quite reversible: ice sublimates as subzero temperature increases and the sublimated vapor is reabsorbed by the membrane.

Ice formed inside the electrodes leads to irreversible damages which have a direct impact on FC performance losses at rated power. The average order of magnitude in FC performance loss is around 0.03% by each freezing–thaw cycle. In order to prevent these negative effects, it is highly recommended to remove water before the freezing conditions are reached.

The typical electrical behavior of a PEMFC during an isothermal cold start is mainly dependent on water management in MEA layers. If the FC has been dried prior to start-up, water produced from the ORR at the cathode side is absorbed by the membrane. This leads to an increase in the membrane conductance as the water content grows and therefore an increase of the FC performance. Afterwards, once the membrane at the cathode side/interface is saturated, produced water cannot be absorbed anymore and thus begins to move and to freeze in the CCL; the FC performance reaches its maximal value. Then, the freezing water plugs CCL pores thus hindering oxygen transport and reducing the ECSA. And it obviously leads to the FC starving and the FC shutdown. After such cold start with total starvation, performance losses at rated conditions are clearly higher than the ones observed after freezing–thaw cycles of a nondried cell. The average order of magnitude of voltage loss is around 1% cycles^{-1}, which is about more than 30 times higher than the one observed for freezing–thaw cycles of a fully hydrated cell. This is probably due to the fact that during cold start ice is produced in the core of the CCL and near the catalytic sites.

In order to prevent impact of ice formation, two kinds of solution can be chosen:

1. Keep the FC warm.
2. Remove water at FC shutdown and heat the FC stack either by using waste heat from FC operation (can take several minutes to reach operational temperature) or by using an external energy source (can enable rapid start-up but the system is more complex, more voluminous, and heavier).

It is worth noting that if a dried FC (dried prior to cold start) operates only in the first stage of the start-up, no damage occurs as the water produced by the electrochemical reaction is absorbed by the membrane. Moreover, when the FC operates at its rated temperature, if the ambient temperature suddenly drops down to subzero temperature, almost no effect is observed on FC stack performances. Thermal losses occurring inside the stack are usually sufficient to maintain its rated temperature.

Finally, even though substantial research efforts have been made for the last five years on PEMFC freezing and PEMFC cold start, results from the open literature still remain scarce. Moreover, the degradation level and origin are sometimes conflicting, and discrepancies among different researcher are observed. This may probably be ascribed to different MEA microstructures (porosity, layer thickness, etc.), compositions (hydrophobicity, etc.), manufacturing process of MEAs, FC assembling processes, and testing procedures used by investigators.

References

Alink, R., Gerteisen, D., and Oszcipok, M. 2008. Degradation effects in polymer electrolyte membrane fuel cell stacks by sub-zero operation—An *in situ* and *ex situ* analysis. *J. Power Sources* 182: 175–187.

Bégot, S., Harel, F., Candusso, D. et al. 2010. Fuel cell climatic tests designed for new configured aircraft application. *Energy Conversion and Management* 51: 1522–1535.

Bégot, S., Harel, F., and Kauffmann, J. M. 2008. Experimental studies on the influence of operational parameters on the cold start of a 2 kW fuel cell. *Fuel Cells* 8: 138–150.

Chacko, C., Ramasamy, R., Kim, S. et al. 2008. Characteristic behavior of polymer electrolyte fuel cell resistance during cold start. *J. Electrochem. Soc.* 155: B1145–B1154.

Ge, S. and Wang, C. Y. 2007a. Cyclic voltammetry study of ice formation in the PEFC catalyst layer during cold start. *J. Electrochemical Soc.* 154: B1399–B1406.

Ge, S. and Wang, C.-Y. 2007b. Characteristics of subzero start up and water/ice formation on the catalyst layer in a polymer electrolyte fuel cell. *Electrochimica Acta* 52: 4825–4835.

Guillermo, A., Gebel, G., Mendil-Jakani, H. et al. 2009. NMR and pulsed field gradient NMR approach of water sorption properties in Nafion at low temperature. *J. Phys. Chem. B* 113: 6710–6717.

Guo, Q. and Qi, Z. 2006. Effect of freeze–thaw cycles on the properties and performance of membrane-electrode assemblies. *J. Power Sources* 160: 1269–1274.

Hou, J., Song, W., Yu, H. L. et al. 2007. Electrochemical impedance investigation of proton exchange membrane fuel cells experienced subzero temperature, *J. Power Sources* 165(1): 287–292.

Hou, J., Yu, H., Zhang, S. et al. 2006. Analysis of PEMFC freeze degradation at −20°C after gas purging, *J. Power Sources* 162: 513–520.

Kim, S. and Mench, M. M. 2007. Physical degradation of membrane electrode assemblies undergoing freeze/thaw cycling: Micro-structure effects. *J. Power Sources* 174: 206–220.

Luo, M., Huang, C., Liu, W. et al. 2010. Degradation behaviors of polymer electrolyte membrane fuel cell under freeze/thaw cycles. *International Journal of Hydrogen Energy* 35: 2986–2993.

Meier, F. and Eigenberger, G. 2004. Transport parameters for the modelling of water transport in ionomer membranes for PEM fuel cell. *Electrochim. Acta* 49: 1731–1742.

Oszcipok, M., Hakenjos, A., Riemann, D. et al. 2007. Start up and freezing processes in PEM fuel cells. *Fuel Cells* 7: 135–141.

Oszcipok, M., Riemann, D., Kronenwett, U. et al. 2005, Statistic analysis of operational influences on the cold start behaviour of PEM fuel cells, *J. Power Sources* 145: 407–415.

Oszcipok, M., Zedda, M., Riemann, D. et al. 2006. Low temperature operation and influence parameters on the cold start ability of portable PEMFCs. *J. Power Sources* 154: 404–411.

Padfield, T. 1999, What happens to water absorbent materials below freezing? http://www.natmus.dk/cons/tp/cool/suprcool.htm

Park, G.-G., Lim, S.-J., Park, J.-S. et al. 2010. Analysis on the freeze/thaw cycled polymer electrolyte fuel cells. *Current Applied Physics* 10: S62–S65.

Peseran, A., Kim, G., and Gonder, D. 2005, PEM fuel cell freeze and rapid start up investigation, US National Renewable Energy Laboratory, *NREL/MP-540-38760.*

Pineri, M., Gebel, G., Davies, R. J. et al. 2007. Water sorption–desorption in Nafion˙ membranes at low temperature, probed by micro x-ray diffraction. *J. Power Sources* 172: 587–596.

Pineri, M., Volino, F., and Escoubes, M. 1985. Evidence for sorption–desorption phenomena during thermal cycling in highly hydrated perfluorinated. *Journal Polymer Sci., Polymer Phys.* 23: 2009–2020.

Pinton, E., Antoni, L., Fourneron, Y. et al. 2009a. Cold start and freeze/thaw cycles effect on PEFMC performances. *ECS Transaction* 17/1: 251–261.

Pinton, E., Fourneron, Y., and Guillermo, A. 2009b. Method for storing a fuel cell at negative temperature, Patent WO2009133274.

Pinton, E., Fourneron, Y., Rosini, S. et al. 2009c. Experimental and theoretical investigations on a proton exchange membrane fuel cell starting up at subzero temperatures. *J. Power Sources* 186: 80–88.

Schießwohl, E., von Unwerth, T., Seyfried, F. et al. 2009. Experimental investigation of parameters influencing the freeze start ability of a fuel cell system, *J. Power Sources* 193: 107–115.

Song, W., Hou, J., Yu, H. et al. 2009. Sub-freezing endurance of PEM fuel cells with different catalyst-coated membranes. *J. Applied Electrochemistry* 39: 609–615.

Springer, T. E., Zawodzinski, T. A., and Gottesfeld, S. 1991. Proton electrolyte fuel cell model. *J. Electrochem. Soc.* 138: 2334–2341.

Tajiri, K., Tabuchi, Y., Kagami, F. et al. 2007. Effects of operating and design parameters on PEFC cold start. *J. Power Sources* 165: 279–286.

Thompson, E. L., Capehart, T. W., Fuller, T. J. et al. 2006. Investigation of low temperature proton transport in Nafion using direct current conductivity and differential scanning calorimetry. *J. Electrochem. Soc.* 153: A2351–A2362.

Thompson, E. L., Jorne, J., Gu, W. et al. 2008a. PEM fuel cell operation at –20°C. I. Electrode and membrane water (charge) storage. *J. Electrochem Soc.* 155: B625–B634.

Thompson, E. L., Jorne, J., Gu, W. et al. 2008b. PEM fuel cell operation at –20°C. II. Ice formation dynamics, current distribution, and voltage losses within electrodes. *J. Electrochemical Soc.* 155: B887–B896.

Yan, Q., Toghiani, H., Lee, Y.-W. et al. 2006. Effect of sub-freezing temperatures on a PEM fuel cell performance, start up and fuel cell components. *J. Power Sources* 160: 1242–1250.

10

Operation-Induced Degradation

Pucheng Pei
Tsinghua University

10.1 Introduction

Powering vehicles is an important application of fuel cells. In such engines the operating conditions change frequently, and current research shows that fuel cells for automobiles have a much shorter lifetime than fuel cells in applications with stationary working conditions. Such fuel cells can encounter many potentially destructive conditions, such as feed starvation and low relative humidity (RH).

In this book's earlier chapters, the various components of fuel cell performance decay mechanisms were introduced. This chapter will cover fuel cell performance degradation under different operating cycles, and describe various mitigation strategies under actual application conditions.

While previous chapters have provided theoretical analyses, the experimental results described in this chapter, which primarily relate to transportation conditions, provide feedback on these theories. The following operating conditions will be covered, followed by a discussion of mitigation strategies:

- Operating cycles
- Feed starvation

- Low RH
- High temperature
- Dynamic response
- Water management

10.2 Operating Cycles

To date, while comprehensive experimental results and reviews have been published in an attempt to understand the degradation mechanisms of fuel cell components such as electro-catalysts, membranes, and bipolar plates, only a relatively small number of studies aimed at real proton exchange membrane (PEM) fuel cell lifetimes have been conducted, due to the high costs and prolonged testing periods required. For example, more than 4.5 years of uninterrupted testing is needed to reach the 40,000-h lifetime requirement for a fuel cell system for stationary applications. To test a fuel cell bus system (275 kW) for 20,000 h, the fuel expense alone would be approximately US $2 million (3.8 billion liters of hydrogen at US $5.3 m^{-3}). To increase sample throughput and reduce the experimental time required, several fuel cell developers and companies, such as Ballard Power Systems, DuPont, Gore, and General Motors, have proposed and implemented different accelerated stress tests (ASTs) to determine the durability and performance of current fuel cell components. This section summarizes papers published in the last decade on PEM fuel cell degradation and lifetimes. Tables 10.1 and 10.2 present work on steady-state and accelerated lifetime tests, respectively. Although most experiments on fuel cell lifetime under steady-state operation have demonstrated acceptable results, with a degradation rate between 2 and 10 V h^{-1}, they were conducted for far less than 40,000 h. In ASTs, almost all degradation rates were greater than 10 V h^{-1}. Prior to commercializing fuel cell technology, more thorough studies of components and analyses of system failure modes are imperative (Wu, 2008).

Pei et al. studied fuel cell performance degradation results under different operating conditions, and have presented an arithmetic equation for fuel cell lifetime that relates to load changing cycles, start–stop cycles, idling time, high-power load conditions, and air pollution factors (Pei, 2008). Based on the practical data gathered from a fuel cell bus and the test results of a fuel cell stack in a laboratory, the calculated lifetime fits the bus real running lifetime very well. Results show that automotive fuel cell lifetime heavily depends on driving cycles.

Figure 10.1 presents the results of a fuel cell bus trial run at an average speed of 32 km h^{-1}. To protect the fuel cell, during the first minute of a trial the electrical current must not exceed 0.7 V. After the fuel

TABLE 10.1   Summary of Steady-State Lifetime Tests in the Literature

Authors	Test Time (h)	Degradation Rate
Ralph	5000	4 μVh^{-1}
St-Pierre et al.	5000	1 μVh^{-1}
Washington	4700	6 μVh^{-1}
	8000	2.2 μVh^{-1}
Endoh et al.	4000	2 μVh^{-1}
Yamazaki et al.	8000	2–3 μVh^{-1}
St-Pierre and Jia	11,000	2 μVh^{-1}
Fowler et al.	1350	11 μVh^{-1}
Ahn et al.	1800	>4 mVh^{-1}
Cheng et al.	4000	3.1 μVh^{-1}
Scholta et al.	2500	20 μVh^{-1}
Cleghorn et al.	26,300	4–6 μVh^{-1}

TABLE 10.2 Summary of Accelerated Durability Tests in the Literature

Authors	Test Time (h)	Degradation Rate	Operating Conditions
Sishtla et al.	5100	$6\,\mu Vh^{-1}$	Reformate fuel
Nakayama	4000	$4.3\,\mu Vh^{-1}$	Reformate fuel
Isono et al.	2000	$10\,\mu Vh^{-1}$	Reformate fuel
Maeda et al.	5000	$6\,\mu Vh^{-1}$	Reformate fuel
Sakamoto et al.		$50-90\,\mu V$	Per start/stop cycles
Fowler et al.	600	$120\,\mu Vh^{-1}$	Humidity cycles
Cho et al.		$4200\,\mu V$	Per thermal cycles
Knights et al.	13,000	$0.5\,\mu Vh^{-1}$	Methane reformate fuel
			Low humidification
Oszcipok et al.		$22,500\,\mu V$	Per cold start-up
Xie et al.	1916	$60\,\mu Vh^{-1}$	Over-saturated humidification
	1000	$54\,\mu Vh^{-1}$	
Yu et al.	2700	$21\,\mu Vh^{-1}$	Low humidification
Endoh et al.	3500	$3\,\mu Vh^{-1}$	High temperature
			Low humidification
Du et al.	1900	$70-800\,\mu Vh^{-1}$	Cold start and hot stop
Xu et al.	1000	$<10\,\mu Vh^{-1}$	High temperature
			Low humidification
Owejan et al.		$0.212\,mV$	Per start/stop cycles

Source: Reprinted from *J. Power Sources*, 184, J. Wu et al. A review of PEM fuel cell durability: Degradation mechanisms and mitigation strategies, 104–119, Copyright (2008), with permission from Elsevier.

cell bus ran 35,000 km, which took about 1100 h, with the cell voltage decreased by 10% below the limiting electrical current, the fuel cell stack power decreased more rapidly (Pei, 2008).

Figure 10.2 shows the results from 1800 h of continuous stack operation at gas utilization of 0.5 for H_2 and 0.25 for air. Although nonuniform voltage distribution in the unit cells was observed, continuous operation of the stack up to 1800 h was possible—specifically, for 300 h under full load and 1500 h under partial load. Continuous operation of the stack was composed of three parts. The first period, from (I) to (III), was a warming-up step for the stack and the system for continuous operation under partial load. The second period, from (IV) to (V), was operation at 70 A under full load. At that time, the power of the stack was about 1.75 kW. During the third period, the stack was operated under partial load

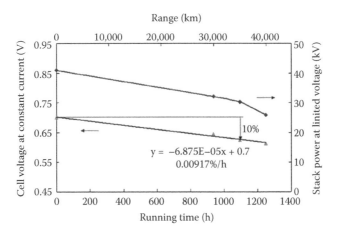

FIGURE 10.1 Fuel cell bus trial results. (Reprinted from *Int. J. Hydrogen Energy*, 33, P. Pei, Q. Chang, T. Tang. A quick evaluating method for automotive fuel cell lifetime, 3829–3836, Copyright (2008), with permission from Elsevier.)

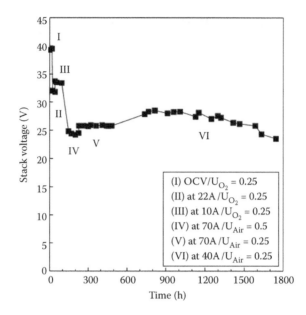

FIGURE 10.2 Performance of a stack during continuous operation. (Reprinted from *J. Power Sources*, 106, S. Y. Ahn, S.-J. Shin, H. Y. Ha. 2002. Performance and lifetime analysis of the kW-class PEMFC stack, 295–303, with permission from Elsevier.)

at 40 A for 12 h a day, and then for the other 12 h at OCV. As can be seen in Figure 10.2, the stack performance did not decrease during the second period, but slowly dropped during the third period at an average decay rate of 4.37 mV h^{-1}; then, after continuous operation for 1800 h, the stack showed rapid performance decay. It was postulated that careless operation of the stack might have caused the rapid decay in performance, because the stack was exposed to flooding (Ahn, 2002). Several factors can be at work in stack performance degradation after continuous operation, and these have been investigated with various analytical tools, as will be described in the following subsections.

10.2.1 Load Cycling

A load cycling test was set up, as in Figure 10.3, where the load changed from an idling condition to a rated power condition, and the fuel cell performance was measured after every 200 loading cycles. Each day, the researchers used 30 min idling time to warm the fuel cell to above 40°C, then carried out 2000 loading cycles. Figure 10.4 shows all test points in order, with a decay rate of 0.0000606%/cycle. In Figures 10.4 and 10.5, the fuel cell performance tends to steady after 1000 loading cycles. Thus, the performance decay rate may be determined in a much shorter time if the test is performed continuously after start-up, without a mid-test shutdown (Pei, 2008).

A driving cycle is a specific kind of load changing cycle under simulated vehicle driving conditions. Driving cycles are designed by mimicking real vehicle internal engines, including start-up, idling, constant load running, variable load acceleration, full power running, and overload running. Overloading in particular normally occurs during actual vehicle operation and is known to play a role in accelerating degradation. In one experiment, every cycle consisted of six periods (simulating cold-start, idling, full power running, continuous loading running, and overload running) and lasted 1184 s. The overload capacity was set at 35 A; the driving cycle protocol consisted of continuously running the above cycles. The evolution of the current in one cycle is presented in Figures 10.6 and 10.7 (Li, 2009).

In the load changing process, it is important to know how the automotive fuel cell with a large area behaves in various parts of the flow field. To measure the membrane drying out time under dynamic

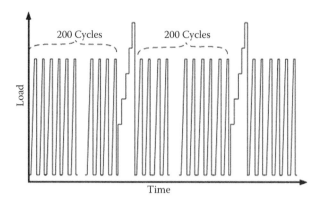

FIGURE 10.3 Load cycling test mode. (Reprinted from *Int. J. Hydrogen Energy*, 33, P. Pei, Q. Chang, T. Tang. A quick evaluating method for automotive fuel cell lifetime, 3829–3836, Copyright (2008), with permission from Elsevier.)

current loading, high-frequency resistance (HFR) has been measured under a constant current. Fluctuation in the HFR was observed in each segment, as shown in Figure 10.8. The HFR was higher in segment 1 due to drying out of the membrane, which occurred because less water was present to humidify it. When the gas shifted downstream, the amount of water increased, the membrane was humidified, and the HFR decreased. It can also be observed that when the loading changed from a large to a small current, the HFR was lower due to the low temperature of the membrane, and the water content was sufficient in segments 1–4. The HFR value was found to increase with increasing temperature. In segments 5–8, the HFR was steady because there was sufficient water from upstream to humidify the membrane. The HFR increased in segments 1–7 when the loading was changed from a large current to a small one, because the membrane was still at a higher temperature and water production decreased, and so the membrane's water content dropped. When the gas shifted downstream it took up more water, which humidified the membrane. This phenomenon caused the HFR to overshoot then ease up in the last segments. There was also flooding in segment 8 because the water produced in prior segments passed to

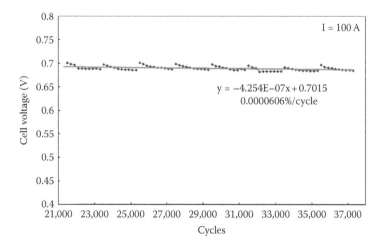

FIGURE 10.4 Performance change in load cycling. (Reprinted from *Int. J. Hydrogen Energy*, 33, P. Pei, Q. Chang, T. Tang. A quick evaluating method for automotive fuel cell lifetime, 3829–3836, Copyright (2008), with permission from Elsevier.)

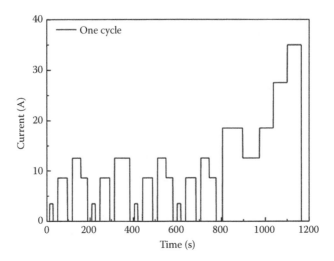

FIGURE 10.5 Driving cycle protocol in one cycle, with evolution of current over time during 20 min; the overloading capacity was set at 35 A. (Reprinted from R. Lin, B. Li, Y. P. Hou. 2009. *Int. J. Hydrogen Energy* 34: 2369–2376. With permission.)

segment 8 when the system shifted from a large to a small load, causing the HFR to undershoot due to that flooding. When the flooding was removed, the HFR stabilized (Weng, 2010).

The voltage fluctuation in segments 1–4 under dynamic current loading is shown in Figure 10.9. In segments 5–8, the gas in the electrode could not be supplied quickly enough when the load suddenly increased from 70 to 700 mA cm^{-2} (Weng, 2010).

An aging experiment under current cycling conditions comprised 450 cycles in total, taking about 150 h. Polarization test and linear sweep voltammetry (LSV) were conducted every 50 cycles. Figure 10.10a and b show the polarization curves at the 140th and 450th cycles, respectively (Weng, 2010).

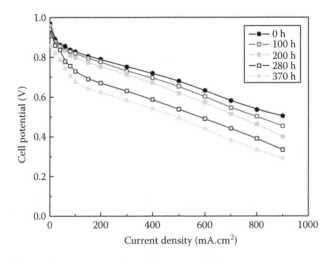

FIGURE 10.6 Current voltage characteristics of an electrode with 50 m^2 of different driving cycles. (Reprinted from R. Lin, B. Li, Y. P. Hou. 2009. *Int. J. Hydrogen Energy* 34: 2369–2376. With permission.)

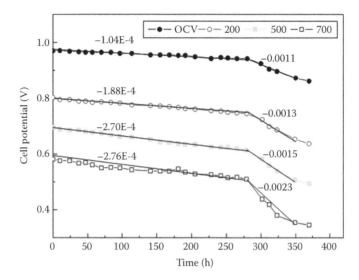

FIGURE 10.7   Evolution of voltage versus time under driving cycles at densities of 0, 200, 500, and 700 mA/cm². (Reprinted from R. Lin, B. Li, Y. P. Hou. 2009. *Int. J. Hydrogen Energy* 34: 2369–2376. With permission.)

10.2.2 Idle Conditions

When the fuel cell works in an idling cycle, the power generated is not large so not much water is produced from the reaction of hydrogen and oxygen. When the gas flows in the channels, the membrane electrode assembly (MEA) dries out easily. In the previously described fuel cell bus, the idling voltage had to be set low, no more than 0.9 V, to prolong fuel cell lifetime. The idling current density was set to 10 mA cm^{-2}, and an automatic program recorded the fuel cell performance every 15 min, as shown in Figure 10.11. During the testing period, the operating temperature was no more than 60°C, with air and hydrogen stoichiometric ratios of 2.5 and 1.2, respectively. Since the fuel cell bus operated during the day and rested at night, the test was performed 5 h every day from morning to afternoon, with one start and one stop. Figure 10.12 shows the 50 h test results, in which the fuel cell performance is almost fully recovered at each start-up, with a slight decay rate beyond. From this figure the voltage decay rate can be obtained. It is significant that although measurements were taken irregularly for 10 h after the 25 h mark, the results still showed the same decay rate. Nonetheless, to enhance the accuracy of decay rate measurements, it is important to follow a regular testing and measurement procedure (Pei, 2008).

When working under different idling currents, the fuel cell has a different degradation rate, as shown in Figure 10.13. With idling currents of 8.4, 2.8, and 1.4 A, the voltages of all three currents could be obtained and fit into the corresponding equation. From Figure 10.13 it is evident that the PEMFC performance was recoverable at a low idling current.

10.2.3 High-Power Conditions

A steady high-power load cycling test has been carried out according to a strict process, in which the first step was start-up and then warming to above 40°C, and the second step was to provide a steady high-power load and run a performance test every 15 min. The high-power load conditions were set to match the limited voltage in a fuel cell bus. Testing was performed every day for 5 h, with a 30 min warming time and one start–stop cycle. Figure 10.14 shows the orderly patterns of test points for all of the days; from these can be calculated the voltage decay rate caused by high-power load cycles (Pei, 2008).

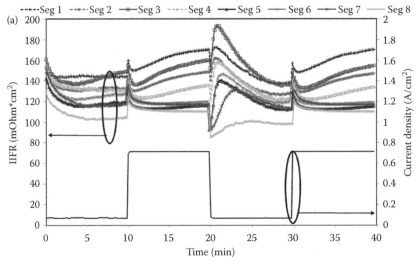

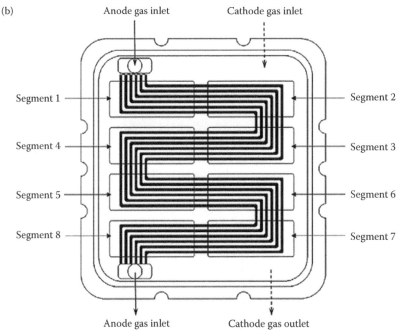

FIGURE 10.8 (a) The HFR in each segment of a fuel cell in a dynamic loading cycle versus time. RH 60%, constant density 0.07–0.7A/cm²; (b) schematic of the eight segments in the flow field. (Reprinted from *Int. J. Hydrogen Energy*, 33, F.-B. Weng, C.-Y. Hsu, C.-W. Li. Experimental investigation of PEM fuel cell aging under current cycling using segmented fuel cell, 3664–3675, Copyright (2010), with permission from Elsevier.)

10.2.4 Start–Stop Cycles

A fuel cell stack start–stop cycling test has been carried out, also following a strict process: start-up, idling for 1 min at a constant current of 10 mA cm⁻², stop, purging of the hydrogen with nitrogen gas, waiting until the stack voltage falls to zero, and then repeating the cycle. For every ten start–stop cycling tests, the fuel cell stack performance was recorded. Testing was conducted for 80 h to obtain the fuel cell performance

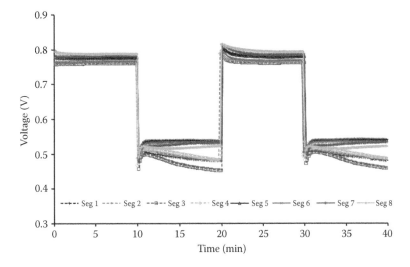

FIGURE 10.9 Voltage versus time in a dynamic loading cycle. RH 60%, constant density 0.07–0.7A/cm². (Reprinted from *Int. J. Hydrogen Energy*, 33, F.-B. Weng, C.-Y. Hsu, C.-W. Li. Experimental investigation of PEM fuel cell aging under current cycling using segmented fuel cell, 3664–3675, Copyright (2010), with permission from Elsevier.)

deterioration history. As shown in Figure 10.15, the cell voltage decayed 0.00196% every start–stop cycle, indicating that if not enough test cycles are performed, the decay rate precision will be poor (Pei, 2008).

10.2.5 Analysis after Operating Cycles

After a long period of operating cycles, many microscale changes occur in a fuel cell. Figure 10.16 shows the H_2 crossover rate as a function of time for different membranes (Wu, 2008).

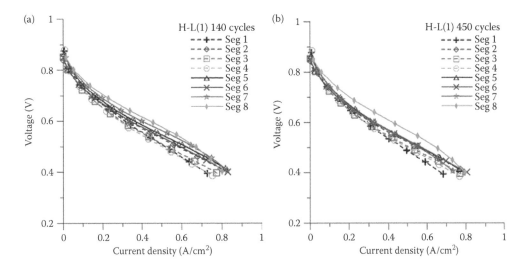

FIGURE 10.10 Polarization curves of the dynamic loading cycles of a multisegment PEMFC: (a) after 140 cycles, and (b) after 450 cycles. (Reprinted from *Int. J. Hydrogen Energy*, 33, F.-B. Weng, C.-Y. Hsu, C.-W. Li. Experimental investigation of PEM fuel cell aging under current cycling using segmented fuel cell, 3664–3675, Copyright (2010), with permission from Elsevier.)

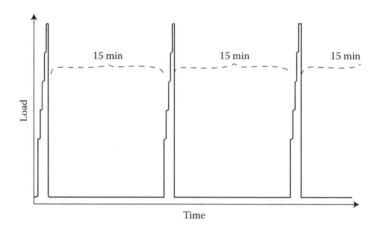

FIGURE 10.11 Idling test cycles. (Reprinted from *Int. J. Hydrogen Energy*, 33, P. Pei, Q. Chang, T. Tang. A quick evaluating method for automotive fuel cell lifetime, 3829–3836, Copyright (2008), with permission from Elsevier.)

Figure 10.17 shows Pt load changes for an original catalyst-coated membrane (CCM) and a modified CCM under accelerated testing. After 100 h of treatment, the Pt catalyst loss from the modified CCM was radically less than the loss from the original CCM: 0.1 mg cm^{-2} versus 0.32 mg cm^{-2} (Pei, 2010).

Figures 10.18 through 10.20 show fluoride emissions from CCMs in hydrogen peroxide. Both the original and the modified CCM had increasing fluoride emissions over time, but the modified CCM emitted more. Fluoride in a fuel cell solution can originate from the degradation of Nafion in the catalyst layers and/or in the Nafion membranes (Li, 2010).

After 500 h accelerated lifetime testing on an automotive PEM fuel cell stack with idling cycles, load changing cycles, high-power cycles, and start–stop cycles, the stack performance clearly deteriorated. As shown in Figure 10.21, the average cell voltage curve of every cell from 1 to 14 cells is lower than for cells 15–100.

Figure 10.22 expresses the polarization curve of different small cells within primary cells. The small cells taken from the air-outlet parts of primary cells display the worst performance overall and have the

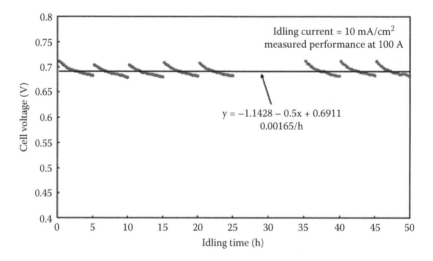

FIGURE 10.12 Performance change in idling test cycles. (Reprinted from *Int. J. Hydrogen Energy*, 33, P. Pei, Q. Chang, T. Tang. A quick evaluating method for automotive fuel cell lifetime, 3829–3836, Copyright (2008), with permission from Elsevier.)

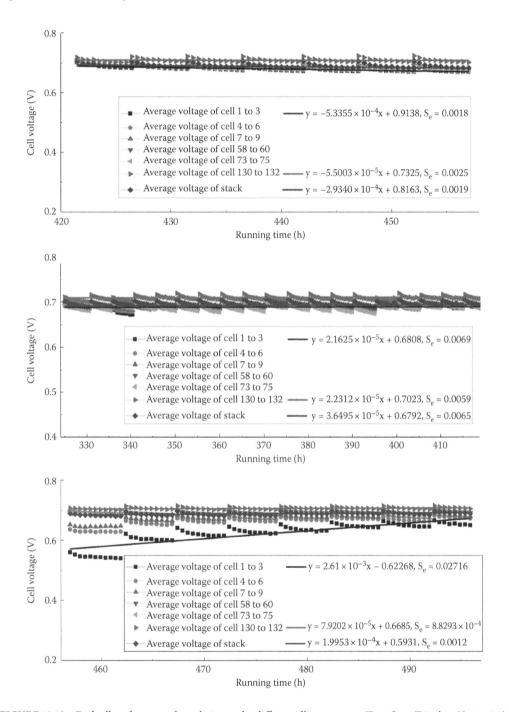

FIGURE 10.13 Fuel cell performance degradation under different idling currents. (Data from Tsinghua University.)

lowest open-circuit potentials. Their polarization curves drop rapidly in the ohmic potential loss region, and their maximum current density is less than 300 mA cm^{-2}. The small cells taken from the hydrogen-outlet parts of primary cells also perform poorly, sometimes even worse than the air-outlet parts, such as cell 4 (Figure 10.22b). The small cells taken from the air-inlet, middle parts display average performance, while the small cells taken from the hydrogen-inlet parts have the best performance, especially

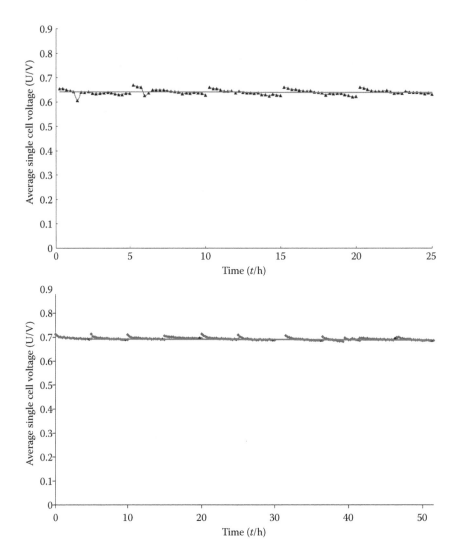

FIGURE 10.14 Voltage degradation by high-power cycles. (Reprinted from *Int. J. Hydrogen Energy*, 33, P. Pei, Q. Chang, T. Tang. A quick evaluating method for automotive fuel cell lifetime, 3829–3836, Copyright (2008), with permission from Elsevier.)

cell 8 (Figure 10.22c), which performs even better than a new cell, with an open-circuit potential of more than 0.85 V. The polarization curves of the hydrogen-inlet cells drop gently in the ohmic potential loss region. Their maximum current density is the highest, more than 1000 mA cm^{-2}.

Figure 10.23a shows the electrochemically active surface area coefficients of the different parts in every cell. The coefficients of the hydrogen-inlet parts are the largest, indicating that the catalysts in these areas are the most active after a long working time. Figure 10.23b compares the particle diameters of three differently positioned, representative cells in the 100-cell stack, which worked more than 500 h. As shown in Figure 10.23c, the contact angles in different parts of cell range from 124° to 130°, which makes it clear that the hydrophobic nature of any part of an MEA surface experiences no obvious change after more than 500 h of cell operation. The cell drainage is steady, indicating that the probability of flooding in the flow channel of the old cell is almost the same as in a new cell. As shown in Figure 10.23d, the impedances are largest for the air-outlet and hydrogen-outlet cells, while those for the hydrogen-inlet cells are the

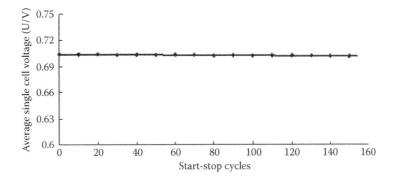

FIGURE 10.15 Voltage degradation by start–stop cycles. (Reprinted from *Int. J. Hydrogen Energy*, 33, P.Pei, Q. Chang, T. Tang. A quick evaluating method for automotive fuel cell lifetime, 3829–3836, Copyright (2008), with permission from Elsevier.)

smallest, but all are bigger than that of a new MEA (120 mΩ cm^2). The impedances of cell 1 are the largest while those of cell 8 are the smallest.

10.3 Feed Starvation

Starvation, which describes the operating conditions of fuel cells in substoichiometric fuel or oxidant feeding, is a potential cause of fuel cell failure. Many factors will lead to starvation in a fuel cell: poor cell design or machining, leading to uneven mass distribution in the flow fields; poor stack design or assembly, causing uneven flux distribution between cells; poor water management, resulting in channel blockage by flooding; poor heat management during cold start-up, causing ice blockage; and incorrect operation, inducing substoichiometric gas feeding.

10.3.1 Feed Starvation at the Anode Side

It is possible for one or more MEAs in a stack, or even a complete stack in a multistack system, to show a reverse in polarity during fuel cell operation. Various circumstances can result in a fuel cell being

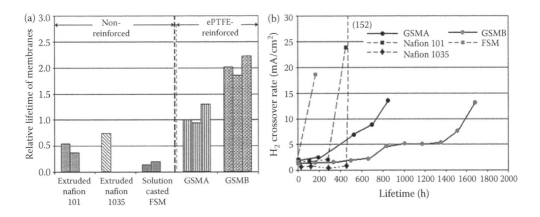

FIGURE 10.16 Comparison of Gore reinforced membranes and nonreinforced membranes: (a) lifetime of various membranes in accelerated fuel cell conditions; (b) H$_2$ crossover rate as a function of time. (Reprinted from *J. Power Sources*, 184, J. Wu et al. A review of PEM fuel cell durability: Degradation mechanisms and mitigation strategies, 104–119, Copyright (2008), with permission from Elsevier.)

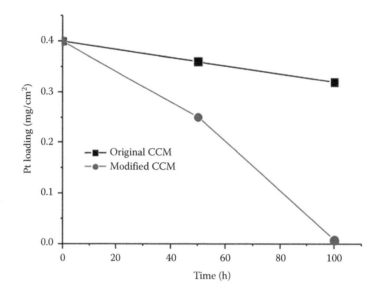

FIGURE 10.17   Changes in Pt loading in catalyst layers undergoing accelerated durability testing. (Reprinted from *Int. J. Hydrogen Energy*, 35, S. Mu. Accelerated durability tests of catalyst layers with various pore volumes for catalyst coated membranes applied in PEM fuel cells, 2872–2876, Copyright (2010), with permission from Elsevier.)

driven into voltage reversal by other cells in the series stack, and severe damage to the MEAs may result. The most likely reason for cell reversal is reactant starvation of the MEA at the anode, due to inadequate fuel supply. Fuel and oxidant starvation can occur during a sudden change in reactant demand, such as start-up and load change. Taniguchi et al. (Taniguchi, 2004) analyzed electro-catalyst degradation during fuel starvation and found anode ruthenium dissolution and cathode platinum sintering. Sanyo

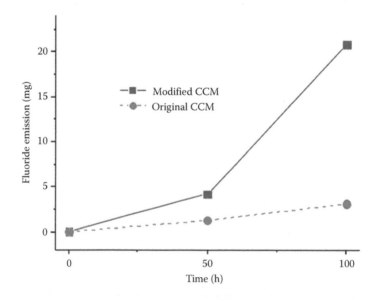

FIGURE 10.18   Fluoride emissions of CCMs during accelerated durability testing. (Reprinted from *Int. J. Hydrogen Energy*, 35, S. Mu. Accelerated durability tests of catalyst layers with various pore volumes for catalyst coated membranes applied in PEM fuel cells, 2872–2876, Copyright (2010), with permission from Elsevier.)

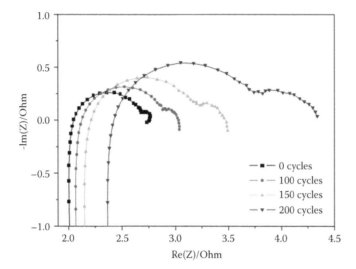

FIGURE 10.19 Electrode impedance plot with varying cycles (0, 100, 150, and 200). (Reprinted from *Int. J. Hydrogen Energy*, 35, B. Li, R. Lin, D. Yang. Effect of driving cycle on the performance of PEM fuel cell and microstructure of membrane electrode assembly, 2814–2819, Copyright (2010), with permission from Elsevier.)

Electric reported performance degradation caused by reactant starvation, in a national R&D project on PEMFCs (Sakamoto, 2000). Kim et al. reported a vacuum effect during anode starvation operation, which could cause fuel to be drawn from the stack manifold or ambient air to enter a laboratory-scale cell (Kim, 2004). Using cyclic voltammetry (CV) measurements, Kang et al. also detected cell reversal when the fuel supply was inadequate (Kang, 2010). According to the findings of Wang, the cell potential decreases if there is insufficient moisture at the anode side, resulting in a momentary loss of power (Wang, 2010).

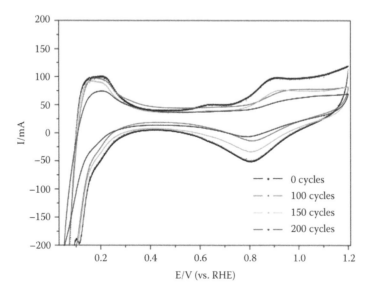

FIGURE 10.20 Cyclic voltammograms of an electrode with varying driving cycles (0, 100, 150, and 200). (Reprinted from *Int. J. Hydrogen Energy*, 35, B. Li, R. Lin, D. Yang. Effect of driving cycle on the performance of PEM fuel cell and microstructure of membrane electrode assembly, 2814–2819, Copyright (2010), with permission from Elsevier.)

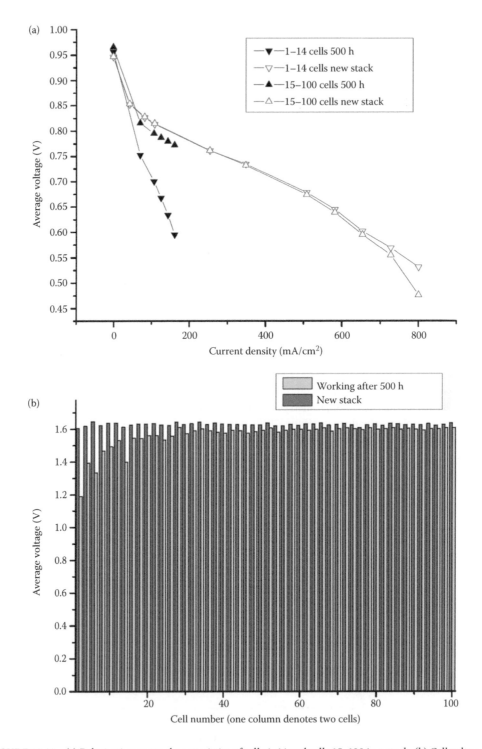

FIGURE 10.21 (a) Polarization curve characteristics of cells 1–14 and cells 15–100 in a stack. (b) Cell voltages of the stack after 500 h of operation, compared to a new stack with the same load. (Reprinted from *Int. J. Hydrogen Energy*, 35, P. Pei, X. Yuan, P. Chao. Analysis on the PEM fuel cells after accelerated life experiment, 3147–3151, Copyright (2010), with permission from Elsevier.)

Cell reversal occurs when the fuel cell stack is loaded and not enough fuel is supplied to the anode. Drawing excessive current from any single cell—that is, more than its fuel delivery can produce—can lead to cell reversal. Taniguchi et al. carried out an experiment using a single cell that mimics cell reversal in a stack. After cell reversal degradation under conditions of 100% fuel utilization, using external direct-current power sources, various characterizations were carried out. The potential change in the cell terminal voltage and individual electrodes during the cell reversal experiment was measured using cell connected to a reference electrode by an electrolyte junction (Taniguchi, 2004).

Figure 10.24 shows the change in cell terminal voltage with time, as well as time-dependent changes in the anode and cathode potentials. According to their results, as soon as the experiment started, the cell terminal voltage rapidly dropped and became negative, and the MEA changed polarity due to cell reversal. After this initial rapid drop, the cell terminal voltage showed a steady decrease with time. Cell reversal occurred when the anode potential increased—which it did rapidly increase to nearly 1.5 V—and became more positive than the cathode potential. This result indicates that the anode potential increased as soon as the experiment started, until water electrolysis occurred because the anode was starved of fuel.

They also found that ruthenium dissolution from individual catalyst particles occurred in the anode catalyst layer, the most severe degradation occurring in the anode plane and the area close to the outlet region.

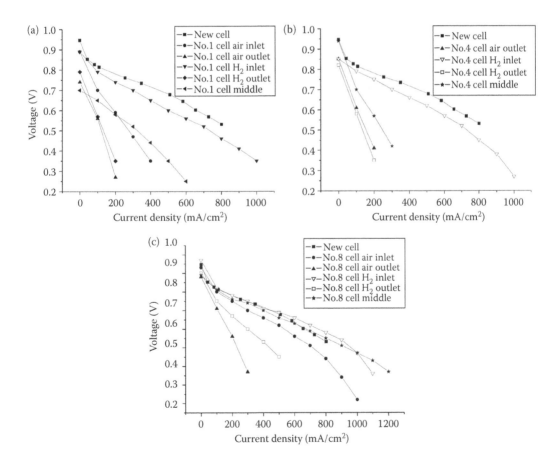

FIGURE 10.22 Polarization curves of small cells in different parts of a primary stack: (a) No. 1 cell; (b) No. 4 cell; (c) No. 8 cell. (Reprinted from *Int. J. Hydrogen Energy*, 35, P. Pei, X. Yuan, P. Chao. Analysis on the PEM fuel cells after accelerated life experiment, 3147–3151, Copyright (2010), with permission from Elsevier.)

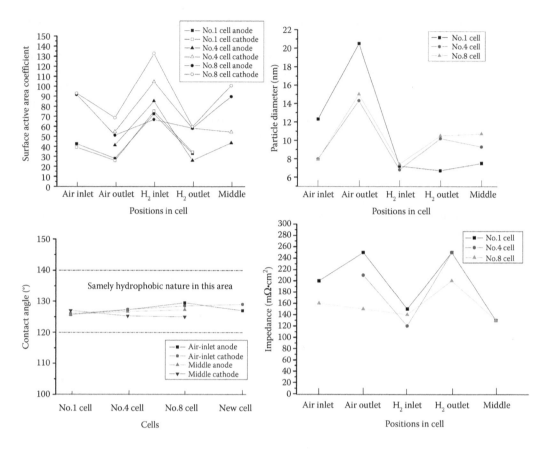

FIGURE 10.23 Results comparison of all parts in every cell. (a) Electrochemically active surface area coefficient. (b) Average particle diameter values. (c) Contact angle values. (d) Impedance values. (Reprinted from *Int. J. Hydrogen Energy*, 35, P. Pei, X. Yuan, P. Chao. Analysis on the PEM fuel cells after accelerated life experiment, 3147–3151, Copyright (2010), with permission from Elsevier.)

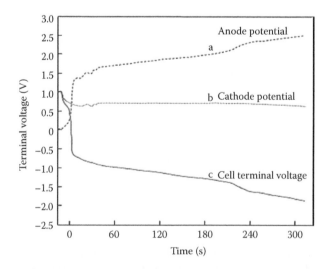

FIGURE 10.24 Changes in cell terminal voltage vs. time.

Fuel starvation causes severe and permanent damage to PEMFC electrocatalysts, and even momentary starvation must be absolutely avoided.

10.3.2 Feed Starvation at the Cathode Side

Oxidant (air) starvation can also occur during operation. Since water is produce at the PEMFC cathode via three processes—transport by humidified reactants, water generation from the cathode reaction, and transport via electroosmotic drag associated with proton transport across the PEM—condensed liquid water present within the cathode tends to block oxygen from accessing the electrocatalyst. A sudden change in oxygen demand, such as at start-up and load change, and water accumulation during long-term operation can cause air starvation. Rao et al. reported air starvation through simulation, finding that it is not enough to control a lumped oxygen excess ratio (OER) parameter to avoid oxygen starvation, and that variation in the process gains suggest that a nonlinear control approach is most likely needed to solve the problem of oxygen starvation (Rao, 2006).

Taniguchi et al. found degradation caused by air starvation to be much smaller than under fuel starvation because of the large difference in anode degradation. However, long-term operation (for a few tens of minutes) under the air-starved condition resulted in accelerated degradation in the electrode and in PEMFC performance (Taniguchi, 2008). Therefore, air starvation also has to be avoided to prevent a dramatic decrease in PEMFC performance.

Figure 10.25 shows their experimental results. The cell terminal voltage dropped rapidly to a negative value, and the MEA changed polarity due to cell reversal as soon as the experiment started. After the initial drop, the cell terminal voltage remained constant. Cell reversal occurred when the cathode potential decreased (to slightly under 0 V) and became more negative than the anode potential. This result indicates that as soon as the experiment started, the anode potential decreased until proton reduction to hydrogen occurred, since the cathode was starved of air. The anode potential was at about 0.1 V, which is higher than the anode potential of a normally operated PEMFC fueled with pure hydrogen under the same current density. This suggests that the local current density might have shown significant variation across the cell plane. The local current density in the region near the fuel inlet may have increased considerably if the oxygen concentration had been sufficient. Therefore, the anode overpotential increased as a result of the current concentration in a small region of the cell plane.

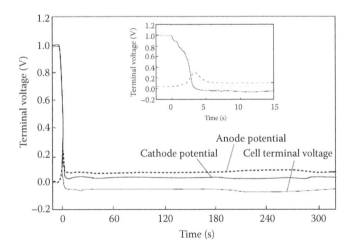

FIGURE 10.25 Time-dependent changes in the anode and cathode potential during a cell-reversal experiment. (Reprinted from *Int. J. Hydrogen Energy*, 33, A. Taniguchi, T. Akita, K. Yasuda. Analysis of degradation in PEMFC caused by cell reversal during air starvation, 2323–2329, Copyright (2008), with permission from Elsevier.)

Shen et al. found that with increased current density and decreased air stoichiometry, the voltage difference between the inlet and outlet increased. When loaded dynamically in load following mode, the fuel cell would suffer a temporary voltage fluctuation due to air starvation because the air response rate lagged behind the loading rate (Shen, 2008). They suggested that if the presupplied air stoichiometry was 1.5, the fuel cell would not suffer from any air starvation at the loading transient.

When starvation occurs at the cathode side, the pressure difference on either side of the MEA is too large, and consequently so is the force on the MEA, which decreases its lifetime.

10.3.3 Mitigation Strategies to Feed Starvation

When the fuel cell loads rapidly, the cathode gas starvation readily occurs. The usual approach is to provide enough air before loading but the drawback is that this can lead to increased fuel cell parasitic power and decreased efficiency. Qiang Shen et al. (2008) have researched this topic and found that when the excess air coefficient is 1.5, cathode gas starvation will not occur.

10.4 Low Relative Humidity

10.4.1 Impact on Fuel Cell Performance

In the course of PEMFC operation, hydrogen ions (protons) formed from the anode will conduct to the cathode through the PEM. Protons in the membrane transfer in the form of hydrated protons carried by water. The greater the number of protons through the membrane, the greater the amount of water moves from the anode to the cathode. Water content in the membrane largely determines the electrical conductivity and the efficiency of fuel cell power generation. As water penetrates the membrane, the amount of water decreases on the anode side and increases on the cathode side due to water infiltration and reactions. Because of this concentration difference, the water molecules should also transfer to the anode through back diffusion. In practice, water always passes from the anode to the cathode, which may lead to membrane dehydration on the anode side.

When water content in the membrane is moderate, the best fuel cell conductivity can be achieved, as the ohmic resistance is relatively small and the power efficiency is maximized. Too little water in the PEM will lead to membrane dehydration and to the disadvantage of proton transfer, greatly reducing its conductivity, and increasing ohmic resistance and decreasing power generation efficiency. For maintaining the membrane sufficiently hydrated, ways need to be found to add water on the anode side due to the water loss caused by the electric osmotic drag (EOD), and anode hydrogen must be humidified. If air, instead of pure oxygen, is used at the cathode, the oxygen concentration at the cathode will be much less. To obtain a higher oxygen concentration it is generally required to increase air flow rate. With water migrating from the anode to the cathode, the water loss at the anode side of the membrane is more serious. Therefore, the cathode air must also be humidified.

If the PEM membrane water content is too high, liquid water will form from saturated water vapor. Diffused liquid water will then dilute the concentration of the reactive gas and block the cathode gas diffusion layer (GDL) channel, reducing the speed of the oxygen transport and the efficiency of fuel cell power generation. A large amount of liquid water can submerge the electrodes and stop the fuel cell from working. Maintaining appropriate water content is one of the most effective ways to improve fuel cell power generation efficiency and service life.

Figure 10.26 describes seven means of water transmission in the PEM: production of water from cathode reaction; removal with air; back diffusion; humidification of the air supply; humidification of the hydrogen supply; and removal by circulating hydrogen (Yan, 2006).

In the case of membrane electrode dehydration, membrane electrical conductivity decreases and ohmic resistance increases, and so the internal fuel cell voltage loss will be even more serious, leading

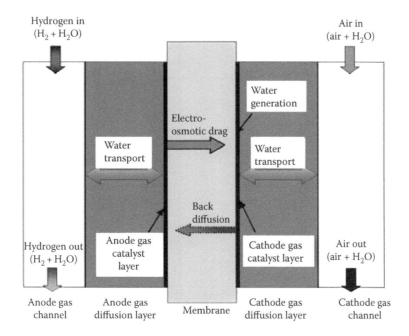

FIGURE 10.26 Schematic of the water transport process in a typical hydrogen PEMFC. (Reprinted from *J. Power Sources*, 158, Q. Yan, Toghiani, Hossein. Investigation of water transport through membrane in a PEM fuel cell by water balance experiments, 316–325, Copyright (2006), with permission from Elsevier.)

the fuel cell polarization curve to decrease with the drop in RH. Figures 10.27 and 10.28 are fuel cell polarization and resistance curves for different RHs.

Low RH will also make thermal management of the fuel cell difficult. As the membrane resistance increases, the output voltage drops. To achieve the same power, the output current must increase. Thus, the fuel cell temperature rises, further decreasing the RH and making the membrane more hydrophobic. The result is continuous deterioration in fuel cell performance.

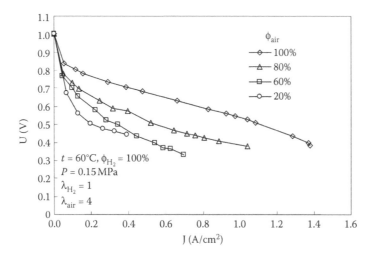

FIGURE 10.27 Fuel cell polarization curve under different RH. (Reprinted from handouts of P. Pei. With permission.)

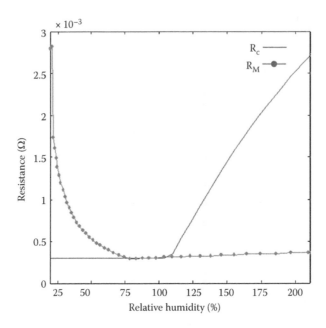

FIGURE 10.28 Effect of RH on fuel cell internal resistances (RC = electrode resistance, RM = membrane resistance). (Reprinted from *J. Power Sources*, 184, L. A. M. Riascos. Relative humidity control in polymer electrolyte membrane fuel cells without extra humidification, 204–211, Copyright (2008), with permission from Elsevier.)

10.4.2 The Effects of Different Anode and Cathode Gas Humidity

10.4.2.1 The Role of Air-Side Humidity and Low-RH Effects

To reduce the effects of air drying on the membrane, the air side needs to be humidified. Different humidity levels, as presented in Figure 10.29, have varying impacts on the fuel cell.

A moist membrane will have increased electrical conductivity and yield better fuel cell performance. When RH is constant in the anode inlet, water diffusion from the cathode to the anode can increase the membrane's water content. But under working conditions with a high current density, the amount of water carrying protons will be greater than the water diffusing from anode to cathode. Therefore, if the anode has low humidity, the fuel cell performance improves when the cathode RH is increased. When the anode humidity increases, this humidification can to some extent compensate for the loss of membrane water content high RH at the cathode can cause a decline in the partial pressure of oxygen in air, and even a flooding phenomenon, resulting in decreased fuel cell performance (Yan, 2006).

On the other hand, because the fuel cell's electrochemical reaction produces water, the RH is generally high (see Figure 10.30) for oxygen export. If the cathode RH is low, the membrane moisture content will not be uniform, resulting in lower RH in the gas intake area and higher RH in the gas export area, and making the current density uneven.

10.4.2.2 The Role of Hydrogen-Side Humidity and Low-RH Effects

Similarly, if the anode RH is very low, for example, 10%, membrane performance will increase with anode humidification. If the cathode RH is high, the membrane can be humidified through water back diffusion from cathode to anode when the hydrogen humidification is low. On the other hand, if the hydrogen humidity increases, the hydrogen partial pressure drops, and under this situation the hydrogen humidity has little effect on cell performance (see Figure 10.31).

In general, to improve fuel cell performance, keeping cathode humidity 100% while reducing anode humidity is more effective than keeping anode humidity 100% while reducing cathode humidity. This is

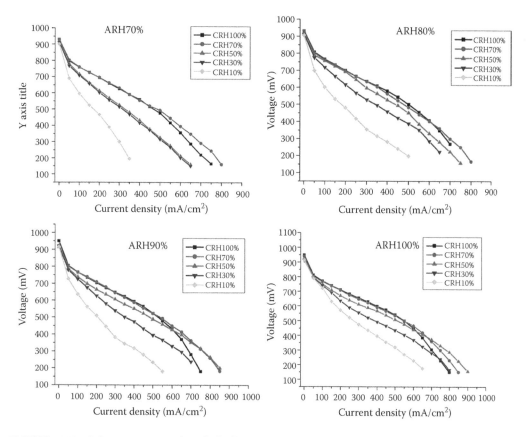

FIGURE 10.29 Polarization curves for a fuel cell operated at different air humidities. (Reprinted from *J. Power Sources*, 158, Q. Yan, Toghiani, Hossein. Investigation of water transport through membrane in a PEM fuel cell by water balance experiments, 316–325, Copyright (2006), with permission from Elsevier.)

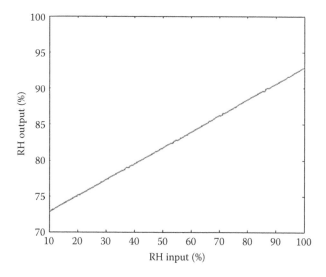

FIGURE 10.30 Variation in RH output vs. RH input (cathode). (Reprinted from *J. Power Sources*, 184, L. A. M. Riascos. Relative humidity control in polymer electrolyte membrane fuel cells without extra humidification, 204–211, Copyright (2008), with permission from Elsevier.)

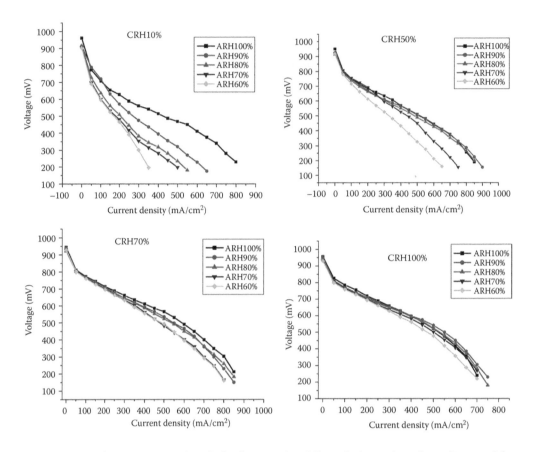

FIGURE 10.31 Polarization curves for a fuel cell operated at different hydrogen humidities. (Reprinted from *J. Power Sources*, 158, Q. Yan, Toghiani, Hossein. Investigation of water transport through membrane in a PEM fuel cell by water balance experiments, 316–325, Copyright (2006), with permission from Elsevier.)

mainly because maintaining 100% RH at the cathode side can basically ensure the membrane humidification. Particularly with a low operating voltage, the cathode produces large amounts of liquid water, and water from back diffusion wetting membrane at the anode side, at which point water flooding becomes the key issue. Reducing the anode humidity can direct more water to the anode by back diffusion, which alleviates water flooding at the cathode and reduces mass transfer loss, thereby increasing fuel cell performance (Yan, 2006).

10.4.3 Fuel Cell Humidification Methods

When the fuel cell is working under regular conditions, the membrane requires adequate moisture so it needs a humidification system, which consists of hydrogen humidifiers, air humidifiers, and a humidifier water system. Bringing moistened air and hydrogen into the fuel cell ensures its performance. The humidifier water system can be separate or can utilize water from the cooling system, the latter method reducing the system's complexity and using heat generated by the fuel cell (Xu, 2006).

PEMFC humidification technology primarily involves external, internal, and self-humidification.

10.4.3.1 External Humidification

External humidification uses additional equipment. Common methods include warm-up, dew point, permeable membrane, and liquid water spray.

10.4.3.1.1 Warm-Up Humidifier

In this method, hydrogen or oxygen is led into the bottom of a container through a pipe, and then bubbled through glass pearls and water. The glass pearls provide a large evaporation surface area, and ensure the export gas is nearly saturated. A set of temperature and humidity required by the fuel cell operation can be maintained for the gases that flow into the fuel cell. Adjusting the gas flow, temperature, and humidity levels changes the bubbling humidifier size (see Figure 10.32).

A small flow of gas with high humidity is achieved by this method, but controlling temperature and humidity is difficult. In addition, excess liquid is removed after gas bubbling through the water container, resulting in liquid water aggregation at the container's outlet, which is a problem.

10.4.3.1.2 Dew Point Humidifier

The basic principle is the same as for the warm-up humidification method, as a dew point humidifier also depends on water evaporation. However, its structure is much more complex. First, water is evaporated to steam, becoming the humidifier gas; then this high-temperature gas is cooled to saturation. Next, the water content of the gas can be accurately calculated according to the saturation vapor pressure at this temperature, after which the gas is heated to the fuel cell temperature and supplied to the fuel cell. The dew point method can accurately control gas humidity, but it consumes a huge amount of energy and requires a complex system.

10.4.3.1.3 Permeable Membrane Humidifiers

In this method, water is delivered through a pipeline to the top of the humidifier. Under gravity, water filters down along the wet membrane material, which absorbs it to create a uniform wet membrane. When dry air gets through the wet membrane material, water molecules absorb the air's heat and vaporize, thereby increasing the air humidity and forming moist air. The temperature drops but the air enthalpy remains unchanged. Adjusting the gas flow, wet membrane thickness, and water temperature can change the level of humidification.

10.4.3.1.4 Liquid Water Spray Humidifier

A liquid water spray humidifier includes an injection room, a section of pipeline for evaporating excess water, and an expansion chamber for recovery. First, water droplets are atomized through a nozzle using high-pressure gas, and sprayed into the gas. Then, the atomized water droplets are fully evaporated into the gas using a sufficiently long pipe. The gas is then passed through an expansion chamber and the excess water flow is filtered out. The gas temperature will drop due to the evaporation process, and so a heating wire type heater is added to the liquid water spray humidifier system to facilitate the evaporation

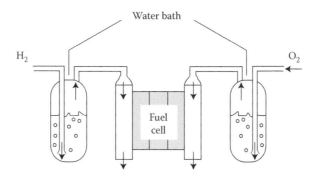

FIGURE 10.32 Schematic of a warm-up humidifier. (Reprinted from, S. Xu et al. 2006. *Journal of China Science and Technology*. No. 2. With permission.)

of liquid water. By adjusting the gas flow, water pressure, and water temperature to change the liquid water spray humidifier size, a certain temperature and humidity for the gas required by fuel cell operation are achieved (see Figure 10.33).

This system has two issues. First, liquid water cannot be fully injected for evaporation and so liquid water accumulation is a problem. Second, the dynamic response of the system is slow.

10.4.3.2 Internal Humidifier

The internal humidifier takes external liquid into the cell interior through a pump. Liquid water is then allocated according to the internal structure. Usually, the bipolar plate structure or the diffusion layer designs perform this function.

10.4.3.3 Self-Humidifier

A fuel cell by itself can also carry out humidification by changing its internal structure. The water generated by the fuel cell is used to achieve humidification.

Technology has simplified the use of a self-humidifying stack system, enhancing the stack power and the weight-to-power ratio. But this method is limited to low-power stacks. The major difficulties with self-humidification are insufficient water intake for wetting, and its slow dynamic response (see Figure 10.34).

10.4.4 Membrane Dehydration Solutions

10.4.4.1 Reduce the Temperature of the Fuel Cell

As shown from the gas temperature and RH curves in Figure 10.35, the gas RH range is not large. When membrane dehydration occurs, reducing the fuel cell temperature will increase the gas RH.

10.4.4.2 Increase the Inlet Temperature or the Difference in Temperature between Inlet and Outlet

Based on the relationship between saturated water vapor and temperature, the higher the inlet temperature, the more water vapor will be taken into the fuel cell, supplying more water to the dehydrated membrane. Similarly, increasing the temperature difference between inlet and outlet can bring more water into the fuel cell and allow less out.

10.4.4.3 Reduce the Inlet Gas Flow Rate

Membrane dehydration usually occurs in several cells at the front. The gas velocity on both sides of the MEA is larger at these cells than at the back cells. Thus, a great deal of water vapor is taken away by the

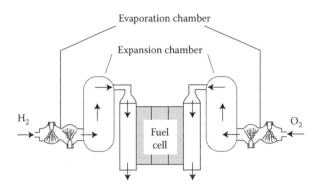

FIGURE 10.33 Schematic of a liquid water spray humidifier. (Reprinted from S. Xu et al. 2006. *Journal of China Science and Technology*. No. 2. With permission.)

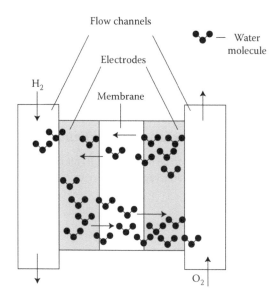

FIGURE 10.34 Schematic of self-humidification. (Reprinted from S. Xu et al. 2006. *Journal of China Science and Technology.* No. 2. With permission.)

reaction gases. Reducing the flow rate at the inlet can efficiently decrease the reactant velocity on both sides of the MEA and thereby mitigate membrane dehydration.

10.5 High Temperature

Usually, PEM fuel cell operating temperature is 60–80°C. But in practical applications, the internal temperature will increase quickly if there is a sudden load change, and if the system's cooling capacity

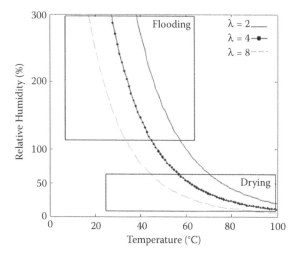

FIGURE 10.35 Temperature and RH for constant stoichiometries (2, 4, 8). (Reprinted from *J. Power Sources,* 184, L. A. M. Riascos. Relative humidity control in polymer electrolyte membrane fuel cells without extra humidification, 204–211, Copyright (2008), with permission from Elsevier.)

is insufficient and the fuel cell stack's internal temperature will become too high. In addition, if a pressurized intake system is cooled inadequately, the intake air temperature will increase. Running at higher temperatures can improve the fuel cell electrode kinetic parameters (Beattie, 1999; Parthasarathy, 1992; Xu, 2005; Zhang, 2006), and the fuel and oxidant mass transfer rates will increase, which is helpful in improving fuel cell performance. However, high-temperature operation will accelerate PEMFC aging and failure, for reasons to be explored in the following subsections.

10.5.1 Material Shortages and Flooding Phenomena

The fuel and oxidant gases of PEMFCs need to be humidified. RH is related to the saturation vapor pressure of water. The formula for calculating water vapor pressure is

$$\ln P_{sat} = A - \frac{B}{T}$$

(10.1)

where A and B are constants, and T is the temperature (K).

From Equation 10.1 it is evident that water vapor pressure increases exponentially with rising temperature. When gas enters the fuel cell stack, the gas temperature will increase. Under certain circumstances, if a constant inlet pressure and RH are maintained, the water vapor intake will increase and that the amount of reactive gas will decrease, which may cause fuel cell material deficiencies as well as flooding phenomena; the latter will lead to a decline in fuel cell performance.

10.5.2 Membrane Dehydration

Typically, a low-temperature PEMFC using a proton conductive membrane has a water-based PEM, and the proton conductivity strongly depends on the RH. When the fuel cell stack's internal temperature is too high, the proton conductive membrane will lose water and conductivity will drop. In addition, if the humidification of a hot gas is insufficient when entering the fuel cell stack, the dry and hot gas will lead to dehydration of the proton conductive membrane. In the case of dehydration, the PEM may shrink and crack, which will accelerate the crossover of fuel and oxidant; the resulting exothermic reaction will cause local hot spots, leading to membrane pinholes that further accelerate gas crossover, creating a vicious circle. Figure 10.36 presents a Nafion® membrane TGA diagram (Samms, 1996). It is evident that when the temperature reached ~300°C, the membrane started to decompose; a continued rise in temperature would likely cause membrane softening, followed by combustion.

10.5.3 Gas Crossover

When the fuel cell is running, hydrogen or air will always penetrate across the proton conductive membrane. With increases in temperature or pressure, the crossover rate also increases. Figure 10.37 presents hydrogen crossover in a Nafion112-based MEA.

The crossover hydrogen reacts with oxygen in the cathode catalyst layer, reducing the fuel cell potential and increasing the membrane electrode temperature, leading to rapid aging of the fuel cell. Oxygen penetration to the anode side forms active peroxide free radicals such as HOO• and HO• that can chemically attack the proton conductive membrane, resulting in membrane degradation (Buchi, 1995; Wang, 1998). Therefore, increasing the fuel cell working temperature will accelerate performance degradation.

10.5.4 Accelerated Material Aging

To improve Pt catalyst utilization and reduce Pt loading, Pt is usually dispersed on a carbon support that has a large surface area. The fuel cell cathode maintains a relatively high oxidation potential. After

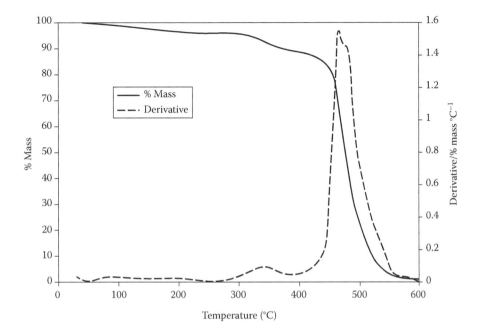

FIGURE 10.36 A Nafion membrane TGA diagram. (Reprinted from S. R. Samms, S. Wasmus, R. F. Savinell. 1996. *J. Electrochem. Soc.* 143: 1498–1504. With permission from ECS.)

adsorption of oxygen atoms on the catalyst, the carbon support becomes corroded and generates CO or CO_2, which leads to catalyst loss and shortens the fuel cell lifetime. This process accelerates with higher temperatures. During fuel cell operation, the Pt particles will agglomerate, decreasing the active surface area; increasing the temperature will speed up this process (Wilson, 1993).

High humidity will accelerate the degradation or oxidation of fuel cell materials such as seals, gaskets, bipolar plates, and so forth. Silicone rubber is often used in gaskets. Figure 10.38 presents the degradation characteristics of silicone rubber. As the test time progresses, the silicone rubber weight loss ratio

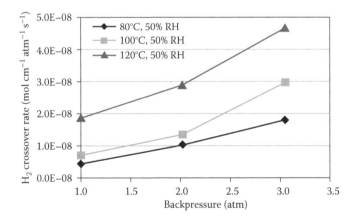

FIGURE 10.37 H_2 crossover rate as a function of fuel cell back pressure at 50% RH and three different temperatures, marked in the graph. (Reprinted from *J. Power Sources*, 167, X. Cheng, J. L. Zhang, Y. H. Tang. Hydrogen crossover in high-temperature PEM fuel cells, 25–31, Copyright (2007), with permission from Elsevier.)

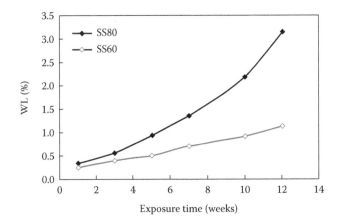

FIGURE 10.38 Weight loss with exposure time from test samples of silicone rubber at 60°C and 80°C. (Reprinted from *J. Power Sources*, 172, J. Z. Tan, Y. J. Chao, X. D. Li. Degradation of silicone rubber under compression in a simulated PEM fuel cell environment, 782–789, Copyright (2007), with permission from Elsevier.)

at 80°C rapidly increases, illustrating how operating at higher temperatures will negatively impact the fuel cell system's engineering materials.

10.5.5 Mitigation Measures

To avoid the many issues arising from high internal temperature and high intake air temperature, the first steps are to avoid sudden increases in the fuel cell load and improve the system's cooling capacity. Preventing high intake air temperatures resulting from pressurization and other factors, and ensuring proper intake RH are also important. Second, in order to adapt PEM fuel cells to a wider range of working temperatures, a proton conductive membrane that has an RH-independent proton conductivity and low gas permeability should be used. In addition, a catalyst with high activity and stability, and new catalyst synthesis process should be employed to reduce free radicals.

10.6 Dynamic Response

Dynamic response is an important aspect of evaluating PEMFCs. For different parameters, the dynamic response of a fuel cell varies widely. If the external load changes, the fuel cell will have a transient fluctuation process, which in turn affects parameters such as gas excessive coefficient, temperature, pressure, and humidity. Such fluctuation will also ultimately lead to fuel cell decay.

10.6.1 Dynamic Response Characteristics and Influencing Factors

Many factors affect the dynamic response of fuel cells, including the loading speed, loading rate, excess coefficient, temperature, humidity, and pressure. This section will introduce them in detail.

10.6.1.1 Loading Range

Figure 10.39 presents a single fuel cell dynamic response curve of voltage under different loadings, with an excess coefficient of $A/C = 3.0{:}3.0$. This is a typical dynamic response curve that cannot meet the load requirements due to restricted air flow (Hsu, 2009). The negative voltage pulse process can be observed from the group loading curves. The larger the load, the greater the voltage negative pulse amplitude, and considerable time is required to reach a new steady-state value.

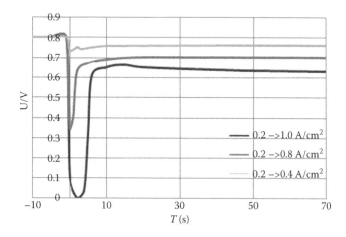

FIGURE 10.39 Dynamic response curves of voltage under different loadings. (Reprinted from *Renewable Energy*, 34, C.-Y. Hsu, F.-B. Weng. Transient phenomenon of step switching for current or voltage in PEMFC, 1979–1985, Copyright (2009), with permission from Elsevier.)

10.6.1.2 Loading Speed

Figure 10.40 shows a group of dynamic response curves for a single fuel cell stack with different loading times and average voltage (Yan, 2007). The shorter the time is and the faster the voltage load, the greater is the magnitude of the voltage negative pulse. When accelerates, the electrode potential and overall fuel cell voltage drop, sometimes even leading to a negative voltage because the cathode lacks oxygen. Therefore, to improve the fuel cell stack's stability and longevity, we should try to reduce the loading speed and extend the loading time.

10.6.1.3 Excess Coefficient

Shown in Figure 10.41 are six groups of data from an experiment looking at the effects of the gas excess coefficient on dynamic response (Cho, 2008). When the excess coefficient of H_2 and air change, the fuel cell voltage undergoes basically the same process of change, and as the gas excess coefficient increases, the dynamic response and voltage change accelerate.

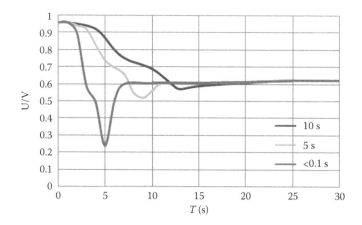

FIGURE 10.40 The dynamic response curves of a single fuel cell. (Reprinted from *J. Power Sources*, 163, X. Yan, M. Hou, L. Sun. The study on transient characteristic of proton exchange membrane fuel cell stack during dynamic loading, 966–970, Copyright (2007), with permission from Elsevier.)

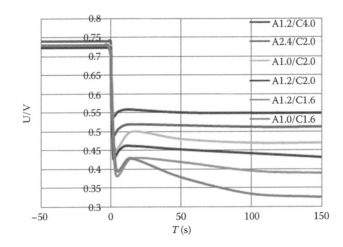

FIGURE 10.41 Gas excess coefficient effect on the dynamic response. (Reprinted from *J. Power Sources*, 185, J. Cho, H.-S. Kim. Transient response of a unit proton-exchange membrane fuel cell under various operating conditions, 118–128, Copyright (2008), with permission from Elsevier.)

10.6.1.4 Temperature

Temperature has an important effect on the performance of fuel cells, directly affecting the reactant gas humidity and water balance. Increasing the temperature will accelerate water evaporation and reduce the fuel cell humidity. However, increasing the temperature can also enhance the reactant gas diffusion and the reaction kinetics. Therefore, there exists the most suitable working temperature for fuel cells.

10.6.1.5 Humidity

Humidity is another important factor affecting a fuel cell's dynamic response. Figure 10.42 is a 100–200 A dynamic voltage response curve obtained by changing the gas humidity. At 100% humidity, the

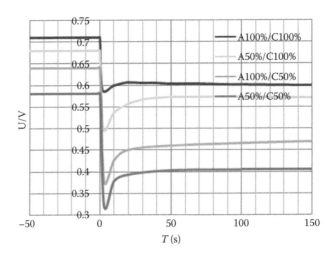

FIGURE 10.42 Dynamic voltage response curves. (Reprinted from *J. Power Sources*, 185, J. Cho, H.-S. Kim. Transient response of a unit proton-exchange membrane fuel cell under various operating conditions, 118–128, Copyright (2008), with permission from Elsevier.)

dynamic response is best. If the upload process is changed and there is insufficient water supply, membrane dehydration and decreased membrane ionic conductivity will result, with a corresponding increase in voltage drop. In Figure 10.42, it is evident that the cathode gas humidity has a slightly larger impact on the fuel cell than the anode gas humidity (Cho, 2008).

10.6.2 Effect of Dynamic Response on Fuel Cell Performance

10.6.2.1 Effect on Temperature

Figure 10.43 presents temperature fluctuations in relation to dynamic load, for different regions of a fuel cell. Evidently, the temperature differs according to the location and air excess coefficient. The temperature change is most obvious at the air entrance, resulting in overheating that accelerates membrane degradation and fuel cell performance degradation (Yan, 2007).

10.6.2.2 Impact on Pressure

Pathapati et al. proposed a fuel cell system integrated model, in which both sides of the gas pressure can cause transient changes in the dynamic process. In Figure 10.44, the hydrogen pressure changes significantly while the air side changes only slightly, which will create a difference in membrane pressure and produce fluctuations in the dynamic process, potentially shortening the membrane's lifetime (Pathapati, 2004).

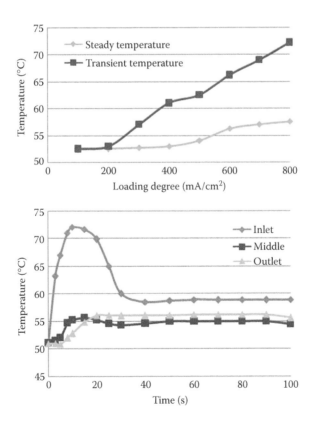

FIGURE 10.43 Temperature fluctuation in relation to dynamic load, for different regions of a fuel cell. (Reprinted from *J. Power Sources*, 163, X. Yan, M. Hou, L. Sun. The study on transient characteristic of proton exchange membrane fuel cell stack during dynamic loading, 966–970, Copyright (2007), with permission from Elsevier.)

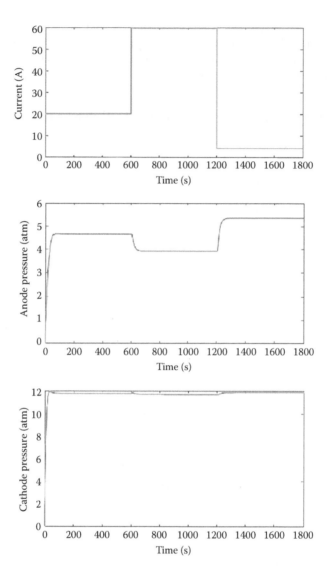

FIGURE 10.44 The hydrogen and air pressure change with load changing. (Reprinted from *J. Renewable Energy*, 30, P. R. Pathapati, X. Xue, J. Tang. A new dynamic model for predicting transient phenomena in a PEM fuel cell system, 1–22, Copyright (2005), with permission from Elsevier.)

10.6.2.3 Effect on Reactant Gas Distribution

In the dynamic change process, lack of reactant gas is a common problem, which occurs because the gas load response lags behind the current response. When the gas excess coefficient and the load speed are low, lack of gas will become particularly obvious. This occurs mostly in the cathode exit, and thus the cathode current at the entrance will be smaller than at the exit.

10.6.3 Measures to Improve the Dynamic Response

When fuel cell loads change rapidly from low to high, cathode gas shortage quickly occurs. The usual approach to mitigate this is to provide enough air before loads, but the drawback is that this increases fuel cell parasitic power and decreases the system overall efficiency.

Pathapati has conducted research in this area. Figure 10.45 describes the current density from 50 to 500 mA cm^{-2} and the fuel cell voltage changes with the excess air coefficients of 1.1, 1.3, and 1.5 respectively. With the presupply excess air coefficient increasing, the dynamic response of the fuel cell performance improves, becoming more stable and faster. When the excess air coefficient is 1.5, starvation does not occur.

10.7 Water Management

Water management, which can significantly influence catalyst and membrane degradation, is of vital importance for a PEMFC's lifetime.

Excess water is a common issue in PEMFC operation. As reactant gas humidity increases, Pt dissolution–precipitation and carbon support corrosion accelerate. Too much water can block the flow channels and pores of the GDL and lead instantly to reactant starvation, which can induce catalyst support degradation (see Chapter 3).

Dehydration is another important issue. Dehydration can cause mechanical stress in the membrane that leads to failures (tearing and cracking), and it can also accelerate the chemical degradation of PEMs (Endoh, 2004). RH cycling (rather than steady-state operation) can further accelerate membrane degradation. Severe dehydration will cause irreversible membrane degradation (delamination, pinholes) within about 100 s (Canut, 2006).

10.7.1 Flooding

Flooding is the accumulation of excess water at the anode or cathode side of the membrane, but especially at the cathode.

Flooding immediately increases the mass transport losses, which means that the transport rate of the reactants is significantly reduced. Water blocks the GDL pores, preventing the reactants from reaching the catalysts and thus leading to gas starvation and an immediate drop in cell potential.

Flooding also has a negative effect on PEMFC durability. Too much water accelerates corrosion of the electrodes, the catalyst layers, the gas diffusion media, and the membrane (Pierre, 2000). Corrosion in turn increases ohmic losses. Dissolved catalyst particles and impurities can also be transported into the membrane, replacing H$^+$ ions and reducing the proton conductivity over time, eventually leading to cell failure (Ge, 2007).

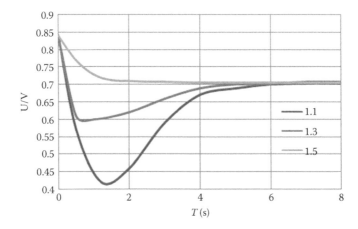

FIGURE 10.45 Voltage change with the excess air coefficients of 1.1, 1.3, and 1.5. (Reprinted from *J. Power Sources*, 179, Q. Shen, M. Hou, X. Yan. The voltage characteristics of proton exchange membrane fuel cell under steady and transient states, 292–296, Copyright (2008), with permission from Elsevier.)

It takes much longer to accumulate water at the anode (Nguyen, 1993), but the accumulated water in the anode side can also cause fuel starvation with subsequent carbon corrosion in the catalyst layer, which is a very serious consequence.

10.7.2 Membrane Dehydration

Dehydration of the membrane, mainly caused by a shortage of water, usually occurs on the anode side. It can lead to membrane degradation, either instant or long term. With decreasing water content the conductivity drops, which leads to higher ionic resistance and higher ohmic losses (Canut, 2006). That in turn leads to a decline in cell voltage. The decline caused by temporary dehydration can usually be recovered by humidification, but long-term dehydration will cause irreversible damage to the membrane (see Figure 10.46).

As water is generated in the fuel cell reaction, dehydration is more serious at the flow field inlet. When membranes are exposed to dry conditions over a longer period, they can become brittle and develop crazes or cracks (Huang, 2006). This causes gas leakage or crossover, and then leads to the exothermic reaction of hydrogen and oxygen. Eventually, hot spots form, which will accelerate membrane degradation. Therefore, the drier the conditions, the shorter will be the life of the cell.

Dehydration can be caused by insufficient humidification, high temperature, and electroosmosis. At high current density, the electroosmotic force is strong, and the water present at the anode side by back diffusion is not sufficient to keep the membrane wet. It has been shown that during a step increase in the current density, the electroosmotic force will immediately pull water molecules from the anode to the cathode (Wang, 2006).

10.7.3 Water Management Strategies

Since water management significantly affects the lifetime of PEMFCs, efficient management strategies are needed. These consist of failure diagnosis and mitigation.

Failure diagnosis for water management focuses mainly on flooding and membrane dehydration.

As mentioned in Section 10.5.1, the easiest way to implement flooding diagnosis is to monitor pressure drop. Before beginning the diagnosis, the normal pressure drop curve should be tested for comparison; this test will be the baseline for the diagnosis. If the pressure drop is seriously larger than normal, flooding is supposed to have occurred.

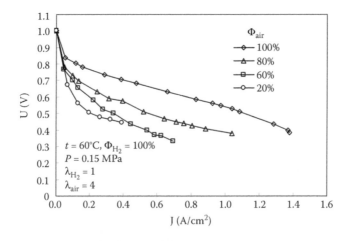

FIGURE 10.46   Effect of dehydration on fuel cell performance. (Data from Tsinghua University.)

The two main methods to solve the flooding problem are: (1) opening the pulse purge valve to allow a pressure wave of hydrogen to push the water out of the flow field; and (2) raising the operating temperature to decrease the RH, turning the liquid water into stream that is then expelled with the reactant gas as exhaust.

To diagnose membrane dehydration, membrane resistance measurements can be used to show the level of dehydration, as it has been proven that membrane resistance is higher when dehydration occurs. Three resistance measurement methods present a good compromise between simplicity, invasiveness, and reliability (Steiner, 2008): current interrupt (CI), AC resistance (ACR) and high-frequency resistance (HFR). All are based on the cell's response to certain solicitations.

If membrane dehydration occurs, raising the humidification temperature will help to bring more water into the flow field, and lowering the fuel cell temperature can prevent too much water from flowing away. Another method is to increase the current density to generate more water for humidifying the membrane. The extra power that results from increasing the current density can be stored in batteries.

10.8 Summary

The purpose of this chapter is to understand PEMFC performance degradation under different operating conditions, and to extend cell lifetime by mitigating those factors that degrade the performance of fuel cells. The main points include:

- During various operating cycles (load cycling, idling conditions, high-power conditions, start–stop cycles), PEMFC performance decreases linearly. Under load cycling, PEMFC performance declines most sharply.
- PEMFC performance can be recovered by idling with a small current.
- PEMFC performance decreases rapidly during feed starvation.
- In the case of membrane dehydration, membrane electrical conductivity decreases and ohmic resistance increases. Low RH will make thermal management of the fuel cell difficult. The result is continuous deterioration in fuel cell performance.
- Humidification involves external, internal, and self-humidifiers.
- The main reasons for PEMFC performance degradation at high temperatures are material shortages, memrane dehydration, gas crossover through the membrane, and accelerated material aging.
- Dynamic response is an important factor in evaluating the PEMFC. Rapid loading is very likely to cause feed starvation.
- Water management, which can significantly influence the degradation of catalyst and membrane, is of vital importance for a PEMFC's lifetime.
- Water flooding can be diagnosed by anodic pressure drop, and mitigated by raising the cell temperature, reducing the gas temperature, and purging.
- Membrane dehydration can be diagnosed from internal resistance measurement and mitigated by reducing the cell temperature and raising the gas temperature.
- To mitigate feed starvation, the usual approach is to provide enough air before loading but the drawback is that this can lead to increased fuel cell parasitic power and decreased overall efficiency.
- To mitigate high temperature, sudden increases in the fuel cell load should be avoided, and the cooling system capacity should be improved. Additional mitigation strategies include using proton conductive membrane that has RH-independent conductivity and low gas permeability, using catalysts with high activity and stability, and new catalyst synthesis processes to reduce free radicals.
- Mitigation strategies for water management include controlling reactant humidity, controlling reactant flow rate, controlling temperature, controlling pressure, and modifying current density.

References

A. Parthasarathy, S. Srinivasan, A. J. Appleby. 1992. Temperature dependence of the electrode kinetics of oxygen reduction at the platinum/Nafion® interface—A microelectrode investigation. *J. Electrochem. Soc* 139: 2530–2537.

A. Taniguchi, T. Akita, K. Yasuda. 2004. Analysis of electrocatalyst degradation in PEMFC caused by cell reversal during fuel starvation. *J. Power Sources* 130: 42–49.

A. Taniguchi, T. Akita, K. Yasuda. 2008. Analysis of degradation in PEMFC caused by cell reversal during air starvation. *Int. J. Hydrogen Energy* 33: 2323–2329.

B. Li, R. Lin, D. Yang. 2010. Effect of driving cycle on the performance of PEM fuel cell and microstructure of membrane electrode assembly. *Int. J. Hydrogen Energy* 35: 2814–2819.

C.-Y. Hsu, F.-B. Weng. 2009. Transient phenomenon of step switching for current or voltage in PEMFC. *Renewable Energy* 34: 1979–1985.

E. Endoh, S. Terazono, H. Widjaja. 2004. Degradation study of MEA for PEMFCs under low humidity conditions. *Electrochemical Solid State Letters* 7: A209–A211.

F.-B. Weng, C.-Y. Hsu, C.-W. Li. 2010. Experimental investigation of PEM fuel cell aging under current cycling using segmented fuel cell. *Int. J. Hydrogen Energy* 33: 3664–3675.

F. N. Buchi, B. Gupta, O. Haas. 1995. Study of radiation-grafted FEP-G-polystyrene membranes as polymer electrolytes in fuel cells. *J. Electrochim. Acta* 40: 345–353.

H. Wang, G. A. Capuano. 1998. Behavior of Raipore radiation-grafted polymer membranes in H_2/O_2 fuel cells. *J. Electrochemist. Soc.* 145: 780–784.

H. Xu, Y. Song, H. R. Kunz. 2005. Effect of elevated temperature and reduced relative humidity on ORR kinetics for PEM fuel cells. *J. Electrochem. Soc* 152(9): A1828–A1836.

J. Wu, X. Z. Yuan, J. J. Martin et al. 2008. A review of PEM fuel cell durability: Degradation mechanisms and mitigation strategies. *J. Power Sources* 184: 104–119.

J. Kang, D. W. Jung, S. Park. 2010. Accelerated test analysis of reversal potential caused by fuel starvation during PEMFCs operation. *J. Hydrogen Energy* 35: 3727–3735.

J. L. Zhang, Z. Xie, J. Zhang. 2006. High temperature PEM fuel cells. *J. Power Sources* 160: 872–891.

J. St Pierre, D. Wilkinson, S. Knights. 2000. Relationships between water management, contamination and lifetime degradation in PEMFC. *J. New Mater. Electrochem. Systems* 3: 99–106.

J. Cho, H.-S. Kim. 2008. Transient response of a unit proton-exchange membrane fuel cell under various operating conditions. *J. Power Sources* 185: 118–128.

J. Z. Tan, Y. J. Chao, X. D. Li. 2007. Degradation of silicone rubber under compression in a simulated PEM fuel cell environment. *J. Power Sources* 172: 782–789.

L. A. M. Riascos. 2008. Relative humidity control in polymer electrolyte membrane fuel cells without extra humidification. *J. Power Sources* 184: 204–211.

Le Canut, J. M. Abouatallah. 2006. Detection of membrane drying, fuel cell flooding, and anode catalyst poisoning on PEMFC stacks by electrochemical impedance spectroscopy. *J. Electrochemical Society* 153: A857–A864.

M. P. Rodgers, R. Agarwal, B. P. Pearman et al., 2009. Accelerated durability testing of perfluorosulfonic acid MEAs for PEMFCs using different relative humidities. *ECS Transactions* 25: 1861–1871.

M. S. Wilson, F. H. Garzon, K. E. Sickafus. 1993. Surface area loss of supported platinum in polymer electrolyte fuel cells. *J. Electrochem.Soc* 140: 2872–2877.

N. Yousfi-Steiner, P. Mootguy, D. Candusso. 2008. A review on PEM voltage degradation associated with water management: Impacts, influent factors and characterization. *J. Power Sources* 183: 260–274.

P. D. Beattie, V. I. Basura, S. Holdcroft. 1999. Temperature and pressure dependence of O_2 reduction at Pt | Nafion® 117 and Pt | BAM® 407 interfaces. *J. Electroanal. Chem* 468: 180–192.

P. R. Pathapati, X. Xue, J. Tang. 2005. A new dynamic model for predicting transient phenomena in a PEM fuel cell system. *J. Renewable Energy* 30: 1–22.

P. Pei. 2008. Lifetime evaluate and the effects of operation conditions on Automotive fuel cells. *Chinese Journal of Mechanical Engineering* 23: 66–71.

P. Pei, Q. Chang, T. Tang. 2008. A quick evaluating method for automotive fuel cell lifetime. *Int. J. Hydrogen Energy* 33: 3829–3836.

P. Pei, X. Yuan, P. Chao. 2010. Analysis on the PEM fuel cells after accelerated life experiment. *Int. J. Hydrogen Energy* 35: 3147–3151.

Q. Shen, M. Hou. 2008. The voltage characteristics of proton exchange membrane fuel cell under steady and transient states. *J. Power Sources* 179: 292–296.

Q. Shen, M. Hou, X. Yan. 2008. The voltage characteristics of proton exchange membrane fuel cell (PEMFC) under steady and transient states. *J. Power Sources* 179: 292–296.

Q. Yan, Toghiani, Hossein. 2006. Investigation of water transport through membrane in a PEM fuel cell by water balance experiments. *J. Power Sources* 158: 316–325.

R. Lin, B. Li, Y. P. Hou. 2009. Investigation of dynamic driving cycle effect on performance degradation and micro-structure change of PEM fuel cell. *Int. J. Hydrogen Energy* 34: 2369–2376.

R. Madhusudana Rao, R. Rengaswamy. 2006. A distributed dynamic dynamic model for chronoamperometry, chronopotentiometry and gas starvation studies in PEM fuel cell cathode. *J. Chem. Eng. Sci.* 61: 7393–7409.

S. Ge, C. Y. Wang. 2007. Liquid water formation and transport in the PEFC anode. *J. Electrochem. Soc.* 154: B998–B1005.

S. Mu. 2010. Accelerated durability tests of catalyst layers with various pore volumes for catalyst coated membranes applied in PEM fuel cells. 35: 2872–2876.

S. Xu, Q. Chen, T. Ma. 2006. The research and development prospect of fuel cell vehicles with humidifiers. *China Sci. Technol. Inf.* 2: 60–63.

S. Kim, S. Shimpalee, J. W. van Zee. 2004. The effect of stoichiometry on dynamic behavior of a proton exchange membrane fuel cell (PEMFC) during load change. *J. Power Sources* 135: 110–121.

S. R. Samms, S. Wasmus, R. F. Savinell. 1996. Thermal stability of Nafion® in simulated fuel cell environments. *J. Electrochem.Soc* 143: 1498–1504.

S. Sakamoto, M. Karakane, H. Maeda. 2000. Study of the factors affecting PEMFC life characteristic. In: *Abstracts Fuel cell seminar*: 141–144.

S.-Y. Ahn, S.-J. Shin, H. Y. Ha. 2002. Performance and lifetime analysis of the kW-class PEMFC stack. *J. Power Sources* 106: 295–303.

T. Nguyen, R. White. 1993. A water and heat management model for PEMFC. *J. Electrochem. Soc.* 140: 2178–2186.

X. Cheng, J. L. Zhang, Y. H. Tang. 2007. Hydrogen crossover in high-temperature PEM fuel cells. *J. Power Sources* 167: 25–31.

X. Huang, R. Solasi, Y. Zou. 2006. Mechanical endurance of polymer electrolyte membrane and PEM fuel cell durability. *J. Polymer Science part B-Polymer Phys.* 44: 2346–2357.

X. Yan, M. Hou, L. Sun. 2007. The study on transient characteristic of proton exchange membrane fuel cell stack during dynamic loading. *J. Power Sources* 163: 966–970.

Y. Wang, C. Y. Wang. 2006. Dynamics of polymer electrolyte fuel cells undergoing load changes. *Electrochem Acta* 51: 3924–3933.

Design-Related
Durability Issues

Kui Jiao
University of Waterloo

Xianguo Li
University of Waterloo

11.1 Introduction

Durability is one of the key issues that hinder the commercialization of proton exchange membrane (PEM) fuel cells. As presented in the previous chapters, degradations occur in all the components of a PEM fuel cell with different mechanisms under various operating conditions. The design of PEM fuel cell therefore must be optimized to enhance the durability while maintaining/improving the performance and reducing the cost before its successful commercialization. The design improvement is needed from each single component to system level. On the single component level, development of novel materials and optimization of geometrical structures of the cell components are the key issues. On the stack level, manifold design and material and design capabilities among different components need to be carefully considered. On the system level, fuel/oxidant conditioning, thermal management, and monitoring/controlling of the operating conditions are critically important. In this chapter, the design-related durability issues of membrane, catalyst layer (CL), gas diffusion layer (GDL), membrane electrode assembly (MEA), bipolar plate, flow channel, PEM fuel cell stack, and PEM fuel cell system are reviewed in different sections, and finally a summary follows to provide an overview of the chapter.

11.2 Membrane Design

Membrane is one of the key components of a PEM fuel cell, and the state-of-the-art membrane at present is the perfluorinated sulfonic acid (PFSA) membrane. The degradation modes of membrane can be classified into physical degradation and chemical degradation. The physical degradation is caused by time-dependent deformation under compression, thermal expansion/contraction, water absorption/desorption, pressure difference, etc; and the chemical degradation is caused by the change of the membrane chemical structure through chemical/electrochemical reactions. Since the detailed mechanisms of membrane degradation have been discussed in Chapter 4, this section only focuses on the effects of design parameters on the membrane durability. Generally, the durability of the state-of-the-art PFSA membranes is only acceptable under continuous (i.e., constant load), intermediate current density, fully humidified, and low-temperature (e.g., <80°C) operations. However, fully humidified condition requires complicated water management strategies, and practical operations often involve load changes (especially for automotive applications). Developing more durable membranes is therefore needed for the commercialization of PEM fuel cells. The general design objectives are to improve the durability and proton conductivity of membrane under the corrosive and oxidative conditions with rapid and large load changes, high temperature, and low humidity.

11.2.1 General Aspects of Membrane Design

Although the PFSA membrane, Nafion, was first used as the PEM in General Electric (GE) designed PEM fuel cells more than forty years ago, it remains the archetypical membrane in PEM fuel cells under development today. This is despite well-recognized durability-related drawbacks that include: limitations in thermal stability (thermal expansion/contraction and membrane transition temperature of about 100°C), the requirement that the material remains fully hydrated during operation of the device (the expansion/contraction of membrane due to water absorption/desorption), and gas crossover (mainly causing chemical degradation). In addition to the durability issues, high proton conductivity and no electronic conductivity are also the critical design requirements of membrane. However, the design requirements are mutually conflicting, and the practical membranes of today are a result of compromise satisfying these requirements. For example, high proton conductivity for low ohmic polarization (or low internal resistance) suggests that the membrane structure be porous, thin and with high fixed charge concentration; while the ability to separate reactant gases from intermixing requires that the membrane be of low gas permeability (i.e., low porosity and small pore sizes) and thick; sufficient mechanical strength manifests that the membrane be reasonably thick as well; and the fixed charge concentration must be reasonably low to maintain the mechanical and thermal stabilities. Moreover, PEM fuel cell performance in general improves at high temperatures due to higher proton conductivity of membrane, faster-reaction kinetics, and easier thermal and water management, yet many membranes are generally not stable in hot corrosive and oxidative environments of PEM fuel cell operating condition.

Different designs of membrane have been explored to improve the durability while maintaining or improving the performance. The design approaches can be generally categorized as (1) modification of the present PFSA membranes (i.e., modification of backbone, side chain or protogenic group); (2) addition of inorganic particles (e.g., silicon oxides) to various membranes; (3) blends of polymers with oxo-acids (replacing water with another proton conduction assisting solvent) such as polybenzimidazole (PBI) membrane doped with phosphoric acid; and (4) synthesis of entirely novel membranes (Paddison et al., 2007). Even though great efforts have been made, development of membrane superior to Nafion has not emerged yet. Since all the durability-related design efforts tend to solve the physical degradation, chemical degradation or both of them, in this section, the different designs are presented in two subsections: those primarily related to physical degradation are presented first, followed by the designs mainly related to chemical degradation.

11.2.2 Physical Degradation-Related Design

Improving the mechanical and thermal strengths of membrane is critical to solve the physical degradation problems, and this has been mainly achieved by reducing the polymer side chain length of PFSA membranes or by using the reinforced/composite membranes.

The polymer side chains of PFSA membranes are the proton conductors, which are usually hydrophilic and therefore can attract water to further enhance the proton conductivity. The hydrophilic side chains are supported by the hydrophobic polymer backbones. The combination of the hydrophobic backbones and hydrophilic side chains makes the PFSA membranes both physically strong and proton conductive. The famous PFSA membranes such as Nafion, Flemion, and Aciplex membranes possess long side chains, for example, Nafion membrane has two ether oxygen atoms in each side chain (between the sulfonic acid group and the backbone), as shown in Figure 11.1 (Li, 2005). Since the hydrophilic side chains are not physically strong, reducing the side chain length becomes an effective way to solve the physical degradation problem. The first PFSA membrane with short side chains is the Dow membrane developed in the 1980s. The Dow membrane has similar backbones as the Nafion membrane, but only one ether oxygen atom at each side chain, as shown in Figure 11.1 (Li, 2005). The durability and other characteristics of Dow membrane have been extensively studied. Tant et al. (1989) found that this short side-chain membrane has a transition temperature of about 165°C, which is much higher than the Nafion membrane (around 100°C), indicating that the thermal stability is significantly improved with short side chains. They also found that the Dow membrane is able to have a higher fixed charge concentration (lower equivalent weight (EW)) than the Nafion, and a similar conclusion was also reported in Moore et al. (1989). The Dow membrane has also been found to feature other favorable properties for PEM fuel cell operation, such as low gas permeability and low electro-osmotic drag coefficient (Ren et al., 2001; Ghielmi et al., 2005). The low gas permeability prevents gas crossover, therefore preventing the chemical degradation and potentially allowing lower thickness. The low electroosmotic drag coefficient simplifies the water management since the drying out of the membrane close to the anode side can be abated. With these attractive features, other types of short side chain PFSA membranes have also been developed (Arcella et al., 2003; Rivard et al., 2003; Hamrock et al., 2006; Borup et al., 2007). 3 M has developed its own short side chain membrane. In comparison with the Nafion membranes, the 3 M short side chain membranes have both better mechanical (7% higher modulus and 9% higher break

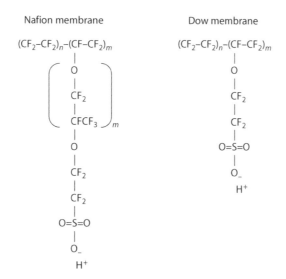

FIGURE 11.1 The chemical structures of Nafion and Dow membranes ($m \geq 1$, $n = 6$–10). (Adapted from Li, X. 2005. *Principles of Fuel Cells*. New York, NY: Taylor & Francis.)

stress at ambient conditions) and thermal (transition temperature of 125°C) strengths than the Nafion membrane (Rivard et al., 2003; Borup et al., 2007). It has also been reported that the relative lifetime of this membrane is about 4 times longer than the Nafion membrane at 90°C (Rivard et al., 2003; Borup et al., 2007).

Another way to solve the physical degradation problem is to use reinforced membranes. The reinforced membranes usually use physically strong porous membranes (e.g., polytetrafluoroethylene (PTFE)) as the reinforcing agent, combined with proton conductive materials (e.g., Nafion). Such membranes are also called the reinforced composite membranes or composite membranes. The most popular reinforced membrane at present is perhaps the Gore membrane, which uses porous PTFE membrane as the reinforced agent combined with Nafion as the proton conductor (Bahar et al., 1996, 1997). Generally, by comparing with Nafion membrane, the Gore membrane (or other PTFE-reinforced membranes) has lower proton conductivity but higher physical strength, and therefore it is usually thinner (e.g., around 35 μm) (Liu et al., 2003; Yu et al., 2004; Cleghorn et al., 2006). Single cell life tests have been conducted by using 35 μm Gore membranes at different current densities (Cleghorn et al., 2006). Figure 11.2 (Cleghorn et al., 2006) shows the polarization curves at various periods in time (0, 500, 5348, 10,100, 15,000, 20,000, and 26,330 h) during the life test at a current density of 0.8 A cm^{-2}, under fully humidified operating condition, and at 70°C. This test is under continuous operation with interruptions at every 500 or 1000 h for measurement purposes. As shown in Figure 11.2 (Cleghorn et al., 2006), the life test had run for up to 26,334 h with a performance degradation rate of about 6 μV h^{-1}. The Gore membrane greatly improves the mechanical strength by comparing with conventional PFSA membranes; however, the transition temperature is not improved, and PEM fuel cells with such membrane still have to run at temperatures lower than 100°C.

Rather than using the PTFE as the reinforcing agent, researchers have also used carbon nanotubes for reinforced membranes, mainly by mixing and casting the carbon nanotubes with Nafion to enhance the mechanical strength (Liu et al., 2006; Thomassin et al., 2007; Wang et al., 2008). The experimental studies in Liu et al. (2006), Thomassin et al. (2007), and Wang et al. (2008) all demonstrated improved mechanical strength with reasonable proton conductivity. Generally, reinforced membranes using PTFE and carbon nanotubes as the reinforcing agent seem to behave similarly, and they are still to be used as low-temperature membranes (e.g., <100°C).

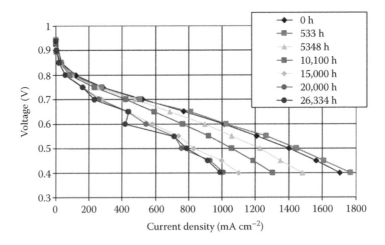

FIGURE 11.2 Polarization curves at various periods in time (0, 500, 5348, 10,100, 15,000, 20,000, and 26,330 h) during the life test. Cell temperature 70°C. Air: 2.0 stoichiometry, ambient pressure, 100% relative humidity. Hydrogen: 1.2 stoichiometry, ambient pressure, and 100% relative humidity. (Reprinted from *Journal of Power Sources*, 158, Cleghorn, S.J.C. et al., A polymer electrolyte fuel cell life test: 3 years of continuous operation, 446–454, Copyright (2006), with permission from Elsevier.)

The addition of inorganic particles (e.g., silicon oxides) as the reinforcing agents to PFSA membranes has been adopted to mainly maintain the mechanical strength at elevated temperatures (e.g., up to about 120°C to avoid liquid water flooding in PEM fuel cells) (Deng et al., 1998; Adiemian, 2002; Kim et al., 2003; Yang et al., 2004; Chalkova et al., 2005; Hill et al., 2006; Statterfield et al., 2006; Woo et al., 2006). Not only for PFSA membranes, other mechanically weak membranes (e.g., hydrocarbon membrane) have also been reinforced by adding reinforcing agents (Kerres, 2005; Fu et al., 2006) to improve the mechanical strength. However, these methods still need further life durability test under PEM fuel cell operating conditions to confirm their feasibilities.

As mentioned previously, PEM fuel cell performance is generally improved at high temperatures (e.g., higher than 100°C), therefore, development of membranes with high mechanical strength at the temperatures higher than 100°C has attracted many attentions in the last decade. One approach is to add inorganic particles to PFSA membranes, as just mentioned in the previous paragraph. However, since the PFSA membranes need to be well hydrated to provide high proton conductivity, and membrane hydration at elevated temperatures is very difficult due to the significantly decreased relative humidity (increased saturation pressure of water), developing high-temperature membranes with high proton conductivity in an anhydrous environment is demanded. The second approach to high-temperature membranes relies on the sulfonation of thermally resistant polymers (Zhang et al., 2006), such as sulfonated polyetheretherketones (SPEEK) (Shibuya et al., 1992; Kobayashi et al., 1998; Bonnet et al., 2000), polyimides (PI) (Guo et al., 2002), polysulfones (PSF) (Hasiotis et al., 2001), and poly(p-phenylene) (Child et al., 1994). Some of these membranes are less dependent on humidity than the PFSA membranes if their associated H_2O/SO_3 values are lower, which allows for good proton conductivity at elevated temperatures (Zhang et al., 2006). The third approach to high-temperature membranes is to replace water with another proton transport assisting solvent that possesses a higher boiling point (e.g., phosphoric acid or imidazoles) (Staiti et al., 2001; Asensio et al., 2004; He et al., 2004; Li et al., 2004; Zhang et al., 2006; Schmidt et al., 2008). The proton conductivities of such membranes therefore mainly rely on the new proton transport assisting solvent and are less dependent on the hydration level. Among such membranes, phosphoric acid-doped PBI membrane (may also be incorporated with inorganic proton conductors) has attracted most of the attentions due to its high proton conductivity and mechanical strength. The proton conductivity of such membrane mainly depends on the phosphoric acid doping level rather than the hydration level, and good proton conductivity and cell performance have been demonstrated in nonhumidified condition (Li et al., 2004; Jiao et al., 2010a). Promising durability of the phosphoric acid-doped PBI membrane has also been demonstrated (Schmidt et al., 2008). Figure 11.3 (Schmidt et al., 2008) shows the cell potentials at 0.2 A cm^{-2} as a function of lifetime (up to about 6000 h) of a constant load test with and without start/stop cycling for a PEM fuel cell with phosphoric acid-doped PBI membrane. The cell operated with H_2 and air without humidification at 160°C and 1 bar. The continuous operation shows a degradation rate of 5 μV h^{-1}, and it is 11 μV h^{-1} for the 240 start/stop cycling operation.

11.2.3 Chemical Degradation-Related Design

The chemical degradation of PFSA membranes is mainly caused by the gas crossover, forming hydrogen peroxide, which reacts with impurities (e.g., iron contaminations) producing hydroxyl and hydroperoxy radicals that could attack (decompose) the membrane (refer to Chapter 4 for detailed degradation mechanisms).

By treating Nafion with fluorine gas for 50 h, Curtin et al. (2004) found that the number of hydrogen-containing end groups can be reduced, and this improves the chemical stability against the hydroxyl and hydroperoxy radicals (a 56% decrease in released fluoride ions in the Fenton test).

Endoh et al. (2009) and Endoh (2006, 2008) have developed new polymer composites (NPCs), which have a highly durable PFSA base under high-temperature and low-humidity conditions. The NPC membrane showed excellent stability that approximated 3000 h in an open-circuit voltage (OCV) test at 120°C and 18% relative humidity, and the continuous operation using an NPC membrane could be

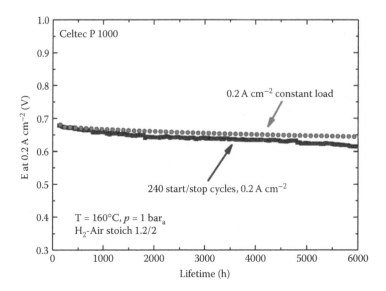

FIGURE 11.3 Cell potentials at 0.2 A cm^{-2} as a function of lifetime of a constant load test with and without start/stop cycling for a PEM fuel cell with phosphoric acid-doped PBI membrane. H$_2$–air (stoichiometry 1.2/2), $T = 160°C$, $p = 1$ bar. (Reprinted from *Journal of Power Sources,* 176, Schmidt, T.J. and J. Baurmeister, Properties of high-temperature PEFC Celtec®-P 1000 MEAs in start/stop operation mode, 428–434, Copyright (2008), with permission from Elsevier.)

operated for more than 6000 h at 120°C, 200 kPa, 0.2 A cm^{-2}, and 50% relative humidity, as shown in Figure 11.4 (Endoh, 2008). Even though promising durability has been demonstrated, the detailed membrane structure has not been disclosed yet.

Steck et al. (1997), Wei et al. (1995), Borup et al. (2007) claimed that their sulfonated polytrifluorostyrene-based BAM3G membrane has excellent chemical stability and can operate for several tens of thousands hours for continuous and fully humidified operation. The BAM3G membrane has demonstrated that the fluorination of the backbone containing styrene sulfonic acid is an effective way to enhance the chemical stability against attacking radicals (Wei et al., 1995; Steck et al., 1997; Borup et al., 2007). Gubler et al. (2005) also demonstrated that a trifluorostyrene-grafted ethylene tetrafluoroethylene

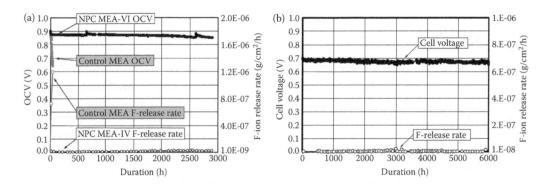

FIGURE 11.4 (a) OCV durability of the NPC MEA-VI at 120°C and 18% relative humidity; (b) durability of the NPC MEA-VII at 120°C, 50% relative humidity, 0.2 A cm^{-2} and 200 kPa. (Reprinted from Endoh, E. 2008. *ECS Transactions* 12: 41–50. With permission from The Electrochemical Society.)

(ETFE)-based membrane showed better chemical stability than a styrene sulfonic acid-grafted PTFE membrane, due to the poor oxidative stability of the styrene group.

Xing et al. (2006) found that for the sulfonated polyarylene ether membranes, the chemical stability, thermal stability, and proton conductivity all depend on the location of the sulfonic acid (the sulfonated position). Three sulfonated positions have been studied, which are the meta-sulfone sulfonated poly-phenyl sulfone (SPSU), ortho-sulfone SPSU, and ortho-ether SPSU, as shown in Figure 11.5a (Xing et al., 2006). The synthesis of the three locations followed the procedures published in the open litera-ture. They showed that the sulfonic acid group at the meta-sulfone position has better chemical and thermal stability and higher proton conductivity than those at the ortho-ether and ortho-sulfone posi-tions. Figure 11.5b (Xing et al., 2006) shows the weight losses of the three different SPSU membranes in H_2O_2 solution, indicating that the meta-sulfone SPSU has the best chemical stability.

Since the chemical degradation is mainly caused by the reactant gas crossover that forms attacking radicals, developing membranes with lower gas crossover rate or lower radical formation rate has attracted people's attention. To reduce the amount of gas crossover, the membranes have been modified by introducing cross-linking, increasing crystallinity, and increasing membrane thickness (Buchi et al., 1995; Chen et al., 2006; Borup et al., 2007). However, the proton transport resistance often increases in these cases. To reduce the amount of radical formation, additional catalyst layers for H_2O_2 decomposi-tion or trapping radicals have been placed inside the membranes or on both sides of the membranes

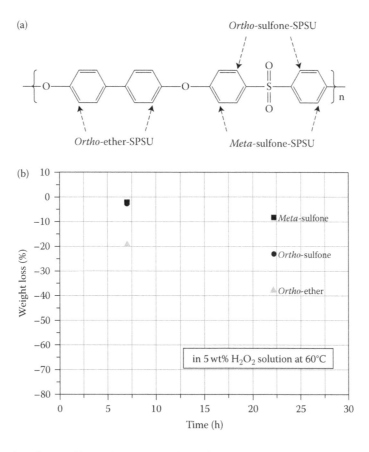

FIGURE 11.5 (a) Different sulfonated position on the sulfonated polyphenylsulfone; (b) the weight losses of different SPSU membranes in H_2O_2 solution. (Xing, D. and J. Kerres. Improved performance of sulfonated poly-arylene ethers for proton exchange membrane fuel cells. *Polymers for Advanced Technologies*. 2006. 17: 591–597. Copyright Wiley-VCH Verlag GmbH & Co. KGaA. Reprinted with permission.)

(Borup et al., 2007). Both methods have demonstrated improvement in chemical stability; however, since the catalyst layers are mainly based on rare earth metals such as platinum, palladium, iridium, etc, this method has not attracted many attentions.

11.3 Catalyst Layer Design

The CL is a critical component of PEM fuel cell where the electrochemical reactions take place. The present state-of-the-art CL microstructure is perhaps still the platinum–carbon particles mixed with electrolyte. Due to the acid nature of the membrane electrolyte and the low-temperature operation of the PEM fuel cells, platinum generally remains the most effective catalyst for the facilitation of both the hydrogen oxidation (HOR) and oxygen reduction reactions (ORR) in the PEM fuel cells. The carbon particles have been used to reduce the platinum loading by increasing the electrochemically active surface area. PTFE can also be mixed in the CL for water rejection and as a binder. The degradation of the platinum–carbon CL is mainly caused by the platinum dissolution, platinum particle growth, and carbon corrosion, which all lead to the loss of the electrochemically active surface area. Since the detailed degradation mechanisms in CL have been presented in Chapters 2 and 3, this section only focuses on the effects of design on the CL durability. The general design objectives are to reduce the catalyst cost, maintain/improve the reaction kinetics, and improve the durability of CL under the corrosive, oxidative, and acidic conditions with rapid and large load change and high temperature.

11.3.1 General Aspects of Catalyst Layer Design

The present platinum–carbon CL offers acceptable durability under continuous intermediate/high-current density operations with sufficient and even reactant gas distributions. However, practical operations often involve load changes (especially for automotive applications), and it is difficult to maintain sufficient reactants everywhere in the fuel cell, especially under fully humidified condition under which water flooding is often an issue. Moreover, due to the cost consideration and the limited amount of platinum resource, significant reduction in platinum loading is required to construct a practical fuel cell. Reduction of platinum loading has been achieved by mixing the platinum particles with carbon particles in early days, and recent researches mainly focused on platinum alloys and novel catalyst supports for further reduction of platinum loading and improved durability. Substantial reduction of platinum catalyst loading has been achieved in the past two decades, from about 2 mg cm^{-2} of the flat active area to below 0.5 mg cm^{-2} without significant impact on cell performance and lifetime (Gesteiger et al., 2005; Borup et al., 2007). However, from January 1999 to January 2010, the average monthly price of platinum has been increased from \$355.72 per ounce to \$1563.3 per ounce (Johnson Matthey Precious Metals Marketing, 2010). The increased price overcomes all the efforts in reducing the platinum loading, and due the limited platinum resource on the earth, a platinum-free CL with good performance and durability is necessary for PEM fuel cell commercialization. Furthermore, replacing platinum with other precious metals such as palladium and ruthenium suffers similar or even more severe price increment and durability issue (Borup et al., 2007). As a result, nonprecious catalysts are becoming more attractive to the PEM fuel cells. However, almost all the nonprevious catalysts developed in the past several decades suffer from low activity and poor stability in the acidic environment of PEM fuel cells. In addition, development of support materials to both increase the electrochemically active surface area and solve the carbon corrosion issue has also attracted many attentions. In the next three subsections, the platinum and other precious metal-based catalysts are discussed first, followed by nonprecious catalyst and support materials.

11.3.2 Precious Catalyst

Since other precious metals such as palladium and ruthenium suffer similar or even more severe price increment than platinum, and these metals are often less durable than platinum, most of the precious catalyst layers are still based on platinum–carbon or mixing with other metals (platinum alloy). Jalan and Taylor (Jalan et al., 1983) showed that the Pt–Pt distance changes when platinum is mixed with other substances, and the ORR activity changes with the Pt–Pt interatomic distance. Figure 11.6 shows the specific activity for ORR versus the nearest-neighbor distance determined from x-ray diffraction for the various platinum alloy catalysts (Jalan et al., 1983). A linear relationship is shown and the Pt–Cr alloy electrocatalyst exhibits the highest electrochemical activity for ORR and the smallest nearest-neighbor distance. The Pt–C catalyst also shows a higher activity than pure platinum. Catalysts based on platinum alloys on carbon (e.g., Pt–Co–C) in acid electrolytes have shown higher crystallinity and enhanced exchange current densities than pure platinum on carbon (Pt–C) (Stamenkovic et al., 2002; Benesch et al., 2005; Borup et al., 2007; Koh et al., 2007; Sasaki et al., 2009). However, the main reason of the enhanced performance is still not fully understood.

Even though improvement on the reaction activity has been obtained, the stability of the new catalysts still needs to be considered. It has been generally summarized that Pt–Cr and Pt–Co tend to exhibit greater stability than Pt–V, Pt–Ni, and Pt–Fe (Antolini et al., 2006; Borup et al., 2007). The preparation procedure also affects the stability. The Pt–Co alloy has been demonstrated to have significantly improved reaction activity in acid electrolyte (Stamenkovic et al., 2002; Koh et al., 2007); however, it has also been found that Pt–Co formed at high annealing temperature has doubled specific activity and is more stable than conventional Pt–C, and Pt–Co formed at low annealing temperature has tripled activity but is less stable than Pt–C (Koh et al., 2007). Pt–Co–C has been demonstrated to improve both the activity and stability of the Pt–C when it is properly prepared (Yu et al., 2005). Figure 11.7 shows the H$_2$/air performance curves of Pt–C and Pt–Co–C after potential cycling for every 400 cycles between 0.87 and 1.2 V at 65°C (Yu et al., 2005), showing a significant improvement in the stability, and improved cell performance as well. It has also been observed that the Pt–Co–C catalyst has less catalyst migration into membrane than Pt–C after load cycling (Ball et al., 2006). However, no apparent improvement in performance and durability by using Pt–Co–C has been observed in Colon-Mercado et al. (2006), this is perhaps due to the different preparation procedures. Other platinum alloys such as Pt–Au has also been demonstrated to improve the catalyst durability (Sasaki et al., 2009). Even though some platinum alloys have shown improved durability, many platinum alloys are less stable than pure platinum. One example is that the Pt – Ru catalyst helps enhance the CO tolerance, but is less stable than pure platinum (Piela et al., 2004; Borup et al., 2007). Generally, the mechanisms behind the improved durability and

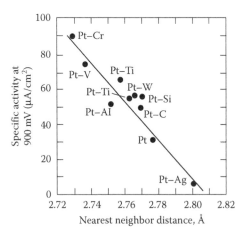

FIGURE 11.6 Specific activity for oxygen reduction vs. electrocatalyst nearest-neighbor distance. (Reprinted from Jalan, V. and E.J. Taylor. 1983. *Journal of Electrochemical Society* 130: 2299–2302. With permission from The Electrochemical Society.)

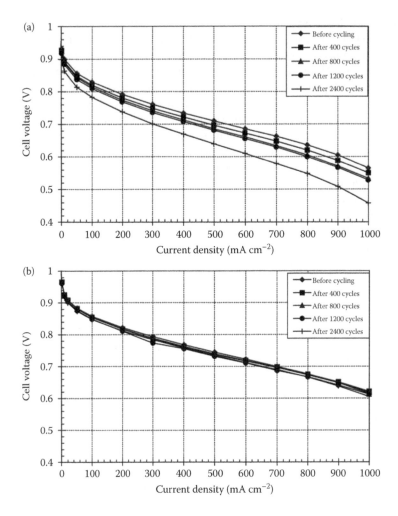

FIGURE 11.7 Cell H$_2$/air performance curves of (a) Pt–C and (b) Pt–Co–C after potential cycling for every 400 cycles between 0.87 and 1.2 V. (Reprinted from *Journal of Power Sources,* 144, Yu, P., M. Pemberton and P. Plasse, PtCo/C cathode catalyst for improved durability in PEMFCs, 11–20, Copyright (2005), with permission from Elsevier.)

performance by using the platinum alloys have not been fully understood yet, which is critically important to develop better catalyst.

11.3.3 Nonprecious Catalyst

As mentioned earlier, since the platinum (and other precious metals) loading reduction cannot overcome the price increment, the development of a nonprecious catalyst is perhaps the "most" feasible solution for PEM fuel cell commercialization. The nonprecious catalysts generally include carbides, nitrides, oxides, carbonitrides, oxynitrides, materials derived from macrocycles and porphyrins, and composites of these materials (Borup et al., 2007). However, unlike the platinum alloy catalysts that have already shown both the improved reaction activity and stability, none of the nonprecious catalysts demonstrated both good reaction activity and stability to the best of the authors' knowledge. Some nonprecious catalysts have reasonable reaction activity but poor stability (e.g., some transition metal carbides), and some have good stability but low-reaction activity (e.g., some transition metal oxides). Even though the nonprecious catalysts are still not comparable with the state-of-the-art platinum-based catalysts, recent development has demonstrated the opportunities for further improvement.

Transition metal carbides, such as tungsten carbide and its alloys, tantalum carbide, titanium carbide, and molybdenum carbide (Cowling et al., 1970, 1971; Voorhies et al., 1972; Scholl et al., 1992, 1994; Borup et al., 2007), have been studied as catalysts for electrochemical reactions. However, it has been found that these transition metal carbides are unstable under high potentials and in acid solution, and this limits their application as PEM fuel cell catalysts (Borup et al., 2007). Transition metal nitrides have been studied as electrochemical catalysts in PEM fuel cell environments, and Zhong et al. (2006) showed that molybdenum nitride supported on carbon powder resulted in a cell performance of about 0.3 V at 0.2 A cm^{-2}, and the catalyst was stable for 60 h of cell operation. However, the long-term performance durability is still questionable.

Unlike most carbides and nitrides, many oxides, carbonitrides, oxynitrides, and composites of them have good stability in acidic condition. Some transition metal oxides, oxynitrides, and carbonitrides have been found to have high stability in acidic condition, but low-reaction activity (Liu et al., 2005; Doi et al., 2006; Ishihara et al., 2006; Kim et al., 2007; Shibata et al., 2007). The heat treatment temperature has been found to have strong effects on the reaction activity. Shibata et al. (2007) showed that the TaO_xN_y/Ti catalyst heat treated at 900°C had the highest activity.

Macrocycle complexes of transition metals with high-temperature treatment have been studied as catalyst (Gouerec et al., 1999; Sawai et al., 2004; Wang et al., 2005). Even though heat treatment is required for most of the nonprecious metals to improve the stability, some materials also showed better stability without heat treatment (Posudiesky et al., 2004; Bashyam et al., 2006; Borup et al., 2007). The long-term performance of these catalysts needs to be investigated further. Cobalt- and iron-based catalysts obtained via the pyrolysis process (e.g., cobalt tetraphenyl porphyrin (CoTTP) or iron tetraphenyl porphyrin (FeTTP)) have also been investigated (Lalande et al., 1995; Faubert et al., 1996; Zelenay et al., 2006; Borup et al., 2007). However, even well-prepared CoTTP and FeTTP are not stable in PEM fuel cell environments, and Zelenay et al. (2006) developed a cobalt–polypyrrole (CoPPy) catalyst with better durability. As shown in Figure 11.8 (Zelenay et al., 2006), CoPPy composite supported on Vulcan XC-72 results in a much better cell durability than the well-prepared CoTTP. Even though the nonprecious

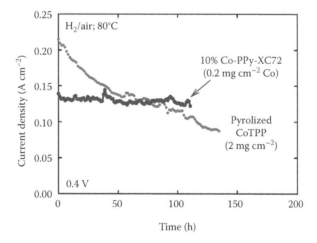

FIGURE 11.8 Life test of an H$_2$-air fuel cell operating with a 10% Co-PPY-C cathode. Life test data for a cell with a CoTTP cathode (the most active pyrolized-porphyrin cathode catalyst) are shown for comparison. Cell voltage: 0.4 V. Cell temperature: 80°C. (Reprinted from Zelenay, P. et al. 2006. *FY 2006 Annual Progress Report.* http://www.hydrogen.energy.gov/pdfs/progress06/v_c_7_zelenay.pdf. With permission.)

catalysts are still not comparable with the precious catalysts in terms of reaction activity and stability, progress has been made toward the success.

11.3.4 Catalyst Support

The catalyst support enhances the electrochemically active surface area and provides pathways for electron transport. The state-of-the-art catalyst support is carbon particle. The carbon particle exhibits degradation in acidic and humid condition, mainly due to the electrochemical reduction reaction especially at fuel starvation. The loss of carbon particle results in aggregation of catalyst particles, leading to decreased electrochemical active surface area, and hinders the reactant transport. Even though the carbon corrosion proceeds slowly, it does affect the long-term durability of PEM fuel cell, and due to the increased durability and performance requirement, developing more durable catalyst support is therefore demanded. Developing new support materials is also motivated by the performance requirement. The improvement includes using graphitized carbon support and using other new materials such as carbon nanotubes, metal oxides, silicon, conducting polymer, conductive diamond, and nonconductive whiskers (Borup et al., 2007; Yu et al., 2009).

Graphitized carbon provides higher corrosion resistance than conventional carbon support in PEM fuel cells but at a higher cost (Yu et al., 2009). Due to the increased stability demand, graphitized carbon has drawn more attention (Makharia et al., 2006; Yu et al., 2007, 2009). Yu et al. (2007, 2009) studied various nongraphitized and graphitized carbon supports through accelerated test. The carbon weight loss was estimated by the CO_2 concentration at the exit of the working electrode. As shown in Figure 11.9 (Yu et al., 2009), the graphitized carbon support shows much better durability. Since a cell voltage loss of approximately 50 mV at 0.8 A cm^{-2} occurs at 11 ± 2% of carbon weight loss for a variety of graphitized and nongraphitized carbon supports, while the same voltage loss can be observed at only 9 ± 2% of

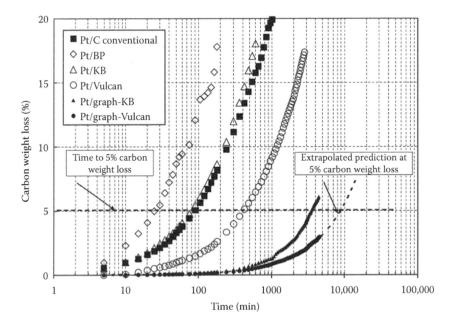

FIGURE 11.9 Integrated carbon weight loss (integrated from CO_2 concentration at the exit of the working electrode) of approximately 50 wt% platinum catalysts supported on various carbon blacks as a function of time at 95°C, 1.2 V, and 80% inlet relative humidity. (With kind permission from Springer Science+Business Media: *Polymer Electrolyte Fuel Cell Durability*, Carbon-support requirements for highly durable fuel cell operation, 2009, pp. 29–53, Yu, P.T. et al.)

carbon weight loss at the higher current density of 1.2 A cm^{-2} (Makharia et al., 2006). Therefore, a 5% of carbon weight loss was used as a conservative indication for performance degradation. Using the 5% indication shows a 35-fold improvement in corrosion resistance for graphitized Ketjenblack versus conventional Ketjenblack, and it is by two orders of magnitude for Vulcan XC-72C (Yu et al., 2009).

Carbon nanotubes, metal oxides and nonconductive whiskers have all been studied as promising catalyst supports. The carbon nanotubes are essentially nanoscale cylinders of rolled-up grapheme sheets. If the cylinder wall is only one sheet thick, it is a single-walled nanotube; more layers form a multiwalled nanotube; and a nonhollow cylinder is a nanofiber (Borup et al., 2007). The nanometer structure increases the electrochemically active surface area of the catalyst, and most of the previous studies of carbon nanotube in cathode showed improved cell performance (Wang et al., 2004; Yuan et al., 2004; Rajalakshmi et al., 2005; Li et al., 2006). The experimental studies in Shao et al. (2006) and Yuan et al. (2006) also showed that multiwalled nanotubes were more oxidation resistant than carbon black at 0.9 and 1.2 V. The carbon nanotubes are not cost competitive versus carbon powders, and further durability investigations are still needed. Metal oxides, especially Ti$_x$O$_y$, have attracted attentions due to their higher corrosion resistance than carbon (Dieckmann et al., 1998; Chen et al., 2003; Ioroi et al., 2005). However, the resulted reaction activity is lower than carbon support, perhaps due to the decreased active area (Dieckmann et al., 1998; Chen et al., 2003; Ioroi et al., 2005). Since such materials have good electron conductivity and stability, further improvement of the microstructure to enhance the reaction activity is the key to success. The nonconductive whiskers developed by 3M showed both better performance and durability than carbon (Parsonage et al., 1994; Debe et al., 1999; Borup et al., 2007). Even though the whisker material is not electron conductive and forms lower active areas for platinum (larger platinum particles were used), the improved reaction kinetics (high platinum utilization) was found to be able to overcome this loss. Further studies of platinum alloys and other nonprecious catalyst on the whiskers would be of interest.

Silicon (Hayase et al., 2004; Yeom et al., 2005), conducting polymer (Lefebvre et al., 1999), and conductive diamond (Montilla et al., 2003; Bennett et al., 2005) have also been studied as catalyst support materials in PEM fuel cells. However, all of these materials result in lower platinum utilization than carbon support. In addition, the durability of silicon and conducting polymer in PEM fuel cell environment is also questionable.

11.4 Gas Diffusion Layer Design

The GDL of PEM fuel cells connects between the bipolar plate/flow channel and catalyst layer. It provides the pathways for mass, heat, and electron transport, and also acts as the physical support of CL and membrane. The present state-of-the-art GDL consists of both a macroporous substrate and a microporous layer (MPL). The macroporous substrate is a porous network of graphite fibers, which is usually carbon paper, carbon felt or carbon cloth. The substrate is often coated with hydrophobic materials such as PTFE for better liquid water removal. The MPL is coated on one side of the substrate which faces the catalyst layer. A typical MPL is a composite layer mixed by carbon particles and PTFE. The MPL usually has lower porosity and permeability but higher hydrophobicity and electrical/thermal conductivity than its substrate layer. The MPL enhances the water management and thermal/electricity connection between the substrate layer and CL to better cell performance. Unlike the membrane and catalyst layer, researches focusing on developing new materials and/or revolutionary design are scarce. Therefore, this subsection only focuses on the macroporous substrate/MPL GDL.

The degradation modes of GDL mainly involve hydrophobicity loss and carbon corrosion. The hydrophobicity loss is caused by the loss of the hydrophobic material (e.g., PTFE) coated on the GDL during operation, especially under fully humidified and high current density operation (Borup et al., 2007). The loss of the hydrophobic coating results in increased hydrophilicity of GDL, which leads to worse water management (water flooding). The carbon corrosion is caused by the carbon electrochemical oxidation (mainly due to fuel starvation) and the impurities created through membrane degradation (e.g., HF) and gas crossover (e.g., H_2O_2) (Healy et al., 2005; Borup et al., 2007; Wood et al., 2009). The mechanical degradation of GDL has also attracted some attentions, mainly focused on the effect of compression force. Refer to Chapter 5 for detailed GDL degradation mechanisms.

The substrate layer (carbon paper, felt, and cloth) is made of carbon fibers in textile form. The raw fibers have to undergo polyacrylonitrile oxidation and carbonization steps to form the polyacrylonitrile form and remove the hydrogen, oxygen, and nitrogen atoms. Then the carbon fibers are graphitized at about 1750–2700°C to enhance the corrosion resistance, and graphitization at higher temperatures enhances the corrosion resistance, but increases the manufacturing cost (Wood et al., 2009). Refer to Mathias et al. (2003), and Wood et al. (2009) for more detailed manufacturing procedures. The different substrate layers (carbon paper, felt and cloth) are intrinsically same materials but in different forms. Both the paper-type substrate and cloth-type substrate have been commercially available for PEM fuel cells, such as SGL, Toray and Spectracorp carbon papers and Zolteck, Gore, and BASF carbon cloths. Generally, the carbon cloth has low porosity and permeability than the carbon paper, but better thermal and electrical conductivities. The selection of the porosity and permeability depends on the design of every component of the fuel cell, as well as fabrication procedure (e.g., compression force) and operating condition. Other structures of GDL has also been proposed through numerical simulations (Jiao et al., 2007, 2008), but not tested in real experiments. Generally, the carbon corrosion in the substrate layer is less a concern than the carbon particles in MPL and CL, because the carbon electrochemical oxidation requires proton-conductive materials, and the MPL and CL are also closer to the corrosive substances produced in membrane and CL (e.g., H_2O_2, HF).

Carbon fiber itself is a hydrophilic material. The hydrophobic coating materials used for PEM fuel cells are typically PTFE or fluorinated ethylene propylene (FEP). Recently, poly (dimethylsiloxane) (PDMS) with good flexibility, thermal stability and nontoxicity has also been used for liquid water removal in PEMFCs; however, its durability has not been tested in an operating cell yet (Wang et al., 2011). Unlike the conventional hydrophobic coating to mainly change the static contact angle, the PDMS coating in Wang et al. (2011) also considered another important parameter called the sliding angle, representing the "stickiness" of the surface. It has been demonstrated that both the static contact angle and sliding angle are needed to quantify the surface dynamic wettability (Jiao et al., 2010; Wang et al., 2011). The hydrophobic coating of the substrate layer is usually obtained by submersion or spraying by using dilutions of the hydrophobic materials (Wood et al., 2010). The hydrophobic coating enhances liquid water transport, but also occupies the pore volumes of the substrate layer, thus decreasing the porosity and permeability. Low level of hydrophobic coating facilitates liquid water transport and does not occupy the pore volume as much as the high level; however, the hydrophobicity loses quickly. On the other hand, high-level coating makes the substrate layer remain hydrophobic for a long period of time. Figure 11.10 (Wood et al., 2010) shows the single fiber contact angles with different levels of hydrophobic coating aged in accelerated fashion in different liquid–water environments, showing that increasing the coating level makes the fiber remain longer as hydrophobic.

The MPL often suffers more severe carbon corrosion and hydrophobicity loss than the substrate layer, because it is closer to the corrosive substances produced in membrane and CL (e.g., H_2O_2, HF). Similar to the carbon powders in the catalyst layer, graphitization of carbon powders in MPL also enhances the corrosion resistance. The hydrophobic material (e.g., PTFE) can also protect the carbon from corrosion. Wood and Borup (Wood et al., 2009) found that no significant structure change occurred in MPL after 664 h of durability test at 1 A cm^{-2}. In order to enhance the durability, Gore CARBEL CL MPL uses PTFE as substrate filled with carbon black (like the concept of reinforced membrane), and it showed good durability in a PEM fuel cell life test (Cleghorn et al., 2006; Wood et al., 2009).

The compression force distribution can affect the GDL durability significantly. Severe compression may result in substrate fibers puncture into the membrane, and light compression may result in high contact resistance between the cell components and poor sealing effect. The optimized compression is a compromise between durability and performance. The compression force also changes the GDL properties under the land (e.g., porosity, permeability, and effective thermal and electron conductivities). The

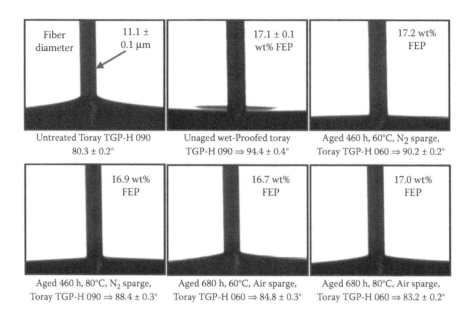

| Fiber diameter | 11.1 ± 0.1 µm | 17.1 ± 0.1 wt% FEP | 17.2 wt% FEP |

Untreated Toray TGP-H 090
80.3 ± 0.2°

Unaged wet-Proofed toray
TGP-H 090 ⟹ 94.4 ± 0.4°

Aged 460 h, 60°C, N₂ sparge,
Toray TGP-H 060 ⟹ 90.2 ± 0.2°

16.9 wt% FEP

16.7 wt% FEP

17.0 wt% FEP

Aged 460 h, 80°C, N₂ sparge,
Toray TGP-H 090 ⟹ 88.4 ± 0.3°

Aged 680 h, 60°C, Air sparge,
Toray TGP-H 060 ⟹ 84.8 ± 0.3°

Aged 680 h, 80°C, Air sparge,
Toray TGP-H 060 ⟹ 83.2 ± 0.2°

FIGURE 11.10 Single fiber contact angles of Toray TGP-H materials (hydrophobized to 17 wt% fluorinated ethylene propylene, FEP) aged in accelerated fashion in different liquid-water environments. (With kind permission from Springer Science+Business Media: *Polymer Electrolyte Fuel Cell Durability*, Durability aspects of gas-diffusion and microporous layers, 2009, pp. 159–195, Wood, D.L. and Borup R.L.)

experimental study in Jiao et al. (2010) showed significantly increased pressure drop along the flow channel when increasing the compression level, and the numerical study in Qi et al. (2011) tried to find the optimized compression level of GDL.

11.5 Membrane Electrode Assembly Design

Membrane, CL and GDL together fabricate the MEA. Since the design aspects of the three components are discussed in the previous sections, this section only focuses on the MEA level for the unaddressed issues in the previous sections. The MEA forms the basic structure for a single PEM fuel cell and it is the central component for a PEM fuel cell stack because it is the sole place that electric energy is produced in a PEM fuel cell stack, and all other stack components are just put in place to facilitate and harness the continual and steady production of electric energy here. The fabrication process of MEA can be generally categorized into bonding the CL to membrane or to GDL (Li, 2005). The platinum and carbon particles are often mixed with ionomer and PTFE to enhance the three-phase contact and platinum utilization. The methods to apply catalyst onto the membrane or GDL include paint brushing, spraying, screen printing, and so on (Li, 2005). The catalyzed membrane with plain GDL or catalyzed GDL with plain membrane can be hot pressed together to decrease the contact resistance, or they can just be put together during the fabrication process. There is no direct evidence that the above-mentioned fabrication processes affect the MEA durability.

Since the membrane in the middle of the MEA expands and contracts with the change of temperature and hydration level, it is important to control the MEA thickness to avoid possible physical failure. As mentioned in the GDL design section, this is usually achieved by proper control of the compression force. Other methods to control the MEA thickness after compression include using proper seal (Steck et al., 1995; Barton et al., 2000; Bonk et al., 2002), bonded layer (Schmid et al., 2000), plastic spaces (Kelland et al., 1992), and metal shims (Jiao et al., 2010). These methods all provide physical supports against the compression between bipolar plates to optimize the compression level of the GDL.

11.6 Bipolar Plate Design

In PEM fuel cells, each of the MEA is interposed between two fluid-impermeable, electrically conductive plates, commonly referred to as the anode and the cathode plates, respectively. The plates serve as current collectors, provide structural support for the thin and mechanically weak MEAs, provide means for supplying the fuel and oxidant to the anode and cathode, respectively, and provide means for removing water of the anode and cathode plates. The plates are normally referred to as fluid flow-field plates. When the flow channels are formed on both sides of the same plate, one side serves as the anode plate and the other side as the cathode plate for the adjacent cell, the plate is normally referred to as bipolar (separator) plate. It is more often that one of the reactants flows on one side of such a plate, while cooling fluid flows on the other side of the same plate, and these plates collectively have to keep the fuel, oxidant, and cooling fluid apart. The bipolar plate also serves as the connections between unit cells to a stack to achieve higher voltage and power output, and therefore needs to have high electrical conductivity, good mechanical strength, and strong stability in the PEM fuel cell environment. Both the graphite-based and metal-based materials have been widely used to make bipolar plates. In the following subsection, these two types of bipolar plates are discussed.

11.6.1 Graphite-Based Bipolar Plate

Graphite has high electron conductivity, good corrosion resistance, and low density. However, graphite is quite brittle, making it difficult to handle mechanically and has a relatively lower electrical and thermal conductivity compared with metals. Furthermore, graphite is quite porous, making it virtually impossible to make very thin gas-impermeable plates, which is desirable for low-weight, low-volume, and low-internal-resistance fuel cell stacks. Normally graphite is resin-impregnated to improve mechanical strength and fluid impermeability and to reduce the cost to form the so-called graphite composite bipolar plates (Li, 2005; Mitani et al., 2009). The graphite composite bipolar plates are made of graphite and polymers, and conventional processing methods such as compression molding (Kuan et al., 2004; Radhakrishnan et al., 2006; Yin et al., 2007) and injection molding (Heinzel et al., 2004; Mighri et al., 2004; Muller et al., 2006) can be used. These methods save the cost and processing time significantly than engraving or milling the flow channels. The injection molding also has lower cost and shorter processing time than the compression molding. However, these methods are more difficult to precisely control the plate dimensions due to the thermal deformation and distortion. The graphite composite is made by mixing graphite powders in polymers, and the polymers for composites are classified into two types: thermosetting resin and thermoplastic resin (Mitani et al., 2009). The compression molding method is often used with thermosetting resin and the injection molding method is usually used with thermoplastic resin, refer to Mitani et al. (2009) for more detailed manufacturing processes. The optimized design of graphite composite is a compromise between mechanical strength and electron conductivity. Increasing the graphite power concentration enhances the electron conductivity but makes the plate more brittle. The polymer matrix makes the plate mechanically strong, but they are often electrically insulating.

Generally, for the graphite-based bipolar plates, the corrosion and release of contaminants under normal PEM fuel cell operating condition is not a problem (de Bruijn et al., 2008). Stable cell performance for 1750 h was reported in Davies et al. (2000) by using the POCO graphite bipolar plates. Both the compression-molded and injection-molded graphite composite bipolar plates were tested in diluted sulfuric acid at 85°C and in concentrated phosphoric acid at 200°C for more than 2000 h without significant corrosion by Mueller et al. (2006). The long-term performance of a PEM fuel cell stack using the graphite composite bipolar plate for more than 3000 h was also tested by Mitani et al. (2009), and as shown in Figure 11.11a (Mitani et al., 2009) no significant performance loss was observed at the end of the testing. Figure 11.11b (Mitani et al., 2009) shows that both the machined and compression molded bipolar plates still had sufficient coolant resistance (to prevent leakage of coolant) at the end of the test.

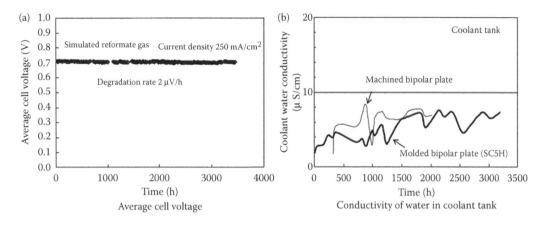

FIGURE 11.11 Long-term performance of a polymer electrolyte fuel cell stack using the graphite composite bipolar plate. (a) Average cell voltage and (b) conductivity of water in coolant tank. (With kind permission from Springer Science+Business Media: *Polymer Electrolyte Fuel Cell Durability*, Durability of graphite composite bipolar plates, 2009, pp. 257–268, Mitani, T. and K. Mitsuda.)

The major concern of graphite-based bipolar plate is still the mechanical strength and water resistance, which requires sufficiently thick bipolar plates. This feature makes the graphite-based material less favorable than the metal-based materials, especially for automotive applications which require high-power density. Another type of carbon-based bipolar plates, the carbon foam bipolar plates, was studied by Kim et al. (2010). The carbon foam is porous and without flow channel on it. However, this design requires high pumping energy for reactant delivery and also has sealing difficulties, and the long-term performance has not been investigated yet.

11.6.2 Metal-Based Bipolar Plate

Metals have much higher mechanical strength and are fluid impervious, hence bipolar plates can be made thin and easy for handling and stack assemblage; metals have higher thermal and electron conductivity, potentially improving the stack performance. Bipolar plates can be made by three-piece metallic components assembled together or by utilizing metallic sheet stamped into shape. A metallic sheet can be made from corrosion-resistant or even corrosion-susceptible metals with corrosion-resistant coatings. Particularly effective corrosion-resistant metals for PEM fuel cell operating environment include titanium, chromium, stainless steel, and so on. These metals develop a dense, passive oxide barrier layer over the surface to resist corrosion and prevent dissolution into the coolant flow. However, these oxide layers have low electrical conductivity, thus increasing the internal resistance of the fuel cell. Furthermore, the metal bipolar plates are not easy to make with the flow channels with sufficient flatness that is required in order to avoid uneven compression of the MEAs in the stack. Metal-based bipolar plates are regarded as the primary pathway to reduce overall stack cost and increase the power density, especially in automotive applications, in the transition to mass manufacturing.

Even though metal-based bipolar plates have so many advantages, it suffers corrosion and releases impurities that may attack other components (e.g., membrane). For example, in the acid and humid environment of PEM fuel cells, most of the stainless steels are corrosion active at the potentials between 0.1 to over 1 V, indicating that the corrosion happens at the cathode in all the operating loads and the anode is also corrosion active at high current densities (Scherer et al., 2009). Moreover, the impurities released due to membrane degradation (e.g., HF) also support the corrosion (Scherer et al., 2009). The different metals also have different features. For example, aluminum and titanium have light weight and low stability in acid condition and are not easy to be stamped; and copper- and nickel-based alloys provide excellent electron conductivity, but are also not stable (Tawfik et al., 2007). In order to enhance the corrosion resistance and formability, it has been reported that it is essential to increase the contents of chromium, nickel, and molybdenum and partially of nitrogen, forming a cubic face-centered structure of austenite (Scherer et al., 2009). However, higher corrosion resistance results in a thicker insulating oxide layer, resulting in higher contact resistance. To reduce the contact resistance, surface treatment or coatings are often needed (Tawfik et al., 2007). The surface coatings include thermal nitration of metal surface (Brady et al., 2004; Tian et al., 2007), metal nitride coating (e.g., chromium nitride, titanium nitride) (Li et al., 2004; Cho et al., 2005), carbide-based amorphous metallic coating (Tawfik et al., 2007), iron-based amorphous coating (Jayaraj et al., 2005), conductive polymers (Joseph et al., 2005) and precious metal coating (e.g., gold) (Hentall et al., 1999; Wind et al., 2002). The precious metal, gold, coating has been found to improve the contact resistance significantly; however, the cost consideration, especially when at least several hundred nanometer-thick coating is needed for sufficient protection, makes this method unfavorable (Hentall et al., 1999; Wind et al., 2002; Scherer et al., 2009). The durability of polymer conductive coating has not been approved under PEM fuel cell operating condition yet (Joseph et al., 2005; Tawfik et al., 2007). The inexpensive coatings, such as metal surface nitration (Brady et al., 2004; Tian et al., 2007), metal nitride coating (e.g., chromium nitride, titanium nitride) (Li et al., 2004; Cho et al., 2005), carbide-based amorphous metallic coating (Tawfik et al., 2007), and iron-based amorphous coating (Jayaraj et al., 2005) all showed promising improvement on the durability with reduced long-term contact resistance.

Perforated or foamed metal has also been explored as the flow fields on bipolar plates (Murphy et al., 1998). However, similar to the carbon foam bipolar plate (Kim et al., 2010), this design requires high pumping energy for reactant delivery, and the durability has not been approved yet.

11.7 Flow-Field Design

Flow channels are formed on the bipolar plates for reactant gas delivery and product water removal, and in the often absence of dedicated cooling plates, cooling channels are usually formed on the bipolar plates as well. The topologies of flow channels are significant for PEM fuel cell performance and durability. Proper flow channel design requires even reactant gas distribution and effective water removal while maintaining a hydrated membrane for proton transport (Jiao et al., 2010). Improper design of flow channel may cause fuel/oxidant starvation, local dry out of membrane that leads to local heating, and so on. Therefore, proper flow channel design is of paramount importance. The design of the flow channel also has to be such that it is easy to be manufactured, and provide sufficient contact area between the bipolar plate and GDL.

The flow channels are typically rectangular in cross-section, especially for graphite-based bipolar plates, and other configurations such as trapezoidal, triangular, semicircular, and so on have also been explored (Li et al., 2005). Normally the rectangular design is not good to absorb cell compression/expansion due to membrane hydration/dehydration change/cycling, and metal bipolar plate with trapezoidal cross-section has a spring-like function that can easily absorb/accommodate the cell compression/ expansion in a stack. For the rectangular cross-section, the flow channel dimensions range from 1 to 2 mm in width and depth as a low limit for a reasonable fluid pressure loss due to friction losses. Values of channel depth, width, and land width have also been suggested close to 1.5, 1.5, and 0.5 mm, respectively (Li et al., 2005). As to the geometrical configurations of the flow channel, a variety of different designs are known and the conventional designs typically comprise either pin, parallel, serpentine, interdigitated, integrated, or combined design (Li et al., 2005).

The pin flow-field network is formed by many pins arranged in a regular pattern, and these pins can be of any shape, although cubical and circular pins are most often used in practice (Reiser et al., 1988; Reiser, 1989; Li et al., 2005). Normally, both the cathode and the anode flow-field plates have an array of regularly spaced cubical or circular pins protruding from the plates, and the reactant gases flow across the plates through the intervening grooves formed by the pins. The actual fluid flow thus goes through a network of series and parallel flow paths. As a result, pin design flow fields result in low reactant pressure drop. However, reactants flowing through such flow fields tend to follow the path of least resistance across the flow field, which may lead to channeling and the formation of stagnant areas, and thus uneven reactant distribution, inadequate product water removal and poor cell performance.

The integrated design (Chow et al., 1999; Ernst et al., 1999) possesses both reactant flow field and cooling flow field on the same plate surfaces. The gas flow field directly faces the electrochemically active area of the adjacent MEA, while the cooling flow field surrounds or is in between the gas flow field. This integrated reactant and coolant flow-field plate design eliminates the need for a separate cooling layer in a stack, thus significantly improves the stack power density. However, it is difficult for these designs to maintain a uniform temperature distribution over the entire fuel cell surface.

The interdigitated flow fields have been explored to provide convection velocity normal to the electrode surface for better mass transfer, and convection flow in the porous backing layer for enhanced water removal capability (Li et al., 2005). An interdigitated flow field consists of dead-ended flow channels built on the flow distribution plates. The flow channels are not continuous from the stack inlet manifold to the exit manifold, as the reactant flow is forced under pressure to go through the porous electrode backing layer to reach the flow channels connected to the stack exit manifold, thus developing the convection velocity toward the CL and convection flow in the backing layer itself. Such flow-field design can remove water effectively from the electrode structure, preventing water flooding phenomenon and providing enhanced performance at high current density operation. However, a large pressure loss occurs for the reactant gas flow, especially the oxidant air stream. The parasitic power required for air compression may limit the application of this flow-field design to smaller stack sizes.

The parallel flow-field plate includes a number of separate parallel flow channels connected to the gas inlet and exhaust headers, which are parallel to the edges of the plate. This design provides short paths for reactant transport from inlet to outlet, and therefore the pressure loss is low. It is found that low and unstable cell voltages occur after extended periods of operation, because of cathode gas flow distribution and water flooding (Li et al., 2005). During PEM fuel cell operation, the water formed at the cathode accumulates in the flow channels, the channels become wet, and the water thus tends to cling to the bottom and the sides of the channels. The water droplets also tend to coalesce forming larger droplets. Since the number and size of the water droplets in the parallel channels are likely different, the reactant gas then flows preferentially through the least obstructed channels (Jiao et al., 2006a,b). Water thus tends to collect in the channels in which little or no gas is passing. Accordingly, stagnant areas tend to form at various areas throughout the plate. This problem is similar to the one that occurs in the pin-type flow field, as discussed earlier.

To resolve the problem of water flooding resulting from the inadequate water removal from the cells, Watkins et al. (1991) proposed using a continuous flow channel that had an inlet at one end and an outlet at the other, and typically followed a serpentine path. Such a single serpentine flow-field forces the reactant flow to traverse the entire active area of the corresponding electrode thereby eliminating areas of stagnant flow. However, this channel layout results in a relatively long reactant flow path, hence a substantial pressure drop and significant concentration gradients from the flow inlet to outlet. In addition, the use of a single channel to collect all the liquid water produced from the electrode reaction may promote flooding of the single serpentine, especially at high current densities. Watkins et al. (1992) pointed out that several continuous separate flow channels might be used in order to limit the pressure drop and thus minimize the parasitic power required to pressurize the air, which can be as much as over 30% of the stack power output. This combined parallel and serpentine design is thus called the parallel serpentine design. In fact, most practical PEM fuel cells combine the flow channel designs mentioned above for the optimized performance (Cavalca et al., 1997; Rock, 2001; Li et al., 2005). The parallel serpentine design on a graphite bipolar plate is shown in Figure 11.12 (Marvin et al., 2002).

A further problem with the flow design is the possible and often-arisen nonuniform distribution of a compressive load carried across the fuel cells within the stack when the flow channels on the anode and the cathode plates are not properly aligned. The contact area, as defined by the overlap of the ribs on the anode and the cathode plates, depends on the manufacturing tolerances affecting the width of the ribs,

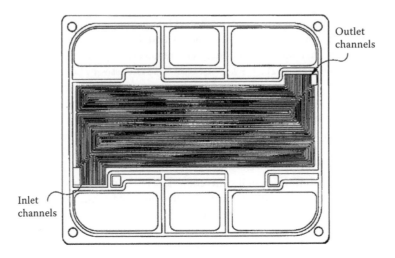

FIGURE 11.12 Parallel serpentine channel design on graphite plate. (Reprinted from Marvin, R.H. and C.M. Carlstrom. 2002. *Fuel cell fluid flow plate for promoting fluid service.* U.S. Patent 6500580. With permission.)

the smoothness of the rib surface, the exact location of the ribs, rib-edge machining and assembly alignments (plate to plate), etc. The variation in the contact areas of the ribs results in variation in the local stress and the associated cell strain. A minimum local stress is necessary to maintain minimum electrical (as well as thermal) contact resistance, whereas a significantly high local stress may lead to the damage and premature failure of cell components. To ensure uniform compression load across the cell, it is necessary to have even distribution of the ribs between the anode and cathode bipolar plates.

The porous flow-field design, such as the carbon foam (Kim et al., 2010) or metal foam (Murphy et al., 1998) discussed previously, requires high pumping energy for reactant delivery, and may also result in uneven distribution of reactant gases. The "Biomimetic" technology developed by Morgan fuel cell drew its inspiration from the natural world (Chapman et al., 2003; Li et al., 2005). It mimics the structure as seen in animal lungs and plant tissues to allow the gases to flow through the plate in an efficient way. Looking at how animal lungs and plant leaves "breathe," a structure consisting of large distribution channels feeding progressively to smaller capillaries is the most efficient way to distribute reactants. This structure reduces the pressure drop found in the industry-standard serpentine design of flow field and ensures a more even delivery of reactants, so that more power can be extracted from the fuel cell.

11.8 Stack Design

Single cells are often connected in series to form a PEM fuel cell stack to increase the output voltage and power. Since the design of each component of a PEM fuel cell stack is discussed in each previous section, this section only focuses on those not covered in previous sections. The manifold design is discussed first, followed by the seal design. The manifold design is important to evenly distribute reactant gases in each single cell of the stack and effectively remove product water to avoid fuel starvation and other problems that may lead degradation. The seal design is important to prevent possible leakage and provide mechanical support against the compression force between bipolar plates to avoid possible physical damage of the MEA.

11.8.1 Manifold Design

For PEM fuel cell stacks, the reactant supply and exhaust manifolds are typically built inside the stack in what is called the internal manifold design. Therefore, the stack contains the manifolds and inlet ports for directing the fuel and oxidant streams to the anode and cathode flow channels and exhaust manifolds and outlet ports for the expulsion of the unreacted fuel and oxidant exhaust streams. The cooling and humidification water flow manifolds are also built inside the stack. Therefore, the stack includes another set of manifolds and inlet port for distributing the coolant fluid, to interior cooling channels built within the stack to absorb heat generated within the stack and an outlet port for the coolant fluid exiting the stack. The inlet and outlet ports of the intake and exhaust manifolds in the stack can be located at the same end of the stack, the opposite ends, and on the side around the middle of the stack or any combination of the arrangements, which are typically parallel (Z configuration), reverse (U configuration), mixed (L configuration), and centered (I configuration), as shown in Figure 11.13 (Li, 2005). The manifold for reactant gas is to distribute the reactant gas evenly, to avoid fuel/oxidant starvation or flooding in each single cell. The manifold structure and the location of the inlet and outlet ports for a stack can have significant influence on the stack performance, especially for the performance variation from one cell to another. The manifold channels can also be tapered or made with baffle plate to achieve the optimized design. It has been suggested that the Z configuration produced the best performance (Baschuk et al., 2004). However, this really depends on the flow condition and relative sizes of the manifold and flow channels on the plates. In fact, each of the configurations shown should produce a more uniform and better performance for the active cells in the stack under a specific operating and design condition. This may explain the variety of the design arrangements in practice.

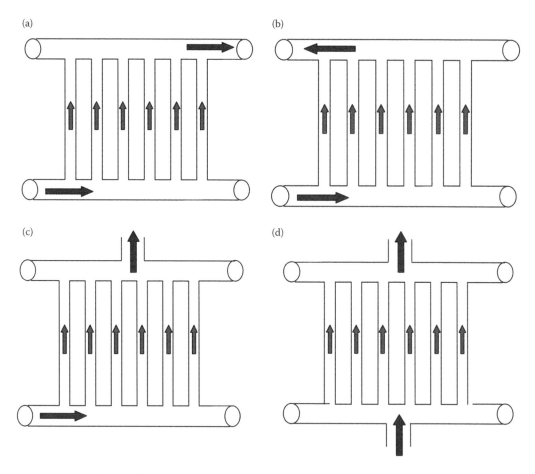

FIGURE 11.13 The manifold arrangement along with the various possible locations of the inlet and outlet ports in the stack. (a) Parallel (Z configuration); (b) reverse (U configuration); (c) mixed (L configuration); and (d) centered (I configuration). (Adapted from Li, X. 2005. *Principles of Fuel Cells.* New York, NY: Taylor & Francis.)

11.8.2 Seal Design

Beside the general sealing function, to avoid the mechanical overload on the MEA and bipolar plate while allowing the proper contact between the GDL and bipolar plate to reduce the contact resistance, the proper material and design (profile) of seal have to be chosen. The sealing materials have to be stable in the PEM fuel cell environment. For conventional PEM fuel cells, the sealing materials have to be able to tolerate the operating temperature and seal air, hydrogen, water, coolant, and other impurities (e.g., HF released from membrane degradation). The sealing materials also have to be able to tolerate higher temperature when used in high-temperature PEM fuel cells. The sealing materials also cannot be corrosive to the other components which may cause degradation problems. It has been reported that elastic materials (e.g., silicone rubber) are generally more favorable than stiff and hard materials (e.g., metal), and profiled seals are also generally better than flat seals (Bieringer et al., 2009).

Elastomeric materials with good relaxation behavior and low hardness are most favorable at present, while other harder materials (e.g., metal) are of less interest (Bieringer et al., 2009). Elastomeric materials are usually made of base rubbers with added fillers (e.g., carbon black, silica) to improve the material strength and plasticizers to reduce the compound viscosity, and antidegradants and vulcanizing agents can also be added to improve the durability and elasticity, respectively (Bieringer et al., 2009). The actual formulation of the material and the required manufacturing process must be carefully chosen to meet all the design requirements. Refer to (Bieringer et al., 2009) for more detailed material information. They have summarized that VMQ (silicone rubber), IIR (butyl rubber), EPDM (ethylene propylene diene rubber), FKM (fluoro rubber), and hydrocarbon-based rubbers are perhaps the only materials suitable in PEM fuel cell environment. Among these materials, the VMQ (silicone rubber) is widely employed for PEM fuel cell sealing because of its softness, simple processability and widespread availability; however, hydrocarbon-based rubbers and FKM (fluoro rubber) often have a longer lifetime than the VMQ (silicone rubber) (Tan et al., 2007; Bieringer et al., 2009). It has also been shown that properly designed profiled seal provides better prevention of overcompression than flat seal (Bieringer et al., 2009). As shown in Figure 11.14 (Bieringer et al., 2009), the seal with a "W" shape results in more even compression force distributions than the flat seal. The "W" design mitigates the unfavorable large compressions at the edges of the seal.

Even though the elastomeric materials are more favorable for PEM fuel cell sealing than the hard materials, the thickness of the MEA with the elastomeric seal is difficult to be predicted after fabrication (compressed between bipolar plates). In order to achieve the optimized compression level of both the MEA and the elastomeric seal, Jiao et al. (2010) used metal shims in parallel with silicone rubber seal. The thin metal shim is almost "incompressible," therefore the thickness of the metal shim is expected to be the thickness of the elastomeric seal after compression.

11.9 System Design

A PEM fuel cell system usually has at least one or multiple PEM fuel cell stacks for the generation of DC electric power. If a number of stacks are employed, they may be connected electrically either in series or in parallel, or a combination of them. Series connections provide higher voltage output while the electric current is relatively low, leading to lower ohmic losses both inside and outside the stacks and among all the electrical connections. However, a series connection can result in a system breakdown if any of the cell or stack components breaks down, thereby increasing maintenance requirements and potentially limiting the reliability and lifetime of the entire system. On the other hand, if all the stacks are connected electrically in parallel, the system can still provide partial power if any one of the stacks breaks down, thus providing time for response and avoiding potential accidents or even catastrophic events. But a parallel connection among the stacks yields a relatively high electric current for the system, thus increasing the ohmic losses. In addition, a single PEM fuel cell stack may also involve both series and

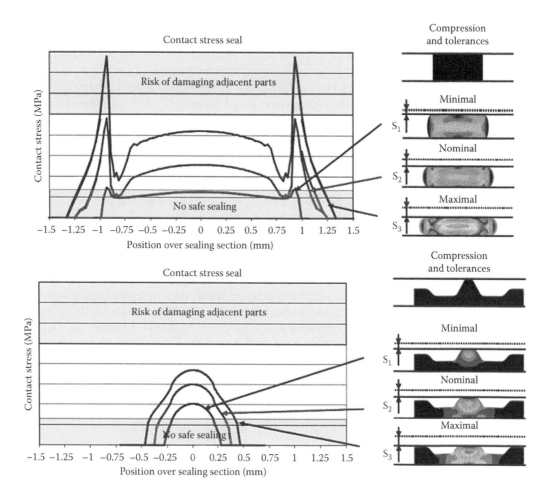

FIGURE 11.14 Comparison of flat (top) and profiled (bottom) gaskets: range of compression between secure sealing and damage of adjacent parts. (With kind permission from Springer Science+Business Media: *Polymer Electrolyte Fuel Cell Durability*, Gaskets: Important durability issues, 2009, 271–281, Bieringer, R. et al.)

parallel connections of MEA, and the parallel connection of MEA can be achieved by using a pair of large bipolar plates with multiple small MEAs between them. In this case, the flow fields for the MEAs on the same bipolar plate are isolated.

In addition to stack(s), a PEM fuel cell system includes a fuel and oxidant supply and conditioning subsystem, thermal and water management subsystem, control and monitoring subsystem, and possibly a DC-to-AC inverter, which may be avoided depending on the application for which whether DC or AC power is needed. The basic construction and materials of a PEM fuel cell stack is not so different between stationary and automotive applications. The main difference is the operating condition. The operating temperature is slightly lower (<75°C) for stationary system and the operating pressure is normally the ambient pressure, while for automotive system it is higher than 80°C and often pressurized for higher power density (Tabata et al., 2009; Yamamoto et al., 2009). The automotive applications also require more often startup/shutdown, idling and large and rapid load changes, and stationary applications need relatively stable operating condition with small and slow load changes. Therefore, the targeted lifetime for stationary system (about 90,000 h) is significantly higher than automotive system (about 5000 h) (Tabata et al., 2009; Yamamoto et al., 2009).

The fuel and oxidant conditioning subsystem involves purification, preheating, and humidification devices. As mentioned in the previous chapters, PEM fuel cell durability can be improved if pure hydrogen and oxygen are supplied. However, hydrogen is usually produced from hydrocarbon fuels and ambient air is often used as the oxidant. Both of them contain impurities such as CO, SO_2, NH_3, and so on (Du et al., 2009). Therefore, impurity filters are important to improve the durability. Different filters have been developed to remove the different impurities (Lakshmanan et al., 2002; Donaldson Company Inc. 2003; Du et al., 2009). However, most filters work on the principle of chemical adsorption and dusty filtration, and have to be replaced or serviced periodically. This adds additional service cost to the system (Du et al., 2009). Preheating and humidification of the fuel and oxidant are also very important, not only to improve the cell performance, but also to enhance the durability of membrane.

Depending on the stack size (power output), and the specific applications, waste heat removal from the PEM fuel cell stack can be accomplished in a variety of ways by using cooling flow, attaching the entire stack to a cooler or fixing extended heat transfer surfaces (fines) to the stack's outside surfaces (Li et al., 2005). However, cooling circulation is most practical for commercial applications, with liquid coolant for large stacks and air coolant for small ones. Precise control of the entire stack is critically important to improving the durability, since local hot spots may result in local overheating and local cold spots may lead to extra water condensation. Proper water management is also important to avoid water flooding that may cause fuel starvation and other problems. With the control and monitoring subsystem, the oxidant flow rate and humidification level can be controlled according to the operating condition (e.g., power output and pressure drop etc.) to effectively hydrate or dehydrate the stack for optimized performance and durability (Jiao et al., 2010), and of course, the control and monitoring subsystem also adds the additional cost.

To improve the durability on the system level, multiple stacks can be connected in parallel, and only some of the stacks operate at partial load conditions to reduce the frequency of load changes and operating time of each stack. A hybrid system can also be used by adding extra batteries to mitigate the large and rapid load changes of the stacks that are required by automotive applications.

11.10 Summary

This chapter is devoted to the design related durability issues of PEM fuel cells, from each single component to system level. Optimization of the chemical structure of the current state-of-the-art PFSA membrane and development of completely new materials have all been carried out to improve the membrane durability, especially under high-temperature and dry conditions, and significant progresses have been made. Even though the present platinum alloy catalyst with graphitized carbon support or other novel support materials may present promising durability and performance, the platinum reduction cannot overcome the rapid increase of the platinum price; and the similar condition also applies to other precious metals. Development of nonprecious catalyst is therefore perhaps the only way to commercialization. Although significant progresses have been made, the currently available nonprecious catalysts still cannot satisfy both the durability and performance requirements. Using graphitized carbon fibers and graphitized carbon powders in the substrate layer (carbon paper or carbon cloth coated with hydrophobic material) and MPL (mixture of carbon powder and hydrophobic material) of GDL is still the state-of-the-art design, and the durability is of less a concern for this kind of GDL compared with membrane and catalyst layer. Both the graphite-based and metal-based bipolar plates have been developed. The durability of sufficiently thick graphite-based bipolar plate is often not a problem, but it increases the system volume and mass as well as the electron transport resistance. The metal-based bipolar plates can be made thin, but it often needs corrosion-resistant coatings to improve the durability. The flow channel also needs to be properly designed to avoid fuel starvation, improper compression on the MEA and other problems. On the stack level, manifold design, proper sealing and material and design capabilities among the different components are the critical issues for optimized design. For the system design, purification, preheating, and humidification of the supplied fuel/oxidant, proper design of the water and thermal management subsystem, and monitor/control of the operating condition are the important design considerations.

Acknowledgments

The financial support by the Natural Sciences and Engineering Research Council of Canada (NSERC) via a strategic Project Grant (Grant No. 350662-07) and by Auto21 is greatly appreciated. Kui Jiao also would like to thank the Chinese Government Award for Outstanding Self-financed Students Abroad and the Next Energy Inc. Graduate Scholarship in Sustainable Energy.

References

Adiemian, K.T., S.J. Lee, S. Srinivasan et al. 2002. Silicon oxide Nafion composite membranes for proton-exchange membrane fuel cell operation at 80–140°C. *Journal of Electrochemical Society* 149: A256–A261.

Antolini, E., J.R.C. Salgado, and E.R. Gonzalez. 2006. The stability of Pt–M (M = first row transition metal) alloy catalysts and its effect on the activity in low temperature fuel cells: A literature review and tests on a Pt–Co catalyst. *Journal of Power Sources* 160: 957–968.

Arcella V., A. Ghielmi, and G. Tommasi. 2003. High performance perfluoropolymer films and membranes. *Annals of the New York Academy of Sciences* 984: 226–244.

Asensio, J.A., S. Borros, and P. Gomez-Romero. 2004. Proton-conducting membranes based on poly(2,5-benzimidazole) (ABPBI) and phosphoric acid prepared by direct acid casting. *Journal of Membrane Science* 241: 89–93.

Bahar, B., A.R. Hobson, and J.A. Kolde. 1997. *Integral composite membrane.* U.S. Patent 5599614.

Bahar, B., A.R. Hobson, J.A. Kolde et al. 1996. *Ultra-thin integral composite membrane.* U.S. Patent 5547551.

Ball, S.C., S. Hudson, B. Theobald et al. 2006. The effect of dynamic and steady state voltage excursions on the stability of carbon supported Pt and PtCo catalysts. *ECS Transactions* 3: 595–605.

Barton, R.H., P.R. Gibb, J.A. Ronne et al. 2000. *Membrane electrode assembly for an electrochemical fuel cell and a method of making an improved membrane electrode assembly.* U.S. Patent 6057054.

Baschuk, J.J. and X. Li. 2004. Modeling of polymer electrolyte membrane fuel cell stacks based on a hydraulic network approach. *International Journal of Energy Research* 28: 697–724.

Bashyam, R. and P. Zelenay. 2006. A class of non-precious metal composite catalysts for fuel cells. *Nature* 443: 63–66.

Benesch, R. and T. Jacksier. 2005. Determination of O[H] and CO coverage and adsorption sites on PtRu electrodes in an operating PEM fuel cell. *Journal of American Chemical Society* 127: 14607–14615.

Bennett, J.A., Y. Show, S.H. Wang et al. 2005. Pulsed galvanostatic deposition of Pt particles on microcrystalline and nanocrystalline diamond thin-film electrodes. *Journal of Electrochemical Society* 152: E184–E192.

Bieringer, R., M. Adler, S. Geiss et al. 2009. Gaskets: Important durability issues. In *Polymer Electrolyte Fuel Cell Durability*, eds. F.N. Buchi, M. Inaba, and T.J. Schmidt, 271–281. New York, NY: Springer.

Bonk, S.P., M. Krasij, and C.A. Reiser. 2002. *Fuel cell stack assembly with edge seal.* U.S. Patent 6399234.

Bonnet, B. D.J. Jones, J.J. Roziere et al. 2000. Hybrid organic–inorganic membranes for a medium temperature fuel cell. *Journal of New Materials for Electrochemical Systems* 3: 87–92.

Borup, R., J. Meyers, B. Pivovar et al. 2007. Scientific aspects of polymer electrolyte fuel cell durability and degradation. *Chemical Reviews* 107: 3904–3951.

Brady, M.P., K. Weisbrod, I. Paulauskas et al. 2004. Preferential thermal nitridation to form pin-hole free Cr-nitrides to protect proton exchange membrane fuel cell metallic bipolar plates. *Scripta Materialia* 50: 1017–1022.

Buchi, F.N., B. Gupta, O. Hass et al. 1995. Performance of differently cross-linked, partially fluorinated proton exchange membranes in polymer electrolyte fuel cells. *Journal of Electrochemical Society* 142: 3044–3048.

Cavalca, C., S.T. Homeyer, and E. Walsworth. 1997. *Flow field plate for use in a proton exchange membrane fuel cell.* U.S. Patent 5686199.

Chalkova, E., M.V. Fedkin, D.J. Wesolowski et al. 2005. Effect of TiO_2 surface properties on performance of Nafion-based composite membranes in high temperature and low relative humidity PEM fuel cells. *Journal of Electrochemical Society* 152: A1742–A1747.

Chapman, A. and I. Mellor. 2003. Development of biomimetic flow field plates for PEM fuel cells. *Eighth Grove Fuel Cell Symposium*, September 2003, London, UK.

Chen, G.Y., C.C. Waraksa, H.G. Cho et al. 2003. EIS studies of porous oxygen electrodes with discrete particles. *Journal of Electrochemical Society* 150: E423–E428.

Chen, J., M. Asano, T. Yamaki et al. 2006. Effect of crosslinkers on the preparation and properties of ETFE-based radiation-grafted polymer electrolyte membranes. *Journal of Applied Polymer Science* 100: 4565–4574.

Child, A.D. and J.R. Reynolds. 1994. Water-soluble rigid-rod polyelectrolytes: A new self-doped, electro-active sulfonatoalkoxy-substituted poly(*p*-phenylene). *Macromolecules* 27: 1975–1977.

Cho, E.A., U.S. Jeon, S.A. Hong et al. 2005. Performance of a 1 kW-class PEMFC stack using TiN-coated 316 stainless steel bipolar plates. *Journal of Power Sources* 142: 177–183.

Chow C.Y., B. Wozniczka, and J.K.K. Chan. 1999. *Integrated reactant and coolant fluid flow field layer for a fuel cell with membrane electrode assembly.* Canadian Patent 2274974.

Cleghorn, S.J.C., D.K. Mayfield, D.A. Moore et al. 2006. A polymer electrolyte fuel cell life test: 3 years of continuous operation. *Journal of Power Sources* 158: 446–454.

Colon-Mercado, H.R. and B.N. Popov. 2006. Stability of platinum based alloy cathode catalysts in PEM fuel cell. *Journal of Power Sources* 155: 253–263.

Cowling, R.D. and H.E. Hintermann. 1970. The corrosion of titanium carbide. *Journal of Electrochemical Society* 117: 1447–1449.

Cowling, R.D. and H.E. Hintermann. 1971. The anodic oxidation of titanium carbide. *Journal of Electrochemical Society* 118: 1912–1916.

Curtin, D.E., R.D. Lousenberg, T.J. Henry et al. 2004. Advanced materials for improved PEMFC performance and life. *Journal of Power Sources* 131: 41–48.

Davies, D.P., P.L. Adcock, M. Turpin et al. 2000. Bipolar plate materials for solid polymer fuel cells. *Journal of Applied Electrochemistry* 30: 101–105.

Debe, M.K., R.J. Poirier, M.K. Wackerfuss et al. 1999. *Membrane electrode assembly.* U.S. Patent 5879828.

de Bruijn, F.A., V.A.T. Dam, and G.J.M. Janssen. 2008. Review: Durability and degradation issues of PEM fuel cell components. *Fuel Cells* 8: 3–22.

Deng, Q., R.B. Moore, and K.A. Mauritz. 1998. Nafion®/(SiO_2, ORMOSIL, and dimethylsiloxane) hybrids via *in situ* sol–gel reactions: Characterization of fundamental properties. *Journal of Applied Polymer Science* 68: 747–763.

Dieckmann, G.R. and S.H. Langer. 1998. Comparisons of Ebonex® and graphite supports for platinum and nickel electrocatalysts. *Electrochimica Acta* 44: 437–444.

Doi, S., Y. Liu, A. Ishihara et al. 2006. Zirconium nitride and oxynitride for new cathode of polymer electrolyte fuel cell. *ECS Transactions* 1: 17–25.

Donaldson Company, Inc. 2003. Point of use (POU) filtration for optical elements in semiconductor lithography tools. http://www.donaldson.com/en/semiconductor/support/datalibrary/001802.pdf

Du, B., R. Pollard, and J.F. Elter. 2009. Performance and durability of a polymer electrolyte fuel cell operating with reformatted: Effects of CO, CO_2, and other trace impurities. In Polymer electrolyte fuel cell durability, eds. F.N. Buchi, M. Inaba and T.J. Schmidt, pp. 341–366. New York, NY: Springer.

Endoh, E. 2006. Highly durable MEA for PEMFC under high temperature and low humidity conditions. *ECS Transactions* 3: 9–18.

Endoh, E. 2008. Progress of highly durable MEA for PEMFC under high temperature and low humidity conditions. *ECS Transactions* 12: 41–50.

Endoh, E. and S. Hommura. 2009. Improvement of membrane and membrane electrode assembly durability. In *Polymer Electrolyte Fuel Cell Durability*, eds. F.N. Buchi, M. Inaba, and T.J. Schmidt, New York, NY: Springer, pp. 119–132.

Ernst, W.D. and G. Mittleman. 1999. *PEM-type fuel cell assembly having multiple parallel fuel cell sub-stacks employing shared fluid plate assemblies and shared membrane electrode assemblies.* U.S. Patent 5945232.

Faubert, G., G. Lalande, R. Cote et al. 1996. Heat-treated iron and cobalt tetraphenylporphyrins adsorbed on carbon black: Physical characterization and catalytic properties of these materials for the reduction of oxygen in polymer electrolyte fuel cells. *Electrochimica Acta* 41: 1689–701.

Fu, Y.Z., A. Manthiram, and M.D. Guiver. 2006. Blend membranes based on sulfonated poly(ether ether ketone) and polysulfone bearing benzimidazole side groups for proton exchange membrane fuel cells. *Electrochemistry Communications* 8: 1386–1390.

Gasteiger, H.A., S. Kocha, B. Sompalli et al. 2005. Activity benchmarks and requirements for Pt, Pt-alloy, and non-Pt oxygen reduction catalysts for PEMFCs. *Applied Catalysis B: Environmental* 56: 9–35.

Ghielmi, A., P. Vaccarono, C. Troglia et al. 2005. Proton exchange membranes based on the short-side-chain perfluorinated ionomer. *Journal of Power Sources* 145: 108–115.

Gouerec, P. and M. Savy. 1999. Oxygen reduction electrocatalysis: Ageing of pyrolyzed cobalt macrocycles dispersed on an active carbon. *Electrochimica Acta* 44: 2653–2661.

Gubler, L., S.A. Gursel, and G.G. Scherer. 2005. Radiation grafted membranes for polymer electrolyte fuel cells. *Fuel Cells* 5: 317–335.

Guo, X., J. Fang, T. Watari et al. 2002. Novel sulfonated polyimides as polyelectrolytes for fuel cell application. 2. Synthesis and proton conductivity of polyimides from 9,9-bis(4-aminophenyl)fluorene-2,7-disulfonic acid. *Macromolecules* 35: 6707–6713.

Hamrock, S.J. and M.A. Yandrasits. 2006. Proton exchange membranes for fuel cell applications. *Journal of Macromolecular Science Part C: Polymer Reviews* 46: 219–244.

Hasiotis, C., V. Deimede, and C. Kontoyannis. 2001. New polymer electrolytes based on blends of sulfonated polysulfones with polybenzimidazole. *Electrochimica Acta* 46: 2401–2406.

Hayase, M., T. Kawase, and T. Hatsuzawa. 2004. Miniature 250 μm thick fuel cell with monolithically fabricated silicon electrodes. *Electrochemical and Solid-State Letters* 7: A231–A234.

He, R.H., Q.F. Li, G. Xiao et al. 2004. Proton conductivity of phosphoric acid doped polybenzimidazole and its composites with inorganic proton conductors. *Journal of Membrane Science* 226: 169–184.

Healy, J., C. Hayden, T. Xie et al. 2005. Aspects of the chemical degradation of PFSA ionomers used in PEM fuel cells. *Fuel Cells* 5: 302–308.

Heinzel, A., F. Mahlendorf, and O. Niemzig. 2004. Injection moulded low cost bipolar plates for PEM fuel cells. *Journal of Power Sources* 131: 35–40.

Hentall, P.L., J.B. Lakeman, G.O. Mepsted et al. 1999. New materials for polymer electrolyte membrane fuel cell current collectors. *Journal of Power Sources* 80: 235–241.

Hill, M.L., Y.S. Kim, B.R. Einsla et al. 2006. Zirconium hydrogen phosphate/disulfonated poly(arylene ether sulfone) copolymer composite membranes for proton exchange membrane fuel cells. *Journal of Membrane Science* 283: 102–108.

Ioroi, T., Z. Siroma, N. Fujiwara et al. 2005. Sub-stoichiometric titanium oxide-supported platinum electrocatalyst for polymer electrolyte fuel cells. *Electrochemistry Communications* 7: 183–188.

Ishihara, A., S. Doi, Q.Y. Liu et al. 2006. Tantalum nitride and oxynitride for new cathode of polymer electrolyte fuel cell. *ECS Transactions* 1: 51–60.

Jalan, V. and E.J. Taylor. 1983. Importance of interatomic spacing in catalytic reduction of oxygen in phosphoric acid. *Journal of Electrochemical Society* 130: 2299–2302.

Jayaraj, J., Y.C. Kim, K.B. Kim et al. 2005. Corrosion studies on Fe-based amorphous alloys in simulated PEM fuel cell environment. *Science and Technology of Advanced Materials* 6: 282–289.

Jiao, K. and X. Li. 2010a. A three-dimensional non-isothermal model of high temperature proton exchange membrane fuel cells with phosphoric acid doped polybenzimidazole membranes. *Fuel Cells* 10: 351–362.

Jiao, K. and X. Li. 2010b. Effect of surface dynamic wettability in proton exchange membrane fuel cells. *International Journal of Hydrogen Energy* 35: 9095–9103.

Jiao, K. and X. Li. 2010c. Water transport in polymer electrolyte membrane fuel cells. *Progress in Energy and Combustion Science* 37: 221–291.

Jiao, K. and B. Zhou. 2007. Innovative gas diffusion layers and their water removal characteristics in PEM fuel cell cathode. *Journal of Power Sources* 169: 296–314.

Jiao, K. and B. Zhou. 2008. Effects of electrode wettabilities on liquid water behaviours in PEM fuel cell cathode. *Journal of Power Sources* 175: 106–119.

Jiao, K., J. Park, and X. Li. 2010. Experimental investigations on liquid water removal from the gas diffusion layer by reactant flow in a PEM fuel cell. *Applied Energy* 87: 2770–2777.

Jiao, K., B. Zhou, and P. Quan. 2006a. Liquid water transport in parallel serpentine channels with manifolds on cathode side of a PEM fuel cell stack. *Journal of Power Sources* 154: 124–137.

Jiao, K., B. Zhou, and P. Quan. 2006b. Liquid water transport in straight micro-parallel-channels with manifolds for PEM fuel cell cathode. *Journal of Power Sources* 157: 226–243.

Johnson Matthey Precious Metals Marketing, 2010. http://www.platinum.matthey.com/prices/price_charts.html.

Joseph, S., J.C. McClure, R. Chianelli et al. 2005. Conducting polymer-coated stainless steel bipolar plates for proton exchange membrane fuel cells (PEMFC). *International Journal of Hydrogen Energy* 30: 1339–1344.

Kelland, J.W. and S.G. Braun. 1992. *Unitized fuel cell structure.* U.S. Patent 5187025.

Kerres, J.A. 2005. Blended and cross-linked ionomer membranes for application in membrane fuel cells. *Fuel Cells* 5: 230–247.

Kim, J. and N. Cunningham. 2010. Development of porous carbon foam polymer electrolyte membrane fuel cell. *Journal of Power Sources* 195: 2291–2300.

Kim, J.H., A. Ishihara, S. Mitsushima et al. 2007. Catalytic activity of titanium oxide for oxygen reduction reaction as a non-platinum catalyst for PEFC. *Electrochimica Acta* 52: 2492–2497.

Kim, Y.S., F. Wang, M. Hickner et al. 2003. Fabrication and characterization of heteropolyacid (H3PW12O40)/directly polymerized sulfonated poly(arylene ether sulfone) copolymer composite membranes for higher temperature fuel cell applications. *Journal of Membrane Science* 212: 263–282.

Kobayashi, T., M. Rikukawa, K. Sanui et al. 1998. Proton-conducting polymers derived from poly(ether-etherketone) and poly(4-phenoxybenzoyl-1,4-phenylene). *Solid State Ionics* 106: 219–225.

Koh, S., J. Leisch, M.F. Toney et al. 2007. Structure–activity–stability relationships of Pt – Co alloy electro-catalysts in gas-diffusion electrode layers. *Journal of Physical Chemistry C* 111: 3744–3752.

Kuan, H.C., C. Ma, K. Chen et al. 2004. Preparation, electrical, mechanical and thermal properties of composite bipolar plate for a fuel cell. *Journal of Power Sources* 134: 7–17.

Lakshmanan, B. and Weidner, J.W. 2002. Electrochemical CO filtering of fuel-cell reformate. *Electrochemical and Solid-State Letters* 5: A267–A270.

Lalande, G., R. Cote, and G. Tamizhmani. 1995. Physical, chemical and electrochemical characterization of heat-treated tetracarboxylic cobalt phthalocyanine adsorbed on carbon black as electrocatalyst for oxygen reduction in polymer electrolyte fuel cells. *Electrochimica Acta* 40: 2635–2646.

Lefebvre, M.C., Z.G. Qi, and P.G. Pickup. 1999. Electronically conducting proton exchange polymers as catalyst supports for proton exchange membrane fuel cells. Electrocatalysis of oxygen reduction, hydrogen oxidation, and methanol oxidation. *Journal of Electrochemical Society* 146: 2054–2058.

Li, M., S. Luo, C. Zeng et al. 2004. Corrosion behavior of TiN coated type 316 stainless steel in simulated PEMFC environments. *Corrosion Science* 46: 1369–1380.

Li, Q., R. He, J.O. Jensen et al. 2004. PBI-based polymer membranes for high temperature fuel cells—Preparation, characterization and fuel cell demonstration. *Fuel Cells* 4: 147–159.

Li, X. 2005. *Principles of Fuel Cells.* New York, NY: Taylor & Francis.

Li, X. and I. Sabir. 2005. Review of bipolar plates in PEM fuel cells: Flow-field designs. *International Journal of Hydrogen Energy* 30: 359–371.

Li, X. and I.M. Hsing. 2006. The effect of the Pt deposition method and the support on Pt dispersion on carbon nanotubes. *Electrochimica Acta* 51: 5250–5258.

Liu, F.Q., B.L. Yi, D.M. Xing et al. 2003. Nafion/PTFE composite membranes for fuel cell applications. *Journal of Membrane Science* 212: 213–223.

Liu, L., A. Ishihara, S. Mitsushima et al. 2005. Zirconium oxide for PEFC cathodes. *Electrochemical Solid-State Letters* 8: A400–A402.

Liu, Y.H., B. Yi, Z.G. Shao et al. 2006. Carbon nanotubes reinforced nafion composite membrane for fuel cell applications. *Electrochemical Solid-State Letters* 9: A356–A359.

Makharia, R., S.S. Kocha, P.T. Yu et al. 2006. Durability PEM fuel cell electrode materials: Requirements and benchmarking methodologies. *ECS Transactions* 1: 3–18.

Makkus, R.C., A.H.H. Janssen, F.A. de Bruijn et al. 2000. Use of stainless steel for cost competitive bipolar plates in the SPFC. *Journal of Power Sources* 86: 274–282.

Marvin, R.H. and C.M. Carlstrom. 2002. *Fuel cell fluid flow plate for promoting fluid service.* U.S. Patent 6500580.

Mathias, M.F., J. Roth, J. Fleming et al. 2003. Diffusion media materials and characterization. In *Handbook of Fuel Cells: Fundamentals, Technology and Applications*, eds. W. Vielstich, A. Lamm, and H.A. Gasteiger, vol. 3, pp. 517–537. Chichester: Wiley.

Mighri, F., M.A. Huneault, and M.F. Champagne. 2004. Electrically conductive thermoplastic blends for injection and compression molding of bipolar plates in the fuel cell application. *Polymer Engineering, and Science* 44: 1755–1765.

Mitani, T. and K. Mitsuda. 2009. Durability of graphite composite bipolar plates. In *Polymer Electrolyte Fuel Cell Durability*, eds. F.N. Buchi, M. Inaba, and T.J. Schmidt, pp. 257–268. New York, NY: Springer.

Montilla, F., E. Morallon, I. Duo et al. 2003. Platinum particles deposited on synthetic boron-doped diamond surfaces. Application to methanol oxidation. *Electrochimica Acta* 48: 3891–3897.

Moore, R.B. and C.R. Martin. 1989. Morphology and chemical properties of the Dow perfluorosulfonate ionomers. *Macromolecules* 22: 3594–3599.

Mueller, A., P. Kauranen, A. von Ganski et al. 2006. Injection moulding of graphite composite bipolar plates. *Journal of Power Sources* 154: 467–471.

Muller, A., P. Kauranen, A. von Ganski et al. 2006. Injection moulding of graphite composite bipolar plates. *Journal of Power Sources* 154: 467–471.

Murphy, O.J., A. Cisar, and E. Clarke. 1998. Low cost light weight high power density PEM fuel cell stack. *Electrochimica Acta* 43: 3829–3840.

Paddison, S.J. and J.A. Elliott. 2007. Selective hydration of the 'short-side-chain' perfluorosulfonic acid membrane. An ONIOM study. *Solid State Ionics* 178: 561–567.

Parsonage, E.E. and M.K. Debe. 1994. *Nanostructured electrode membranes.* U.S. Patent 5338430.

Piela, P., C. Eickes, E. Brosha et al. 2004. Ruthenium crossover in direct methanol fuel cell with Pt–Ru black anode. *Journal of Electrochemical Society* 151: A2053–A2059.

Posudievsky, O.Y., Y.I. Kurys, and V.D. Pokhodenko. 2004. 12-Phosphormolibdic acid doped polyaniline–V_2O_5 composite. *Synthetic Metals* 144: 107–111.

Qi, L., K. Jiao, A. Pereira et al. 2011. Numerical investigation of the effects of GDL deformation on water transport in PEM fuel cells. *International Journal of Energy Research*, in press.

Radhakrishnan, S., B.T.S. Ramanujam, A. Adhikari et al. 2006. High-temperature, polymer–graphite hybrid composites for bipolar plates: Effect of processing conditions on electrical properties. *Journal of Power Sources* 163: 702–707.

Rajalakshmi, N., H. Ryu, M.M. Shaijumon et al. 2005. Performance of polymer electrolyte membrane fuel cells with carbon nanotubes as oxygen reduction catalyst support material. *Journal of Power Sources* 140: 250–257.

Reiser C.A. 1989. *Water and heat management in solid polymer fuel cell stack.* U.S. Patent 4826742.

Reiser, C.A. and R.D. Sawyer. 1988. *Solid polymer electrolyte fuel cell stack water management system.* U.S. Patent 4769297.

Ren, X.M. and S.J. Gottesfeld. 2001. Electro-osmotic drag of water in poly(perfluorosulfonic acid) membranes. *Journal of Electrochemical Society* 148: A87–A93.

Rivard, L.M., D. Pierpont, H.T. Freemeyer et al. 2003 Development of a new electrolyte membrane for PEM fuel cells. *Fuel Cell Seminar,* p. 73. Miami Beach, FL—Abstract Book.

Rock, J.A. 2001. *Serially-linked serpentine flow channels for PEM fuel cell.* U.S. Patent 6309773.

Sasaki, K., M. Shao, and R. Adzic. 2009. Dissolution and stabilization of platinum in oxygen cathodes. In *Polymer Electrolyte Fuel Cell Durability,* eds. F.N. Buchi, M. Inaba, and T.J. Schmidt, pp. 7–27. New York, NY: Springer.

Sawai, K. and N. Suzuki. 2004. Heat-treated transition metal hexacyanometallates as electrocatalysts for oxygen reduction insensitive to methanol. *Journal of Electrochemical Society* 151: A682–A688.

Scherer, J., D. Munter, and R. Strobel. 2009. Influence of metallic bipolar plates on the durability of polymer electrolyte fuel cells. In *Polymer Electrolyte Fuel Cell Durability,* eds. F.N. Buchi, M. Inaba, and T.J. Schmidt, pp. 243–255. New York, NY: Springer.

Schmidt, T.J. and J. Baurmeister. 2008. Properties of high-temperature PEFC Celtec®-P 1000 MEAs in start/stop operation mode. *Journal of Power Sources* 176: 428–434.

Scholl, H., B. Hofman, and A. Rauscher. 1992. Anodic polarization of cemented carbides of the type [(WC,M): M = Fe, Ni or Co] in sulphuric acid solution. *Electrochimica Acta* 37: 447–452.

Scholl, H., B. Hofman, J. Kupis et al. 1994. Anodic dissolution of cemented carbides of the type [(WC, M); M = Co, Ni OR Fe] in sulphuric acid solution. Electrochemical impedance spectroscopy. *Electrochimica Acta* 39: 115–117.

Schmid, O. and J. Einhart. 2000. *Polymer electrolyte membrane fuel cells and stacks with adhesively bonded layers.* U.S. Patent 6080503.

Shao, Y., G. Yin, J. Zhang et al. 2006. Comparative investigation of the resistance to electrochemical oxidation of carbon black and carbon nanotubes in aqueous sulfuric acid solution. *Electrochimica Acta* 51: 5853–5857.

Shibata, Y., A. Ishihara, S. Mitsushima et al. 2007. Effect of heat treatment on catalytic activity for oxygen reduction reaction of TaO_xN_y/Ti prepared by electrophoretic deposition. *Electrochemical and Solid-State Letters* 10: B43–B46.

Shibuya, N. and R.S. Porter. 1992. Kinetics of PEEK sulfonation in concentrated sulfuric acid. *Macromolecules* 25: 6495–6499.

Staiti, P. and M. Minutoli. 2001. Influence of composition and acid treatment on proton conduction of composite polybenzimidazole membranes. *Journal of Power Sources* 94: 9–13.

Stamenkovic, V., T.J. Schmidt, P.N. Ross et al. 2002. Surface composition effects in electrocatalysis: Kinetics of oxygen reduction on well-defined Pt3Ni and Pt3Co alloy surfaces. *Journal of Physical Chemistry* B 106: 11970–11979.

Statterfield, M.B., P.W. Majsztrik, H. Ota et al. 2006. Mechanical properties of Nafion and titania/Nafion composite membranes for polymer electrolyte membrane fuel cells. *Journal of Polymer Science Part B: Polymer Physics* 44: 2327–2345.

Steck, A.E. and C. Stone. 1997. New materials for fuel cell and modern battery systems II. *Proceedings of the 2nd International Symposium on New Materials for Fuel Cell and Modern Battery Systems,* Montreal, July 6–10, 1997, pp. 792–807.

Steck, A.E. and J. Wei. 1995. *Gasketed membrane electrode assembly for electrochemical fuel cells.* U.S. Patent 5464700.

Tabata, T., O. Yamazaki, H. Shintaku et al. 2009. Degradation factors of polymer electrolyte fuel cells in residential cogeneration systems. In *Polymer Electrolyte Fuel Cell Durability,* eds. F.N. Buchi, M. Inaba, and T.J. Schmidt, pp. 447–463. New York, NY: Springer.

Tan, J., Y.J. Chao, J.W. Van Zee et al. 2007. Degradation of elastomeric gasket materials in PEM fuel cells. *Materials Science and Engineering A* 445–446: 669–675.

Tant, M.R., K.P. Darst, K.D. Lee et al. 1989. Structure and properties of short-side-chain perfluorosulfonate ionomers. *ACS Symposium Series* 395: 370–400.

Tawfik, H., Y. Hung, and D. Mahajan. 2007. Metal bipolar plates for PEM fuel cell—A review. *Journal of Power Sources* 163: 755–767.

Thomassin, J.M., J. Kollar, G. Caldarella et al. 2007. Beneficial effect of carbon nanotubes on the performances of Nafion membranes in fuel cell applications. *Journal of Membrane Science* 303: 252–257.

Tian, R.J., J.C. Sun, and L. Wang. 2007. Effect of plasma nitriding on behavior of austenitic stainless steel 304L bipolar plate in proton exchange membrane fuel cell. *Journal of Power Sources* 163: 719–724.

Voorhies, J.D. 1972. Electrochemical and chemical corrosion of tungsten carbide (WC). *Journal of Electrochemical Society* 119: 219–222.

Wang, B. 2005. Recent development of non-platinum catalysts for oxygen reduction reaction. *Journal of Power Sources* 152: 1–15.

Wang, C., M. Waje, X. Wang et al. 2004. Proton exchange membrane fuel cells with carbon nanotube based electrodes. *Nano Letters* 4: 345–348.

Wang, L., D.M. Xing, H.M. Zhang et al. 2008. MWCNTs reinforced Nafion® membrane prepared by a novel solution-cast method for PEMFC. *Journal of Power Sources* 176: 270–275.

Wang, Y., S. Al Shakhshir, X. Li et al. 2011. Development of superhydrophobic flow channel surface for improved PEM fuel cell performance. *International Journal of Hydrogen Energy* in press.

Watkins, D.S., K.W. Dircks, and D.G. Epp. 1991. *Novel fuel cell fluid flow field plate*. U.S. Patent 4988583.

Watkins, D.S., K.W. Dircks, and D.G. Epp. 1992. *Fuel cell fluid flow field plate*. U.S. Patent 5108849.

Wei, J., C. Stone, and A.E. Steck. 1995. *Trifluorostyrene and substituted trifluorostyrene copolymeric compositions and ion-exchange membranes formed therefrom*. U.S. Patent 5422411.

Wind, J., R. Spah, and W. Kaiser. 2002. Metallic bipolar plates for PEM fuel cells. *Journal of Power Sources* 105: 256–260.

Woo, M.H., O. Kwon, S.H. Choi et al. 2006. Zirconium phosphate sulfonated poly (fluorinated arylene ether)s composite membranes for PEMFCs at 100–140°C. *Electrochimica Acta* 51: 6051–6059.

Wood, D.L. and Borup R.L. 2009. Durability aspects of gas-diffusion and microporous layers. In *Polymer Electrolyte Fuel Cell Durability*, eds. F.N. Buchi, M. Inaba, and T.J. Schmidt, pp. 159–195. New York, NY: Springer.

Xing, D. and J. Kerres. 2006. Improved performance of sulfonated polyarylene ethers for proton exchange membrane fuel cells. *Polymers for Advanced Technologies* 17: 591–597.

Yamamoto, S., S. Sugawara, and K. Shinohara. 2009. Fuel cell stack durability for vehicle application. In *Polymer Electrolyte Fuel Cell Durability*, eds. F.N. Buchi, M. Inaba, and T.J. Schmidt, pp. 467–482. New York, NY: Springer.

Yan, W., H.S. Chu, C.P. Wang et al. 2006. Transient behavior of CO poisoning of the anode catalyst layer of a PEM fuel cell. *Journal of Power Sources* 159: 1071–1077.

Yang, C., S. Srinivasan, A.B. Bocarsly et al. 2004. A comparison of physical properties and fuel cell performance of Nafion and zirconium phosphate/Nafion composite membranes. *Journal of Membrane Science* 237: 145–161.

Yeom, J., G.Z. Mozsgai, B.R. Flachsbart et al. 2005. Microfabrication and characterization of a silicon-based millimeter scale, PEM fuel cell operating with hydrogen, methanol, or formic acid. *Sensors and Actuators B: Chemical* 107: 882–891.

Yin, Q., A.J. Li, W.Q. Wang et al. 2007. Study on the electrical and mechanical properties of phenol formaldehyde resin/graphite composite for bipolar plate. *Journal of Power Sources* 165: 717–721.

Yu, P., M. Pemberton, and P. Plasse. 2005. PtCo/C cathode catalyst for improved durability in PEMFCs. *Journal of Power Sources* 144: 11–20.

Yu, P.T., W. Gu, and R. Makharia. 2007. The impact of carbon stability on PEM fuel cell startup and shutdown voltage degradation. *ECS Transactions* 3: 797–809.

Yu, P.T., W. Gu, J. Zhang et al. 2009. Carbon-support requirements for highly durable fuel cell operation. In *Polymer Electrolyte Fuel Cell Durability*, eds. F.N. Buchi, M. Inaba, and T.J. Schmidt, pp. 29–53. New York, NY: Springer.

Yu, T.L., H.L. Lin, K.S. Shen et al. 2004. Nafion/PTFE composite membranes for fuel cell applications. *Journal of Polymer Research* 11: 217–224.

Yuan, F.L., H.K. Yu, and H.J. Ryu. 2004. Preparation and characterization of carbon nanofibers as catalyst support material for PEMFC. *Electrochimica Acta* 50: 685–691.

Zelenay, P., R. Bashyam, E. Brosha et al. 2006. Non-platinum cathode catalysts. *FY 2006 Annual Progress Report*. http://www.hydrogen.energy.gov/pdfs/progress06/v_c_7_zelenay.pdf

Zhang, J., Z. Xie, J. Zhang et al. 2006. High temperature PEM fuel cells. *Journal of Power Sources* 160: 872–891.

Zhong, H., H. Zhang, G. Liu et al. 2006. A novel non-noble electrocatalyst for PEM fuel cell based on molybdenum nitride. *Electrochemistry Communications* 8: 707–712.

Index

Note: n = Footnote

A

AAEMs. *See* Alkaline anion exchanges membranes (AAEMs)
AAS. *See* Atomic adsorption spectroscopy (AAS)
AC resistance (ACR), 301
Accelerated durability test. *See* Accelerated stress test (AST)
Accelerated stress test (AST), 6, 49
    applications, 49–50
    catalyst ionomer resistance, 97
    conditions and test systems, 6
    for FC components, 266
Accelerated testing (AT), 124, 128
    methods, 124
    product life estimation, 125
    Pt load changes for CCM, 274, 278
Acid-type membranes, 75
ACL. *See* Anode catalyst layer (ACL)
ACR. *See* AC resistance (ACR)
Adhesives, hot melt, 189
Adiabatic conditions, 117
AFM. *See* Atomic force microscope (AFM)
Air contaminants in FCs, 200, 202
    on Pt catalyst layer, 215
Air-side humidity, 286
    polarization curves, 286, 287
    RH output vs. input, 286, 287
Alkaline anion exchanges membranes (AAEMs), 229
Alkaline FCs, 229
American Society for Testing and Materials (ASTM), 190
Anion, 15
    chloride, 200
    contamination, 15
Anode and cathode gas humidity
    air-side humidity, 286
    hydrogen-side humidity, 286, 288

low-RH effects, 286
Anode catalyst layer (ACL), 251
    composition effect, 220
    ionomer degradation effect, 224
    ruthenium dissolution, 281
Anode deactivation modeling, 221–222
Asahi Kasei advanced membrane, 93
AST. *See* Accelerated stress test (AST)
ASTM. *See* American Society for Testing and Materials (ASTM)
AT. *See* Accelerated testing (AT)
Atomic adsorption spectroscopy (AAS), 9
Atomic force microscope (AFM), 48, 162
ATR-FTIR. *See* Attenuated total reflection Fourier transform infrared (ATR-FTIR)
Attenuated total reflection Fourier transform infrared (ATR-FTIR), 195

B

BAM3G membrane, 310
BET. *See* Brunauer–Emmett–Teller (BET)
Biomimetic technology, 324
Bipolar plate design (BPP design), 139, 187, 320. *See also* Membrane design
    carbon-filled composite, 145
    chemical analysis, 161–162
    compression molding, 145–146
    corrosion research results, 155–157
    degradation, 153, 165
    DOE targets for, 154
    elemental analysis, 162
    epoxy-based and phenolic-bonded, 173
    ex situ test, 153
    global manufacturers, 150
    graphite-based, 143, 320–321
    ICR measurement, 153–158
    immersion tests, 160–161